DOMINATION IN GRAPHS

MONOGRAPHS AND TEXTBOOKS IN
PURE AND APPLIED MATHEMATICS

54. *J. Cronin*, Differential Equations (1980)
55. *C. W. Groetsch*, Elements of Applicable Functional Analysis (1980)
56. *I. Vaisman*, Foundations of Three-Dimensional Euclidean Geometry (1980)
57. *H. I. Freedan*, Deterministic Mathematical Models in Population Ecology (1980)
58. *S. B. Chae*, Lebesgue Integration (1980)
59. *C. S. Rees et al.*, Theory and Applications of Fourier Analysis (1981)
60. *L. Nachbin*, Introduction to Functional Analysis (R. M. Aron, trans.) (1981)
61. *G. Orzech and M. Orzech*, Plane Algebraic Curves (1981)
62. *R. Johnsonbaugh and W. E. Pfaffenberger*, Foundations of Mathematical Analysis (1981)
63. *W. L. Voxman and R. H. Goetschel*, Advanced Calculus (1981)
64. *L. J. Corwin and R. H. Szczarba*, Multivariable Calculus (1982)
65. *V. I. Istrățescu*, Introduction to Linear Operator Theory (1981)
66. *R. D. Järvinen*, Finite and Infinite Dimensional Linear Spaces (1981)
67. *J. K. Beem and P. E. Ehrlich*, Global Lorentzian Geometry (1981)
68. *D. L. Armacost*, The Structure of Locally Compact Abelian Groups (1981)
69. *J. W. Brewer and M. K. Smith, eds.*, Emmy Noether: A Tribute (1981)
70. *K. H. Kim*, Boolean Matrix Theory and Applications (1982)
71. *T. W. Wieting*, The Mathematical Theory of Chromatic Plane Ornaments (1982)
72. *D. B. Gauld*, Differential Topology (1982)
73. *R. L. Faber*, Foundations of Euclidean and Non-Euclidean Geometry (1983)
74. *M. Carmeli*, Statistical Theory and Random Matrices (1983)
75. *J. H. Carruth et al.*, The Theory of Topological Semigroups (1983)
76. *R. L. Faber*, Differential Geometry and Relativity Theory (1983)
77. *S. Barnett*, Polynomials and Linear Control Systems (1983)
78. *G. Karpilovsky*, Commutative Group Algebras (1983)
79. *F. Van Oystaeyen and A. Verschoren*, Relative Invariants of Rings (1983)
80. *I. Vaisman*, A First Course in Differential Geometry (1984)
81. *G. W. Swan*, Applications of Optimal Control Theory in Biomedicine (1984)
82. *T. Petrie and J. D. Randall*, Transformation Groups on Manifolds (1984)
83. *K. Goebel and S. Reich*, Uniform Convexity, Hyperbolic Geometry, and Nonexpansive Mappings (1984)
84. *T. Albu and C. Năstăsescu*, Relative Finiteness in Module Theory (1984)
85. *K. Hrbacek and T. Jech*, Introduction to Set Theory: Second Edition (1984)
86. *F. Van Oystaeyen and A. Verschoren*, Relative Invariants of Rings (1984)
87. *B. R. McDonald*, Linear Algebra Over Commutative Rings (1984)
88. *M. Namba*, Geometry of Projective Algebraic Curves (1984)
89. *G. F. Webb*, Theory of Nonlinear Age-Dependent Population Dynamics (1985)
90. *M. R. Bremner et al.*, Tables of Dominant Weight Multiplicities for Representations of Simple Lie Algebras (1985)
91. *A. E. Fekete*, Real Linear Algebra (1985)
92. *S. B. Chae*, Holomorphy and Calculus in Normed Spaces (1985)
93. *A. J. Jerri*, Introduction to Integral Equations with Applications (1985)
94. *G. Karpilovsky*, Projective Representations of Finite Groups (1985)
95. *L. Narici and E. Beckenstein*, Topological Vector Spaces (1985)
96. *J. Weeks*, The Shape of Space (1985)
97. *P. R. Gribik and K. O. Kortanek*, Extremal Methods of Operations Research (1985)
98. *J.-A. Chao and W. A. Woyczynski, eds.*, Probability Theory and Harmonic Analysis (1986)
99. *G. D. Crown et al.*, Abstract Algebra (1986)
100. *J. H. Carruth et al.*, The Theory of Topological Semigroups, Volume 2 (1986)
101. *R. S. Doran and V. A. Belfi*, Characterizations of C*-Algebras (1986)
102. *M. W. Jeter*, Mathematical Programming (1986)
103. *M. Altman*, A Unified Theory of Nonlinear Operator and Evolution Equations with Applications (1986)
104. *A. Verschoren*, Relative Invariants of Sheaves (1987)
105. *R. A. Usmani*, Applied Linear Algebra (1987)
106. *P. Blass and J. Lang*, Zariski Surfaces and Differential Equations in Characteristic $p > 0$ (1987)
107. *J. A. Reneke et al.*, Structured Hereditary Systems (1987)
108. *H. Busemann and B. B. Phadke*, Spaces with Distinguished Geodesics (1987)
109. *R. Harte*, Invertibility and Singularity for Bounded Linear Operators (1988)

Additional Volumes in Preparation

DOMINATION IN GRAPHS
Advanced Topics

edited by
Teresa W. Haynes
East Tennessee State University
Johnson City, Tennessee

Stephen T. Hedetniemi
Clemson University
Clemson, South Carolina

Peter J. Slater
University of Alabama in Huntsville
Huntsville, Alabama

MARCEL DEKKER, INC.   NEW YORK · BASEL · HONG KONG

ISBN 0-8247-0034-1

The publisher offers discounts on this book when ordered in bulk quantities. For more information, write to Special Sales/Professional Marketing at the address below.

This book is printed on acid-free paper.

MARCEL DEKKER, INC.
270 Madison Avenue, New York, New York 10016
http://www.dekker.com

Current printing (last digit):
10 9 8 7 6 5 4 3 2 1

PRINTED IN THE UNITED STATES OF AMERICA

Preface

Within the last twenty-five years, concurrent with the growth of computer science, graph theory has seen explosive growth. Perhaps the fastest growing area within graph theory is the study of domination in graphs.

Cockayne and Hedetniemi's survey paper on domination appeared in 1977 and contained 20 references. This survey paper seems to have set in motion the modern study of dominating sets in graphs. In 1990, Hedetniemi and Laskar edited an issue of *Discrete Mathematics* devoted entirely to domination. The 1990 bibliography revealed an impressive increase in thirteen years from 20 to approximately 400 references. Seven years later more than 1,000 research papers had been published on dominating sets and related sets in graphs, and the field is steadily growing.

Noting the wide interest and the need for comprehensive publications in this field of study, we were motivated to produce this book and its companion, *Fundamentals of Domination in Graphs*. The companion book covers the basics of domination and major research accomplishments on domination in textbook form. It includes the only known comprehensive bibliography on the subject. While writing *Fundamentals of Domination in Graphs*, it became apparent to us that there was a collection of topics that ought to be covered in greater depth than a textbook would allow. Thus *Domination in Graphs: Advanced Topics* was conceived.

For this book, we invited leading researchers in domination to contribute chapters. We have been very gratified by the fact that so many of them accepted and have contributed to the project. This book contains 17 chapters, each of which is intended to provide readers with a survey of the known results and to bring them to the forefront of research in a particular aspect of graph domination. These topics are on the frontiers of research and many unsolved problems are presented. Hence it is an important reference work in domination for those interested in research in the area, those wanting to delve more deeply into a specific topic presented here, those interested in a sampling of proof techniques in domination, and those from diverse areas interested in becoming acquainted with the material.

Definitions, Terminology and Notation

For the most part, terminology and notation for graph theory domination concepts are fairly well standardized. The authors of the chapters in this book all graciously consented to be as consistent as possible with the descriptions presented here. Thus the readers can easily move from chapter to chapter and repetitive definitions are avoided.

We acknowledge our hope that what follows becomes the standard throughout the literature.

Basics

For graph $G = (V, E)$ we let n and m denote the order and size of G, respectively. That is, the number of vertices is $n = |V|$ and the number of edges is $m = |E|$. We use $\langle S \rangle$ (or $< S >$) to denote the subgraph induced by the set S.

$N(v)$ and $N[v]$

The *open neighborhood* of $v \in V(G)$ is the set of vertices adjacent to v, $N(v) = \{w | vw \in E(G)\}$, and the *closed neighborhood* of v is $N[v] = N(v) \cup \{v\}$.

$N(S)$ and $N[S]$

For a set $S \subseteq V(G)$, $N(S) = \cup_{s \in S} N(s)$ and $N[S] = \cup_{s \in S} N[s]$.

$d(u, v)$

The *distance* $d(u, v)$ between vertices u and v is the minimum number of edges in a $u - v$ path.

$ecc(u)$

The *eccentricity* of $v \in V(G)$ is $ecc(u) = MAX\{d(u, v) : v \in V(G)\}$.

$rad(G)$ and $diam(G)$

The *radius* of G is $rad(G) = MIN\{ecc(u) : u \in V(G)\}$ and the *diameter* of G is $diam(G) = MAX\{ecc(u) : u \in V(G)\}$.

$deg(v)$ or *deg* v

The *degree of* $v \in V(G)$ is $deg(v) = |N(v)|$.

$\delta(G)$ and $\Delta(G)$

The minimum degree of G is $\delta(G) = MIN\{deg(v) : v \in V(G)\}$ and the maximum degree of G is $\Delta(G) = MAX\{deg(v) : v \in V(G)\}$.

Parameters

A generic parameter $\mu(G)$ is sometimes denoted simply as μ when the graph under consideration is clear from the context. That is, we let $\mu = \mu(G)$ and $\overline{\mu} = \mu(\overline{G})$.

$\gamma(G)$ and $\Gamma(G)$

Set $S \subseteq V(G)$ is a *dominating set* if for each $v \in V(G)$ either $v \in S$ or v is adjacent to some $w \in S$. That is, S is a dominating set if and only if $N[S] = V$. The *domination number* $\gamma(G)$ and the *upper domination number* $\Gamma(G)$ are the minimum and maximum cardinalities, respectively, of minimal dominating sets.

$\boxed{i(G) \text{ and } \beta(G)}$

Set $S \subseteq V(G)$ is an *independent set* if no two vertices in S are adjacent. The *independence number* $\beta(G)$ (or $\beta_0(G)$) and *lower independence number* $i(G)$ are the maximum and minimum cardinalities, respectively, of maximal independent sets. Because an independent set is maximally independent if and only if it is dominating, $i(G)$ is also called the *independent domination number*.

$\boxed{ir(G) \text{ and } IR(G)}$

Set $S \subseteq V(G)$ is *irredundant* if for each $s \in S$ there is a vertex w (possibly $w = s$) such that $N[w] \cap S = \{s\}$. The *irredundance number* $ir(G)$ and the *upper irredundance number* $IR(G)$ are the minumum and maximum cardinalities, respectively, of maximal irredundant sets.

$\boxed{\rho(G)}$

Set $S \subseteq V(G)$ is a *packing* if for any two vertices u and v in S we have $d(u,v) \geq 3$. That is, S is a packing if and only if for any vertex $u \in V$, $|N[u] \cap S| \leq 1$. The *packing number* $\rho(G)$ is the maximum cardinality of a packing in G.

$\boxed{\alpha(G) \text{ and } \Lambda(G)}$

Set $S \subseteq V(G)$ is a *vertex cover* if for any edge $uv \in E(G)$, u or v is in S. The *vertex covering number* $\alpha(G)$ (or $\alpha_0(G)$) and *upper vertex covering number* $\Lambda(G)$ are the minimum and maximum cardinalities of minimal vertex covers.

$\boxed{\psi(G) \text{ and } \Psi(G)}$

For $v \in S \subseteq V(G)$ the vertex v is an *enclave* of S if $N[v] \subseteq S$, and S is *enclaveless* if it has no enclaves. The *enclaveless number* $\Psi(G)$ and *lower enclaveless number* $\psi(G)$ are the maximum and minimum cardinalities, respectively, of maximal enclaveless sets.

$\boxed{\gamma_t(G)}$

Set $S \subseteq V(G)$ is a *total dominating set* if every vertex $v \in V(G)$ is adjacent to a vertex in S. In this context a vertex does not dominate itself. Equivalently, the union of the open neighborhoods of the vertices in S is all of $V(G)$, $\cup_{s \in S} N(s) = V$, and, hence, such sets are also called *open dominating* sets. The *total/open domination number* of G, denoted $\gamma_t(G)$, is the minimum cardinality of a total dominating set.

$\boxed{\gamma_c(G)}$

For a connected graph G the *connected domination number* $\gamma_c(G)$ is the minimum cardinality of a dominating set S whose induced subgraph $< S >$ is connected.

$\boxed{F(G)}$

If $S \subseteq V(G)$ is a packing and also a dominating set, then S is called an *efficient dominating set*. Not all graphs have efficient dominating sets, and the *efficient domination number* $F(G)$ is the maximum number of vertices that can be dominated given that no vertex gets dominated more than once. That is, $F(G) = MAX\{|\cup_{s \in S} N[s]| : S \text{ is a packing}\}$. Note that $F(G) = MAX\{\sum_{s \in S}(1 + deg(s)) : S \text{ is a packing}\}$.

$\boxed{R(G)}$

If we consider the minimum amount of domination that can be done given that every vertex gets dominated at least once, we have the *redundance number* $R(G) = MIN\{\sum_{s \in S}(1 + deg(s)) : S \text{ is a dominating set}\}$.

$\boxed{\alpha_1(G) \text{ or } \alpha'(G)}$

Set $B \subseteq E(G)$ is an *edge cover* if every vertex is incident with at least one edge in B. The *edge covering number*, denoted $\alpha_1(G)$ or $\alpha'(G)$, is the minimum cardinality of an edge cover.

$\boxed{\beta_1(G) \text{ or } \beta'(G)}$

A *matching* is an independent set B of edges, i.e., no two edges in B have a vertex in common. The *matching number* of G, denoted by $\beta_1(G)$ or $\beta'(G)$, is the maximum cardinality of a matching.

We note that a minimum (maximum, respectively) dominating set is called a γ-set (Γ-set, respectively). Similar notation is used to denote sets achieving any of the parameters, e.g. γ_t-set, i-set, or ir-set.

Y-valued domination parameters

$\boxed{w(f)}$

For a subset Y of the reals $\Re$, $Y \subseteq \Re$, a function $f : V(G) \to Y$ is called a *Y-valued function*. For $S \subseteq V(G)$, we let $f(S) = \sum_{s \in S} f(s)$. The *weight* of f is $w(f) = f(V(G))$. Function f is a (*Y-valued*) *dominating function* if $f(N[v]) \geq 1$ for every $v \in V(G)$.

$\boxed{\gamma_Y(G)}$

The *Y-domination number* of G is the minimum weight of a Y-dominating function. Note that $\gamma(G) = \gamma_{\{0,1\}}(G)$. Perhaps the three other most studied cases are the *fractional domination number* $\gamma_f(G) = \gamma_{[0,1]}(G)$, the *minus domination number* $\gamma^-(G) = \gamma_{\{-1,0,1\}}(G)$, and the *signed domination number* $\gamma_s(G) = \gamma_{\{-1,1\}}(G)$.

It is to be noted that Y-valued parameters have been defined for concepts in addition to domination, such as independence, packing, and covering. Furthermore, many other domination parameters in addition to the ones defined here have been studied. A comprehensive listing of all the known domination related parameters has been produced in T. W. Haynes, S. T. Hedetniemi, and P. J. Slater, *Fundamentals of Domination in Graphs*, Marcel Dekker Inc., New York, 1998.

Acknowledgments

We are extremely grateful to the following people for reviewing these chapters.

David Bange	Dieter Kratsch	Bob Brigham
Christine Mynhardt	Ernie Cockayne	Mike Plummer
Wayne Goddard	Doug Rall	Bert Hartnell
Johan Hattingh	Brooks Reid	Kelly Schultz
Mike Henning	Chris Smart	Dave Jacobs
Doug Weakley		

We also want to thank Srini Madella and Ben Phillips for their assistance with the preparation of this book.

Teresa W. Haynes
Stephen T. Hedetniemi
Peter J. Slater

Contents

Contributors

Robert C. Brigham University of Central Florida, Orlando, Florida

Julie R. Carrington Rollins College, Winter Park, Florida

E. J. Cockayne University of Victoria, Victoria, British Columbia, Canada

Gayla S. Domke Georgia State University, Atlanta, Georgia

Jean E. Dunbar Converse College, Spartanburg, South Carolina

Gerd H. Fricke Wright State University, Dayton, Ohio

J. Ghoshal Clemson University, Clemson, South Carolina

Bert Hartnell Saint Mary's University, Halifax, Nova Scotia, Canada

Johannes H. Hattingh Rand Afrikaans University, Johannesburg, South Africa

Teresa W. Haynes East Tennessee University, Johnson City, Tennessee

Sandra M. Hedetniemi Clemson University, Clemson, South Carolina

Stephen T. Hedetniemi Clemson University, Clemson, South Carolina

Michael A. Henning University of Natal, Pietermaritzburg, South Africa

Dieter Kratsch Friedrich-Schiller-Universität, Jena, Germany

Renu R. Laskar Clemson University, Clemson, South Carolina

Aniket Majumdar Purdue University, Fort Wayne, Indiana

Alice A. McRae Appalachian State University, Boone, North Carolina

C. M. Mynhardt University of South Africa, Pretoria, South Africa

Dolores A. Parks Appalachian State University, Boone, North Carolina

D. Pillone Clemson University, Clemson, South Carolina

Douglas F. Rall Furman University, Greenville, South Carolina

Robert Reynolds Data General Corporation, Research Triangle Park, North Carolina

E. Sampathkumar University of Mysore, Mysore, India

Peter J. Slater The University of Alabama in Huntsville, Huntsville, Alabama

David P. Sumner University of South Carolina, Columbia, South Carolina

Ulrich Teschner RWTH Aachen, Aachen, Germany

Lutz Volkmann RWTH Aachen, Aachen, Germany

Ewa Wojcicka University of Charleston, Charleston, South Carolina

Bohdan Zelinka Technical University, Liberec, Czech Republic

DOMINATION IN GRAPHS

Chapter 1

LP-Duality, Complementarity, and Generality of Graphical Subset Parameters

Peter J. Slater

Mathematical Sciences Department
The University of Alabama in Huntsville
Huntsville, Alabama 35899 USA

Abstract. This chapter describes in matrix form many graph theoretic subset problems such as domination, packing, covering, and independence. Graph theoretic minimization (respectively, maximization) problems expressed as linear programming problems have dual maximization (respectively, minimization) problems. Many of them also have interesting "complementary" problems, as described herein. As is the case for duality, complementarity involves one minimization and one maximization problem. The complementation theorem applied to specific pairs of complementary problems produces results including Gallai's covering/independence theorem $\alpha(G) + \beta(G) = |V(G)|$ and the domination/enclaveless theorem $\gamma(G) + \Psi(G) = |V(G)|$. Also, the matrix formulations suggest "generalized" parameters for which vertices can be assigned values in some arbitrarily specified subset Y of the reals. A generalized theory is introduced.

1.1   Introduction-Matrix Representations

Throughout this chapter graph G of order n and size m has vertex set $V(G) = \{v_1, v_2, ..., v_n\}$ and edge set $E(G) = \{e_1, e_2, ..., e_m\}$, and the focus is on matrices which can be used not only to represent G in several ways but also to define many interesting graph-theoretic parameters in linear programming (LP) and integer programming (IP) formats. The concepts of LP-duality and of complementarity are used both to demonstrate relations among such parameters and to actually define interesting (and perhaps some not so interesting) new

1

parameters. In the first four sections primary emphasis will be on integer (indeed, $\{0,1\}$) programming problems. However, the duals of these problems and even just the LP-format for these problems naturally suggest the extension to generalized graphical parameters. Generalized graph parameters are the focus of Section 1.5.

There are four basic matrices I think to use in representing graph G, namely T, H, A and N. The *adjacency matrix* A is the n-by-n binary matrix $A = [a_{i,j}]$ with $a_{i,j} = 1$ if $v_i v_j \in E(G)$, and $a_{i,j} = 0$ if $v_i v_j \notin E(G)$. The *closed neighborhood matrix* N is the n-by-n binary matrix $N = A + I_n$ where I_n is the n-by-n identity matrix. Thus $N = [n_{i,j}]$ with $n_{i,j} = 1$ if v_i and v_j are in the closed neighborhoods of each other, and $n_{i,j} = 0$ otherwise. Note that the open and closed neighborhoods of v_i are identified by the i-th columns of A and N, respectively, with $N(v_i) = \{v_t : a_{t,i} = 1\}$ and $N[v_i] = \{v_t : n_{t,i} = 1\}$. Here H denotes the n-by-m binary *incidence matrix* $H = [h_{i,j}]$ with $h_{i,j} = 1$ if e_j is incident with v_i ($e_j = v_i v_k$ for some v_k in the open neighborhood $N(v_i)$), and $h_{i,j} = 0$ otherwise. Matrix $T = [t_{i,j}]$ is the n-by-(n+m) *total matrix* obtained by juxtaposing N and $-H$. Thus $T = [N | - H]$ satisfies $t_{i,j} = 1$ if $v_i \in N[v_j]$ and $1 \leq j \leq n$, and $t_{i,j} = -1$ if v_i is incident with e_{j-n} and $n + 1 \leq j \leq n + m$, and $t_{i,j} = 0$ otherwise. Obviously, one can define additional representation matrices for G other than T,H,A,N.

Three particular n-tuples for G are the following. The column n-vector D corresponds to the degree sequence of G with $D = [d_i]$ where d_i is the degree of v_i , $d_i = deg(v_i) = |N(v_i)|$ is the number of ones in row i of A. Similarly $D^* = [d_i^*]$ where $d_i^* = 1 + d_i = 1 + deg\ v_i$ equals the number of ones in row i of N. Let $\bar{1}_n$ denote the column n-vector of all ones. So $D^* = D + \bar{1}_n$.

For a vertex subset $S \subseteq V(G)$ the characteristic function $f_S : V(G) \rightarrow \{0,1\}$ satisfies $f_S(v_i) = 1$ if $v_i \in S$ and $f_S(v_i) = 0$ if $v_i \notin S$. The characteristic function column vector $X_S = [x_i]$ for $1 \leq i \leq n$ satisfies $x_i = f_S(v_i)$. Now S is a dominating set if $|N[v_i] \cap S| \geq 1$ for $1 \leq i \leq n$ or, equivalently, if $N \cdot X_S \geq \bar{1}_n$. Thus we have the following integer programming formulation for the domination number $\gamma(G)$.

(1) $$\gamma(G) = MIN \sum_{i=1}^{n} x_i$$

subject to $N \cdot X \geq \bar{1}_n$

with $x_i \in \{0,1\}$

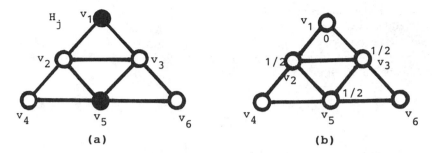

Figure 1.1: Hajós graph H_j.

For example, consider the Hajós graph H_j in Figure 1.1(a). For $S = \{v_1, v_5\}$ we have $X_S = [1, 0, 0, 0, 1, 0]^t$ and $N \cdot X_s = [1, 2, 2, 1, 1, 1]^t \geq \bar{1}_n$, so S is a dominating set. In fact, clearly $\gamma(H) = |S| = 2$. For the linear programming version of domination problem (1) we only require $x_i \geq 0$, which implies that $x_i \in [0, 1]$ because $x_i > 1$ is never necessary, and we have the formulation of "fractional" domination γ_f in Figure 1.2. Farber [11] investigated the problem of determining when the linear programming formulation of γ_f would provide an integer solution to (1), and he showed this to be the case for strongly chordal graphs. See also [7, 16]. The graph theoretic model for γ_f is to call a function $f : V(G) \to [0, \infty)$ a *dominating function* if for each $v \in V(G)$ we have $\sum_{u \in N[v]} f(u) \geq 1$, and the weight of function f is $w(f) = \sum_{u \in V(G)} f(v)$. For γ_f we seek the minimum weight of a dominating function. The function f illustrated in Figure 1.1(b) with $f(v_2) = f(v_3) = f(v_5) = 1/2$ and $f(v_1) = f(v_4) = f(v_6) = 0$ has $w(f) = \gamma_f(H_j) = 3/2$.

The constraint system for domination is $N \cdot X \geq \bar{1}_n$, and we seek to minimize the weight of a function that achieves this. On the other hand, assume that each vertex can be dominated at most once and we wish to achieve as much domination as possible. For the three-cube Q_3 vertex 000 dominates itself, 100, 010, and 001, while $N[111] = \{111, 110, 101, 011\}$. Hence for $S = \{000, 111\}$ we have $N \cdot X_s = \bar{1}_n$ and every vertex gets dominated exactly once. Mauldoon [22] and Zaremba [34] have shown that every cube Q_{2^k-1} has a vertex set S with $N \cdot X_S = \bar{1}_n$. In general, coding theorists (see, for example, Biggs [5]) would call a graph G with $S \subseteq V(G)$ satisfying $N \cdot X_S = \bar{1}_n$ a *perfect graph* and S a *perfect code*. Clearly cycle C_n on n vertices is perfect if and only if n = 3k. The concept of efficiency measures how close to perfect a graph is. Specifically, the *efficient domination number* $F(G)$ equals the maximum number of vertices that can be dominated given that no vertex is dominated more than once. So $F(C_{3k+1}) = F(C_{3k+2}) = 3k$ and $F(H_j) = 5$. Noting that v_i dominates $1 + deg\ v_i = d_i^*$ vertices, the integer programming formulation for F is given in (2) below with the constraint system being $N \cdot X \leq \overline{1}_n$. Returning to $N \cdot X \geq \overline{1}_n$, for "redundance" we seek to minimize the total amount of domination done, given that every vertex gets dominated at least once. See (3) below. Studies of efficiency and redundance include [1, 2, 3, 14, 16, 17, 20, 21], and these topics are treated in detail in [18].

(2)
$$F(G) = MAX \sum_{i=1}^{n} (1 + deg\ v_i)x_i$$

subject to $N \cdot X \leq \overline{1}_n$
with $x_i \in \{0,1\}$

(3)
$$R(G) = MIN \sum_{i=1}^{n} (1 + deg\ v_i)x_i$$

subject to $N \cdot X \geq \overline{1}_n$
with $x_i \in \{0,1\}$

The linear programming formulations for F_f and R_f are given in Figure 1.2. Note that if G is regular of degree r and we define $f : V(G) \to [0,1]$ by $f(v_i) = 1/(1+r)$ for $1 \leq i \leq n$, then $\sum_{u \in N[v_i]} f(u) = 1$.

Proposition 1.1 *If G is regular of degree r, then $F_f(G) = R_f(G) = n$.*

For "open" or "total" domination as in Cockayne, Dawes and Hedetniemi [6] a vertex v is considered to dominate the vertices in $N(v)$ but not itself. Thus $S \subseteq V(G)$ is a total dominating set if $\bigcup_{s \in S} N(s) = V(G)$, equivalently $A \cdot X_S \geq \overline{1}_n$. The integer programming formulation of total domination follows.

(4)
$$\gamma_t(G) = MIN \sum_{i=1}^{n} x_i$$

subject to $A \cdot X \geq \overline{1}_n$
with $x_i \in \{0,1\}$

While N and A are symmetric, H is not. The condition for a vertex set $S \subseteq V(G)$ to be independent is that $H^t \cdot X_S \leq \overline{1}_m$, and for edge set $S \subseteq E(G)$ to be independent (a matching) we have $H \cdot X_S \leq \overline{1}_n$. Thus the IP-formulations for the (vertex) independence number and the edge independence (matching) number are as follows.

(5) $\beta(G) = MAX \sum_{i=1}^{n} x_i$

subject to $H^t X \leq \bar{1}_m$
with $x_i \in \{0,1\}$

(6) $\beta^1(G) = MAX \sum_{i=1}^{m} x_i$

subject to $HX \leq \bar{1}_n$
with $x_i \in \{0,1\}$

The choice of $-H$ rather than H in $T = [N| - H]$ was specifically to accommodate independent domination. The independent domination number $i(G)$ is the minimum number of vertices in an independent set $S \subseteq V(G)$ such that S also dominates $V(G)$. That is, $H^t \cdot X_S \leq \bar{1}_m$ and $N \cdot X_S \geq \bar{1}_n$. Let $\bar{1}_{n+m}^*$ denote the $n + m$ column vector whose first n entries are 1 and whose last m entries are -1.

(7) $i(G) = MIN \sum_{i=1}^{n} x_i$

subject to $N \cdot X \geq \bar{1}_n$ and $H^t X \leq \bar{1}_m$
with $x_i \in \{0,1\}$

$i(G) = MIN \sum_{i=1}^{n} x_i$

subject to $T^t \cdot X \geq \bar{1}_{n+m}^*$
with $x_i \in \{0,1\}$

In (1), ..., (7) there are seven integer $\{0,1\}$-programming problems with constraint matrices $N, N, N, A, H^t, H,$ and T^t, respectively. Figure 1.2 contains the LP-formulations of the first three involving N and their duals. A generic expression for maximization/minimization LP/IP problems follows. Let Y denote a subset of the reals R and, specifically, $R^+ = [0, \infty)$ and $Z^+ = \{0, 1, 2, 3, ...\}$. Let M denote an arbitrary k-by-h real matrix; let MIN and MAX indicate minimization and maximization; let $C = [c_1, c_2, ..., c_h]^t$ be the vector of coefficients for the objective function; and let $B = [b_1, b_2, ..., b_k]^t$ be the vector of constraint values. Then (M, MIN, C, B, Y) and (M, MAX, C, B, Y) represent the following minimization and maximization problems. (As well, they will also be used to denote the actual optimal values.)

Domination

$$\gamma_f(G) = \text{MIN} \sum_{i=1}^{n} x_i$$

subject to $N \cdot X \geq \vec{1}_n$
$$x_i \geq 0$$

Packing

$$\rho_f(G) = \text{MAX} \sum_{i=1}^{n} x_i$$

subject to $N \cdot X \leq \vec{1}_n$
$$x_i \geq 0$$

$\gamma(H) = 2$ $\rho(H) = 1$

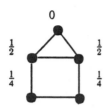

$$\gamma_f(H) = \tfrac{3}{2} = \rho_f(H)$$

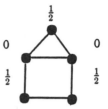

CLOD

$$W_f(G) = \text{MIN} \sum_{i=1}^{n} x_i$$

subject to $N \cdot X \geq D_n^{\bullet}$
$$x_i \geq 0$$

Efficient Domination

$$F_f(G) = \text{MAX} \sum_{i=1}^{n} (1 + \deg v_i) x_i$$

subject to $N \cdot X \leq \vec{1}_n$
$$x_i \geq 0$$

$W(H) = 5$ $F(H) = 4$

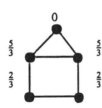

$$W_f(H) = \tfrac{14}{3} = F_f(H)$$

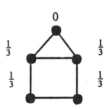

Redundance

$$R_f(G) = \text{MIN} \sum_{i=1}^{n} (1 + \deg v_i) x_i$$

subject to $N \cdot X \geq \vec{1}_n$
$$x_i \geq 0$$

CLOP

$$P_f(G) = \text{MAX} \sum_{i=1}^{n} x_i$$

subject to $N \cdot X \leq D_n^{\bullet}$
$$x_i \geq 0$$

$R(H) = 6$ $P(H) = 5$

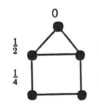

$$R_f(H) = \tfrac{11}{2} = P_f(H)$$

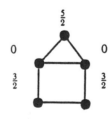

$$n \leq P \leq P_f = R_f \leq R$$

Figure 1.2: Linear programming formulations.

(8)
$$(M, MIN, C, B, Y) = MIN \sum_{i=1}^{h} c_i x_i$$

subject to $MX \geq B$
with $x_i \in Y$

(9)
$$(M, MAX, C, B, Y) = MAX \sum_{i=1}^{h} c_i x_i$$

subject to $MX \leq B$
with $x_i \in Y$

When matrix M is N, A, H, or T associated with a graph G, then one can also write $(M, MIN, C, B, Y)(G)$ and $(M, MAX, C, B, Y)(G)$. So $\gamma(G) = (N, MIN, \bar{1}_n, \bar{1}_n, \{0, 1\})(G)$, $F(G) = (N, MAX, D^*, \bar{1}_n, \{0, 1\})(G)$, $R(G) = (N, MIN, D^*, \bar{1}_n, \{0, 1\})(G)$, $\gamma_t(G) = (A, MIN, \bar{1}_n, \bar{1}_n, \{0, 1\})(G)$, $\beta(G) = (H^t, MAX, \bar{1}_n, \bar{1}_m, \{0, 1\})(G)$, $\beta_1(G) = (H, MAX, \bar{1}_m, \bar{1}_n, \{0, 1\})(G)$, $i(G) = (T^t, MIN, \bar{1}_n, \bar{1}_{n+m}^*, \{0, 1\})(G)$, and an LP-problem is $\gamma_f(G) = (N, MIN, \bar{1}_n, \bar{1}_n, R^+)(G)$.

The LP-problems whose IP-formulations are the (maximum) vertex independence number $\beta(G)$ and the (minimum) edge cover number $\alpha^1(G)$ are LP-duals of each other, where $\alpha^1(G) = (H, MIN, \bar{1}_m, \bar{1}_n, \{0, 1\})(G)$ denotes the minimum number of edges in a set $S_1 \subseteq E(G)$ that covers $V(G)$ (that is, $\bigcup_{uv \in S_1} \{u, v\} = V(G)$). Recalling Gallai's Theorem [12] $|V(G)| = \alpha(G) + \beta(G)$, and while the dual of β_f is α_f^1, the "complementary parameter" for β_f is the minimum vertex cover parameter α_f. Section 1.2 discusses duality and the Automorphism Class Theorem (A.C.T.). The A.C.T. shows how one can frequently greatly reduce the computations for graphically based LP problems. Section 1.3 is concerned with complementarity, and the Complementation Theorem extends Gallai's result to a very general situation. Section 1.4 treats the problems resulting from (8) and (9) when $M = N$, A, H, or T. Section 1.5 treats the extension from subset problems involving $f : V(G) \to \{0, 1\}$ to generalized parameters with $f : V(G) \to Y$ for an arbitrary subset of the reals, $Y \subseteq R$, and a further extension to the Matrix Complementation Theorem is presented.

1.2 Duality and the Automorphism Class Theorem

The following two linear programs are duals of each other.

(10) Primal $MIN\ C^t \cdot X$ Dual $MAX\ B^t \cdot X$
            subject to $MX \geq B$ subject to $M^tX \leq C$
            with $x_i \geq 0$ with $x_i \geq 0$

Theorem 1.2 (LP-duality) *If an optimal solution exists to either the primal or its dual program, then the other program also has an optimal solution and the two objective functions have the same optimal value.*

In the notation here (M, MIN, C, B, R^+) and (M^t, MAX, B, C, R^+) are duals, and all of the pairs of graph theoretic LP-dual problems in this chapter have optimal solutions. If $f : V(G) \to R^+$ with $f(v_i) = x_i$ and $S \subseteq V(G)$, let $f(S) = \sum_{v \in S} f(v)$ and, in particular, then $w(f) = f(V(G))$. As illustrated in Figure 1.2, $\gamma_f(G) = (N, MIN, \overline{1}_n, \overline{1}_n, R^+)(G)$ and $\rho_f(G) = (N, MAX, \overline{1}_n, \overline{1}_n, R^+)(G)$ are duals. For the Hajós graph H_j of Figure 1.1 let $f(v_2) = f(v_3) = f(v_5) = 1/2$ and $f(v_1) = f(v_4) = f(v_6) = 0$, then $f(N[v_i]) \geq 1$ for $1 \leq i \leq 6$ and $w(f) = 3/2$ shows $\gamma_f(H_j) \leq 3/2$. Also, $g(v_1) = g(v_4) = g(v_6) = 1/2$ and $g(v_2) = g(v_3) = g(v_5) = 0$ satisfies $g(N[v_i]) \leq 1$ for $1 \leq i \leq 6$ and $w(g) = 3/2$, so $\rho_f(H_j) \geq 3/2$. Thus, $3/2 \leq \rho_f(H_j) = \gamma_f(H_j) \leq 3/2$ implies $\rho_f(H_j) = \gamma_f(H_j) = 3/2$. We consider (M, MIN, C, B, Z^+) and (M^t, MAX, B, C, Z^+) to be dual IP problems, and $(M, MIN, C, B, \{0, 1\})$ and $(M^t, MAX, B, C, \{0, 1\})$ to be "binary duals", although either pair will usually not have the same optimal value. In general, call (M, MIN, C, B, Y) and (M^t, MAX, B, C, Y) *Y-duals* of each other.

Corollary 1.3 $(M^t, MAX, B, C, Z^+)(G) \leq (M^t, MAX, B, C, R^+)(G) = (M, MIN, C, B, R^+)(G) \leq (M, MIN, C, B, Z^+)(G)$ *for* $M = N, A, H,$ *or* T.

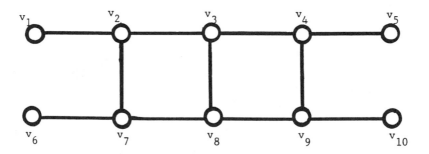

Figure 1.3: Closed neighborhood order domination/packing.

As noted in [27], although the closed neighborhood order domination function W_f, the dual of efficient domination F_f, and the closed neighborhood order packing function P_f, the dual of R_f, have the following interesting interpretations/applications, their study [19, 27, 28, 32, 33] began with the theoretical consideration of them as the duals of F_f and R_f. Assume each of the vertices in graph L of Figure 1.3 represents a town, each of which supports a security service such as a fire station. Further assume that each town v can be serviced (for multiple or multi-alarm fires) by any service station in its closed neighborhood $N[v]$. Note for the graph L that $N \cdot \overline{1}_{10} = [2, 4, 4, 4, 2, 2, 4, 4, 4, 2]^t = B_1$. If we place two service stations at each of v_2, v_4, v_7, and v_9, then $N \cdot [0, 2, 0, 2, 0, 0, 2, 0, 2, 0]^t \geq B_1$. (In fact, it equals B_1.) Thus every town is serviced as well with the eight stations so arranged as it is with the ten arranged one per town. Thus the closed neighborhood order domination function $(N, MIN, \overline{1}_n, D^*, Z^+)(G) = W(G)$ represents the minimum number of service facilities one can place on $V(G)$ so that each vertex location is serviced by as many facilities as it is by the all-ones function $f : V(G) \to Z^+$ with $f(v_i) = 1$ for $1 \leq i \leq n$, that is, $f(N[v_i]) \geq |N[v_i]|$ for $1 \leq i \leq n$. On the other hand, suppose each town supports an obnoxious facility such as a garbage dump or a noisy airport. For the latter example one assumes the traffic density at each airport is equivalent and the noise affects that location and its immediate neighbors. Note for the graph L that $N \cdot [2, 0, 2, 0, 2, 2, 0, 2, 0, 2]^t \leq B_1$. (Again, in fact, it equals B_1.) The dual of redundance is closed neighborhood order packing $(N, MAX, \overline{1}_n, D^*, Z^+)(G) = P(G)$ in which the objective is to place as many facilities on $V(G)$ as possible with each $N[v_i]$ containing at most $|N[v_i]|$ of them.

Corollary 1.4 [27] *For any graph G with $|V(G)| = n$, we have $\rho(G) \leq \rho_f(G) = \gamma_f(G) \leq \gamma(G) \leq F(G) \leq F_f(G) = W_f(G) \leq W(G) \leq n \leq P(G) \leq P_f(G) = R_f(G) \leq R(G)$.*

The use of the Automorphism Class Theorem [16, 26] to simplify computing LP-problems is illustrated next. Let $g : V(G) \to [0, \infty)$ be any real valued function defined on vertex set $V(G)$ of a graph G. For each $v \in V(G)$ let $[v]$ denote the set of vertices in the automorphism class of v. Define $g^* : V(G) \to [0, \infty)$ by $g^*(v) = \sum_{x \in [v]} g(x)/|[v]|$.

Theorem 1.5 (A.C.T. [16, 27]) *If $g : V(G) \to [0, \infty)$ is a ρ_f, γ_f, F_f, W_f, P_f, or R_f function, then so is g^*.*

Thus each of these n variable LP-problems with an n-by-n constraint matrix can be reduced to a system of k variables and a k-by-k constraint matrix where k is the number of automorphism classes. For example, the Hajós graph H_j has only two equivalence classes. The A.C.T. guarantees an optimum solution f with $f(v_1) = f(v_4) = f(v_6) = a$ and $f(v_2) = f(v_3) = f(v_5) = b$.

$$
(11) \qquad R_f(H_j) = MIN \; 9a + 15b
$$
$$
\text{Subject to} \quad
\begin{aligned}
a + 2b &\geq 1 \\
2a + 3b &\geq 1 \\
a, b &\geq 0
\end{aligned}
$$

An optimum solution with $a = 0$ and $b = 1/2$ shows $P_f(H_j) = R_f(H_j) = 15/2$. (A P_f-function g has $g(v_1) = g(v_4) = g(v_6) = 5/2$ and $g(v_2) = g(v_3) = g(v_5) = 0$.) An integer valued CLOP-function f showing $P(H_j) = 7$ is $f(v_1) = f(v_4) = 2$, $f(v_6) = 3$, and $f(v_2) = f(v_3) = f(v_5) = 0$, and an integer valued redundance function showing $R(H_j) = 8$ is $g(v_1) = g(v_5) = 1$ and $g(v_2) = g(v_3) = g(v_4) = g(v_6) = 0$.

For the complete t-partite graph $K_{n1,n2,...,nt}$ on $n = n1 + n2 + ... + nt$ vertices the LP-formulation for γ_f is a system of n variables with n constraints. The A.C.T. reduces this to a system of only t variables and t constraining inequalities. With full details in [16], $K_{n1,n2,...,nt}$ is actually fractionally efficiently dominatable. Specifically, for $K_{4,6,9}$ one has a system of only three variables and three equations. A solution consists of letting $f(u) = 40/319$, $f(v) = 24/319$, and $f(w) = 15/319$ for every u, v, w of degree 15, 13, and 10, respectively. Then $f(N[u]) = (40 + 6 \cdot 24 + 9 \cdot 15)/319 = 1$, $f(N[v]) = (4 \cdot 40 + 24 + 9 \cdot 15)/319 = 1$, and $f(N[w]) = (4 \cdot 40 + 6 \cdot 24 + 15)/319 = 1$ and $\gamma_f(K_{4,6,9}) = 439/319$.

1.3   Complementarity and Duality: The 8-Cycle

The paradigm to be used for the concept of complementarity is the (vertex) covering number and the (vertex) independence number. The (vertex) covering number, denoted $\alpha(G)$, is the minimum number of vertices in a set $S \subseteq V(G)$ such that for every edge $uv \in E(G)$ we have $|S \cap \{u,v\}| \geq 1$; and the vertex independence number, denoted $\beta(G)$, is the maximum cardinality of a vertex set $S \subseteq V(G)$ such that for $uv \in E(G)$ we have $|S \cap \{u,v\}| \leq 1$. Thus $\alpha(G) = (H^t, MIN, \overline{I}_n, \overline{I}_m, \{0,1\})(G)$ and $\beta(G) = (H^t, MAX, \overline{I}_n, \overline{I}_m, \{0,1\})(G)$.

Obviously, α and β are not binary-duals because the dual for each problem involves H and not H^t. The binary-dual of $\alpha(G) = (H^t, MIN, \overline{I}_n, \overline{I}_m, \{0,1\})(G)$ is $(H, MAX, \overline{I}_m, \overline{I}_n, \{0,1\})(G) = \beta^1(G)$, the maximum size of an independent edge set (the matching number), and the binary dual of $\beta(G) = (H^t, MAX, \overline{I}_n, \overline{I}_m, \{0,1\})(G)$ is $(H, MIN, \overline{I}_m, \overline{I}_n, \{0,1\})(G) = \alpha^1(G)$, the edge covering number.

The relation between α and β is simply "complementation". Namely, if $S \subseteq V(G)$ then S covers $E(G)$ if and only if $V(G) - S$ is independent. In general, let $\mathcal{F}$ be any family of subsets of some set X. Define $M(X, \mathcal{F})$ and $m(X, \mathcal{F})$ as follows.

(12) $M(X, \mathcal{F}) = \mathcal{MAX}\{|\mathcal{S}| : \mathcal{S} \in \mathcal{F}\}$ and $m(X, \mathcal{F}) = \mathcal{MIN}\{|\mathcal{S}| : \mathcal{S} \in \mathcal{F}\}$

Families $\mathcal{F}_1$ and $\mathcal{F}_2$ of subsets of X will be called complement-related if $S \in \mathcal{F}_1$ if and only if $X - S \in \mathcal{F}_2$. Suppose $\mathcal{F}_1$ and $\mathcal{F}_2$ are complement-related. Since the complement of any set in $\mathcal{F}_1$ is in $\mathcal{F}_2$, $m(X, \mathcal{F}_2) \leq |X| - M(X, \mathcal{F}_1)$; since the complement of any set in $\mathcal{F}_2$ is in $\mathcal{F}_1$, $M(X, \mathcal{F}_1) \geq |X| - m(X, \mathcal{F}_2)$. Thus

$M(X, \mathcal{F}_1) + m(X, \mathcal{F}_2) = |X|$. Note that one could let $\mathcal{F}_1$ and $\mathcal{F}_2$ be the complement-related families of independent sets and covering sets, respectively. Then $M(V(G), \mathcal{F}_1) = \beta(G)$ and $m(V(G), \mathcal{F}_2) = \alpha(G)$ implies $\beta(G) + \alpha(G) = n$. Recall that $i(G)$, the lower independence number (or the independent domination number), is the minimum cardinality of a maximal independent set. In general, let $\mathcal{F}^+$ denote the family of those members of $\mathcal{F}$ which are set-theoretically maximal with respect to membership, and $\mathcal{F}^-$ those which are minimal. It is easily seen that if $\mathcal{F}_1$ and $\mathcal{F}_2$ are complement-related, then so are $\mathcal{F}_1^+$ and $\mathcal{F}_2^-$. Hence $m(X, \mathcal{F}_1^+) + M(X, \mathcal{F}_2^-) = |X|$.

Theorem 1.6 (Set Complementation [26]) *If families $\mathcal{F}_1$ and $\mathcal{F}_2$ of subsets of X are complement-related, then $M(X, \mathcal{F}_1) + m(X, \mathcal{F}_2) = |X| = m(X, \mathcal{F}_1^+) + M(X, \mathcal{F}_2^-)$.*

As noted, independence and covering are complementary properties. As in [24] let $\Lambda(G)$ denote the maximum cardinality of a minimal vertex cover of $E(G)$. If $S \subseteq V(G)$ then vertex v is an enclave of S if $N[v] \subseteq S$, and S is enclaveless if it has no enclaves. Observe that S is enclaveless if and only if $V(G) - S$ is a dominating set, and as in [26] let $\Psi(G)$ and $\psi(G)$ denote the maximum and minimum order of maximally enclaveless sets. Hence, $\Psi(G) = (N, MAX, \overline{1}_n, D, \{0, 1\})(G)$ and $\gamma(G) = (N, MIN, \overline{1}_n, \overline{1}_n, \{0, 1\})(G)$ are complementary properties.

Corollary 1.7

    (a) [12] $\alpha(G) + \beta(G) = |V(G)|$ *(covering/independence)*

    and [24] $\Lambda(G) + i(G) = |V(G)|$.

    (b) [26] $\gamma(G) + \Psi(G) = |V(G)|$ *(dominating/enclaveless)*

    and [26] $\Gamma(G) + \psi(G) = |V(G)|$.

Clearly Theorem 1.6 has any number of corollaries. For one example with $X = E(G)$ consider the dual of $\beta(G) = (H^t, MAX, \overline{1}_n, \overline{1}_m, \{0, 1\})(G)$, namely $\alpha^1(G) = (H, MIN, \overline{1}_m, \overline{1}_n, \{0, 1\})(G)$. Now consider for α^1 its complementary property. Set $S_1 \subseteq E(G)$ is an edge cover of $V(G)$ if and only if $E(G) - S_1$ has the property that for each $v \in V(G)$ at least one edge incident with v is not in $E(G) - S_1$, or, equivalently, $E(G) - S_1$ has at most $deg(v) - 1$ of the edges incident with v for each $v \in V(G)$. Thus, the complementary property for $\alpha^1 = (H, MIN, \overline{1}_m, \overline{1}_n, \{0, 1\})$ is $(H, MAX, \overline{1}_m, D - \overline{1}_n, \{0, 1\})$. Thinking about this in matrix terms, an edge of $E(G)$ corresponds to a column of the n-by-m binary matrix H. Thus an edge cover S_1 corresponds to a collection C of columns of H with the property that for each row of H one of the columns in C has a one in that row. Assuming the i-th row has r_i ones in it for $1 \leq i \leq n$, the complementary collection C^c of columns contains at most $r_i - 1$ columns with ones in row i. For H we had $r_i = deg(v_i)$.

This motivates the following definition of the complementary parameter for $(M, MIN, C, B, \{0, 1\})$. As noted, in Section 1.5 set $\{0,1\}$ will be generalized to subsets $Y \subseteq R$.

Let M be a k-by-h matrix $M = [m_{j,i}]$ with $1 \leq j \leq k$ and $1 \leq i \leq h$, $C = [c_1, c_2, ..., c_h]^t$ and $B = [b_1, b_2, ..., b_k]^t$. Denote the row sum $r_j = \sum_{i=j}^{h} m_{j,i}$, and let $L_M = [r_1, r_2, ..., r_k]^t$. A binary h-vector $X = [x_1, x_2, ..., x_h]^t$ is to be considered to be the characteristic function of a set S of columns of M, those columns with $x_j = 1$. Clearly $\overline{1}_h - X = [1 - x_1, 1 - x_2, ..., 1 - x_h]^t$ is the characteristic function of the complement of S. (For $S \subseteq \{1, 2, ..., h\}$ the characteristic function of S is binary vector $X_S = [e_1, e_2, ..., e_h]^t$ with $e_i = 1$ if and only if $i \in S$.)

(13) $(M, MIN, C, B, \{0, 1\})$

$$MIN \sum_{i=1}^{h} c_i x_i$$
Subject to $MX \geq B$
with $x_i \in \{0, 1\}$

Now X is a feasible solution for (13) if and only if $\overline{1}_h - X$ is a feasible solution for (14).

(14) $(M, MAX, C, L_M - B, \{0, 1\})$

$$MAX \sum_{i=1}^{h} c_i x_i$$
Subject to $M \cdot X \leq L_M - B$
with $x_i \in \{0, 1\}$

Definition 1.8 *Problems $(M, MIN, C, B, \{0, 1\})$ and $(M, MAX, C, L_M - B, \{0, 1\})$ are complements of each other, where $L_M = [r_1, r_2, ..., r_k]^t$ and $r_j = \sum_{i=1}^{h} m_{j,i}$ is the j-th row-sum of M.*

In Sections 1.4.1, 1.4.2, 1.4.3 and 1.4.4, respectively, $M = N$, A, H, and T will be considered. Note that $L_N = \overline{1}_n + D$, $L_A = D$, $L_H = D$, $L_{H^t} = [2, 2, ..., 2]^t = \overline{2}_m$ and $L_T = \overline{1}_n$. Y-duals have been defined and even after Y-complements are defined for suitable $Y \subseteq R$ the process of successively taking the duals and complements of a problem cannot result in a string of arbitrary length of distinct problems. The next result is easily verified by applying the definitions of dual and complementary problems.

Theorem 1.9 (Dual/Complementary 8-Cycle) *The process of successive complementations and dualizations starting with (M, MN, C, B, Y) produces at most eight distinct parameters, as illustrated in (15).*

(15) (M, MIN, C, B, Y) $\qquad$ (M^t, MAX, B, C, Y)

$\quad (M^t, MIN, L_M - B, C, Y)$ $\qquad$ $(M, MAX, C, L_M - B, Y)$

$\quad (M, MIN, L_{M^t} - C, L_M - B, Y)$ $\qquad$ $(M^t, MAX, L_M - B, L_{M^t} - C, Y)$

$\quad (M^t, MIN, B, L_{M^t} - C, Y)$ $\qquad$ $(M, MAX, L_{M^t} - C, B, Y)$

Recall that even when $Y = \{0, 1\}$ for an IP-problem, restricting Y to $\{0, 1\}$ for its dual might not be appropriate/interesting. For example, the dual of efficient domination $F(G) = (N, MAX, D^*, \bar{I}_n, \{0, 1\})(G)$ is closed neighborhood domination $W(G)$, but the only feasible solution for $W(G) = (N, MIN, \bar{I}_n, D^*, \{0, 1\})(G)$ is the trivial solution in which one uses all of $V(G)$. However, $W(G) = (N, MIN, \bar{I}_n, D^*, Z^+)(G)$ is quite interesting. As promised, generalized cases with suitable $Y \subseteq R$ will be considered in Section 1.5. Until then, whenever reasonable $Y = \{0, 1\}$ will be considered.

To motivate Theorem 1.10, a generalization of the set-complementation Theorem 1.6, consider $\alpha(G) = (H^t, MIN, \bar{I}_n, \bar{I}_m, \{0, 1\})(G)$ and assume that if we use v_i then there is a cost of c_i rather than having every $c_i = 1$. For $S \subseteq V(G)$ let $c(S) = \sum_{v_i \in S} c_i$, and so $c(V(G)) = \sum_{i=1}^{n} c_i$. Recall that S is a cover of $E(G)$ if and only if $V(G) - S$ is independent and that $\alpha(G) + \beta(G) = (H^t, MIN, \bar{I}_n, \bar{I}_m, \{0, 1\})(G) + (H^t, MAX, \bar{I}_n, \bar{I}_m, \{0, 1\})(G) = n$. More generally, we have $(H^t, MIN, C, \bar{I}_m, \{0, 1\})(G) + (H^t, MAX, C, \bar{I}_m, \{0, 1\})(G) = \sum_{i=1}^{n} c_i$.

Theorem 1.10 (Complementation Theorem) *Let M be any k-by-h matrix, C an h-tuple, B a k-tuple, and let $L_M = [r_1, r_2, ..., r_k]^t$ with row sum $r_j = \sum_{i=1}^{h} m_{j,i}$. Assume $(M, MIN, C, B, \{0, 1\})$ has a feasible solution S, that is, for some $S \subseteq \{1, 2, ..., h\}$ one has $M \cdot X_S \geq B$. Then $(M, MIN, C, B, \{0, 1\}) + (M, MAX, C, L_M - B, \{0, 1\}) = \sum_{i=1}^{h} c_i$.*

Proof. For each $S \subseteq \{1, 2, ..., h\}$ let $S^c = \{1, 2, ..., h\} - S$. Now $L_M = M \cdot \bar{I}_h = M(X_S + X_{S^c}) = M \cdot X_S + M \cdot X_{S^c}$. Simply observe that $M \cdot X_S \geq B$ if and only if $M \cdot X_{S^c} \leq L_M - B$. Hence, if S is an optimal feasible solution for $(M, MIN, C, B, \{0, 1\})$ with $c(S) = (M, MIN, C, B, \{0, 1\})$, then $(M, MAX, C, L_M - B, \{0, 1\}) \geq c(S^c) = (\sum_{i=1}^{h} c_i) - c(S)$. Thus $(M, MIN, C, B, \{0, 1\}) + (M, MAX, C, L_M - B, \{0, 1\}) \geq \sum_{i=1}^{h} c_i$. And if S is a feasible solution

for $(M, MAX, C, L_M - B, \{0,1\})$ and $c(S) = (M, MAX, C, L_M - B, \{0,1\})$, then
$$(M, MIN, C, B, \{0,1\}) \leq c(S^c) = (\sum_{i=1}^{h} c_i) - c(S). \text{ Thus } (M, MIN, C, B, \{0,1\}) +$$
$$(M, MAX, C, L_M - B, \{0,1\}) \leq \sum_{i=1}^{h} c_i. \ \square$$

This Complementation Theorem has any number of corollaries, several of which are presented throughout Section 1.4. An even more general Matrix Complementation Theorem is presented in Section 1.5.

1.4   Representation Matrices and Graphical Parameters

In this section $M = N$, A, H, and T will be considered and some of the graphical parameters that evolve from these matrix representations will be presented. Samples of theorems one can derive using complementarity and duality will also be presented. In the figures dual problems will be placed horizontally (for duality one assumes $Y = R^+$), and complementary problems are connected by diagonals. Set Y will be interpreted as is appropriate.

1.4.1   Closed neighborhood matrix N

The matrix N is an n-by-n matrix with $N^t = N$ and $L_N = D^* = [1 + deg v_1, ..., 1 + deg \ v_n]^t$. So when $B = \bar{1}_n$ then $L_N - \bar{1}_n = D$, and when $B = D$ then $L_n - B = \bar{1}_n$. Note, for example, that closed neighborhood domination W_f has $B = D^*$, $W_f(G) = (N, MIN, \bar{1}_n, D^*, R^+)$. Its complement with $L_n - D^* = \bar{0}$ is $(N, MAX, \bar{1}_n, \bar{0}, R^+)$ with only a trivial solution assigning every v_i a weight of zero, and the dual of $(N, MAX, \bar{1}_n, \bar{0}, R^+)$ is meaningless. When $B = L_M$ Theorem 1.9 will not produce the eight parameters as in (15).

One of the central points of this chapter is that new interesting/applicable parameters will be observed in the complement/dual strings suggested by Theorem 1.9. Such was the case for closed neighborhood order domination and packing $W(G)$ and $P(G)$ [19, 27, 28, 32, 33]. Let us first think about the choice of $\bar{1}_n$, D or D^* for vector C. Assume $Y = \{0,1\}$ and $f : V(G) \rightarrow \{0,1\}$. If every $c_i = 1$ (that is, $C = \bar{1}_n$) think about putting one token on each v_i with $f(v_i) = 1$. Then f is a dominating function if every $N[v_i]$ contains at least one token. Alternatively, for each v_i with $f(v_i) = 1$ place one token on each $v \in N[v_i]$, so the resulting number of tokens on v_j is $|\{v_i \in N[v_j] : f(v_i) = 1\}|$. Then f is a dominating function if every vertex has at least one token. For $Y = Z^+$ and each $c_i = 1 + deg(v_i)$ (that is, $C = D^*$) consider putting $f(v_i)$ tokens on each $v \in N[v_i]$. Then f is a closed neighborhood order dominating function if every v_i has at least $|N[v_i]|$ tokens on it. With $Y = \{0,1\}$ and $c_i = deg(v_i)$, for domination (such as modelling total domination) one might place a token on each $v \in N(v_i)$ when $f(v_i) = 1$. Alternatively, thinking of covering rather than dominating, consider placing one token on each edge incident with v_i when $f(v_i) = 1$. Specifically, the dual of $\Psi_f = (N, MAX, \bar{I}_n, D, R^+)$ is $(N, MIN, D, \bar{I}_n, R^+)$. Now we can interpret the $(N, MIN, D, \bar{I}_n, \{0,1\})$-problem to be to minimize the amount of coverage done by a dominating set. Call $(N, MIN, D, \bar{I}_n, \{0,1\})(G) = \eta(G)$ the *domination-coverage number* of G. The study of η [31] and the ten unlabelled N-parameters in Figure 1.4 is just beginning.

$$
\begin{array}{ll}
\gamma_Y & (N, MIN, \bar{I}_n, \bar{I}_n, Y) \\
\eta_Y & (N, MIN, D, \bar{I}_n, Y) \\
& (N, MIN, D, D, Y) \\
& (N, MIN, \bar{I}_n, D, Y) \\
W_Y & (N, MIN, \bar{I}_n, D^*, Y) \\
& (N, MIN, D^*, D, Y) \\
R_Y & (N, MIN, D^*, \bar{I}_n, Y) \\
& (N, MIN, D, D^*, Y) \\
& (N, MIN, D^*, D^*, Y)
\end{array}
\qquad
\begin{array}{ll}
(N, MAX, \bar{I}_n, \bar{I}_n, Y) & \rho_Y \\
(N, MAX, \bar{I}_n, D, Y) & \Psi_Y \\
(N, MAX, D, D, Y) & \\
(N, MAX, D, \bar{I}_n, Y) & \\
(N, MAX, D^*, \bar{I}_n, Y) & F_Y \\
(N, MAX, D, D^*, Y) & \\
(N, MAX, \bar{I}_n, D^*, Y) & P_Y \\
(N, MAX, D^*, D, Y) & \\
(N, MAX, D^*, D^*, Y) &
\end{array}
$$

Figure 1.4: 18 MIN/MAX closed neighborhood matrix problems with C and B being $\bar{I}_n$, D, and D^*.

As corollaries of Theorem 1.2 we have the next result.

Proposition 1.11

a.) $(N, MAX, \bar{I}_n, \bar{I}_n, \{0,1\}) \le (N, MAX, \bar{I}_n, \bar{I}_n, R^+) = (N, MIN, \bar{I}_n, \bar{I}_n, R^+) \le (N, MIN, \bar{I}_n, \bar{I}_n, \{0,1\})$.
That is, $\rho(G) \le \rho_{R^+}(G) = \gamma_{R^+}(G) \le \gamma(G)$.

b.) $(N, MAX, \bar{I}_n, D, \{0,1\}) \le (N, MAX, \bar{I}_n, D, R^+) = (N, MIN, D, \bar{I}_n, R^+) \le (N, MIN, D, \bar{I}_n, \{0,1\})$.
That is, $\Psi(G) \le \Psi_{R^+}(G) = \eta_{R^+}(G) \le \eta(G)$.

c.) $(N, MAX, D, D, \{0,1\}) \le (N, MAX, D, D, R^+) = (N, MIN, D, D, R^+) \le (N, MIN, D, D, \{0,1\})$.

d.) $(N, MAX, D, \bar{I}_n, \{0,1\}) \le (N, MAX, D, \bar{I}_n, R^+) = (N, MIN, \bar{I}_n, D, R^+) \le (N, MIN, \bar{I}_n, D, \{0,1\})$.

e.) $(N, MAX, D^*, \bar{I}_n, \{0,1\}) \leq (N, MAX, D^*, \bar{I}_n, R^+) =$
$(N, MIN, \bar{I}_n, D^*, R^+) \leq (N, MIN, \bar{I}_n, D^*, \{0,1\})$.
That is, $F(G) \leq F_{R^+}(G) = W_{R^+}(G) \leq W(G)$.

f.) $(N, MAX, D, D^*, \{0,1\}) \leq (N, MAX, D, D^*, R^+) =$
$(N, MIN, D^*, D, R^+) \leq (N, MIN, D^*, D, \{0,1\})$.

g.) $(N, MAX, \bar{I}_n, D^*, \{0,1\}) \leq (N, MAX, \bar{I}_n, D^*, R^+) =$
$(N, MIN, D^*, \bar{I}_n, R^+) \leq (N, MIN, D^*, \bar{I}_n, \{0,1\})$.
That is, $P(G) \leq P_{R^+}(G) = R_{R^+}(G) \leq R(G)$.

h.) $(N, MAX, D^*, D, \{0,1\}) \leq (N, MAX, D^*, D, R^+) =$
$(N, MIN, D, D^*, R^+) \leq (N, MIN, D, D^*, \{0,1\})$.

i.) $(N, MAX, D^*, D^*, \{0,1\}) \leq (N, MAX, D^*, D^*, R^+) =$
$(N, MIN, D^*, D^*, R^+) \leq (N, MIN, D^*, D^*, \{0,1\})$.

And for the six complementary pairs of Figure 1.4, Theorem 1.10 applies as follows. Note that when $C = \bar{I}_n$ then $\sum_{i=1}^{n} c_i = n$, when $C = D$ then $\sum_{i=1}^{n} c_i = \sum_{i=1}^{n} deg(v_i) = 2|E(G)| = 2m$, and when $C = D^*$ then $\sum_{i=1}^{n} c_i = n + 2m$.

Proposition 1.12

a.) $(N, MIN, \bar{I}_n, \bar{I}_n, \{0,1\})(G) + (N, MAX, \bar{I}_n, D, \{0,1\})(G) = n$. That is, $\gamma(G) + \Psi(G) = n$.

b.) $(N, MIN, D, \bar{I}_n, \{0,1\})(G) + (N, MAX, D, D, \{0,1\})(G) = 2m$.

c.) $(N, MIN, D, D, \{0,1\})(G) + (N, MAX, D, \bar{I}_n, \{0,1\})(G) = 2m$.

d.) $(N, MIN, \bar{I}_n, D, \{0,1\})(G) + (N, MAX, \bar{I}_n, \bar{I}_n, \{0,1\})(G) = n$.

e.) $(N, MIN, D^*, D, \{0,1\})(G) + (N, MAX, D^*, \bar{I}_n, \{0,1\})(G) = n + 2m$.

f.) $(N, MIN, D^*, \bar{I}_n, \{0,1\})(G) + (N, MAX, D^*, D, \{0,1\})(G) = n + 2m$.

1.4.2 Adjacency matrix A

The adjacency matrix A is also an n-by-n symmetric matrix, and $L_A = D$. Allowing C and B to be $\bar{I}_n$, D, and $D - \bar{I}_n$, one produces 18 problems precisely paralleling those in Figure 1.4.

γ_Y^{op} $(A, MIN, \bar{I}_n, \bar{I}_n, Y)$ $(A, MAX, \bar{I}_n, \bar{I}_n, Y)$ ρ_Y^{op}
$(A, MIN, D - \bar{I}_n, \bar{I}_n, Y)$ $(A, MAX, \bar{I}_n, D - \bar{I}_n, Y)$
$(A, MIN, D - \bar{I}_n, D - \bar{I}_n, Y)$ $(A, MAX, D - \bar{I}_n, D - \bar{I}_n, Y)$
$(A, MIN, \bar{I}_n, D - \bar{I}_n, Y)$ $(A, MAX, D - \bar{I}_n, \bar{I}_n, Y)$

W_Y^{op} $(A, MIN, \bar{I}_n, D, Y)$ $(A, MAX, D, \bar{I}_n, Y)$ F_Y^{op}
$(A, MIN, D, D - \bar{I}_n, Y)$ $(A, MAX, D - \bar{I}_n, D, Y)$

R_Y^{op} $(A, MIN, D, \bar{I}_n, Y)$ $(A, MAX, \bar{I}_n, D, Y)$ P_Y^{op}
$(A, MIN, D - \bar{I}_n, D, Y)$ $(A, MAX, D, D - \bar{I}_n, Y)$
(A, MIN, D, D, Y) (A, MAX, D, D, Y)

Figure 1.5: MIN/MAX open neighborhood matrix problems.

As noted, $\gamma_t = \gamma_{\{0,1\}}^{op}$ was introduced by Cockayne, Dawes and Hedetniemi [6], and $F_{\{0,1\}}^{op}$ is considered in [13]. To date the other parameters have not been studied.

Proposition 1.13

a.) $(A, MAX, \bar{I}_n, \bar{I}_n, \{0,1\}) \le (A, MAX, \bar{I}_n, \bar{I}_n, R^+) =$
$(A, MIN, \bar{I}_n, \bar{I}_n, R^+) \le (A, MIN, \bar{I}_n, \bar{I}_n, \{0,1\})$.
That is, $\rho^{op}(G) \le \rho_{R+}^{op}(G) = \gamma_{R+}^{op}(G) \le \gamma^{op}(G)$.

b.) $(A, MAX, \bar{I}_n, D - \bar{I}_n, \{0,1\}) \le (A, MAX, \bar{I}_n, D - \bar{I}_n, R^+) =$
$(A, MIN, D - \bar{I}_n, \bar{I}_n, R^+) \le (A, MIN, D - \bar{I}_n, \bar{I}_n, \{0,1\})$.

c.) $(A, MAX, D - \bar{I}_n, D - \bar{I}_n, \{0,1\}) = (A, MAX, D - \bar{I}_n, D - \bar{I}_n, R^+) =$
$(A, MIN, D - \bar{I}_n, D - \bar{I}_n, R^+) \le (A, MIN, D - \bar{I}_n, D - \bar{I}_n, \{0,1\})$.

d.) $(A, MAX, D - \bar{I}_n, \bar{I}_n, \{0,1\}) \le (A, MAX, D - \bar{I}_n, \bar{I}_n, R^+) =$
$(A, MIN, \bar{I}_n, D - \bar{I}_n, R^+) \le (A, MIN, \bar{I}_n, D - \bar{I}_n, \{0,1\})$.

e.) $F^{op}(G) \leq F^{op}_{R+}(G) = W^{op}_{R+}(G) \leq W^{op}(G).$

f.) $(A, MAX, D - \bar{I}_n, D, \{0,1\}) \leq (A, MAX, D - \bar{I}_n, D, R^+) =$
$(A, MIN, D, D - \bar{I}_n, R^+) \leq (A, MIN, D, D - \bar{I}_n, \{0,1\}).$

g.) $P^{op}(G) \leq P^{op}_{R+}(G) = R^{op}_{R+}(G) \leq R^{op}(G).$

h.) $(A, MAX, D, D - \bar{I}_n, \{0,1\}) \leq (A, MAX, D, D - \bar{I}_n, R^+) =$
$(A, MIN, D - \bar{I}_n, D, R^+) \leq (A, MIN, D - \bar{I}_n, D, \{0,1\}).$

i.) $(A, MAX, D, D, \{0,1\}) \leq (A, MAX, D, D, R^+) =$
$(A, MIN, D, D, R^+) \leq (A, MIN, D, D, \{0,1\}).$

Proposition 1.14

a.) $\gamma^{op}(G) + (A, MAX, \bar{I}_n, D - \bar{I}_n, \{0,1\})(G) = n.$

b.) $(A, MIN, D - \bar{I}_n, \bar{I}_n, \{0,1\})(G) + (A, MAX, D - \bar{I}_n, D - \bar{I}_n, \{0,1\})(G) =$
$2m - n.$

c.) $(A, MIN, D - \bar{I}_n, D - \bar{I}_n, \{0,1\})(G) + (A, MAX, D - \bar{I}_n, \bar{I}_n, \{0,1\})(G) =$
$2m - n.$

d.) $(A, MIN, \bar{I}_n, D - \bar{I}_n, \{0,1\})(G) + \rho^{op}(G) = n.$

e.) $(A, MIN, D, D - \bar{I}_n, \{0,1\})(G) + F^{op}(G) = 2m.$

f.) $R^{op}(G) + (A, MAX, D, D - \bar{I}_n, \{0,1\})(G) = 2m.$

Perhaps the most interesting of the as yet unstudied parameters from Figure 1.5 is $(A, MAX, \bar{I}_n, D - \bar{I}_n, \{0,1\})$. Recall for a set $S \subseteq V(G)$ that a vertex v is an enclave of S if $N[v] \subseteq S$, that is, $v \in S$ and $N(v) \subseteq S$. An isolate of S is a vertex $v \notin S$ with $N(v) \subseteq S$. The value of the parameter $(A, MAX, \bar{I}_n, D - \bar{I}_n, \{0,1\})$ is the maximum number of vertices in a set with no isolates or enclaves.

1.4.3   Incidence matrix H

The n-by-m incidence matrix H has row sum vector $L_H = D$, and its transpose H^t is m-by-n with $L_{H^t} = \bar{2}_m$. In this section the only choices considered for C and B are $\bar{1}_n$, $\bar{1}_m$, D and $D - \bar{1}_n$. These cases include the standard vertex and edge independence and vertex and edge covering parameters. Note that the condition $H^t \cdot X \geq \bar{1}_m$ implies that one has a vertex cover of $E(G)$, $H^t \cdot X \leq \bar{1}_m$ implies an independent vertex set, $H \cdot X \geq \bar{1}_n$ implies an edge cover of $V(G)$, and $H \cdot X \leq \bar{1}_n$ implies an independent edge set.

α_Y	$(H^t, MIN, \bar{1}_n, \bar{1}_m, Y)$	$(H, MAX, \bar{1}_m, \bar{1}_n, Y)$	β_Y^1
α_Y^1	$(H, MIN, \bar{1}_m, \bar{1}_n, Y)$	$(H^t, MAX, \bar{1}_n, \bar{1}_m, Y)$	β_Y
	$(H^t, MIN, D - \bar{1}_n, \bar{1}_m, Y)$	$(H, MAX, \bar{1}_m, D - \bar{1}_n, Y)$	
	$(H, MIN, \bar{1}_m, D - \bar{1}_n, Y)$	$(H^t, MAX, D - \bar{1}_n, \bar{1}_m, Y)$	
R_Y^{01}	$(H^t, MIN, D, \bar{1}_m, Y)$	$(H, MAX, \bar{1}_m, D, Y)$	P_Y^{10}
W_Y^{10}	$(H, MIN, \bar{1}_m, D, Y)$	$(H^t, MAX, D, \bar{1}_m, Y)$	F_Y^{01}

Figure 1.6: 12 MIN/MAX incidence matrix problems.

From Theorem 1.2 we have the following.

Proposition 1.15

a.) $\beta^1(G) \leq \beta_{R^+}^1(G) = \alpha_{R^+}(G) \leq \alpha(G)$.

b.) $\beta(G) \leq \beta_{R^+}(G) = \alpha_{R^+}^1(G) \leq \alpha^1(G)$.

c.) $(H, MAX, \bar{1}_m, D - \bar{1}_n, \{0, 1\})(G) \leq (H, MAX, \bar{1}_m, D - \bar{1}_n, R^+)(G) = (H^t, MIN, D - \bar{1}_n, \bar{1}_m, R^+)(G) \leq (H^t, MIN, D - \bar{1}_n, \bar{1}_m, \{0, 1\})(G)$.

d.) $(H^t, MAX, D - \bar{1}_n, \bar{1}_m, \{0, 1\})(G) \leq (H^t, MAX, D - \bar{1}_n, \bar{1}_m, R^+)(G) = (H, MIN, \bar{1}_m, D - \bar{1}_n, R^+)(G) \leq (H, MIN, \bar{1}_m, D - \bar{1}_n, \{0, 1\})(G)$.

e.) $P^{10}(G) \leq P_{R^+}^{10}(G) = R_{R^+}^{01}(G) \leq R^{01}(G)$.

f.) $F^{01}(G) \leq F_{R^+}^{01}(G) = W_{R^+}^{10}(G) \leq W^{10}(G)$.

The last four parameters of Figure 1.6 are quite interesting. $R^{01}(G)$ is the minimum total coverage done by a vertex set that covers every edge at least once; and, $F^{01}(G)$ is the maximum amount of vertex coverage one can have given that no edge is covered more than once (i.e., twice). That is, $R^{01}(G)$ is the redundance-coverage number, and $F^{01}(G)$ is the efficient-coverage number. These parameters have only recently been studied. (See [8].) And W_{R+}^{10} and P_{R+}^{01} parallel W_{R+} and P_{R+} from Figure 1.4. Namely, $W_{R+}^{10}(G)$ is the minimum weight of a function $f : E(G) \to R^{+}$ such that each vertex has as much weight incident with it as it would from $g : E(G) \to R^{+}$ with $g(uv) = 1$ for every edge uv, and P_{R+}^{01} is the maximum weight of an $f : E(G) \to R^{+}$ such that each vertex has no more weight incident with it than it does with the all-ones function g.

Combining Proposition 1.15 e. and f., mimicking Corollary 1.4 we have the following corollary.

Corollary 1.16 *For any graph G of size $|E(G)| = m$, we have $F^{01}(G) \le F_{R+}^{01}(G) = W_{R+}^{10}(G) \le W^{10}(G) \le m \le P^{10}(G) \le P_{R+}^{10}(G) = R_{R+}^{01}(G) \le R^{01}(G)$.*

Note that, unlike F and R, parameters F^{01} and R^{01} are complementary. Theorem 1.10 implies the following complementarity results.

Proposition 1.17

a.) $\alpha(G) + \beta(G) = n$.

b.) $\alpha^{1}(G) + (H, MAX, \overline{1}_m, D - \overline{1}_n, \{0,1\})(G) = m$.

c.) $(H^t, MIN, D - \overline{1}_n, \overline{1}_m, \{0,1\})(G) + (H^t, MAX, D - \overline{1}_n, \overline{1}_m, \{0,1\})(G) = 2m - n$.

d.) $(H, MIN, \overline{1}_m, D - \overline{1}_n, \{0,1\})(G) + \beta^{1}(G) = m$.

e.) $R^{01}(G) + F^{01}(G) = 2m$.

1.4.4   Total matrix T

As noted in Section 1.1, the transpose T^t of n-by-(n+m) total matrix $T = [N| - H]$ is used as the constraint matrix for independent domination, as presented in (7) of Section 1.1. Results like those in the previous three subsections for T are presented in Chris Smart's dissertation [30]. Here I will simply illustrate that taking the dual for parameter $i(G)$ results in an interesting parameter defined on $V(G) \cup E(G)$.

We have $i_{R^+}(G) = (T^t, MIN, \bar{1}_n, \bar{1}^*_{n+m}, R^+)(G)$, and its dual is $\rho^i_{R^+}(G) = (T, MAX, \bar{1}^*_{n+m}, \bar{1}_n, R^+)(G)$.

$$(16) \qquad \rho^i_{R^+}(G) = MAX \sum_{j=1}^{n} y_i + \sum_{k=1}^{m} (-1) z_k$$

$$\text{Subject to } T[y_1, ..., y_n, z_1, ..., z_m]^t \le \bar{1}_n$$
$$\text{with } y_j \ge 0, \, z_k \ge 0.$$

For $\rho^i_{R^+}(G)$ we want the maximum weight of a function $f : V(G) \cup E(G) \to R^+$ such that for each vertex $v \in V(G)$ the sum of the weights on vertices in $N[v]$ minus the sum of the edge weights incident to v is at most one. Let $\rho^i(G) = (T, MAX, \bar{1}^*_{n+m}, \bar{1}_n, \{0, 1\})(G)$.

Proposition 1.18

a.) $\rho^i(G) \le \rho^i_{R^+}(G) = i_{R^+}(G) \le i(G)$

b.) $\rho(G) \le \rho^i(G)$ *and* $\rho_{R^+}(G) \le \rho^i_{R^+}(G)$;

$\gamma(G) \le i(G)$ *and* $\gamma_{R^+}(G) \le i_{R^+}(G)$.

Recall from Proposition 1.11(a.) that $\rho(G) \le \rho_{R^+}(G) = \gamma_{R^+}(G) \le \gamma(G)$.

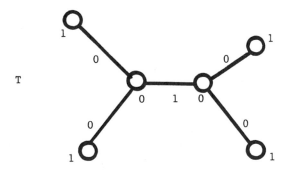

Figure 1.7: Tree T with $\rho(T) = \gamma(T) = 2$ and $\rho^i(T) = i(T) = 3$.

For the five cycle C_5 we have $\rho(C_5) = \rho^i(C_5) = 1$, $\rho_{R^+}(C_5) = \gamma_{R^+}(C_5) = \rho^i_{R^+}(C_5) = i_{R^+}(C_5) = 5/3$, and $\gamma(C_5) = i(C_5) = 2$. The tree T in Figure 1.7 provides an interesting example with $\gamma(T) < \rho^i(T)$. Specifically, $\rho(T) = \gamma(T) = 2$ and $\rho^i(T) = i(T) = 3$.

1.5   Generalized Graph Parameters

The beauty/simplicity of $\{0,1\}$-subset graph theory problems belies the fact that these problems are usually computationally intractable, while the $R^+ - LP$ problems are usually tractable. As noted, Farber investigated when solutions for the LP-domination problem $\gamma_f(G) = (N, MIN, \bar{1}_n, \bar{1}_n, [0,1])$ (which clearly equals $\gamma_{R^+}(G) = (N, MIN, \bar{1}_n, \bar{1}_n, R^+))$ would also be solutions for $\gamma(G) = (N, MIN, \bar{1}_n, \bar{1}_n, \{0,1\})$. Signed domination $\gamma_{\{-1,1\}}(G) = (N, MIN, \bar{1}_n, \bar{1}_n, \{-1,1\})$ and minus domination $\gamma_{\{-1,0,1\}}(G) = (N, MIN, \bar{1}_n, \bar{1}_n, \{-1,0,1\})$, respectively denoted $\gamma_s(G)$ and $\gamma^-(G)$, have been introduced in [10] and [9].

Theorem 1.19 [2] *If G has an efficient dominating set, then the cardinality of any efficient dominating set equals $\gamma(G)$. In particular, all efficient dominating sets have the same cardinality.*

McRae [25] conjectured that all signed (respectively, all minus) dominating functions would have the same weight. This motivated the definition of Y-domination for an arbitrary $Y \subseteq R$ in [4].

Theorem 1.20 [4] *For a graph G there exists a vector W satisfying $N \cdot W = \bar{1}_n$ if and only if $N \cdot X = N \cdot Z$ implies $\sum_i x_i = \sum_i z_i$. In particular, for any $Y \subseteq R$ any two efficient Y-dominating functions have the same weight.*

After Y-domination was first introduced in [4], other generalized graph parameters such as Y-independence, Y-covering, and Y-enclaveless (and Y-edge-independence and Y-edge-covering for $f : E(G) \to Y$) were introduced in [29]. Goddard and Henning [15] are studying Y-domination. The Y-valued minimization and maximization problems (8) and (9) defined in Section 1.1 extend these to arbitrary matrix problems. Complementation Theorem 1.10 will be generalized in this section.

First note that it does not extend for arbitrary Y. For a particular example, let $Y = \{-1,1\}$ and consider $\alpha_{\{-1,1\}} = (H^t, MIN, \bar{1}_n, \bar{1}_m, \{-1,1\})$ and $\beta_{\{-1,1\}} = (H^t, MAX, \bar{1}_n, \bar{1}_m, \{-1,1\})$. For complete graph K_5 we have $\alpha_{\{-1,1\}}(K_5) + \beta_{\{-1,1\}}(K_5) = 5 + (-3) = 2 \neq 5$. Before presenting the generalized complementation theorem, some further examples will be presented.

For the Hajós graph H_j in Figure 1.1, for $k \geq 1$ let $f(v_2) = f(v_3) = f(v_5) = k$ and $f(v_1) = f(v_4) = f(v_6) = (1 - 3k)/2$. Then $f(N[v_1]) = f(N[v_4]) =$

$f(N[v_6]) = (1 + k)/2 \geq 1$, and $f(N[v_2]) = f(N[v_3]) = f(N[v_5]) = 1$. So f is a dominating function of weight $w(f) = 3(1 - k)/2$. When we can find a Y-dominating function for a graph G of arbitrarily small weight, we write $\gamma_Y(G) = (N, MIN, \bar{1}_n, \bar{1}_n, Y)(G) = -\infty$. For example, $\gamma_R(H_j) = \gamma_Z(H_j) = -\infty$.

For the complementary parameter $\Psi_Y(H_j) = (N, MAX, \bar{1}_n, D, Y)(H_j)$, define $g : V(H_j) \to Y$ by $g(v_i) = 1 - f(v_i)$. Specifically, let $g(v_2) = g(v_3) = g(v_5) = 1 - k$ and $g(v_1) = g(v_4) = g(v_6) = (1 + 3k)/2$. Then $g(N[v_1]) = g(N[v_4]) = g(N[v_6]) = (5 - k)/2 = |N[v_1]| - f(N[v_1]) \leq |N[v_1]| - 1$, and $g(N[v_2]) = g(N[v_3]) = g(N[v_5]) = 4 = |N[v_2]| - 1$. So g is an enclaveless function of weight $w(g) = 3(3 + k)/2$, $\Psi_R(H_j) = \Psi_Z(H_j) = \infty$, and $w(f) + w(g) = 6 = |V(H_j)|$.

Again consider the Hajós graph, this time with $Y = (0, 1)$. Define $f : V(H_j) \to (0, 1)$ by $f(v_1) = f(v_4) = f(v_6) = \epsilon$ and $f(v_2) = f(v_3) = f(v_5) = (1 - \epsilon)/2$. One can verify that f is a dominating function of weight $w(f) = 3(1 + \epsilon)/2$. Although there is no $(0,1)$-dominating function f with $w(f) = 3/2$, these functions approach $3/2$ in weight as $\epsilon \to 0$. To handle examples like these we can actually define (8) and (9) in terms of infima and suprema.

Definition 1.21 *For any m-by-n matrix M, n-tuple C, m-tuple B, and $Y \subseteq R$,*

$$(17)\, (M, MIN, C, B, Y) = INF\{\sum_{i=1}^{n} c_i x_i \mid M \cdot X \geq B, x_i \in Y\}, \text{ and}$$

$$(M, MAX, C, B, Y) = SUP\{\sum_{i=1}^{n} c_i x_i \mid M \cdot X \leq B, x_i \in Y\}.$$

The next result is obvious.

Theorem 1.22 *If $Y1 \subseteq Y2 \subseteq R$, then $(M, MIN, C, B, Y2) \leq (M, MIN, C, B, Y1)$ and $(M, MAX, C, B, Y1) \leq (M, MAX, C, B, Y2)$.*

The Matrix Complementation Theorem, a generalization of Theorem 1.6 and of Theorem 8 in [29], will next be developed. In brief, the reason that $\alpha_{\{-1,1\}}(K_5) + \beta_{\{-1,1\}}(K_5) \neq |V(K_5)|$ is that the set $Y = \{-1, 1\}$ is not complementable. Call $Y \subseteq R$ a *complementable set* if $x \in Y$ implies $1 - x \in Y$. If Y is complementable and X is an n-tuple in Y^n, let $1 - X = [1 - x_1, 1 - x_2, ..., 1 - x_n]^t \in Y^n$.

Theorem 1.23 (Matrix Complementation) *For any m-by-n matrix M, n-tuple C, m-tuple B, and complementable set $Y \subseteq R$, let $L = [r_1, r_2, ..., r_m]^t$ be the row sum vector of M with $r_j = \sum_{i=1}^{n} m_{j,i}$. Then either $(M, MIN, C, B, Y) = -\infty$ and $(M, MAX, C, L - B, Y) = \infty$ or else $(M, MIN, C, B, Y) + (M, MAX, C, L - B, Y) = \sum_{i=1}^{n} c_i$.*

Proof. If $(M, MIN, C, B, Y) = k > -\infty$, then given $\epsilon > 0$ there exists an n-tuple X in Y^n with $M \cdot X \geq B$ and $\sum_{i=1}^{n} c_i x_i \leq k + \epsilon$. Now $M \cdot X \geq B$

if and only if for $1 \leq j \leq m$ we have $m_{j,1} x_1 + m_{j,2} x_2 + ... + m_{j,n} x_n \geq b_j$

if and only if for $1 \leq j \leq m$ we have $r_j - (m_{j,1} x_1 + ... + m_{j,n} x_n) \leq r_j - b_j$

if and only if for $1 \leq j \leq m$ we have $m_{j,1}(1 - x_1) + ... + m_{j,n}(1 - x_n) \leq r_j - b_j$

if and only if $M(\overline{1}_n - X) \leq L - B$. By assumption, $\overline{1}_n - X \in Y^n$.

Thus $(M, MAX, C, L-B, Y) \geq \sum_{i=1}^{n} c_i(1 - x_i) = \sum_{i=1}^{n} c_i - \sum_{i=1}^{n} c_i x_i \geq \sum_{i=1}^{n} c_i(k + \epsilon)$.

Hence $(M, MIN, C, B, Y) + (M, MAX, C, L - B, Y) \geq \sum_{i=1}^{n} c_i - \epsilon$.

And if $(M, MAX, C, L-B, Y) = j < \infty$, then given $\epsilon > 0$ there exists an n-tuple X with $M \cdot X \leq L - B$ and $\sum_{i=1}^{n} c_i x_i \geq j - \epsilon$. It follows that $M \cdot (\overline{1}_n - X) \geq B$,

and $(M, MIN, C, B, Y) \leq \sum_{i=1}^{n} c_i(1 - x_i) = \sum_{i=1}^{n} c_i - \sum_{i=1}^{n} c_i x_i \leq \sum_{i=1}^{n} c_i - j + \epsilon$.

Hence $(M, MIN, C, B, Y) + (M, MAX, C, L - B, Y) \leq \sum_{i=1}^{n} c_i + \epsilon$. Consequently,

$$(M, MIN, C, B, Y) + (M, MAX, C, L - B, Y) = \sum_{i=1}^{n} c_i.$$

If $(M, MIN, C, B, Y) = -\infty$ and $-\infty < k$, then there exists an n-tuple X in Y^n with $M \cdot X \geq B$ and $\sum_{i=1}^{n} c_i x_i < k$. As above, $M(\overline{1}_n - X) \leq L - B$ with $\overline{1}_n - X \in Y^n$ by complementarity. Thus $(M, MAX, C, L - B, Y) \geq \sum_{i=1}^{n} c_i - \sum_{i=1}^{n} c_i x_i \geq \sum_{i=1}^{n} c_i - k$. So $(M, MAX, C, L - B, Y) = \infty$. Similarly, $(M, MAX, C, L - B, Y) = \infty$ implies $(M, MIN, C, B, Y) = -\infty$, completing the proof. $\square$

Two examples will illustrate the many corollaries of this theorem: a weighted Gallai Theorem (covering/independence) and a weighted domination/enclaveless theorem. Assume each vertex $v_i \in V(G)$ has a weight c_i. For $S \subseteq V(G)$ define the weight of S to be $w(S) = \sum_{v_i \in S} c_i$. Let $w\alpha_Y$ and $w\gamma_Y$ denote the minimum weights of Y-covering and Y-dominating sets, respectively, using just $w\alpha$ and $w\gamma$ when $Y = \{0, 1\}$. Let $w\beta_Y$ and $w\Psi_Y$ denote the maximum weights of Y-independent and Y-enclaveless sets, respectively, using $w\beta$ and $w\Psi$ for $Y = \{0, 1\}$.

Corollary 1.24 *If Y is complementable, then*

 a.) $w\alpha_Y(G) + w\beta_Y(G) = w(V(G))$, *and*

 b.) $w\gamma_Y(G) + w\Psi_Y(G) = w(V(G))$.

 When $Y = \{0, 1\}$ we get the following.

Corollary 1.25

 a.) $w\alpha(G) + w\beta(G) = w(V(G))$, *and*

 b.) $w\gamma(G) + w\Psi(G) = w(V(G))$.

 With $Y = \{0, 1\}$ and $C = \bar{I}_n$, we get the following.

Corollary 1.26

 a.) $\alpha(G) + \beta(G) = n$, *and*

 b.) $\gamma(G) + \Psi(G) = n$.

 Corollary 1.26.a is Gallai's Theorem [12]; using different notation 1.26.b appeared in [26].

1.6 Future Study

The matrix format presented here suggests many new graph theoretic subset problems and a series of generalizations of those problems already studied. Note that Corollary 1.25 suggests the weighted extension of subset problems (where $Y = \{0, 1\}$). To my knowledge, little has been done to investigate even weighted domination problems.

 Goddard and Henning [15] have some computational complexity results for Y-domination. Results for other generalized parameters are only now being investigated. See Smart and Slater [32, 33] for computational complexity results on F, W, P and R.

Along with γ there is the upper domination parameter Γ, and β has its lower independence number i. Likewise, minimization problems formatted as in (8) have upper parameters, and maximization problems formatted as in (9) have lower parameters.

Theorem 1.23 illustrates a meta-theorem that applies across entire classes of parameters.

Bibliography

[1] D. W. Bange, A. E. Barkauskas, and P. J. Slater, Disjoint dominating sets in trees. Sandia Laboratories Report, SAND78-1087J, (1978).

[2] D. W. Bange, A. E. Barkauskas, and P. J. Slater, Efficient dominating sets in graphs. *Applications of Discrete Math.*, SIAM, Philadelphia (1988) 189-199.

[3] D. W. Bange, A. E. Barkauskas, L. H. Host and P. J. Slater, Efficient near-domination of grid graphs. *Congr. Numer.* 58 (1987) 83-92.

[4] D. W. Bange, A. E. Barkauskas, L. H. Host and P. J. Slater, Generalized domination and efficient domination in graphs. *Discrete Math.* 159 (1996) 1–11.

[5] N. Biggs, Perfect codes in graphs. *J. Combin. Theory Ser. B* 15 (1973) 289-296.

[6] E. J. Cockayne, R. M. Dawes, and S. T. Hedetniemi, Total domination in graphs. *Networks* 10 (1980) 211-219.

[7] G. S. Domke, S. T. Hedetniemi, and R. Laskar, Fractional packings, coverings, and irredundance in graphs. *Congr. Numer.* 66 (1988) 227-238.

[8] J. Dunbar, J. H. Hattingh, A. McRae, and P. J. Slater, Maximum efficient coverage and minimum redundant coverage in graphs. To appear in *Utilitas Math.*

[9] J. Dunbar, S. T. Hedetniemi, M. A. Henning, and A. McRae, Minus domination in graphs. Submitted for publication.

[10] J. Dunbar, S. T. Hedetniemi, M. A. Henning, and P. J. Slater, Signed domination in graphs. *Graph Theory, Combinatorics and Applications*, Eds. Y. Alavi and A. Schwenk, John Wiley and Sons, Inc. (1995) 311-321.

[11] M. Farber, Domination, independent domination, and duality in strongly chordal graphs. *Discrete Appl. Math.* 7 (1984) 115-130.

[12] T. Gallai, Uber extreme Punkt-und Kantenmengen. Ann. Univ. Sci. Budapest, Eotvos Sect. Math. 2 (1959) 133-138.

[13] H. Gavlas, K. Schultz, and P. J. Slater, Efficient open domination in graphs. Submitted for publication.

[14] W. Goddard, O. R. Oellermann, P. J. Slater and H. C. Swart, Bounds on the total redundance and efficiency of a graph. Submitted for publication.

[15] W. Goddard and M. A. Henning, personal communication.

[16] D. L. Grinstead and P. J. Slater, Fractional domination and fractional packing in graphs. *Congr.*

Numer. 71 (1990) 153-172.

[17] D. L. Grinstead and P. J. Slater, A recurrence template for several parameters in series-parallel graphs. *Discrete Appl. Math.* 54 (1994) 151-168.

[18] T. W. Haynes, S. T. Hedetniemi, and P. J. Slater, *Fundamentals of Domination in Graphs*, Marcel-Dekker, Inc., 1997.

[19] M. A. Henning and P. J. Slater, Closed neighborhood order dominating functions. Submitted for publication.

[20] T. W. Johnson and P. J. Slater, Maximum independent, minimally c-redundant sets in graphs. *Congr. Numer.* 74 (1990) 193-211.

[21] T. W. Johnson and P. J. Slater, Maximum independent, minimally redundant sets in series-parallel graphs. *Quaestiones Math.* 16 (1993) 351-370.

[22] J. G. Mauldoon, Covering theorems for groups. *Quart. J. Math. Oxford Ser. (2)* 1 (1950) 284-287.

[23] S. B. Maurer, Vertex colorings without isolates. *J. Combin. Theory Ser. B* 27 (1979) 294-319.

[24] J. D. McFall and R. Nowakowski, Strong independence in graphs. *Congr. Numer.* 29 (1980) 639-656.

[25] A. McRae, personal communication.

[26] P. J. Slater, Enclaveless sets and MK-systems. *J. Res. Nat. Bur. Standards* 82 (1977) 197-202.

[27] P. J. Slater, Closed neighborhood order domination and packing. *Congr. Numer.* 97 (1993) 33-43.

[28] P. J. Slater, Packing into closed neighborhoods. *Bull. Inst. Combin. Appl.* 13 (1995) 23-33.

[29] P. J. Slater, Generalized graph parameters: Gallai Theorems I. *Bull. Inst. Combin. Appl.* 17 (1996) 27-37.

[30] C. B. Smart, *Studies of Graph Based IP/LP Parameters*. Ph.D. Dissertation, University of Alabama in Huntsville (1996).

[31] C. B. Smart and P. J. Slater, On the domination-coverage number of a graph. To appear in *J. Combin. Math. Combin. Comp.*

[32] C. B. Smart and P. J. Slater, Complexity results for closed neighborhood order parameters. To appear in *Congr. Numer.*

[33] C. B. Smart and P. J. Slater, Closed neighborhood order parameters: complexity, algorithms, and inequalities. Submitted for publication.

[34] S. K. Zaremba, A covering theorem for abelian groups. *J. London Math. Soc.* 26 (1950) 71-72.

Chapter 2

Dominating Functions in Graphs

Michael A. Henning
Research Department of Mathematics
University of Natal
Private Bag X01
Pietermaritzburg, 3209 South Africa

Abstract. For an arbitrary subset $\mathcal{P}$ of the reals, we define a function $f: V \to \mathcal{P}$ to be a $\mathcal{P}$-dominating function of a graph $G = (V, E)$ if the sum of its function values over any closed neighbourhood is at least 1. That is, for every $v \in V$, $f(N(v) \cup \{v\}) \geq 1$. The $\mathcal{P}$-domination number of a graph G is defined to be the infimum of $f(V)$ taken over all $\mathcal{P}$-dominating functions f. When $\mathcal{P} = \{0, 1\}$ we obtain the standard domination number. When $\mathcal{P} = [0, 1]$, $\{-1, 0, 1\}$ or $\{-1, 1\}$ we obtain the fractional, minus or signed domination numbers, respectively. In this chapter, we survey some recent results concerning dominating functions in which negative weights are allowed.

2.1 Introduction

For a graph $G = (V, E)$ and for a real-valued function $f : V \to R$, the *weight* of f is $w(f) = \sum_{v \in V} f(v)$, and for $S \subseteq V$ we define $f(S) = \sum_{v \in S} f(v)$, so $w(f) = f(V)$. For a vertex v in V, we denote $f(N[v])$ by $f[v]$ for notational convenience. Let $f : V \to \{0, 1\}$ be a function which assigns to each vertex of a graph an element of the set $\{0, 1\}$. We say f is a *dominating function* if for every $v \in V$, $f[v] \geq 1$. We say f is a minimal dominating function if there does not exist a dominating function $g: V \to \{0, 1\}$, $f \neq g$, for which $g(v) \leq f(v)$ for every $v \in V$. This is equivalent to saying that a dominating function f is minimal if for every vertex v such that $f(v) > 0$, there exists a vertex $u \in N[v]$ for which $f[u] = 1$. Then the domination number and upper domination number of a graph G can be defined as $\gamma(G) = \min\{w(f) \mid f$ is a dominating function on $G\}$ and $\Gamma(G) = \max\{w(f) \mid f$ is a minimal dominating function on $G\}$.

Several authors have suggested changing the allowable weights. Well-known is fractional domination where the weights are allowed to be in the range $[0, 1]$. Reporting on results in [32], at the Eighteenth Southeastern International Conference Combinatorics, Graph Theory and Computing, S.T. Hedetniemi formally defined fractional domination as follows. For a graph $G = (V, E)$, a function $f : V \to [0, 1]$ is called a *fractional dominating function* of G if $f[v] \geq 1$ for each $v \in V$. The *fractional domination number* of G is given by $\gamma_f(G) = \min \{w(f) \mid f$ is a fractional dominating function for $G\}$. For the Hajös graph H in Figure 2.1, let $f(v_2) = f(v_3) = f(v_5) = 1/2$ and $f(v_1) = f(v_4) = f(v_6) = 0$, which makes $\gamma_f(F) \leq 3/2 = w(f)$. In fact, $\gamma_f(H) = 3/2$. For r-regular graphs on n vertices the fractional domination number is $n/(r + 1)$, which is attained by placing weights $1/(r+1)$ on each vertex. This fractional version of domination has been studied in [16, 21, 24, 27, 32] and elsewhere.

Figure 2.1: The Hajös graph H.

Recently, the idea of allowing negative weights was put forward. This resulted in minus domination [18], where f has codomain $\{-1, 0, 1\}$, signed domination [20], where f has codomain $\{-1, 1\}$, and $\mathcal{P}$-domination [1], where f has codomain $\mathcal{P}$ for an arbitrary subset $\mathcal{P}$ (called the *weight set*) of the reals $\mathbf{R}$. In this chapter, we survey some recent results concerning dominating functions in which negative weights are allowed.

2.2   Minus Domination in Graphs

2.2.1   Introduction

A *minus dominating function* is defined in [20] as a function $f : V \to \{-1, 0, 1\}$ such that $f[v] \geq 1$ for all $v \in V$. The *minus domination number* for a graph G is $\gamma^-(G) = \min\{w(f) \mid f$ is a minus dominating function on $G\}$. The *upper minus domination number* for a graph G is $\Gamma^-(G) = \max\{w(f) \mid f$ is a minimal minus dominating function on $G\}$. Minus domination is similar in many ways to ordinary domination, but also has different properties. For the graph H in Figure 2.1, let $f(v_2) = f(v_3) = f(v_5) = 1$ and $f(v_1) = f(v_4) = f(v_6) = -1$, making $\gamma^-(H) \leq 0 = w(f)$. In fact, $\gamma^-(H) = 0$. Minus domination has been studied in [17, 18, 19, 29, 36, 42] and elsewhere. In [18] various properties of the minus domination number are presented.

Proposition 2.1 *A minus dominating function f on a graph G is minimal if and only if for every vertex $v \in V$ with $f(v) \geq 0$, there exists a vertex $u \in N[v]$ with $f[u] = 1$.*

There are several possible applications for this variation of domination. By assigning the values $-1, 0$ or $+1$ to the vertices of a graph we can model such things as networks of positive and negative electrical charges, networks of positive and negative spins of electrons, and networks of people or organizations in which global decisions must be made (e.g. positive, negative or neutral responses or preferences). In such a context, for example, the minus domination number represents the minimum number of people whose positive votes can assure that all local groups of voters (represented by closed neighborhoods) have more positive than negative voters, even though the entire network may have far more people who vote negative than positive. By contrast, the upper minus domination number represents the greatest number of positive voters that may be required to offset a few negative voters, i.e., to insure that all local groups of voters have positive vote totals. Hence this variation of domination studies situations in which, in spite of the presence of negative vertices, the closed neighborhoods of all vertices are required to maintain a positive sum.

2.2.2 Relationships between domination and minus domination

Let D be a minimum dominating set in a graph $G = (V, E)$. Let $f: V \to \{-1, 0, 1\}$ be the characteristic function of D, i.e., $f(v) = 1$ if $v \in D$ and $f(v) = 0$ if $v \notin D$. Then f is a minus dominating function of G and so $\gamma^-(G) \leq w(f) = |D| = \gamma(G)$. Hence we have the following relationship between domination and minus domination.

Proposition 2.2 *For every graph G, $\gamma^-(G) \leq \gamma(G)$.*

The domination and minus domination number of a tree are related as follows.

Proposition 2.3 [18] *If T is a tree of order $n \geq 4$, then $\gamma(T) - \gamma^-(T) \leq (n-4)/5$.*

That the bound in Proposition 2.3 is sharp, may be seen by considering the tree T_k ($k \geq 1$) obtained from a path $v_1, v_2, \ldots, v_{3k+2}$ on $3k+2$ vertices by adding $2(k + 1)$ new vertices $\{u_{3i} + 1: 0 \leq i \leq k\} \cup \{u_{3i} + 2: 0 \leq i \leq k\}$, and joining u_i to v_i with an edge for each i. The tree T_2 is shown in Figure 2.2, together with a minimum minus dominating function for T_2. Then $\gamma(T_k) = 2(k + 1)$ and $\gamma^-(T_k) = k + 2$, and so $\gamma(T_k) - \gamma^-(T_k) = k = (|V(T_k)| - 4)/5$.

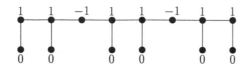

Figure 2.2: The tree T_2.

2.2.3   Graphs with positive or negative minus domination numbers

There exist graphs with minus domination numbers which are positive, negative or zero. Some families of graphs fall into one of these groups. This leads us to examine various classes of graphs in an attempt to classify them according to their minus domination numbers. First we consider the 'Negative Graphs.' In general, graphs with negative minus domination numbers have not been characterized. But we can find various families of graphs with minus domination number less than any negative integer. So far we know (see [18]) that bipartite graphs that contain cycles, chordal graphs, and planar graphs are Negative Graphs. For example, for any positive integer k, the outerplanar graph G_k shown in Figure 2.3 satisfies $\gamma^-(G_k) \leq -k$.

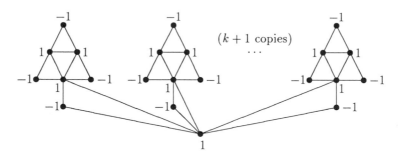

Figure 2.3: An outerplanar graph G_k with $\gamma^-(G_k) \leq -k$.

There are several graph properties which guarantee a nonnegative or positive minus domination number.

Proposition 2.4 [18] *If G is a graph with maximum degree $\Delta(G) \leq 5$, then $\gamma^-(G) \geq 0$.*

Proposition 2.5 [18] *If G is a graph with maximum degree $\Delta(G) \leq 3$, then $\gamma^-(G) \geq 1$.*

Next we consider the 'Positive Graphs.' We show that trees, cycles, regular graphs, and complete multipartite graphs are Positive Graphs. In [18] it is shown that $\gamma(G) = \gamma^-(G) = \lceil n/3 \rceil$ if G is a path P_n on n vertices or a cycle C_n on $n \geq 3$ vertices. Furthermore, for any complete multipartite graph $G \cong K(n_1, n_2, \ldots, n_t)$ $(t \geq 2)$ it is shown that $\gamma^-(G) = 1$ if $min\{n_i\} = 1$ and $\gamma^-(G) = 2$ otherwise.

Proposition 2.6 [18] *For every tree* T, $\gamma^-(T) \geq 1$ *with equality if and only if* T *is a star* $K_{1,n-1}$.

Zelinka [42] established that for every cubic graph G of order n, $\gamma^-(G) \geq n/4$. This result was generalized in [19] to r-regular graphs. The proof follows from simple counting arguments.

Proposition 2.7 [19] *For every* r-*regular graph* $G = (V, E)$ *of order* n, $\gamma^-(G) \geq \frac{n}{r+1}$ *and this bound is sharp.*

It remains an open problem to expand the classification of Positive and Negative graphs, depending on whether their minus domination numbers may be positive or can be arbitrarily negative.

2.2.4   Extremal results

For t an integer, let $p(t, \gamma^-)$ be the smallest order of a connected graph with minus domination number equal to t. For each integer $k \geq 1$, let $I_k = \{k(k + 1)/2,\ k(k+1)/2 + 1,\ \ldots,\ k(k+1)/2 + k\}$. Then the smallest integer in I_k is one larger than the largest integer in I_{k-1} (if $k > 1$), while the largest integer in I_k is one smaller than the smallest integer in I_{k+1}. Hence, each positive integer is contained in a unique interval I_k for some $k \geq 1$. The value of $p(-t, \gamma_s)$, for all integers $t \geq 1$, is as follows.

Proposition 2.8 *Let* $k \geq 1$ *be an integer, and let* $t \in I_k$. *Then* $p(-t, \gamma^-) = 2(k + 3) + t$.

That the bound in Proposition 2.8 is best possible, may be seen by considering the graph G consisting of a complete graph K_{k+3} on $k + 3$ vertices, with an additional $k + t + 3$ vertices of degree 2, each of which is adjacent to a distinct pair of vertices in the complete graph. Then G is a graph of order $2(k + 3) + t$ with $\gamma^-(G) = -t$.

Proposition 2.9 *For any integer* $t \geq 2$, $p(t, \gamma^-) = 2t$.

From the computational point of view, the problem of finding $\gamma^-(G)$ is difficult. Even if we restrict G to being bipartite, the corresponding decision problem is NP-complete (see Section 2.2.7). It is therefore desirable to find good upper bounds on this parameter. In [18], lower bounds on $\gamma^-(G)$ for a bipartite graph G are investigated.

Proposition 2.10 [18] *Let f be a minimum minus dominating function on a bipartite graph G, and let P be the set of vertices in G that are assigned the values $+1$ under f. Then $\gamma^-(G) \geq |P| - \lfloor |P|^2/4 \rfloor$.*

2.2.5 Upper minus domination

The following sequence of inequalities is well-known in domination theory: $ir \leq \gamma \leq i \leq \beta_0 \leq \Gamma \leq IR$. In this chain, ir and IR are the irredundance and upper irredundance numbers, respectively, and i and β_0 are the independent domination and vertex independence numbers. This inequality chain was first presented in [13] in 1978 and has been greatly enriched since then. In [18] it is shown that $\Gamma(G)$ and $\Gamma^-(G)$ are not comparable by showing that $\Gamma(G) < \Gamma^-(G)$ and $\Gamma^-(G) < \Gamma(G)$ are both possible. However for every graph G, it is shown that $\beta_0 \leq \Gamma^-(G)$. This is evident since the characteristic function of a maximum independent set of a graph is a minimal minus dominating function. Hence we have the following string of inequalities.

Proposition 2.11 *For any graph G, $\gamma^-(G) \leq \gamma(G) \leq i(G) \leq \beta_0(G) \leq \Gamma^-(G)$.*

Proposition 2.12 [18] *For any graph G of order $n \geq 3$, $\Gamma^-(G) \leq n - 2$.*

That the bound in Proposition 2.12 is sharp may be seen as follows: Let T be the tree obtained from a star $K_{1,n-1}$ $(n \geq 3)$ by subdividing each edge exactly once. The function that assigns to the central vertex of T the value -1, and to all remaining vertices the value 1, is a minimal minus dominating function of T of weight $|V(T)| - 2$.

2.2.6 Minus domination in cubic graphs

Hedetniemi [28] posed the following question: Does there exist a cubic graph for which $\gamma^-(G) < \gamma(G)$? This question was answered in the affirmative in [36] by constructing a cubic graph G of order 52 satisfying $\gamma^-(G) \leq 14$ and $\gamma(G) = 15$. That there exists an infinite class of cubic graphs in which the difference $\Gamma^- - \beta_0$ can be made arbitrarily large is also shown in [36]. In general, for a cubic graph it is shown in [36] that γ^- and ir are not comparable. We believe that for a cubic graph Γ and Γ^- are not comparable. Although $\Gamma < \Gamma^-$ is possible for a cubic graph, we have yet to settle the question whether there exist a cubic graph G satisfying $\Gamma^-(G) < \Gamma(G)$.

2.2.7 Algorithmic and complexity results

In [17] a variety of algorithmic results on the complexity of minus domination in graphs are presented. It is shown that the decision problem corresponding to the problem of computing $\gamma^-(G)$ is NP-complete, even when restricted to bipartite or chordal graphs, by describing a polynomial transformation from the decision problem **Dominating Set** (DM). The decision problem DM for the domination number of a graph is known to be NP-complete, even when restricted to bipartite graphs (see Dewdney [14]) or chordal graphs (see Booth [4] and Booth and Johnson [5]).

Next we consider the decision problem corresponding to the problem of computing $\Gamma^-(G)$. If a graph G is bipartite or chordal, then it is known that $\beta(G) = \Gamma(G)$, where $\beta(G)$ is the maximum cardinality of an independent set of G (see [11] and [37]). Since the maximum independent set problem can be solved in polynomial time for these two families of graphs, so too can the problem of finding $\Gamma(G)$ for G either bipartite or chordal. However, in [17] it is shown that the decision problem corresponding to the problem of computing $\Gamma^-(G)$ is NP-complete, even when restricted to bipartite or chordal graphs, by describing a polynomial transformation from the known NP-complete decision problem **One-In-Three 3SAT** (see [25]).

In [17] a linear algorithm for finding a minimum minus dominating function in a nontrivial tree T is presented. This algorithm is generalized in Section 2.4.9. As of this writing a linear algorithm for computing $\Gamma^-(T)$ for any tree T has not yet been developed, although we strongly suspect that one exists.

2.3 Signed Domination in Graphs

2.3.1 Introduction

A *signed dominating function* is defined in [20] as a function $f: V \to \{-1, 1\}$ such that $f[v] \geq 1$ for all $v \in V$. The *signed domination number* for a graph G is $\gamma_s(G) = \min\{ w(f) \mid f$ is a signed dominating function on $G \}$. The *upper signed domination number* for a graph G is $\Gamma_s(G) = \max\{ w(f) \mid f$ is a minimal signed dominating function on $G \}$. Signed domination has been studied in [20, 22, 29, 34, 36, 42] and elsewhere.

Proposition 2.13 [20] *A signed dominating function f on a graph G is minimal if and only if for every vertex $v \in V$ with $f(v) = 1$, there exists a vertex $u \in N[v]$ with $f[u] \in \{1, 2\}$.*

Proposition 2.14 [20] *If f is a signed dominating function for a graph G, then each endvertex and each vertex adjacent with an endvertex of G is assigned the value 1 under f.*

Proposition 2.15 [20] *Let G be a graph on n vertices. Then $\gamma_s(G) = n$ if and only if every vertex is either an endvertex or adjacent with an endvertex.*

We know that γ and γ_s are not comparable in general. The Hajös graph shown in Figure 2.1 has $\gamma_s(G) = 0$ while $\gamma(G) = 2$. On the other hand, $\gamma_s(K_2) = 2$ and $\gamma(K_2) = 1$. Since every signed dominating function is also a minus dominating function, we have $\gamma^-(G) \le \gamma_s(G)$ for every graph G. Table 2.1 gives a comparison of $\gamma(G)$ and $\gamma_s(G)$ for several classes of graphs. In these examples γ and γ^- are equal.

	G	$\gamma(G)$	$\gamma_s(G)$
$n \ge 2$	$K_{1,n-1}$	1	n
$n \ge 2$	P_n	$\lceil \frac{n}{3} \rceil$	$n - 2\lfloor \frac{n-2}{3} \rfloor$
$n \ge 3$	C_n	$\lceil \frac{n}{3} \rceil$	$n - 2\lfloor \frac{n}{3} \rfloor$
$r, s \ge 3$	$K_{r,s}$	2	$\begin{cases} 4 & \text{for } r, s \text{ even} \\ 5 & \text{for } r \text{ even, } s \text{ odd} \\ 6 & \text{for } r, s \text{ odd} \end{cases}$
$n \ge 1$	K_n	1	$\begin{cases} 2 & \text{for } n \text{ even} \\ 1 & \text{for } n \text{ odd} \end{cases}$

Table 2.1: A comparison of $\gamma_s(G)$ and $\gamma(G)$ for some families of graphs.

2.3.2   Signed domination for trees

For arbitrary graphs, sharp lower bounds for the signed domination number are not known. For trees, however, a sharp lower bound for γ_s is known.

Proposition 2.16 [20] *Let T be a tree of order $n \ge 2$. Then $\gamma_s(T) \ge (n+4)/3$ with equality if and only if T is a path on $3k + 2$ vertices for some nonnegative integer k.*

Another sharp lower bound for the signed domination number of a tree can be stated in terms of the independent domination number i.

Proposition 2.17 [20] *For every tree T on $n \ge 2$ vertices, $i(T) \le \gamma_s(T) - 1$.*

That the bound given in Proposition 2.17 is sharp may be seen by considering the path P_{3k+2}, $k \ge 0$, on $3k + 2$ vertices. Note that $i(P_{3k+2}) = k + 1 = \gamma_s(P_{3k+2}) - 1$. However for a tree, γ_s and β are not comparable. For any integer $k \ge 2$, let T be the tree obtained from $K_{1,2k}$ by subdividing k edges twice as shown in Figure 2.4. Then $\gamma_s(T) = 2k + 1$ and $\beta(T) = 3k$. This shows that there exist trees T for which $\gamma_s(T) < \beta(T)$. On the other hand, if $T \cong K_{1,n-1}$, then $\gamma_s(T) > \beta(T)$.

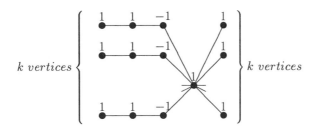

Figure 2.4: A tree with $\gamma_s(T) < \beta(T)$.

2.3.3   Bounds on γ_s

A lower bound on γ_s in terms of the degree sequence of the graph can be given.

Proposition 2.18 [20] *Let G be a graph of order n, the degrees d_i of whose vertices satisfy $d_1 \le d_2 \le \ldots \le d_n$. If k is the smallest integer for which*

$$d_{n-k+1} + d_{n-k+2} + \ldots + d_n - (d_1 + d_2 + \ldots + d_{n-k}) \ge 2(n - k),$$

then $\gamma_s(G) \ge 2k - n$.

To illustrate Proposition 2.18, consider the graph constructed as follows: for each pair of vertices of a complete graph K_5 on five vertices, introduce a new vertex adjacent to these two vertices. Let G denote the resulting graph. Then G has degree sequence $d_1, d_2, \ldots, d_{15}$, where $d_i = 2$ for $1 \le i \le 10$ and $d_i = 8$ for $11 \le i \le 15$. The smallest integer k for which $d_{16-k} + d_{17-k} + \ldots + d_{15} - (d_1 + d_2 + \ldots + d_{15-k}) \ge 30 - 2k$ is $k = 5$. Hence applying Proposition 2.18, we get $\gamma_s(G) \ge -5$. On the other hand, if g is the function on G defined by letting $g(v) = 1$ if $deg\, v = 8$ and letting $g(v) = -1$ if $deg\, v = 2$, then g is a signed dominating function on G of weight -5, so $\gamma_s(G) \le -5$. Hence $\gamma_s(G) = -5$.

Proposition 2.19 [20] *For every k-regular graph G, $\gamma_s(G) \ge n/(k + 1)$.*

For k even, the bound in Proposition 2.19 is sharp. Consider a complete graph on $k + 1$ vertices. Assigning the value -1 to $k/2$ vertices and the value 1 to the remaining $1 + k/2$ vertices produces a signed dominating function of weight $n/(k + 1) = 1$. In [20] an infinite family of k-regular graphs G of order n for which $\gamma_s(G) = n/(k + 1)$ with k even is described. For k odd, the bound on γ_s given in Proposition 2.19 can be improved.

Proposition 2.20 [36] *For every k-regular graph G of order n with k odd, $\gamma_s(G) \ge 2n/(k + 1)$.*

That the bound in Proposition 2.20 is sharp, may be seen by considering a complete graph on $k + 1$ vertices with k odd and assigning to $(k - 1)/2$ vertices the value -1, and to the remaining $(k + 3)/2$ vertices the value 1 to produce a signed dominating function of weight $2n/(k + 1) = 2$. In [36] an infinite family of k-regular graphs G of order n for which $\gamma_s(G) = 2n/(k + 1)$ with k odd is described.

The parameters γ_s and IR, and the parameters γ_s and i, for a cubic graph are related as follows.

Proposition 2.21 [36] *If G is a cubic graph, then $IR(G) \leq \frac{n}{2} \leq \gamma_s(G)$.*

Proposition 2.22 *If G is a cubic graph, then $i(G) \leq \gamma_s(G) - 1$ and this bound is sharp.*

In [36] an infinite family of cubic graphs G for which $IR(G) = \gamma_s(G)$ is established. Also, in [36] the problem of finding a good upper bound on γ_s and Γ_s for a cubic graph is mentioned. Subsequently, Zelinka [42] showed that for every cubic graph G of order n, $\gamma_s(G) \leq \frac{4}{5}n$. Henning [28] generalized this result to r-regular graphs.

Proposition 2.23 [28] *For every r-regular $(r \geq 2)$ graph $G = (V, E)$ of order n,*

$$
\gamma_s(G) \leq
\begin{cases}
\left(\frac{(r+1)^2}{r^2 + 4r - 1} \right) n & \text{for } r \text{ odd} \\
\left(\frac{r+1}{r+3} \right) n & \text{for } r \text{ even.}
\end{cases}
$$

Recently Favaron [22] showed that the bounds in Proposition 2.23 are sharp and improved Zelinka's result.

Proposition 2.24 [22] *Every connected cubic graph of order n different from the Petersen graph has signed domination number at most $3n/4$.*

2.3.4 Algorithmic and complexity results

In [30] a variety of algorithmic results on the complexity of signed domination in graphs are presented. It is shown that the decision problem corresponding to the problem of computing $\gamma_s(G)$ is NP-complete, even when restricted to bipartite or chordal graphs, by describing a polynomial transformation from the decision problem **Dominating Set** (DM).

In [30], it is shown that the decision problem corresponding to the problem of computing $\Gamma_s(G)$ is NP-complete, even when restricted to bipartite or chordal graphs, by describing a polynomial transformation from the known NP-complete decision problem **One-In-Three 3SAT** (see [25]). Furthermore, a linear algorithm for finding a minimum signed dominating function in a nontrivial tree T is presented in [30]. As of this writing a linear algorithm for computing $\Gamma_s(T)$ for any tree T has not yet been developed, although we strongly suspect that one exists.

2.4 Real and Integer Domination

2.4.1 Introduction

Recently, Bange et al. [1] introduced the generalization of $\{0,1\}$-domination to $\mathcal{P}$-domination for an arbitrary subset $\mathcal{P}$ of the reals $\mathbf{R}$. A function $f: V \to \mathcal{P}$ is a $\mathcal{P}$-dominating function if the sum of its function values over every closed neighbourhood is at least 1. That is, for every $v \in V$, $f[v] \geq 1$. The $\mathcal{P}$-domination number of a graph G, denoted $\gamma_{\mathcal{P}}(G)$, is defined to be the infimum of $w(f)$ taken over all $\mathcal{P}$-dominating functions f. Of course this might be $-\infty$. For example, if $\mathcal{P} = \mathbf{Z}$ and α is a positive integer, then for the tree T shown in Figure 2.5, $\gamma_{\mathcal{P}}(T) \leq 2 - 2\alpha$. As we can make α as large as we like, it is evident that $\gamma_{\mathcal{P}}(T) = -\infty$.

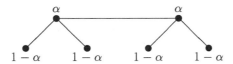

Figure 2.5: A tree T with $\gamma_{\mathbf{Z}} = -\infty$.

When $\mathcal{P} = \{0,1\}$ we obtain the standard domination number. When $\mathcal{P} = [0,1]$, $\{-1,0,1\}$ or $\{-1,1\}$ we obtain the fractional, minus or signed domination numbers, respectively. A trivial observation is that if $\mathcal{P} \subseteq \mathcal{Q}$, then $\gamma_{\mathcal{P}}(G) \geq \gamma_{\mathcal{Q}}(G)$. However, unlike the domination number the $\mathcal{P}$-domination number is not necessarily monotonic. For example, if one removes the central edge from the tree T in Figure 2.5 then the resultant graph has $\gamma_{\mathbf{Z}} = 2$.

The relationship between domination and linear programming has been exploited by several authors. For, the concept of domination can be formulated in terms of solving a $\{0,1\}$-integer programming problem. For $S \subseteq V$, we let $\mathbf{x}(S) = (x_1, x_2, \ldots, x_n)^t$ be the column vector with $x_i = 1$ if $v_i \in S$, and $x_i = 0$ if $v_i \notin S$. Then S is a dominating set if and only if $N \cdot \mathbf{x}(S) \geq \mathbf{1}$, where $\mathbf{1}$ denotes the all 1's column vector in $\mathbf{R}^n$. So $\gamma(G) = \min \sum_{i=1}^{n} x_i$ subject to $N \cdot \mathbf{x} \geq \mathbf{1}$ and $\mathbf{x} = (x_1, x_2, \ldots, x_n)^t$ with $x_i \in \{0,1\}$. For example, for the graph H in Figure 2.1, $\gamma(H) = 2$ and $\mathbf{x} = (0,1,0,0,0,1)^t$ is the characteristic function of the dominating set $\{v_2, v_6\}$. In general a function $f: V \to \mathcal{P}$ may clearly be thought of as a vector $\mathbf{f}$ in $\mathcal{P}^n$. We say that $\mathbf{f}$ is a $\mathcal{P}$-dominating vector if and only if $N\mathbf{f} \geq \mathbf{1}$. Unless otherwise stated, all remaining results in this section are from [26].

2.4.2 Real domination

In this subsection we show that there is a simple solution to the **R**-domination number of a graph. Let $\mathcal{P}$ be a subset of the reals **R**. We say a function $f: V \to \mathcal{P}$ is an *efficient $\mathcal{P}$-dominating function* if for every vertex v it holds that $f[v] = 1$. Equivalently, $N\mathbf{f} = \mathbf{1}$ where $\mathbf{1}$ denotes the all 1's vector in $\mathbf{R}^n$. For example, consider the graph shown in Figure 2.1. The function that assigns 1 to vertices v_2, v_3 and v_5, and -1 to vertices v_1, v_4, and v_6, is an efficient $\{-1, 1\}$-dominating function.

Bange et al. [1] generalized a conjecture of McRae and showed that all efficient $\mathcal{P}$-dominating functions for a graph have the same weight. The following simple proof of this result is provided in [26].

Proposition 2.25 [1] *If f_1 and f_2 are any two efficient $\mathcal{P}$-dominating functions for a graph G, then $w(f_1) = w(f_2)$.*

Proof. Since N is symmetric, $\mathbf{f}_1^t N \mathbf{f}_2 = (\mathbf{f}_1^t N) \mathbf{f}_2 = (N \mathbf{f}_1)^t \mathbf{f}_2 = \mathbf{1}^t \mathbf{f}_2 = w(f_2)$, and, $\mathbf{f}_1^t N \mathbf{f}_2 = \mathbf{f}_1^t (N \mathbf{f}_2) = \mathbf{f}_1^t \mathbf{1} = w(f_1)$. Thus $w(f_1) = w(f_2)$, as required. $\square$

A function is *nonnegative* if all the function values are nonnegative. We denote a function which is both nonnegative and efficient $\mathcal{P}$-dominating as an *NE$\mathcal{P}$D-function*. For example, if G is a regular graph of degree r, then the function f that assigns to each vertex the value $1/(r+1)$ is an NE$\mathcal{P}$D-function for G. If G is a complete bipartite graph of order at least 3 with one partite set $\mathcal{L}$ of cardinality l and the other $\mathcal{R}$ of cardinality r, then the function f that assigns to each vertex of $\mathcal{L}$ the value $(r-1)/(lr-1)$ and to each vertex of $\mathcal{R}$ the value $(l-1)/(lr-1)$ is an NE$\mathcal{P}$D-function for G.

Grinstead and Slater [27] called a graph which has an NE**R**D-function "fractionally efficiently dominatable". Using linear programming duality they observed that if a graph G has a NE**R**D-function g then $\gamma_f(G) = w(g)$. We use related ideas to show that the property of possessing an NE**R**D-function is the key to the real domination number of a graph.

Proposition 2.26 *For any graph G,*

$$\gamma_{\mathbf{R}}(G) = \gamma_{\mathbf{Q}}(G) = \begin{cases} w(f) & \text{if } G \text{ has an } NE\mathbf{Q}D\text{-function } f, \\ -\infty & \text{otherwise.} \end{cases}$$

Proof. Of course, by linear algebra, if a graph has a NE**R**D-function then it has an NE**Q**D-function. The concept of real domination can be formulated in terms of solving the following linear programming problem.

Real Domination $\gamma_{\mathbf{R}}(G)$	**Dual**
$\min \mathbf{1}^t \mathbf{x} \left(= \min \sum_{i=1}^n x_i \right)$	$\max \mathbf{1}^t \mathbf{y} \left(= \max \sum_{i=1}^n y_i \right)$
subject to: $\begin{cases} N \cdot \mathbf{x} \geq \mathbf{1} \\ x_i \text{ unrestricted} \end{cases}$	subject to: $\begin{cases} N \cdot \mathbf{y} &= \mathbf{1} \\ y_i &\geq 0 \end{cases}$

The dual of the above linear programming problem is shown. Since the min problem has a feasible solution (simply take the characteristic function of any dominating set), there are only two possible categories into which solutions to the max and min problems can fall: (1) both problems have feasible solutions, in which case both objective functions have the same solutions; and (2) the max problem has no feasible solution, in which case the objective function for the min problem is unbounded below.

If (1) holds, then the max problem has a feasible solution. However, every feasible solution to the max problem corresponds to an NERD-function for the graph G. Hence the solution to the max problem is an NERD-function of maximum weight. However, by Proposition 2.25, all NERD-functions have the same weight, so the solution to the max problem (and therefore the min problem) is $w(f)$, where f is an arbitrary NERD-function for G; so $\gamma_\mathbf{R}(G) = w(f)$. If (2) holds, then $\gamma_\mathbf{R}(G) = -\infty$. $\square$

If f is an NE$\mathcal{P}$D-function of G, then $w(f) \geq \gamma_{\mathcal{P}}(G) \geq \gamma_\mathbf{R}(G) = w(f)$. Hence we have the following corollary of Proposition 2.26.

Corollary 2.27 *For any subset $\mathcal{P}$ of $\mathbf{R}$, if a graph G has an NE$\mathcal{P}$D-function f, then $\gamma_{\mathcal{P}}(G) = \gamma_\mathbf{R}(G) = w(f)$.*

If there exists a dominating function with total weight less than 1 the $\mathbf{R}$-domination number is $-\infty$. For example, we observed earlier that there is a $\{-1, 1\}$-dominating function of total weight 0 for the graph H shown in Figure 2.1. So $\gamma_\mathbf{R}(H) = -\infty$.

2.4.3   Integer domination

If $\mathcal{P} = \mathbf{R}$ or $\mathbf{Q}$, then the determination of $\gamma_{\mathcal{P}}(G)$ can be formulated in terms of solving a linear programming problem, and so can be computed in polynomial-time (see [38] and [39]). On the other hand, the determination of $\gamma(G)$, $\gamma^-(G)$ and $\gamma_s(G)$ has been shown to be NP-complete even when restricted to bipartite graphs (see [14, 17, 29]) and chordal graphs (see [4, 5, 17, 29]).

Consider $\mathcal{P}$-domination when $\mathcal{P} = \mathbf{Z}$. If $\gamma_\mathbf{Q}(G) = -\infty$ then $\gamma_\mathbf{Z}(G) = -\infty$; this follows since, if one takes a $\mathbf{Q}$-dominating function f and multiplies all the weights by the least common multiple of the weights' dominators, one obtains a $\mathbf{Z}$-dominating function. So if there is a $\mathbf{Q}$-dominating function of arbitrarily negative weight then there is a such a $\mathbf{Z}$-dominating function too.

However, it remains an open problem to determine the complexity of $\mathbf{Z}$-domination. Also we do not know of a graph-theoretic proof that shows that $\mathbf{Z}$-domination is a member of NP. This is, however, a consequence of the general result that integer programming is in NP (see [6]).

Nevertheless, for some families of graphs one can say more. A graph is *chordal* if it contains no cycle of length greater than three as an induced subgraph. A *strongly chordal graph* is a chordal graph that contains no induced trampoline, where a *trampoline* consists of a $2t$-cycle $v_1, v_2 \ldots, v_{2t}, v_1$ in which the vertices v_{2i} of even subscript form a complete graph on t vertices. Since Farber [21] showed that for strongly chordal graphs their fractional domination number is equal to their domination number, it follows from Proposition 2.26 and the above discussion that:

Proposition 2.28 *For any strongly chordal graph G,*

$$\gamma_{\mathbf{Z}}(G) = \left\{ \begin{array}{ll} \gamma_f(G) & \text{if G has an NERD-function,} \\ -\infty & \text{otherwise.} \end{array} \right.$$

2.4.4   NERD-functions and efficient dominating sets

When $\mathcal{P} = \{0, 1\}$ an efficient $\mathcal{P}$-dominating function of a graph G is the characteristic function of a so-called *efficient dominating set* D of G: $|N[v] \cap D| = 1$ for every $v \in V$. (Equivalently, D dominates G and $u, v \in D$ implies $d(u, v) \geq 3$.) Efficient dominating sets were introduced by Bange, Barkauskas, and Slater [2, 3]. As a special case of Corollary 2.27, we have the following result.

Corollary 2.29 *If a graph G has an efficient dominating set D, then $\gamma_{\mathbf{R}}(G) = \gamma(G)$.*

If a graph G has an efficient dominating set D, then G has a NERD-function (simply take the characteristic function of D). However, the converse is not true. Many graphs that do not have efficient dominating sets will have a NERD-function. For example, the graph G shown in Figure 2.6 has a NERD-function as illustrated, but does not have an efficient dominating set. Hence, the existence of a NERD-function does not necessarily imply that G has an efficient dominating set.

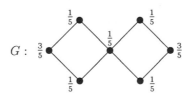

$$G :$$

Figure 2.6: A NERD-function of a graph G.

Proposition 2.30 [26] *For any tree T, T has a NERD-function if and only if it has an efficient dominating set.*

A *block* of a graph is a maximal 2-connected subgraph of the graph. A *complete block graph* is a connected graph in which every block is complete. In particular, if each block is isomorphic to K_2, then the complete block graph is a tree. Henning and Kubicki [35] extended Proposition 2.30 to complete block graphs.

Proposition 2.31 [35] *For any complete block graph G, G has a NERD-function if and only if it has an efficient dominating set.*

For what other classes of graphs does the existence of a NERD-function imply the existence of an efficient dominating set? Every regular graph has a NERD-function. However, for every integer $r \geq 2$, there exist regular graphs G of degree r that do not possess efficient dominating sets. For example, if $r = 2$, then let $G \cong C_n$ where $n \not\equiv 0 \pmod 3$, while for $r \geq 3$, let $G \cong K_{r-1} \times K_2$. The bipartite graph G shown in Figure 2.7(i) has a NERD-function as illustrated, but no efficient dominating set. The unicyclic graph G shown in Figure 2.7(i) has a NERD-function as illustrated, but does not possess an efficient dominating set. The graph H shown in Figure 2.7(ii) is both chordal and outerplanar in which every bounded region is a triangle. Although H has a NERD-function as illustrated, it has no efficient dominating set. Hence we have the following result.

Proposition 2.32 [35] *If a graph is bipartite, chordal, unicyclic, regular, or outerplanar in which every bounded region is a triangle, then the existence of a NERD-function does not necessarily imply the existence of an efficient domi-nating set.*

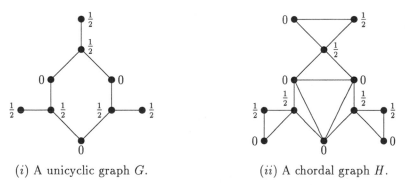

(i) A unicyclic graph G.   (ii) A chordal graph H.

Figure 2.7: NERD-functions of the graphs G and H.

Next we consider the family $\mathcal{U}$ of unicyclic graphs with the property that the vertices of the cycle dominate all other vertices. Equivalently, if $G \in \mathcal{U}$, then every vertex of G either lies on the cycle or is adjacent with a vertex of the cycle. In [35], those graphs G from $\mathcal{U}$ for which the following two conditions are equivalent are characterized:

(1) G has a NE**R**D-function;

(2) G has a efficient dominating set.

Every regular graph has a NE**R**D-function, so if G is a cycle, then it satisfies condition (1). A cycle has an efficient dominating set if and only if its length is a multiple of 3. Therefore, if $G \cong C_n$, then the conditions (1) and (2) are equivalent if and only if $n \equiv 0 \pmod 3$. Hence in what follows we restrict our attention to those graphs of $\mathcal{U}$ that are not cycles. We will refer to the edges that do not belong to the cycle as *legs*. Two nonadjacent legs are called *consecutive legs* if they are joined by a path every internal vertex of which has degree 2.

Proposition 2.33 [35] *Let $G = (V, E)$ be a graph from $\mathcal{U}$ that is not a cycle. Then G has a NE**R**D-function if and only if G has an efficient dominating set, unless G satisfies the following three conditions: (i) G has maximum degree 3, (ii) if G has exactly one leg, then the length of the cycle is congruent to 2 (mod 3), and (iii) if G has at least two legs, then it has an odd number of legs and a path that joins two consecutive legs, the internal vertices of which have degree 2, has length congruent to 2 (mod 3).*

2.4.5   Graphs for which $\gamma_{\mathbf{R}} = \gamma$ implies the existence of an efficient dominating set

As an immediate consequence of Corollary 2.29 and Proposition 2.31 we have the following result.

Corollary 2.34 *For any complete block graph G, $\gamma_{\mathbf{R}}(G) = \gamma(G)$ if and only if G has an efficient dominating set.*

For the class of regular graphs G, $\gamma_{\mathbf{R}}(G) = \gamma(G)$ implies that G has an efficient dominating set, as the following result shows.

Proposition 2.35 [35] *If G is a regular graph of degree r satisfying $\gamma_{\mathbf{R}}(G) = \gamma(G)$, then G has an efficient dominating set.*

As a consequence of Propositions 2.31 and 2.35, we have the following result.

Corollary 2.36 *For any regular graph G, $\gamma_{\mathbf{R}}(G) = \gamma(G)$ if and only if G has an efficient dominating set.*

However, in general a graph G satisfying $\gamma_{\mathbf{R}}(G) = \gamma(G)$ does not necessarily possess an efficient dominating set. For example, the graph G shown in Figure 2.7(i), which is bipartite, unicyclic, and a cactus, has a NE**R**D-function

f as illustrated, and so $\gamma_{\mathbf{R}}(G) = w(f) = 3$. Furthermore, it is evident that $\gamma(G) = 3$. However, the graph G does not possess an efficient dominating set. The graph H shown in Figure 2.7(ii), which is both chordal and outerplanar in which every bounded region is a triangle, has a NERD-function f as illustrated, and so $\gamma_{\mathbf{R}}(H) = w(f) = 3$. Furthermore, it is evident that $\gamma(H) = 3$. However, the graph H does not possess an efficient dominating set. Hence we have the following result.

Proposition 2.37 *If a graph G is bipartite, chordal, unicyclic, or outerplanar in which every bounded region is a triangle, then $\gamma_{\mathbf{R}}(G) = \gamma(G)$ does not necessarily imply that G has an efficient dominating set.*

2.4.6 Graphs for which the existence of a NERD-function implies $\gamma_{\mathbf{R}} = \gamma$

If G is a graph for which $\gamma_{\mathbf{R}}(G) = \gamma(G)$, then G has a NERD-function. This follows since if G has no NERD-function, then, by Proposition 2.26, $\gamma_{\mathbf{R}}(G) = -\infty$. As an immediate consequence of Corollary 2.34 and Proposition 2.31 we have the following result.

Corollary 2.38 *For any complete block graph G, $\gamma_{\mathbf{R}}(G) = \gamma(G)$ if and only if G has a NERD-function.*

Suppose that a strongly chordal graph $G = (V, E)$ has a NERD-function f. Since Farber [21] showed that for strongly chordal graphs their fractional domination number is equal to their domination number, it follows that $w(f) \geq \gamma_f(G) = \gamma(G) \geq \gamma_{\mathbf{R}}(G) = w(f)$. Thus we must have equality throughout. In particular, $\gamma_{\mathbf{R}}(G) = \gamma(G)$. Hence we have the following result.

Corollary 2.39 *For any strongly chordal graph G, $\gamma_{\mathbf{R}}(G) = \gamma(G)$ if and only if G has a NERD-function.*

In [35] it is conjectured that Corollary 2.39 can be extended to the family of all chordal graphs. Corollaries 2.38 and 2.39 may be restated as follows:

Corollary 2.40 *For any complete block graph or strongly chordal graph G,*

$$\gamma_{\mathbf{R}}(G) = \begin{cases} \gamma(G) & \textit{if } G \textit{ has a NERD-function,} \\ -\infty & \textit{otherwise.} \end{cases}$$

However, it is not true in general that if a graph G has a NERD-function then $\gamma_{\mathbf{R}}(G) = \gamma(G)$. We know that every regular graph of order n and of degree r satisfies $\gamma_{\mathbf{R}}(G) = n/(r+1)$. However, for every integer $r \geq 2$, there exist regular graphs G of degree r for which $\gamma(G) \neq n/(r+1)$. For example, if $r = 2$, then let $G \cong C_n$ where $n \not\equiv 0 \pmod 3$, while for $r \geq 3$, let $G \cong K_{r-1} \times K_2$. The bipartite graph G shown in Figure 2.6 has a NERD-function f as illustrated, so $\gamma_{\mathbf{R}}(G) = w(f) = 11/5 > 3 = \gamma(G)$. The unicyclic graph G shown in Figure 2.8 has a NERD-function f as illustrated, so $\gamma_{\mathbf{R}}(G) = w(f) = 7/3$, while $\gamma(G) = 3$. Hence we have the following result.

Proposition 2.41 *If a bipartite, unicyclic or regular graph G has a NERD-function, then it is not necessarily true that $\gamma_{\mathbf{R}}(G) = \gamma(G)$.*

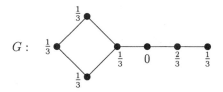

Figure 2.8: A unicyclic graph G with $\gamma_{\mathbf{R}}(G) < \gamma(G)$.

As an immediate consequence of Proposition 2.33, we have the following result.

Corollary 2.42 *If $G \in \mathcal{U}$ is not a cycle, then $\gamma_{\mathbf{R}}(G) = \gamma(G)$ if and only if G has a NERD-function.*

Let $\mathcal{F}$ denote a family of graphs, and let P_1 and P_2 denote the following two properties of the family $\mathcal{F}$.

P_1 : For every $G \in \mathcal{F}$, $\gamma_{\mathbf{R}}(G) = \gamma(G)$ implies that G has an efficient dominating set;

P_2 : For every $G \in \mathcal{F}$, the existence of a NERD-function for G implies that $\gamma_{\mathbf{R}}(G) = \gamma(G)$.

The following table summarizes the results of the last three subsections.

Family of graphs	complete block graphs	unicyclic graphs	bipartite graphs	regular graphs	chordal graphs	strongly chordal graphs
Property P_1 :	yes	no	no	yes	no	?
Property P_2 :	yes	no	no	no	?	yes

The question of whether P_2 implies P_1 has not yet been settled.

2.4.7 Products

Let $G = (V, E)$ and $H = (V', E')$ be two graphs with disjoint vertex sets. The *cartesian product* $G \times H$ has vertex set $V \times V'$ and two vertices (a, b) and (c, d) are adjacent if either $a = c$ and $bd \in E'$ or $b = d$ and $ac \in E$. The *strong product* $G \cdot H$ has vertex set $V \times V'$, and two vertices (a, b) and (c, d) are adjacent if $c \in N[a]$ and $d \in N[b]$.

Fisher et al. [23] observed that for all graphs G and H, $\gamma_f(G \cdot H) = \gamma_f(G)\gamma_f(H)$. From this they deduced that $\gamma_f(G \times H) \geq \gamma_f(G)\gamma_f(H)$, which established the fractional version of Vizing's conjecture.

For real domination there is no result analogous to his conjecture. For example, P_3 has an efficient dominating set, but $P_3 \times P_3$ does not have an NERD-function. So $\gamma_{\mathbf{R}}(P_3 \times P_3) = -\infty$ while $\gamma_{\mathbf{R}}(P_3) = 1$. However, the result on strong direct products does generalize. The reason is that if g and h are dominating functions then the function $f: V \times V' \to \mathbf{R}$ defined by $(a, b) \mapsto g(a)h(b)$ (and denoted by $f = g \otimes h$) is a dominating function of $G \cdot H$, and if g and h are both efficient then so is f. Furthermore, $w(f) = w(g) \times w(h)$.

Proposition 2.43 *For all graph G and H*

$$\gamma_{\mathbf{R}}(G \cdot H) = \begin{cases} \gamma_{\mathbf{R}}(G) \cdot \gamma_{\mathbf{R}}(H), & \text{if } \gamma_{\mathbf{R}}(G) \text{ and } \gamma_{\mathbf{R}}(H) \text{ both positive,} \\ -\infty, & \text{otherwise.} \end{cases}$$

Proof. If G and H have NERD-functions g and h, then $f = g \otimes h$ is an NERD-function for $G \cdot H$ and, by Theorem 2.26, $\gamma_{\mathbf{R}}(G \cdot H) = w(f) = w(g) \cdot w(h) = \gamma_{\mathbf{R}}(G) \cdot \gamma_{\mathbf{R}}(H)$.

On the other hand, suppose $\gamma_{\mathbf{R}}(G)$ say is $-\infty$. Then let h be any dominating function of H with positive total weight (for example the all 1s function). Then since G has a $\mathbf{R}$-dominating function g with arbitrarily negative weight, so does $G \cdot H$: namely, $g \otimes h$. This means that $\gamma_{\mathbf{R}}(G) = -\infty$. □

2.4.8 Hardness results

The following decision problem for the domination number of a graph is known to be NP-complete (see [25]), as was mentioned before.

Dominating Set (DM)

    Instance: A cubic (planar) graph $G = (V, E)$ and a positive integer k.

    Question: Does G have a dominating set of cardinality k or less?

    In this subsection we consider the general version:

$\mathcal{P}$-Dominating Function ($\mathcal{P}DM$)

    Instance: A graph $H = (V, E)$ and a positive integer j.

    Question: Does H have a $\mathcal{P}$-dominating function of weight j or less?

    In this subsection we provide an intractability proof for a large class of $\mathcal{P}$. We will demonstrate a polynomial-time reduction of the normal domination problem to the $\mathcal{P}$-domination problem. To do this, we introduce some notation. We define a *pendant* in a graph G as a subgraph on four vertices, three of which have degree 2 in G, that induce a 4-cycle. Equivalently, a pendant is an end-block that is a 4-cycle. The vertex of the pendant of degree more than 2 in G we call the "*attacher*" and the vertex not adjacent to the attacher we call the "*dangler*."

Proposition 2.44 [26] *Let j be a positive integer and let $\mathcal{P} = \{\, i \in \mathbf{Z} \mid i \leq j\, \}$. For any graph G, there exists a $\mathcal{P}$-dominating function of weight $\gamma_{\mathcal{P}}(G)$ such that for each pendant the two neighbours of the attacher have the same weight as the attacher.*

Using Proposition 2.44, we are now in a position to prove the following result.

Proposition 2.45 *Let j be a positive integer and let $\{0,1\} \subseteq \mathcal{P} \subseteq \{\, i \in \mathbf{Z} \mid i \leq j\, \}$. Then $\mathcal{P}$-Dominating Function is NP-complete (even for planar graphs).*

Proof. It is obvious that $\mathcal{P}$DM is a member of NP since we can, in polynomial time, guess at a function $f : V \to \mathcal{P}$ and verify that f has weight at most w and is a $\mathcal{P}$-dominating function. We next show how a polynomial-time algorithm for $\mathcal{P}$DM could be used to solve DM in polynomial time. Given a cubic graph $G = (V, E)$ on n vertices, and a positive integer k, construct the graph $H = (V', E')$ by attaching $x = \lceil (3j - 1)/2 \rceil$ pendants to each vertex of G. It is easy to see that the construction of the graph H can be accomplished in polynomial time, and that H is planar if G is. Let $\mathcal{Q} = \{\, i \in \mathbf{Z} \mid i \leq j\, \}$.

Claim $\gamma_{\mathcal{Q}}(H) = \gamma_{\mathcal{P}}(H) = \gamma(H) = nx + \gamma(G)$.

Proof. Let D be a minimum dominating set of G, and let D' be the extension of D to H' that includes the dangler of each pendant. Then D' is a dominating set of H of cardinality $nx + \gamma(G)$. Hence $\gamma_{\mathcal{Q}}(H) \leq \gamma_{\mathcal{P}}(H) \leq \gamma(H) \leq nx + \gamma(G)$.

To show that $nx + \gamma(G)$ is a lower bound on $\gamma_{\mathcal{Q}}(H)$, let f be a $\mathcal{Q}$-dominating function of H that satisfies the requirements of Proposition 2.44. Suppose one of the attachers a has a negative weight. Then, by Proposition 2.44, all its $2x$ neighbours in its pendants have negative weight. Thus, since G is cubic, $f[a] \leq -1 \cdot (2x + 1) + 3j \leq 0$, which produces a contradiction. Hence f assigns to all the attachers a nonnegative weight. If $f(a) = 0$, then, by Lemma 1, all its neighbours in the pendants also have weight 0. So one of its neighbours in G must have positive weight. This means that the attachers with positive weight form a dominating set of G. However the closed neighbourhood sum of each dangler can be shown to be 1. So the total weight of f is at least $nx + \gamma(G)$. $\square$

The above claim implies that if we let $w = nx + k$, then $\gamma(G) \leq k$ if and only if $\gamma_{\mathcal{P}}(H) \leq w$. This completes the proof of the proposition. $\square$

2.4.9   Integer intervals

In this subsection, we consider as weight set an interval of integers. Specifically, we let $\mathcal{P} = \{A, A+1, \ldots, B\}$, where A and B are integers satisfying $1 \leq B \leq 1 - A$. For example, $A = -1$ and $B = 1$ correspond to a minus dominating function. We showed in the previous subsection that the $\mathcal{P}$-domination problem is NP-complete for general graphs. Here we present a linear-time algorithm for finding a minimum $\mathcal{P}$-dominating function in a tree T. The algorithm is based on the one for minus domination given in [17].

Let $T_r = (V, E, r)$ be a rooted tree with root r and v a vertex of T. Then the *level number* $\ell(v)$ of v is the length of the unique r-v path in T. The maximum of the level numbers of the vertices of T is called the *height* of T and is denoted by $h(T)$. If a vertex v of T is adjacent to u and $\ell(u) > \ell(v)$, then u is called a *child* of v; if the level numbers of the vertices on the v-w path are monotonically increasing, then w is a *descendant* of v (and v is an *ancestor* of w). The subtree of T induced by v and all of its descendants is called the *maximal subtree* of T rooted at v. We will refer to an endvertex of T as a *leaf*, and the vertex adjacent to an endvertex as a *remote vertex*.

We introduce the following notation. For a rooted tree $T_r = (V, E, r)$ with root r, we call a function $f \colon V \to \mathcal{P}$ an *almost $\mathcal{P}$-dominating function* if the sum of its function values over every closed neighbourhood except that of the root is at least one, and the closed neighbourhood sum of the root is at least $1 - B$. We define $\gamma_{a\mathcal{P}}(T_r)$ to be the minimum weight $w(f)$ of an almost $\mathcal{P}$-dominating function f of the tree T_r. Any such f is called a minimum almost $\mathcal{P}$-dominating function. Note that if f is an almost $\mathcal{P}$-dominating function of T_r then f restricted to any maximal subtree is one also.

Furthermore, we define $\gamma'_{a\mathcal{P}}(T_r, \alpha)$ to be the minimum weight of an almost $\mathcal{P}$-dominating function of the tree T_r with r receiving weight α. We define $\gamma''_{a\mathcal{P}}(T_r, \theta)$ to be the minimum weight of an almost $\mathcal{P}$-dominating function of the tree T_r rooted at r with the sum of the values assigned to the vertices in the closed neighbourhood of r receiving weight θ.

The following recursive linear-time algorithm finds a minimum almost $\mathcal{P}$-dominating function in a tree T. The vertices of T are assigned values from the weight set $\mathcal{P}$ starting with the vertices at the highest level and ending with the root at the lowest level. Hence a vertex v receives a weight only once after all its descendants have received weights. For each vertex v, the algorithm associates a variable $ChildSum(v)$, which is the sum of the values assigned to the children of v.

Algorithm $\mathcal{P}D$:

Input: *A rooted tree $T = (V, E)$ on n vertices with the vertices labelled from 1 to n such that $label(w) > label(y)$ if the level of vertex w is less than the level of vertex y.*

For $i \leftarrow 1$ *to* n **do**

1. $ChildSum(i) \leftarrow$ *(sum of weights of the children of vertex i).*

2. **If** $ChildSum(i) < 1 - 2B$ **then**

   - *Increase the weights of the children of vertex i (so that each weight remains at most B) until $ChildSum(i) = 1 - 2B$.*

   - $f(i) \leftarrow B$.

3. **If** $ChildSum(i) \geq 1 - 2B$ **then**

   - *let $f(i)$ be the minimum weight that may be assigned to vertex i that yields an almost $\mathcal{P}$-dominating function on the maximal subtree rooted at i.*

end for

The following result verifies the validity of Algorithm $\mathcal{P}D$.

Theorem 2.46 [26] *Let $T_r = (V, E, r)$ be a rooted tree of order n with root r, and let f be the function produced by Algorithm $\mathcal{P}D$. Then the following five conditions hold:*

(1) The function f is a minimum almost $\mathcal{P}$-dominating function of the rooted tree T_r; so $\gamma_{a\mathcal{P}}(T_r) = w(f)$.

(2) The root r receives the maximum weight M over all minimum almost $\mathcal{P}$-dominating functions.

(3) The closed neighbourhood sum of r is the maximum value S over all minimum almost $\mathcal{P}$-dominating functions.

(4) $\gamma'_{a\mathcal{P}}(T_r, M + \alpha) = \gamma_{a\mathcal{P}}(T_r) + \alpha$ (for $0 \leq \alpha \leq B - M$).

(5) $\gamma''_{a\mathcal{P}}(T_r, S + \theta) = \gamma_{a\mathcal{P}}(T_r) + \theta$ (for $0 \leq \theta \leq B \cdot (\deg r) - S$).

2.4.10 Upper $\mathcal{P}$-domination

The fractional version of the upper domination number has been studied in [9, 8, 15, 24] and elsewhere. In this subsection we investigate the upper $\mathcal{P}$-domination number of a graph.

We say a $\mathcal{P}$-dominating function f is a *minimal $\mathcal{P}$-dominating function* if there does not exist a $\mathcal{P}$-dominating function h, $h \neq f$, such that $h(v) \leq f(v)$ for every $v \in V$. The *upper $\mathcal{P}$-domination number* for G is $\Gamma_{\mathcal{P}}(G) = \sup\{ f(V) \mid f : V \rightarrow \mathcal{P}$ is a minimal $\mathcal{P}$-dominating function on $G \}$. It is possible that

$\Gamma_\mathcal{P}(G) = \infty$; for example, if $\mathcal{P} = \mathbf{Z}$ and α is a large positive integer in $\mathcal{P}$, then the function that assigns to each leaf of a star $K_{1,n-1}$ ($n \geq 3$) the value $\alpha + 1$ and to the central vertex the value $-\alpha$ is a minimal $\mathcal{P}$-dominating function whose weight tends to $+\infty$ as $\alpha \to +\infty$. A trivial observation is that if $\mathcal{P} \subseteq \mathcal{Q}$, then it does not necessarily follow that $\Gamma_\mathcal{Q} \geq \Gamma_\mathcal{P}$. For example, if $\mathcal{P} = \{2\}$ and $\mathcal{Q} = \{1, 2\}$, then for every graph G of order n, $\Gamma_\mathcal{Q} < \Gamma_\mathcal{P} = 2n$. The next result shows that there is a simple solution to the upper $\mathcal{P}$-domination problem for $\mathcal{P} = \mathbf{Z}, \mathbf{Q}$ or $\mathbf{R}$.

Theorem 2.47 [26] *Let $\mathcal{P} = \mathbf{Z}$, $\mathbf{Q}$ or $\mathbf{R}$. Then for any connected graph G,*

$$\Gamma_\mathcal{P}(G) = \begin{cases} 1, & \text{if } G \text{ is complete,} \\ +\infty, & \text{otherwise.} \end{cases}$$

Corollary 2.48 [26] *Let $\mathcal{P} = \mathbf{Z}$, $\mathbf{Q}$ or $\mathbf{R}$. Then for any graph G, $\Gamma_\mathcal{P}(G)$ equals the number of components of G if every component is complete, and $+\infty$ otherwise.*

2.4.11   Other parameters

Several other parameters can be extended to allow negative weights [41]. In this subsection we consider a few such examples, including the independence number, the total domination number, the covering number, and the matching number of a graph. Unless otherwise stated all results in this section are from [26].

Let $G = (V, E)$ be a graph where $|V| = n$ and $|E| = m$. Let I denote the $n \times m$ incidence matrix of G. We say a function $f: V \to \mathcal{P}$ is a $\mathcal{P}$-*independence function* of G if for every edge e the sum of the values (weights) assigned under f to the two ends of e is at most 1. Or in vector notation, $I^t \mathbf{f} \leq \mathbf{1}$. The $\mathcal{P}$-*independence number* $\beta_\mathcal{P}(G)$ of G is defined to be the supremum of $w(f)$ taken over all $\mathcal{P}$-independence functions f. If $\mathcal{P} = \{0, 1\}$ then one obtains the normal independence number. An obvious lower bound on $\beta_\mathbf{R}$ is $n/2$ attained by assigning to every vertex a weight of $1/2$. The *lower $\mathcal{P}$-independence number* $i_\mathcal{P}(G)$ of G is the infimum of $w(f)$ taken over all maximal $\mathcal{P}$-independence functions f.

We say a function $g: E \to \mathcal{P}$ is an *efficient $\mathcal{P}$-matching function* of G if for every vertex v the sum of the values assigned under g to the edges incident with v is 1. For example, if G is a regular graph of degree r, then G has a nonnegative efficient $\mathbf{R}$-matching function: the function that assigns to each edge the value $1/r$. Note that if $\mathbf{g}$ is an efficient $\mathcal{P}$-matching vector, then the sum of the weights of the edges incident with each vertex is equal to 1, so $w(g) = n/2$. One can obtain a result similar to Proposition 2.26.

Proposition 2.49 *For any connected graph G on n vertices,*

$$\beta_{\mathbf{R}}(G) = \beta_{\mathbf{Q}}(G) = \begin{cases} n/2, & \text{if } G \text{ has a nonnegative efficient} \\ & \mathbf{R}\text{-matching function,} \\ +\infty, & \text{otherwise.} \end{cases}$$

Proof. We use the same argument as in the proof of Theorem 2.46. The real independence problem is to max $\mathbf{1}^t\mathbf{x}$ subject to $I^t\mathbf{x} \le \mathbf{1}$ and $\mathbf{x}$ unrestricted. The linear programming dual is to min $\mathbf{1}^t\mathbf{y}$ subject to $I\mathbf{y} = \mathbf{1}$ and $\mathbf{y} \ge \mathbf{0}$. Since the max problem is always feasible, there are only two possibilities. If the min problem is feasible then the solution to the max is the solution to the min, but we observed above that all efficient matching vectors have the same weight. If the min problem is not feasible then the solution to the max is $+\infty$. $\square$

If $\mathcal{P} = \mathbf{R}$ or $\mathbf{Q}$, then the determination of $\beta_{\mathcal{P}}(G)$ can be formulated in terms of solving a linear programming problem, and so can be computed in polynomial-time (see [38] and [39]). On the other hand, the determination of $\beta(G)$ has been shown to be NP-complete (cf. [25]). Consider $\mathcal{P}$-independence when $\mathcal{P} = \mathbf{Z}$.

Proposition 2.50 *For any connected bipartite graph G on n vertices,*

$$\beta_{\mathbf{Z}}(G) = \begin{cases} n/2, & \text{if } G \text{ has a perfect matching,} \\ +\infty, & \text{otherwise.} \end{cases}$$

Proof. Let G have partite sets $\mathcal{L}$ and $\mathcal{R}$ of cardinalities ℓ and r, respectively. If G has a perfect matching, then $\ell = r$ and the weight of any $\mathbf{Z}$-independence function of G is the sum of the weights of the two ends of the edges of the matching which is at most the number of edges in the matching, namely $n/2$. However the function that assigns to each vertex of $\mathcal{L}$ the weight 1 and to each vertex of $\mathcal{R}$ the weight 0 is an $\mathbf{Z}$-independence function of G of weight $n/2$. Hence $\beta_{\mathbf{Z}}(G) = n/2$.

If G has a perfect matching, then it follows from a well-known result of König [40] and Hall [28] that $|N(S)| < |S|$ for some nonempty subset S of $\mathcal{L}$. For α a positive integer, let f be the function defined as follows: $f(v) = \alpha$ if $v \in S$, $f(v) = 1 - \alpha$ if $v \in N(S)$, $f(v) = 1$ if $v \in \mathcal{L} - S$, and $f(v) = 0$ if $v \in \mathcal{R} - N(S)$. Then f is a $\mathbf{Z}$-independence function of G of weight $w(f) = \alpha|S| - (1-\alpha)|N(S)| + (\ell - |S|) \ge \alpha|S| - (1-\alpha)(1-|S|) + (\ell - |S|) = \alpha + \ell - 1$. As we can make α as large as we like, it is evident that $\beta_{\mathbf{Z}}(G) = +\infty$. $\square$

Corollary 2.51 *For any connected bipartite graph G on n vertices,*

$$\beta_{\mathbf{R}}(G) = \beta_{\mathbf{Q}}(G) = \beta_{\mathbf{Z}}(G).$$

Proof. If G has a perfect matching, then assign to each edge of the matching the weight 1 and to all remaining edges the weight 0 to produce a nonnegative efficient **R**-matching function of G. Hence, by Proposition 2.49, $\beta_{\mathbf{R}}(G) = \beta_{\mathbf{Q}}(G) = n/2$. However, by Proposition 2.50, $\beta_{\mathbf{Z}}(G) = n/2$. Conversely, if G has no perfect matching, then by Proposition 2.50, $\beta_{\mathbf{Z}}(G) = +\infty$. However, $\beta_{\mathbf{R}}(G) = \beta_{\mathbf{Q}}(G) \geq \beta_{\mathbf{Z}}(G)$ and so $\beta_{\mathbf{R}}(G) = +\infty$. $\square$

Corollary 2.52 *For any connected bipartite graph G on n vertices, G has a perfect matching if and only if G has a nonnegative efficient **R**-matching function.*

Another parameter that can be extended is the total domination number which was first introduced and studied in [10]. Let $\mathcal{P}$ be a subset of the reals **R**. A function $f: V \to \mathcal{P}$ is said to be a *total $\mathcal{P}$-dominating function* if the sum of its values over every *open* neighbourhood is at least 1. That is, for every $v \in V$, $f(N(v)) \geq 1$, or in vector notation, $Af \geq 1$ where A is the adjacency matrix. Then $\gamma^t(G)$ is the minimum weight of a total dominating function with range contained in $\{0, 1\}$, and $\gamma^t_{\mathcal{P}}(G)$ is the infimum weight of a total $\mathcal{P}$-dominating function.

We say that $f: V \to \mathcal{P}$ is an *efficient total $\mathcal{P}$-dominating function* if $Af = 1$ where 1 denotes the all 1's vector in $\mathbf{R}^n$. We denote a function which is both nonnegative and efficient $\mathcal{P}$-dominating as an *NET$\mathcal{P}$D-function*. For example, if H is the Hajös graph shown in Figure 2.1, then the function that assigns $1/2$ to vertices v_2, v_3 and v_5, and 0 to vertices v_1, v_4, and v_6, is an NET**R**D-function for G. One can readily establish a result similar to Proposition 2.26.

Proposition 2.53 *For any graph G,*

$$\gamma^t_{\mathbf{R}}(G) = \gamma^t_{\mathbf{Q}}(G) = \begin{cases} w(f) & \text{if } G \text{ has an NET}\mathbf{Q}\text{D-function } f, \\ -\infty & \text{otherwise.} \end{cases}$$

One interesting fact is that although $\gamma(G) \leq \gamma^t(G)$ for all G, no general inequality holds between $\gamma_{\mathbf{R}}(G)$ and $\gamma^t_{\mathbf{R}}(G)$. For the graph G in Figure 2.3 it is readily established that G has no NET**R**D-function, so $\gamma^t_{\mathbf{R}}(G) = -\infty$. However, $\gamma_{\mathbf{R}}(G) = 3/2$. Hence $\gamma^t_{\mathbf{R}}(G) < \gamma_{\mathbf{R}}(G)$ is possible. On the other hand, the Hajös graph H of Figure 2.1 has $\gamma_{\mathbf{R}}(H) = -\infty$, but H has an NET**R**D-function of weight $3/2$ so $\gamma^t_{\mathbf{R}}(H) = 3/2$. This shows that $\gamma_{\mathbf{R}}(G) < \gamma^t_{\mathbf{R}}(G)$ is possible.

Other general $\mathcal{P}$-valued functions have recently been studied by Slater [41]. A function $f: V \to \mathcal{P}$ is a *$\mathcal{P}$-covering* of G if for every edge e the sum of the values (weights) assigned under f to the two ends of e is at least 1. The *$\mathcal{P}$-covering number $\alpha_{\mathcal{P}}(G)$* of G is the infimum of $w(f)$ taken over all $\mathcal{P}$-covering functions f, while the *upper $\mathcal{P}$-covering number $\Lambda_{\mathcal{P}}(G)$* of G is the supremum of $w(f)$ taken over all minimal $\mathcal{P}$-covering functions f.

A function $f\colon E \to \mathcal{P}$ is a $\mathcal{P}$-*matching function* of G if for every $v \in V$ the sum of the values assigned under f to the edges incident with v is at most 1. The $\mathcal{P}$-*matching number* $\beta_{\mathcal{P}}^1(G)$ of G is the supremum of $w(f)$ taken over all $\mathcal{P}$-matching functions f of G.

A function $f\colon E \to \mathcal{P}$ is a $\mathcal{P}$-*edge covering function* of G if for every $v \in V$ the sum of the values assigned under f to the edges incident with v is at least 1. The $\mathcal{P}$-*edge covering number* $\alpha_{\mathcal{P}}^1(G)$ of G is the infimum of $w(f)$ taken over all $\mathcal{P}$-edge covering functions f of G.

A function $f\colon V \to \mathcal{P}$ is a $\mathcal{P}$-*enclaveless function* of G if for every $v \in V$ the sum of the values assigned under f to the vertices in the closed neighborhood of v is at most $|N(v)|$. The $\mathcal{P}$-*enclaveless number* $\psi_{\mathcal{P}}(G)$ of G is the supremum of $w(f)$ taken over all $\mathcal{P}$-enclaveless functions f, while the *lower $\mathcal{P}$-enclaveless number* $\Psi_{\mathcal{P}}(G)$ of G is the infimum of $w(f)$ taken over all maximal $\mathcal{P}$-enclaveless functions f.

In [41], Slater investigates Gallai-type theorems, theorems of the form $\alpha(G) + \beta(G) = n$, for general $\mathcal{P}$-valued functions. A subset $\mathcal{P}$ of the reals is defined to be *complementable* if $x \in \mathcal{P}$ implies $1 - x \in \mathcal{P}$. For example, the signed dominating and minus dominating sets $\{-1, 1\}$ and $\{-1, 0, 1\}$, respectively, are not complementable, but each $P_k = \{1 - k, 2 - k, 3 - k, \ldots, k - 1, k\}$ is a complementable set of order $2k$. The following result connects the above parameters.

Proposition 2.54 [41] *If $\mathcal{P} \subset \mathbf{R}$ is any finite complementable set, then for every graph G of order n,*

$$(a)\ \alpha_{\mathcal{P}}(G) + \beta_{\mathcal{P}}(G) = n \ and \ \Lambda_{\mathcal{P}}(G) + i_{\mathcal{P}}(G) = n,$$

and

$$(b)\ \gamma_{\mathcal{P}}(G) + \Psi_{\mathcal{P}}(G) = n \ and \ \Gamma_{\mathcal{P}}(G) + \psi_{\mathcal{P}}(G) = n.$$

2.5   Summary and Open Problems

In this chapter, we have surveyed some recent results concerning dominating functions in which negative weights are allowed. We have focused our attention on signed, minus, and real dominating functions in graphs. These parameters are often related. For example, Proposition 2.19 is a consequence of Proposition 2.7 which in turn is a consequence of Proposition 2.26. General results such as Proposition 2.54 serve to connect certain $\mathcal{P}$-valued parameters mentioned in this chapter.

The idea of $\mathcal{P}$-dominating functions in graphs provides numerous interesting theoretical and computational questions. Many problems have yet to be settled. We close this survey with a partial listing of some of these problems and open questions.

1. Is it true that, if G is a bipartite graph of order n, then $\gamma^-(G) \geq 4(\sqrt{n+1} - 1) - n$? (This is stated as a conjecture in [18]. If the conjecture is true, then the bound is sharp as shown in [18].)

2. Does there exist a linear algorithm for computing $\Gamma^-(T)$ (respectively, $\Gamma_s(T)$) for any tree T?

3. Does there exist a cubic graph G satisfying $\Gamma^-(G) < \Gamma(G)$?

4. Is it true that if G is a cubic graph, then $\Gamma(G) = IR(G)$?

5. Is it true that if G is a cubic graph, then $\Gamma^-(G) \leq \gamma_s(G)$?

6. For a cubic graph are Γ^- and Γ_s comparable? If so, how do they compare?

7. Does there exist a cubic graph for which $ir < \gamma < i < \beta_0 < \Gamma < IR$?

8. Is it true that if T is a tree of order n, then $(\Gamma_s(G_n) - \gamma_s(G_n))/n \leq 4/7$? If so, then this bound is easily shown to be asymptotically sharp.

9. Is it true that if G is a strongly chordal graph, then $\gamma_{\mathbf{R}}(G) = \gamma(G)$ implies that G has an efficient dominating set?

10. Is it true that for any chordal graph G, $\gamma_{\mathbf{R}}(G) = \gamma(G)$ if and only if G has a NERD-function?

Bibliography

[1] D.W. Bange, A.E. Barkauskas, L.H. Host, and P.J. Slater, Generalized domination and efficient domination in graphs. *Discrete Math.* 159 (1996) 1–11.

[2] D.W. Bange, A.E. Barkauskas, and P.J. Slater, Disjoint dominating sets in trees. Sandia Laboratories Report SAND 78-1087J (Sandia Laboratories, Albuquerque, 1978).

[3] D.W. Bange, A.E. Barkauskas, and P.J. Slater, Efficient dominating sets in graphs. in: R.D. Ringeisen and F.S. Roberts, eds., *Applications of Discrete Mathematics* (SIAM, Philadelphia, 1988) 189–199.

[4] K.S. Booth, Dominating sets in chordal graphs. Research Report CS-80-34, University of Waterloo (1980).

[5] K.S. Booth and J.H. Johnson, Dominating sets in chordal graphs. *SIAM J. Comput.* 11 (1982) 191–199.

[6] I. Borosh and L.B. Trebig, Bounds on positive integral solutions of linear Diopantine equations. *Proc. Amer. Math. Soc.* 55 (1976) 299–304.

[7] G. Chartrand and L. Lesniak, *Graphs & Digraphs.* Third Edition, Wadsworth and Brooks/Cole, Monterey, CA (1996).

[8] G.A. Cheston and G. Fricke, Classes of graphs for which upper fractional domination equals independence, upper domination, and upper irredundance. *Discrete Appl. Math.* 55 (1994) 241–258.

[9] G.A. Cheston, G. Fricke, S.T. Hedetniemi, and D. Pokrass, On the computational complexity of upper fractional domination. *Discrete Appl. Math.* 27 (1990) 195–207.

[10] E.J. Cockayne, R.M. Dawes, and S.T. Hedetniemi, Total domination in graphs. *Networks* 10 (1980) 211–219.

[11] E.J. Cockayne, O. Favaron, C. Payan, and A. Thomason, Contributions to the theory of domination, independence and irredundance in graphs. *Discrete Math.* 33 (1981) 249–258.

[12] E.J. Cockayne and S.T. Hedetniemi, Towards a theory of domination in graphs. *Networks* 7 (1977) 247–261.

[13] E.J. Cockayne, S.T. Hedetniemi, and D.J. Miller, Properties of hereditary hypergraphs and middle graphs. *Canad. Math. Bull.* 21(4) (1978) 461–468.

[14] A.K. Dewdney, Fast Turing reductions between problems in NP. Report 71, University of Western Ontario (1981).

[15] G.S. Domke, G. Fricke, S.T. Hedetniemi, and R. Laskar, Relationships between integer and fractional parameters of graphs. *Graph Theory, Combinatorics, and Applications*, John Wiley & Sons, Inc. 2 (1991) 371–387.

[16] G.S. Domke, S.T. Hedetniemi, and R. Laskar, Fractional packings, coverings and irredundance in graphs. *Congr. Numer.* 66 (1988) 227–238.

[17] J.E. Dunbar, W. Goddard, S.T. Hedetniemi, M.A. Henning, and A. McRae, On the algorithmic complexity of minus domination in graphs. *Discrete Appl. Math.* 68 (1996) 73–84.

[18] J.E. Dunbar, S.T. Hedetniemi, M.A. Henning, and A.A. McRae, Minus domination in graphs. Submitted for publication.

[19] J.E. Dunbar, S.T. Hedetniemi, M.A. Henning, and A.A. McRae, Minus domination in regular graphs. *Discrete Math.* 49 (1996) 311–312.

[20] J.E. Dunbar, S.T. Hedetniemi, M.A. Henning, and P.J. Slater, Signed domination in graphs. *Graph Theory, Combinatorics, and Applications*, John Wiley & Sons, Inc. 1 (1995) 311–322.

[21] M. Farber, Domination, independent domination, and duality in strongly chordal graphs. *Discrete Appl. Math.* 7 (1984) 115–130.

[22] O. Favaron, Signed domination in regular graphs. *Discrete Math.* 158 (1996) 287–293.

[23] D.C. Fisher, J. Ryan, G. Domke, and A. Majumdar, Fractional domination of strong direct products. *Discrete Appl. Math.* 50 (1994) 89–91.

[24] G. H. Fricke, Upper domination on double cone graphs. *Congr. Numer.* 72 (1990) 199–207.

[25] M.R. Garey and D.S. Johnson, *Computers and Intractability: A Guide to the Theory of NP-Completeness.* W.H. Freeman and Company, New York (1979).

[26] W. Goddard and M.A. Henning, Real and integer domination in graphs. Submitted for publication.

[27] D. L. Grinstead and P. J. Slater, Fractional domination and fractional packing in graphs. *Congr. Numer.* 71 (1990) 153–172.

[28] P. Hall, On representation of subsets. *J. London Math. Soc.* 10 (1935) 26–30.

[29] J.H. Hattingh, M.A. Henning, and P.J. Slater, Three-valued k-neighbourhood domination in graphs. *Australas. J. Combin.* 9 (1994) 233–242.

[30] J.H. Hattingh, M.A. Henning, and P.J. Slater, On the algorithmic complexity of signed domination in graphs. *Australas. J. Combin.* 12 (1995) 101–112.

[31] T.W. Haynes, private communication (1996).

[32] S.M. Hedetniemi, S.T. Hedetniemi, and T.V. Wimer, Linear time resource allocation for trees. Technical Report URI-014, Dept. Mathematical Sciences, Clemson Univ. (1987).

[33] S.T. Hedetniemi and R.C. Laskar, Bibliography on domination in graphs and some basic definitions of domination parameters. *Discrete Math.* 86 (1990) 257–277.

[34] M.A. Henning, Domination in regular graphs. *Ars Combin.* 43 (1996) 263–271.

[35] M.A. Henning and G. Kubicki, Real domination in graphs. To appear in *J. Combin. Math. Combin. Comput.*

[36] M.A. Henning and P.J. Slater, Inequalities relating domination parameters in cubic graphs. *Discrete Math.* 158 (1996) 87–98.

[37] M.S. Jacobson and K. Peters, Chordal graphs and upper irredundance, upper domination and independence. *Discrete Math.* 86 (1990) 59–69.

[38] N. Karmarkar, A new polynomial-time algorithm for linear programming. *Combinatorica* 4 (1984) 373–395.

[39] L.G. Khachian, A polynomial algorithm in linear programming. *Dokl. Akad. Nauk SSSR* 244 (1979) 1093–1096.

[40] D. König, Graphen und Matrizen. *Math. Fiz. Lapok.* 38 (1931) 116–119.

[41] P.J. Slater, Generalized graph parameters: Gallai Theorems I. *Bull. Inst. Combin. Appl.* 17 (1996) 27–37.

[42] B. Zelinka, Some remarks on domination in cubic graphs. *Discrete Math.* 158 (1996) 249–255.

Chapter 3

Fractional Domination and Related Parameters

Gayla S. Domke
Department of Mathematics and Computer Science
Georgia State University
Atlanta, GA   30303 USA

Gerd H. Fricke
Department of Mathematics
Wright State University
Dayton, OH   45435 USA

Renu R. Laskar
Department of Mathematical Sciences
Clemson University
Clemson, SC   29634 USA

Aniket Majumdar
Department of Mathematical Sciences
Indiana - Purdue University
Fort Wayne, IN   46805 USA

3.1   Introduction

It is well-known [11] that the following six parameters are related by the following chain of inequalities.

$$ir(G) \leq \gamma(G) \leq i(G) \leq \beta_0(G) \leq \Gamma(G) \leq IR(G).$$

The above parameters and their corresponding sets are defined in the beginning of this book. An alternate way to define the sets is to assign a value of 1 to every vertex in the set S and assign a value of 0 to the vertices not in S. For example, a set S would be a dominating set of vertices if for every vertex

61

$v \in V$, the sum of the values assigned to the vertices in the closed neighborhood of v is greater than or equal to 1. This motivates the definitions of fractional parameters. The values assigned to the vertices would not be restricted to be 0 or 1 but any real value in the closed interval between 0 and 1.

This idea of fractional parameters has been studied before in different contexts. Berge [4] has studied fractional transversals of hypergraphs, Pulleyblank [46] has studied fractional matchings, and Aharoni [2] has studied fractional matchings and covers in infinite hypergraphs. Chung, Furedi, Garey and Graham [9] have studied fractional covers of hypergraphs. The fractional chromatic number has been looked at by Larsen, Propp and Ullman as well as by others (see [40]). It appears that Farber [23] introduced indirectly the concept of fractional domination, whereas Hedetniemi, Hedetniemi and Wimer [37] were the first ones to study fractional domination.

If g is a function mapping the vertex set V into some set of real numbers and $S \subseteq V$, then let $g(S) = \sum_{v \in S} g(v)$ and the *weight* $|g| = g(V) = \sum_{v \in V} g(v)$. A real-valued function $g : V \to [0, 1]$ is a *dominating function* of a graph G if for every $v \in V$, $g(N[v]) \geq 1$. Thus, if S is a dominating set of vertices of G and g is a function where

$$g(v) = \begin{cases} 1 \text{ if } v \in S \\ 0 \text{ otherwise} \end{cases},$$

then g is a *dominating function* of G. A dominating function g is *minimal* if for every $u \in V$ with $g(u) > 0$, there exists a vertex $v \in N[u]$ such that $g(N[v]) = 1$. The *fractional domination number*, denoted $\gamma_f(G)$, and the *upper fractional domination number* of G, denoted $\Gamma_f(G)$, are defined as follows:

$$\gamma_f(G) = \min\{|g| : \text{g is a minimal dominating function of G}\}$$
$$\Gamma_f(G) = \max\{|g| : \text{g is a minimal dominating function of G}\}.$$

Other fractional parameters, such as fractional packing, fractional covering, fractional irredundance and fractional independence have been introduced and studied in many papers including [18], [17], [34] and [46].

A real-valued function $g : V \to [0, 1]$ is an *irredundant function* if for every u with $g(u) > 0$, there exists a vertex $w \in N[u]$ such that $g(N[w]) = 1$. An irredundant function g is *maximal* if there does not exist another irredundant function $h \neq g$, such that $h(v) \geq g(v)$ for every $v \in V$. The *fractional irredundance* number, $ir_f(G)$, and the *upper fractional irredundance* number $IR_f(G)$, are defined as follows:

$$ir_f(G) = \inf\{|g| : \text{g is a maximal irredundant function of G}\}$$
$$IR_f(G) = \max\{|g| : \text{g is a maximal irredundant function of G}\}.$$

The fractional irredundance numbers, $ir_f(G)$ and $IR_f(G)$ were first introduced by Domke, Hedetniemi and Laskar in [18] and later by Cheston, Fricke, Hedetniemi, and Jacobs [8].

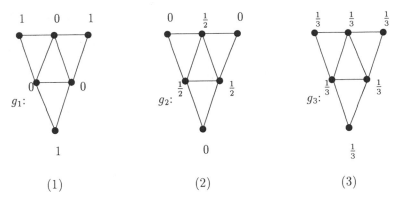

Figure 3.1: Examples of minimal dominating functions.

Fractional packing numbers are defined analogously; a real function $g : V \to [0, 1]$ is a *packing function* if for every $v \in V$, $g(N[v]) \leq 1$. A packing function is maximal if for every $u \in V$ with $g(u) < 1$, there exists a vertex $v \in N[u]$ such that $g(N[v]) = 1$. The *fractional packing number*, $p_f(G)$, and the *upper fractional packing number*, $P_f(G)$, are defined as follows:

$$p_f(G) = \min\{|g| : \text{g is a maximal packing function of G}\}$$
$$P_f(G) = \max\{|g| : \text{g is a maximal packing function of G}\}.$$

The fractional packing numbers were introduced in [18] and [34].

3.2 Known Results

3.2.1 Fractional domination

Consider the graph G with three different minimal dominating functions as shown in (1), (2) and (3) of Figure 3.1. Note that $|g_1| = 3$ and $|g_2| = \frac{3}{2}$ whereas $|g_3| = 2$.

Observation 3.1 *For every graph G,*

$$\gamma_f(G) \leq \gamma(G) \leq \Gamma(G) \leq \Gamma_f(G).$$

For the graph given in Figure 3.1, $\gamma_f(G) = \frac{3}{2}$, $\gamma(G) = 2$ and $\Gamma_f(G) = \Gamma(G) = 3$.

Figure 3.2 gives three different maximal packing functions. Note that $|g_4| = |g_5| = 1$ whereas $|g_6| = \frac{3}{2}$.

A set S of vertices is a *2-packing* of a graph G if for all $x, y \in S$ the distance between x and y is greater than 2. Meir and Moon [45] defined the *2-packing*

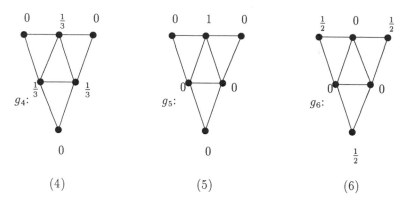

Figure 3.2: Examples of maximal packing functions.

number of G, $P_2(G)$, to be the maximum cardinality of a 2-packing. Domke, Hedetniemi and Laskar prove the following.

Theorem 3.2 [18] *For every graph G, $p_f(G) \leq P_2(G) \leq P_f(G)$.*

For the graph given in Figure 3.2, $p_f(G) = P_2(G) = 1$ and $P_f(G) = \frac{3}{2}$.

The problem of finding the fractional domination number and the upper fractional packing number of a graph can be translated into linear programming problems where $\gamma_f(G)$ is an optimal solution to the primal problem and $P_f(G)$ is an optimal solution to the dual problem. We will discuss these issues in Section 3.6. By the principle of strong duality, the solutions are equal and the following result holds.

Observation 3.3 *For every graph G, $P_f(G) = \gamma_f(G)$.*

This fact is very useful for finding the values of these parameters. If g is a minimal dominating function of a graph G, then $g(N[u]) \geq 1$ for every $u \in V$, and if h is a maximal packing function of G, then $h(N[u]) \leq 1$ for every $u \in V$. By duality, $|g| \geq |h|$. So if a minimal dominating function g of a graph G is found and a maximal packing function h of G is also found where $|g| = |h|$, then the values of $P_f(G)$ and $\gamma_f(G)$ are known. As a matter of fact, $P_f(G) = \gamma_f(G) = |g| = |h|$. It should be noted that it is not necessary for $g(v) = h(v)$ for any $v \in V$. (See Figure 3.1(2) and Figure 3.2(6)).

The following result was obtained independently by Grinstead and Slater and Domke, Hedetniemi and Laskar.

Theorem 3.4 [18, 34] *For every graph G with n vertices, maximum degree $\Delta(G)$ and minimum degree $\delta(G)$, $\frac{n}{\Delta(G)+1} \leq P_f(G) = \gamma_f(G) \leq \frac{n}{\delta(G)+1}$.*

The following results are now easily seen.

Corollary 3.5 [18, 34] *If G is an r-regular graph with n vertices,*

$$\gamma_f(G) = P_f(G) = \frac{n}{r+1}.$$

Corollary 3.6 [18, 34] *For any complete graph K_n or cycle C_n,*

$$\gamma_f(C_n) = P_f(C_n) = \frac{n}{3},$$

$$\gamma_f(K_n) = P_f(K_n) = 1.$$

Meir and Moon showed the following.

Theorem 3.7 [45] *For any tree T, $P_2(T) = \gamma(T)$.*

A graph is a *block graph* if each block (maximal 2-connected subgraph without a cut-vertex) of G is a complete subgraph. Domke, Hedetniemi, Laskar and Allan extended Meir and Moon's result to block graphs.

Theorem 3.8 [19] *For any block graph G, $P_2(G) = \gamma(G)$.*

Lemma 3.9 [18] *For any graph G, $P_2(G) \leq \gamma_f(G) \leq \gamma(G)$.*

From Theorems 3.7, 3.8 and Lemma 3.9, the following results are obvious.

Corollary 3.10 [18] *For any tree T, $\gamma_f(T) = \gamma(T)$.*

Corollary 3.11 [18] *For any block graph G, $\gamma_f(G) = \gamma(G)$.*

The following result gives the fractional domination number of complete bipartite graphs. A more general result for complete multipartite graphs is given by Grinstead and Slater in [34].

Theorem 3.12 [34, 21] *For any complete bipartite graph $K_{r,s}$, r, s $\geq$ 2*

$$\gamma_f(K_{r,s}) = \frac{r(s-1) + s(r-1)}{rs - 1}.$$

Proof. Let $K_{r,s}$ have bipartition of the vertex set V into sets A and B where $|A| = r$ and $|B| = s$. Define $g: A \cup B \rightarrow [0,1]$ by

$$g(v) = \begin{cases} \frac{s-1}{rs-1} & \text{for } v \in A \\ \frac{r-1}{rs-1} & \text{for } v \in B \end{cases}.$$

Now if $v \in A$, v is adjacent to all s vertices in B. Hence, $g(N[v]) = g(v) + \sum_{w \in B} g(w) = \frac{s-1}{rs-1} + s \cdot \frac{r-1}{rs-1} = \frac{s-1+sr-s}{rs-1} = 1$. Similarly if $v \in B$, v is adjacent to all r vertices in A and $g(N[v]) = 1$. Thus, $g(N[v]) = 1$ for every $v \in A \cup B$. So g is both a dominating function and a packing function. Therefore, $\gamma_f(K_{r,s}) = P_f(K_{r,s}) = |g| = \sum_{v \in A} g(v) + \sum_{v \in B} g(v) = \frac{r(s-1)}{rs-1} + \frac{s(r-1)}{rs-1}$. ☐

A real-valued function $g: V \rightarrow [0, 1]$ is a *total dominating function* if, for every $v \in V$, the open neighborhood sum of v satisfies $g(N(v)) \geq 1$. The *upper fractional total domination number*, Γ_f^0, and the *fractional total domination number*, γ_f^0, are defined as follows:

$$\Gamma_f^0(G) = \max\{|g| : g \text{ is a minimal fractional total dominating function of } G\}$$
$$\gamma_f^0(G) = \min\{|g| : g \text{ is a minimal fractional total dominating function of } G\}.$$

A set S is a total dominating set if $N(S) = V$. The upper total domination number, $\Gamma_t(G)$, and the total domination number, $\gamma_t(G)$, are, respectively, the cardinalities of a largest and smallest minimal total dominating set.

On the basis of these definitions we can establish a series of inequalities which relate these parameters to one another.

Proposition 3.13 *For any graph G, $\gamma_f(G) \leq \gamma(G) \leq \gamma_t(G) \leq \Gamma_t(G)$.*

Proposition 3.14 *For any graph G, $\gamma_f(G) \leq \gamma_f^0(G) \leq \gamma_t(G)$.*

Proposition 3.15 *For any graph G, $\gamma_f^0(G) \leq \Gamma_f^0(G)$.*

A dominating function $g: V \rightarrow [0, 1]$ with $\gamma_f(G) = |g|$ and a minimal dominating function $h: V \rightarrow [0, 1]$ and $\Gamma_f(G) = |h|$ are both minimal dominating functions and clearly $|g| \leq |h|$. So, a natural question can be raised concerning interpolation of minimal dominating functions (MDF); specifically: if g and h are MDFs with values $|g| < |h|$, does there exist an MDF k such that $|g| < |k| < |h|$?

A *convex combination* of the dominating functions g_1 and g_2 of G is the function $h_\lambda: V \rightarrow [0, 1]$ where $\lambda \in (0, 1)$ and $h_\lambda(v) = \lambda g_1(v) + (1 - \lambda)g_2(v)$ for each $v \in V$. It is trivial to show that h_λ is a dominating function for each $\lambda \in (0, 1)$. Majumdar [44] noticed that convex combinations of MDFs may

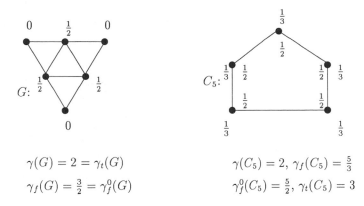

$$\gamma(G) = 2 = \gamma_t(G) \qquad\qquad \gamma(C_5) = 2,\ \gamma_f(C_5) = \tfrac{5}{3}$$
$$\gamma_f(G) = \tfrac{3}{2} = \gamma_f^0(G) \qquad\qquad \gamma_f^0(C_5) = \tfrac{5}{2},\ \gamma_t(C_5) = 3$$

Figure 3.3: Examples of $\gamma(G)$, $\gamma_f(G)$, $\gamma_t(G)$, $\gamma_f^0(G)$.

provide positive answers to the above interpolation question, but such convex combinations are not always minimal.

In a series of papers including [12] - [16], Cockayne, Mynhardt, MacGillivray and Yu have investigated the convexity of extremal domination related functions (e.g. dominating and total dominating functions and packing functions). The reader is referred to Chapter 5 for a complete discussion of these results.

In Figure 3.3, the graph G shows that $\gamma_f(G) < \gamma(G)$ is possible, whereas the graph C_5 shows that $\gamma(G)$, $\gamma_f^0(G)$ and $\gamma_t(G)$ may all be distinct; in particular, $\gamma_f(G) < \gamma(G)$ for G, but $\gamma_f^0(C_5) > \gamma(C_5)$ for C_5.

3.2.2 Fractional irredundance

Recall that a function $g\colon V \to [0,1]$ is irredundant if for every vertex $v \in V$ with $g(v) > 0$ there exists a vertex $w \in N[v]$ such that $g(N[w]) = 1$, and an irredundant function g is maximal if there does not exist another irredundant function $h \neq g$, such that $h(v) \geq g(v)$ for every $v \in V$.

The limit of a convergent sequence of irredundant functions is clearly irredundant. However, maximality may not be preserved under limits. As an example, consider Figure 3.4 and the path P_7 on seven vertices with an irredundant function g that assigns the following values to the vertices. Note that g is maximal irredundant and $g(P_7) = 2 + 2\varepsilon$. But it can be shown that there are no maximal irredundant functions g' with $|g'| = 2$.

As before we define

$$
\begin{aligned}
ir_f(G) &= \inf\{|g| : g \text{ is a maximal irredundant function of G}\}\\
IR_f(G) &= \max\{|g| : g \text{ is a maximal irredundant function of G}\}.
\end{aligned}
$$

Lemma 3.16 [17] *Every minimal dominating function is a maximal irredundant function on a graph G.*

Figure 3.4: An irredundant function on P_7.

Proof. Let g be a minimal dominating function of a graph G. By the minimality of g, for every $v \in V$ with $g(v) > 0$, there is a vertex $u \in N[v]$ such that $g(N[u]) = 1$. Therefore, g is an irredundant function.

Now let $h \colon V \to [0, 1]$ be a function obtained from g by increasing at least one functional value of g. That is, for some $v \in V$, where $g(v) < 1$, define $h(v) = g(v) + \varepsilon$, with $\varepsilon > 0$ and $h(v) \leq 1$.

Note that for any $u \in N[v]$, $h(N[u]) \geq g(N[u]) + \varepsilon > g(N[u]) \geq 1$. Thus h is not irredundant. Consequently, g is maximal irredundant. $\square$

Theorem 3.17 [17] *For any graph G, $ir_f(G) \leq \gamma_f(G) \leq \Gamma_f(G) \leq IR_f(G)$.*

For any graph G, $\Gamma_f(G) \geq \Gamma(G)$. A natural question arises: does there exist a graph G, for which $\Gamma_f(G) > \Gamma(G)$? In search of such a graph, Fricke [29] exhibits an example with 48 vertices and 144 edges called a *double cone graph*.

In light of the fact that $\Gamma_f > \Gamma$ for some graphs it is surprising that for upper irredundance, the fractional and integral versions are identical.

Theorem 3.18 [28] *For any graph G, $IR(G) = IR_f(G)$.*

3.3 Relationships Between Integer and Fractional Parameters

3.3.1 k-domination, k-packings, and k-irredundance

The idea of k-domination was presented by Hare [36] while trying to find values for the fractional domination number. Cheston, Fricke, Hedetniemi, and Jacobs [8] have studied upper k-domination. Also, other k-parameters have been studied. For example, Berge [4] has studied k-transversals of hypergraphs. In this section we will discuss k-packings and k-irredundance and relate these k-parameters as well as the k-domination parameters to fractional parameters and to other packing and covering parameters.

For any fixed positive integer k, a function $g \colon V \to \{0, 1, 2, \ldots, k\}$ is a *k-dominating function* of G if for every $v \in V$, $g(N[v]) \geq k$. Thus, if S is a dominating set of vertices and we define the function g where $g(v) = 1$ if $v \in S$, and $g(v) = 0$ if $v \notin S$, then g is a 1-dominating function of G. A *k-dominating*

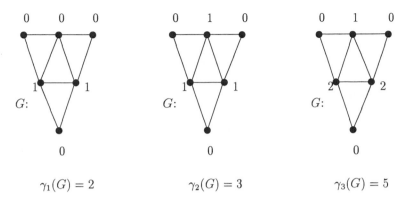

$$\gamma_1(G) = 2 \qquad \gamma_2(G) = 3 \qquad \gamma_3(G) = 5$$

Figure 3.5: Examples of k-domination for $k = 1, 2, 3$.

function is minimal if for every $v \in V$ with $g(v) > 0$, there exists a vertex $u \in N[v]$ such that $g(N[u]) = k$. For each positive integer k, the k-*domination number of G*, denoted $\gamma_k(G)$, and the *upper k-domination number of G*, denoted $\Gamma_k(G)$, are the minimum and maximum values, respectively, of $|g|$, where g is a minimal k-domination function of G. Examples of k-domination are shown in Figure 3.5. The k-domination numbers can be verified with results given in [20].

It should be noted that this definition of k-domination differs from the definition of k-domination given by Fink and Jacobson in [24]. They define a set $S \subseteq V$ to be an k-dominating set of vertices if each vertex $v \in V - S$ is adjacent to at least k vertices in S.

For a fixed positive integer k, a function $g: V \to \{0, 1, 2, \ldots, k\}$ is a k-*packing function of G* if for every $v \in V$, $g(N[v]) \leq k$. A k-packing function is maximal if for every $v \in V$ with $g(v) < k$, there exists a vertex $u \in N[v]$ such that $g(N[u]) = k$. For each positive integer k, the lower k-*packing number of G*, denoted $p_k(G)$, and the upper k-packing number, $P_k(G)$, are the minimum and maximum values of $|g|$ among all maximal k-packing functions g, respectively. Examples of k-packing functions are given in Figure 3.6 for $k = 1, 2, 3$. The (upper) k-packing numbers can be verified with results found in [20].

For a fixed positive integer k, a function $g: V \to \{0, 1, 2, \ldots, k\}$ is a k-*irredundant function of G* if for every vertex $v \in V$, with $g(v) > 0$, there exists a vertex $u \in N[v]$ such that $g(N[u]) = k$. A k-irredundant function is maximal if there does not exist a k-irredundant function h obtained from g by increasing at least one function value of g. For a fixed positive integer k, the k-*irredundance number of G*, denoted $ir_k(G)$, and the *upper k-irredundance number of G*, denoted $IR_k(G)$, are respectively the minimum and maximum values of $|g|$, where g is a maximal k-irredundant function (see Figure 3.7).

It should be noted here that when finding the minimum value of $|g|$, where

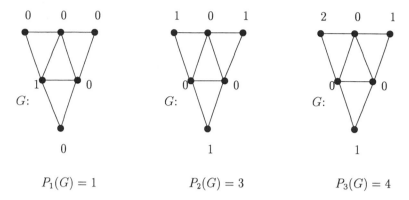

$$P_1(G) = 1 \qquad\qquad P_2(G) = 3 \qquad\qquad P_3(G) = 4$$

Figure 3.6: Examples of k-packings for $k = 1, 2, 3$.

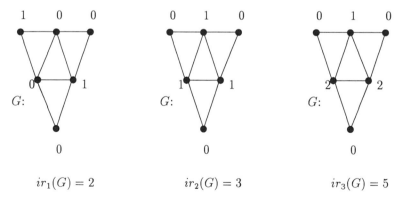

$$ir_1(G) = 2 \qquad\qquad ir_2(G) = 3 \qquad\qquad ir_3(G) = 5$$

Figure 3.7: Examples of k-irredundance for $k = 1, 2, 3$.

g is a *k-dominating function*, it will necessarily be true that g is a minimal k-dominating function. Similarly, if $|g|$ is a maximum, where g is a k-packing (k-irredundant) function, then g is a maximal k-packing (k-irredundant) function. Therefore, $\gamma_k(G) = \min\{|g|: g \text{ is a } k\text{-dominating function}\}$, $P_k(G) = \max\{|g|: g \text{ is a } k\text{-packing function}\}$, and $IR_k(G) = \max\{|g|: g \text{ is a } k\text{-irredundant function}\}$.

Notice that the definitions of k-domination and k-irredundance are closely related to those of domination and irredundance. In particular, if S is a dominating (irredundant) set of vertices of a graph, and if a function value of k is placed on those vertices in S, while a function value of zero is placed on the others, this function will be a k-dominating (k-irredundant) function.

The next results relate the domination and irredundance numbers to their corresponding k-parameters.

Theorem 3.19 [20] *For every graph G and positive integer k,*

$$\gamma_k(G) \leq k \cdot \gamma(G) \leq k \cdot \Gamma(G) \leq \Gamma_k(G).$$

Theorem 3.20 [20] *For every graph G and positive integer k,*

$$ir_k(G) \leq k \cdot ir(G) \leq k \cdot IR(G) \leq IR_k(G).$$

There is always an important relationship between domination and irredundance. As a matter of fact, every minimal dominating set of vertices is a maximal irredundant set. Recall that Cockayne and Hedetniemi [10] showed that $ir(G) \leq \gamma(G)$, for every graph G. As the following shows, an analogous statement is true for minimal k-dominating functions and maximal k-irredundant functions and the string of inequalities holds for k-parameters also.

Lemma 3.21 [20] *If k is a positive integer, every minimal k-dominating function is a maximal k-irredundant function.*

Theorem 3.22 [20] *For every graph G and positive integer k, $ir_k(G) \leq \gamma_k(G) \leq \Gamma_k(G) \leq IR_k(G)$.*

Proof. Let g be a minimal k-dominating function of G where $|g| = \gamma_k(G)$. Since $ir_k(G) = \min\{|h|: h \text{ is a maximal } k\text{-irredundant function}\}$, and g is a maximal k-irredundant function by Lemma 3.21, then $ir_k(G) \leq |g| = \gamma_k(G)$.

Similarly, let g be a minimal k-dominating function where $|g| = \Gamma_k(G)$. Since $IR_k(G) = \max\{|h|: h \text{ is a maximal } k\text{-irredundant function}\}$, then $IR_k(G) \geq |g| = \Gamma_k(G)$, and the result holds. $\square$

It is interesting to note that there is no such relationship between maximal k-packing functions and maximal k-irredundant functions. Figure 3.6 shows examples of maximal k-packing functions which are k-irredundant functions, but which are not maximal. Also, Figure 3.7 gives examples of maximal k-irredundant functions which are not k-packing functions.

The next results show that the sequences of k-domination numbers (similarly k-packing numbers) are strictly increasing. This is helpful to know since once $\gamma_k(G)$ (similarly $P_k(G)$) is known, it gives a bound on $\gamma_{k-1}(G)$ and $\gamma_{k+1}(G)$ ($P_{k-1}(G)$ and $P_{k+1}(G)$).

Theorem 3.23 [20] *For every graph G and positive integer k, $\gamma_k(G) < \gamma_{k+1}(G)$.*

Theorem 3.24 [20] *For every graph G and positive integer k, $P_k(G) < P_{k+1}(G)$.*

It would be interesting to know if similar results are true for the other k-parameters. For example, is it true that $\Gamma_k(G) < \Gamma_{k+1}(G)$? At this point it is not known.

3.3.2   Relating k-parameters and fractional parameters

As we stated in the last section, there is a close relationship between the domination (irredundance) number and the k-domination (k-irredundance) number of a graph. The same is true between fractional parameters and their associated k-parameters. These first results show that any k-dominating (k-packing) function provides a dominating (packing) function when each of its function values is divided by k. If $g: V \to \{0, 1, 2, \ldots, k\}$ is a minimal k-dominating function, then the function $h : V \to [0, 1]$ defined by $h(v) = g(v)/k$ is a minimal dominating function.

Theorem 3.25 [20] *For any graph G and positive integer k, $\gamma_f(G) \le \gamma_k(G)/k \le \Gamma_k(G)/k \le \Gamma_f(G)$.*

If $g: V \to \{0, 1, 2, \ldots, k\}$ is a maximal k-packing function, then the function $h: V \to [0, 1]$ defined by $h(v) = g(v)/k$ is a maximal packing function.

Theorem 3.26 [20] *For any graph G and positive integer k, $p_f(G) \le p_k(G)/k \le P_k(G)/k \le P_f(G)$.*

The above two theorems were used to show that $P_k(G) \le \gamma_k(G)$ for every graph G and positive integer k.

Corollary 3.27 [20] *For any graph G and positive integer k,*

$$P_k(G) \le k \cdot P_f(G) = k \cdot \gamma_f(G) \le \gamma_k(G).$$

Theorem 3.28 [20] *For any graph G and positive integer k, $IR_f(G) > IR_k(G)/k$.*

Proof. Let k be any positive integer, and let $g: V \to \{0, 1, 2, \ldots, k\}$ be a k-irredundant function, where $|g| = IR_k(G)$. Now define a function h where $h(v) = g(v)/k$ for every $v \in V$. Since $0 \le g(v) \le k$, then $0 \le h(v) \le 1$ and $h: V \to [0, 1]$. Since g is a k-irredundant function, to each $v \in V$ with $g(v) > 0$, there corresponds a vertex $u \in N[v]$, where $g(N[u]) = k$. Thus, if $v \in V$, and $h(v) > 0$, and there exists a vertex $u \in N[v]$ where $h(N[u]) = \sum\limits_{w \in N[u]} h(w) = \sum\limits_{w \in N[u]} \frac{g(w)}{k} = \frac{1}{k} g(N[u]) = \frac{1}{k}k = 1$. Therefore, h is an irredundant function of G. Since $IR_f(G) = \max \{|g'|: g'$ is an irredundant function of $G\}$, then $IR_f(G) \ge |h| = \sum\limits_{v \in V} h(v) = \sum\limits_{v \in V} \frac{g(v)}{k} = \frac{1}{k}|g| = \frac{1}{k} IR_k(G)$, and the result holds. $\square$

Since $\gamma_f(G) \le \gamma_k(G)/k$ for every positive integer k, it is natural to ask if equality holds for any value of k (similarly for $P_f(G)$). This is indeed true as shown in the following theorems.

Theorem 3.29 [20] *For any graph G,*

$$\gamma_f(G) = \min\{\gamma_k(G)/k : k \text{ is a positive integer }\}.$$

Theorem 3.30 [20] *For any graph G,*

$$P_f(G) = \max\{P_k(G)/k : k \text{ is a positive integer }\}.$$

3.3.3 Expanding fractional parameter results to k-parameters

In [18] Domke, Hedetniemi, and Laskar show that for any graph G with n vertices, $\gamma_f(G) = 1$ if and only if $\Delta(G) = n - 1$. A similar result holds for $\gamma_k(G)$ as shown by Domke, Hedetniemi, Laskar and Fricke. Note that if g is any k-dominating function, then $|g| \ge g(N[v]) \ge k$, for any vertex $v \in V$. Therefore, $\gamma_k(G) \ge k$ for any graph G and any positive integer k.

Theorem 3.31 [20] *For any graph G with n vertices and any positive integer k, $\gamma_k(G) = k$ if and only if $\Delta(G) = n - 1$.*

For any maximal k-packing function g, if $g(v) < k$, then there is a vertex $u \in N[v]$ where $g(N[u]) = k$. Hence, $|g| \ge g(N[u]) = k$. If $g(v) = k$, then $|g| \ge g(v) = k$. In either case, $|g| \ge k$. Therefore, it is easy to see that $p_k(G) \ge k$ for every positive integer k. Similarly, it is easy to see that $ir_k(G) \ge k$. Now, since $ir_k(G) \le \gamma_k(G)$ and $P_k(G) \le \gamma_k(G)$, the following result can be easily seen.

Corollary 3.32 [20] *For any positive integer k,*

$$ir_k(K_n) = P_k(K_n) = \gamma_k(K_n) = k.$$

Theorem 3.33 [20] *For any graph G and positive integer k,*

$$p_k(G) \le k \cdot P_2(G) \le P_k(G).$$

Corollary 3.34 [20] *For any graph G and positive integer k,*

$$k \le p_k(G) \le k \cdot P_2(G) \le P_k(G) \le \gamma_k(G) \le k \cdot \gamma(G).$$

Recall that a block graph is a graph each of whose blocks is complete. Domke, Hedetniemi, Laskar, and Allan [19] showed that if G is a connected block graph, then $P_2(G) = \gamma(G)$. This result combined with the above corollary gives the following.

Corollary 3.35 [20] *For any positive integer k, if G is a connected block graph, then $k \cdot P_2(G) = P_k(G) = \gamma_k(G) = k \cdot \gamma(G)$.*

3.3.4 Inequalities involving domination parameters

An inequality chain similar to that given in (3.1) at the beginning of this chapter holds for fractional parameters, namely, $ir_f(G) \le \gamma_f(G) \le \Gamma_f(G) \le IR_f(G)$ (see [17]) as well as an inequality chain involving k-parameters, namely $ir_k(G) \le \gamma_k(G) \le \Gamma_k(G) \le IR_k(G)$ (see Theorem 3.22).

Still another inequality holds involving the upper fractional domination number and the upper irredundance number of a graph.

Theorem 3.36 [20] *For any graph G, $\Gamma_f(G) \le IR(G)$.*

Proof. Let $g \colon V \to [0, 1]$ be a minimal dominating function of G where $|g| = \Gamma_f(G)$. Also let $S = \{v_1, v_2, \ldots, v_t\} \subseteq V$ be the set of vertices with $g(N[v_i]) = 1$ for $i = 1, 2, \ldots, t$. Note that since g is minimal, every vertex $u \in V$, with $g(u) > 0$, is adjacent to at least one vertex in S, i.e., S dominates the set P of vertices with positive function values. Let $D \subseteq S$ be a minimal subset of S which dominates P. Now, D is an irredundant set of $< D \cup P >$ and hence of G. Thus, $IR(G) \ge |D|$. Since D is a dominating set of P, every $v \in V$ with $g(v) > 0$ is in $N[v_i]$ for some $v_i \in D$. Also, $g(N[v_i]) = 1$ for every $v_i \in D$. Hence, $|D| = \sum\limits_{v_i \in D} 1 = \sum\limits_{v_i \in D} g(N[v_i]) \ge \sum\limits_{v \in V} g(v) = |g| = \Gamma_f(G)$.
Therefore, $IR(G) \ge \Gamma_f(G)$. □

It should be noted that a similar inequality is not true for the fractional domination number and the irredundance number. For a cycle on four vertices (see Figure 3.8(a)) $\gamma_f(C_4) = 4/3$ and $ir(C_4) = 2$. Hence, $\gamma_f(C_4) < ir(C_4)$. For the tree given in Figure 3.8(b), $\gamma_f(T) = \gamma(T) = 5$ and $ir(T) = 4$. Hence, $\gamma_f(T) > ir(T)$. Therefore, in general there is no strict order relationship between $\gamma_f(T)$ and $ir(G)$.

Combining Theorems 3.19, 3.20, 3.25, 3.26 and 3.28 and other results stated in this section, we obtain the following corollaries.

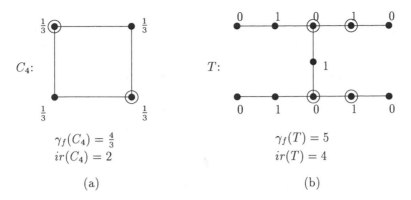

$$\gamma_f(C_4) = \tfrac{4}{3}$$
$$ir(C_4) = 2$$

$$\gamma_f(T) = 5$$
$$ir(T) = 4$$

(a) (b)

Figure 3.8: Examples where $\gamma_f(G) < ir(G)$ and $\gamma_f(G) > ir(G)$.

Corollary 3.37 [20] *For any graph G and positive integer k, $\frac{ir_k(G)}{k} \le ir(G) \le \gamma(G) \le i(G) \le \beta_0(G) \le \Gamma(G) \le \frac{\Gamma_k(G)}{k} \le \Gamma_f(G) \le IR(G) \le \frac{IR_k(G)}{k} \le IR_f(G)$.*

Corollary 3.38 [20] *For any graph G and positive integer k, $1 \le p_f(G) \le \frac{P_k(G)}{k} \le P_2(G) \le \frac{P_k(G)}{k} \le P_f(G) = \gamma_f(G) \le \frac{\gamma_k(G)}{k} \le \gamma(G) \le i(G) \le \beta_0(G) \le \Gamma_f(G) \le \frac{\Gamma_k(G)}{k} \le \Gamma(G) \le IR(G) \le \frac{IR_k(G)}{k} < IR_f(G)$.*

3.4 Fractional Domination and Graph Products

Let G and H be disjoint graphs. We will look at several ways of combining G and H to form a new graph by defining four kinds of products of graphs. The *join* of G and H, denoted $G+H$, has vertex set $V(G+H) = V(G) \cup V(H)$ and edge set $E(G+H) = E(G) \cup E(H) \cup \{(gh)|g \in V(G), h \in V(H)\}$. The *Cartesian product* of G and H, denoted $G \times H$, has vertex set $V(G \times H) = V(G) \times V(H) = \{(g, h)|g \in V(G), h \in V(H)\}$ and edge set $E(G \times H) = \{(g_1, h_1)(g_2, h_2)|$ either $g_1 = g_2$ and $h_1 h_2 \in E(H)$ or $g_1 g_2 \in E(G)$ and $h_1 = h_2\}$. The *weak direct product* of G and H is defined as the graph $G \oplus H$ where $V(G \oplus H) = V(G) \times V(H)$ and $E(G \oplus H) = \{(g_1, h_1)(g_2, h_2)|g_1 g_2 \in E(G)$ and $h_1 h_2 \in E(H)\}$. Finally, the *strong direct product* of G and H, denoted $G \circ H$, has $V(G \circ H) = V(G) \times V(H)$ and $E(G \circ H) = E(G \times H) \cup E(G \oplus H)$.

We will now state several results involving fractional domination of graph products. Clearly, $\gamma_f(G+H) \le \min \{\gamma_f(G), \gamma_f(H)\}$. Laskar, Majumdar, Domke and Fricke give an example that strict inequality is possible for $C_4 + C_4$. They also prove the following result.

Theorem 3.39 [41] *For the join $G + H$, $\Gamma_f(G + H) = max \{\Gamma_f(G), \Gamma_f(H))\}$.*

Vizing conjectured that the domination number of the Cartesian product of graphs satisfies $\gamma(G \times H) \geq \gamma(G) \cdot \gamma(H)$. Fisher, Ryan, Domke and Majumdar proved the following results.

Theorem 3.40 [27] *For the strong direct product $G \circ H$,*

$$\gamma_f(G \circ H) = \gamma_f(G) \cdot \gamma_f(H).$$

Corollary 3.41 [27] *For the Cartesian product $G \times H$,*

$$\gamma_f(G \times H) \geq \gamma_f(G) \cdot \gamma_f(H).$$

Majumdar proved a similar result for weak direct products and fractional total domination.

Theorem 3.42 [43] *For the weak direct product $G \oplus H$,*

$$\gamma_f^0(G \oplus H) = \gamma_f^0(G) \cdot \gamma_f^0(H).$$

Fisher also considered the 2-packing number and proved the following two similar relations for the strong direct product.

Theorem 3.43 [26] *For the strong direct product $G \circ H$,*

$$\gamma(G) \cdot \gamma_f(H) \leq \gamma(G \circ H) \leq \gamma(G) \cdot \gamma(H).$$

Theorem 3.44 [26] *For the strong direct product $G \circ H$,*

$$P_2(G) \cdot P_2(H) \leq P_2(G \circ H) \leq P_2(G) \cdot \gamma_f(H).$$

We conclude this section with a discussion of grid graphs. An $m \times n$ grid graph is the Cartesian product of two paths and is denoted $P_m \times P_n$. Hare [36], and Stewart and Hare [47] proved the following results for fractional domination of $P_m \times P_n$.

Theorem 3.45 [36] *For the Cartesian product $P_2 \times P_n$,*

$$\gamma_f(P_2 \times P_n) = \begin{cases} \frac{n+1}{2} & n \quad \text{odd} \\ \frac{n^2+2n}{2(n+1)} & n \quad \text{even} \end{cases}.$$

Theorem 3.46 [36] *For the Cartesian product $P_m \times P_n$,*

$$\gamma_f(P_3 \times P_3) = \tfrac{5}{2},$$
$$\gamma_f(P_3 \times P_4) = \tfrac{10}{3},$$
$$\gamma_f(P_3 \times P_5) = 4,$$
$$\gamma_f(P_3 \times P_6) = \tfrac{14}{3}, \text{and}$$
$$\gamma_f(P_3 \times P_7) = \tfrac{27}{5}.$$

Theorem 3.47 [47] *For the Cartesian product $P_m \times P_n$,*
$$\gamma_f(P_3 \times P_{10}) = \tfrac{68}{9},$$
$$\gamma_f(P_3 \times P_{11}) = \tfrac{33}{4},$$
$$\gamma_f(P_3 \times P_{14}) = \tfrac{49}{9}.$$

Theorem 3.48 [36] *For the Cartesian product $P_m \times P_n$,*
$$\gamma_f(P_4 \times P_4) = 4 \ and$$
$$\gamma_f(P_5 \times P_6) = \tfrac{22}{3}.$$

Theorem 3.49 [47] *For the Cartesian product $P_m \times P_n$,*
$$\gamma_f(P_4 \times P_8) = 8 \ and$$
$$\gamma_f(P_5 \times P_9) = \tfrac{32}{3}.$$

Theorem 3.45 gives an exact formula for $\gamma_f(P_2 \times P_n)$. Fisher [26] has attempted to find such a formula for $\gamma_f(P_3 \times P_n)$. However, there is no known formula for $\gamma_f(P_m \times P_n)$ for $m > 3$. Hare [35] gives bounds for $\gamma_f(P_m \times P_n)$ for $m, n > 2$ involving the Fibonacci numbers.

3.5   Bondage and Reinforcement

A set F of edges is a *domination alternation set of edges* of G if either $F \subseteq E$ and the removal of the edges in F from G results in a graph G' satisfying $\gamma(G') > \gamma(G)$ or $F \subseteq \bar{E} = E(\bar{G})$ and the addition of the edges in F to G results in a graph G'' satisfying $\gamma(G'') < \gamma(G)$. A graph G is a *domination critical graph* if $\gamma(G + e) < \gamma(G)$ for every edge $e \notin E$. Domination critical graphs were studied by Sumner and Blitch [48]. Also, Acharya and Walikar [1] studied graphs G where $\gamma(G - e) > \gamma(G)$ for every edge $e \in E$.

Fink, Jacobson, Kinch and Roberts [25] defined the *bondage number* of a graph G, denoted $b(G)$, to be the minimum cardinality of a set of edges whose removal from G results in a graph G' for which $\gamma(G') > \gamma(G)$. Similarly, Kok and Mynhardt [39] defined the *reinforcement number* of G, denoted $r(G)$, to be smallest number of edges which must be added to G to form a graph G'' where $\gamma(G'') < \gamma(G)$. If $\gamma(G) = 1$, then define $r(G) = 0$. Ghoshal, Laskar, Pillone and Wallis [32] defined similar parameters involving the bondage and reinforcement numbers associated with the strong domination number. Domke and Laskar [21] defined the *bondage number* of γ_f for a graph G, denoted $b_f(G)$, as the minimum cardinality of a set $F \subseteq E$ of edges whose removal results in a graph $G' = G - F$ satisfying $\gamma_f(G - F) > \gamma_f(G)$ and the *reinforcement number* of γ_f for G, denoted $r_f(G)$, to be the minimum cardinality of a set $K \subseteq \bar{E} = E(\bar{G})$ of edges which when added to G results in a graph $G'' = G + K$ satisfying

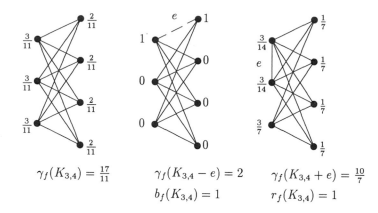

$$\gamma_f(K_{3,4}) = \tfrac{17}{11} \qquad \gamma_f(K_{3,4} - e) = 2 \qquad \gamma_f(K_{3,4} + e) = \tfrac{10}{7}$$
$$b_f(K_{3,4}) = 1 \qquad r_f(K_{3,4}) = 1$$

Figure 3.9: Examples of the bondage and reinforcement numbers of γ_f.

$\gamma_f(G + K) < \gamma_f(G)$. If $\gamma_f(G) = 1$, then define $r_f(G) = 0$. An example of the bondage and reinforcement numbers of γ_f are given in Figure 3.9.

The following values of the bondage number of γ_f for certain classes of graphs are due to Domke and Laskar.

Theorem 3.50 [21] *For the cycle C_n, path P_n, complete graph K_n, and complete bipartite graph $K_{r,s}$,*

$$(i)\ b_f(C_n) = \begin{cases} 1 & \text{if } n \not\equiv 0 \ mod\ 3 \\ 2 & \text{if } n \equiv 0 \ mod\ 3 \end{cases}.$$

$$(ii)\ b_f(P_n) = \begin{cases} 1 & \text{if } n \not\equiv 1 \ mod\ 3 \\ 2 & \text{if } n \equiv 1 \ mod\ 3 \end{cases}.$$

$(iii)\ b_f(K_n) = \lceil \tfrac{n}{2} \rceil.$
$(iv)\ b_f(K_{r,s}) = 1.$

Furthermore, Domke and Laskar found the reinforcement numbers of γ_f for the following classes of graphs.

Theorem 3.51 [21] *For the cycle C_n, path P_n, complete graph K_n, and complete bipartite graph $K_{r,s}$,*

$$(i)\ r_f(C_n) = \begin{cases} 1 & \text{if } n \not\equiv 0 \ mod\ 3 \\ 2 & \text{if } n \equiv 0 \ mod\ 3 \end{cases}, \text{ for } n \geq 4.$$

$$(ii)\ r_f(P_n) = \begin{cases} 1 & if\ n \not\equiv 0\ mod\ 3 \\ 2 & if\ n \equiv 0\ mod\ 3 \end{cases} ,\ for\ n \geq 4.$$

(iii) $r_f(K_n) = 0.$

(iv) $b_f(K_{r,s}) = 1.$

3.6 Fractional Parameters Via Linear Programming

3.6.1 Introduction

In this section we explore how linear programming methods can be used to obtain results about fractional domination and related parameters. As has been pointed out in Section 3.2.1, the principle of strong duality plays a pivotal role in obtaining many of these results.

Let $G = (V, E)$ be a graph with adjacency matrix A and neighborhood matrix $N = A + I$, where I is the $n \times n$ identity matrix, $n = |V(G)|$. We denote all vectors in bold letters. Corresponding to G we can define the domination parameters as follows:

Domination (P1)

$$\gamma(G) = \min \sum_{i=1}^{n} x_i$$

$$s.t.\ N\mathbf{x} \geq \mathbf{1}$$

$$x_i \in \{0,1\};\ \mathbf{1} = \begin{bmatrix} 1 \\ 1 \\ \vdots \\ 1 \end{bmatrix}.$$

Total Domination (P2)

$$\gamma_t(G) = \min \sum_{i=1}^{n} x_i$$

$$s.t.\ A\mathbf{x} \geq \mathbf{1}$$

$$x_i \in \{0,1\}.$$

Fractional Domination (LP1)

$$\gamma_f(G) = \min \sum_{i=1}^{n} x_i$$

$$s.t. \; N\mathbf{x} \geq 1$$

$$0 \leq x_i \leq 1.$$

Fractional Total Domination (LP2)

$$\gamma_f^0(G) = \min \sum_{i=1}^{n} x_i$$

$$s.t. \; A\mathbf{x} \geq 1$$

$$0 \leq x_i \leq 1.$$

Note that (LP1) is the linear programming (LP) relaxation of the integer programming problem (P1) and (LP2) is the LP relaxation of (P2).

Remark: Since both the neighborhood matrix N and the adjacency matrix A are $(0, 1)$-matrices and since the right hand side of our LP formulations is the vector of all $1's$, the solution vectors $\mathbf{x}$ have rational entries. Thus we can state the following.

Theorem 3.52 [41] *For any graph G, there exists a γ_f-function g and a γ_f^0-function h such that $g(v)$ and $h(v)$ are rational for all $v \in V(G)$.*

We shall see later that a similar result is true for the corresponding "upper" parameters.

As is customary for all 0-1 integer programming problems, we ask the question: When does (LP1) yield an optimal solution to (P1) and when does (LP2) yield an optimal solution to (P2)? Farber [23], and Hoffman, Kolen and Sakarovitch [38] have independently investigated such integrality issues in the context of domination and minimum cost covering problems. Following Lovasz [42] they define a $(0, 1)$-matrix to be *totally balanced* if it does not contain a square submatrix which has no two columns identical and whose row and column sums are equal to two. The next two theorems provide examples of graphs for which $\gamma_f = \gamma$ or $\gamma_f^0 = \gamma_t$.

A graph G is *chordal* if every cycle of length greater than three has a chord. A bipartite graph is *chordal bipartite* if every cycle of length greater than four has a chord. (See pp. 261-263 [33]). A chordal graph G is *strongly chordal* if it does not contain any induced trampoline. A *trampoline* is a chordal graph on $2k$ vertices which can be partitioned into sets $X = \{x_1, x_2 \ldots x_k\}$ and $Y = \{y_1, y_2, \ldots, y_k\}$ such that Y is an independent set, $(x_1, x_2 \ldots, x_k, x_1)$ is a cycle and x_i is adjacent to y_j if $i = j$ or $i = j + 1 \bmod k$.

Theorem 3.53 [38] *The adjacency matrix A of a graph G is totally balanced if and only if G is a chordal bipartite graph.*

Theorem 3.54 [22, 23] *The neighborhood matrix N of a graph G is totally balanced if and only if G is a strongly chordal graph.*

The results of Farber assert that if the neighborhood matrix N of a graph is totally balanced then solving (LP1) we get a solution to (P1). In other words, we immediately get the following:

Corollary 3.55 [41] *If G is strongly chordal, then $\gamma_f(G) = \gamma(G)$.*

Corollary 3.56 [41] *If G is chordal bipartite, then $\gamma_f^0(G) = \gamma_t(G)$.*

Since a tree is both strongly chordal and chordal bipartite, Corollary 3.10 and its total domination analogue follow from the above results.

Corollaries 3.55 and 3.56 can be further generalized to graphs whose neighborhood or adjacency matrices are balanced. A $(0, 1)$-matrix is *balanced* if it does not contain incidence matrices of odd cycles (as submatrices); equivalently, if the matrix is the incidence matrix of a balanced hypergraph (see [3]). The class of balanced matrices contains the class of totally balanced matrices, since totally balanced matrices do not contain incidence matrices of cycles.

Brouwer, Duchet and Schrijver have characterized graphs with balanced adjacency and neighborhood matrices.

Theorem 3.57 [5]

  a) The adjacency matrix of a graph G is balanced if and only if G is bipartite with no induced cycles of length $4q + 2$, $q \geq 1$.

  b) The neighborhood matrix of a graph G is balanced if and only if G is odd-sun-free chordal.

The class of *odd-sun-free* chordal graphs (see [5] or [6] for a definition) contains the class of strongly chordal graphs.

Fulkerson, Hoffman and Oppenheim have shown that balanced matrices have the requisite integrality property. Let M be a $(0, 1)$-matrix.

Lemma 3.58 [31] *If M is balanced, and if $\{x: Mx \geq 1, x \geq 0\}$ is not empty, then every vertex of this polyhedron has all coordinates 0 or 1.*

Combining Theorem 3.57 and Lemma 3.58 we get the following result which generalizes Corollaries 3.55 and 3.56.

Theorem 3.59 [5]

  a) If a graph G is odd-sun-free chordal, then $\gamma_f(G) = \gamma(G)$.

  b) If a graph G is bipartite without induced cycles of length $4q + 2$ with $q \geq 1$, then $\gamma_f^0(G) = \gamma_t(G)$.

We now give a proof of Corollary 3.5 and its total domination analogue using linear programming techniques.

Theorem 3.60 [41] *For any r-regular graph G of order n,*

$$\gamma_f(G) = \frac{n}{r+1} \text{ and } \gamma_f^0(G) = \frac{n}{r}.$$

Proof. Consider the dual (DP1) to (LP1)

$$p_f(G) = \max \sum_{i=1}^{n} y_i$$

$$s.t. \; yN \leq 1$$

$$y_i \geq 0, n = |V|,$$

and the dual (DP2) to (LP2)

$$p_f^0(G) = \max \sum_{i=1}^{n} y_i$$

$$s.t. \; yA \leq 1$$

$$y_i \geq 0, n = |V|.$$

Note that $x_i = \frac{1}{r+1} = y_i$ is feasible to both (LP1) and (DP1) and they yield the same objective function value of $\frac{n}{r+1}$. By the duality theorem of linear programming, $\gamma_f(G) = \frac{n}{r+1} = p_f(G)$. Similarly $x_i = \frac{1}{r} = y_i$ is feasible to both (LP2) and (DP2) and yield the optimal objective function value of $\gamma_f^0(G) = \frac{n}{r} = p_f^0(G)$. $\square$

3.6.2   Upper parameters

In the case of upper domination parameters such as $\Gamma(G)$, $\Gamma_t(G)$ and upper fractional parameters such as $\Gamma_f(G)$, $\Gamma_f^0(G)$, linear programming techniques cannot be used so easily, since all these parameters involve finding the maximum cardinality of a set with a minimal property.

Cheston, Fricke, Hedetniemi and Jacobs [8] have shown how to express Γ_f as a solution of a sequence of linear programs. A similar technique has been used by Fricke, Hare, Jacobs and Majumdar [30] to do the same for Γ_f^0. These results imply that both these parameters are computable and are rational.

We now describe the method of computing Γ_f as expounded upon in [8]. The technique to compute Γ_f^0 is essentially the same.

Let f be a minimal dominating function, and let us define the set $S_f = \{v \in V | f(N[v]) = 1\}$. Recall that a dominating function is minimal if and only if whenever $f(v) > 0$, there exists some $u \in N[v]$ such that $f(N[v]) = 1$. As a consequence of this fact we must have $N[N[S_f]] = V$.

This motivates us to define the following class

$$\mathcal{S} = \{S | S \text{ is a set of vertices such that } \mathcal{N}[\mathcal{N}[S]] = \mathcal{V}\}.$$

For each set S in $\mathcal{S}$, consider the problem of finding a minimal dominating function f of maximum weight with the additional restriction that S_f contains S. Now consider the following linear program, where the vertices of the graph G have been labeled by the integers 1, 2, ..., n:

maximize $\sum_{i=1}^{n} x_i$

subject to

$$0 \leq x_i \leq 1, \qquad \text{for all } i \in N[S], \qquad\qquad (1)$$

$$x_i = 0, \qquad \text{for all } i \in V - N[S], \qquad (2)$$

$$\sum_{j \in N[i]} x_j \geq 1, \qquad \text{for all } i \in V - S, \qquad\qquad (3)$$

$$\sum_{j \in N[i]} x_i = 1, \qquad \text{for all } i \in S. \qquad\qquad (4)$$

Note that a feasible solution to the above problem is a dominating function. Given that a solution is dominating, conditions (2) and (4) guarantee that it is minimal dominating. Hence every feasible solution to the LP above is a minimal dominating function. On the other hand, any minimal dominating function f whose weight is equal to $\Gamma_f(G)$ is the solution to this LP for some $S_f \in \mathcal{S}$.

The following theorem on $\Gamma_f(G)$ appears in [8]. The same conclusions are valid for $\Gamma_f^0(G)$ and are implicit in [30].

Theorem 3.61 [8] *For any graph G,*
(i) $\Gamma_f(G)$ is a computable function and is always rational, and
(ii) there is a polynomial p such that for any graph G, there exists a minimal dominating function of weight $\Gamma_f(G)$ having rational values, and such that the length of the representation of this function is bounded by $p(|V(G)|)$.

For our next theorems we again use Lemma 3.58 of Fulkerson, Hoffman and Oppenheim [31] on balanced (0, 1)-matrices.

Theorem 3.62 [30] *If G is a bipartite graph with no induced cycles of length $4q + 2$, $q \geq 1$, then $\Gamma_f^0(G) = \Gamma_t(G)$.*

Proof. If G is such a graph, then, by Theorem 3.57, its adjacency matrix A is balanced. Thus every vertex of the polyhedron $\{\mathbf{x}: A\mathbf{x} \geq \mathbf{1}, \mathbf{x} \geq \mathbf{0}\}$ has all coordinates 0 or 1. Now a minimal total dominating function of maximum weight comes from one such vertex, and the proof is complete.   □

Lemma 3.58 also leads to the following result whose proof is obtained by replacing A by N, the neighborhood matrix, in the above proof. This result is implicit in [8].

Theorem 3.63 [8] *If G is an odd-sun-free chordal graph, then $\Gamma_f(G) = \Gamma(G)$.*

Note that the conclusion of Theorem 3.63 has been subsumed by a more general result that says that $\beta_0(G) = \Gamma(G) = IR(G)$ for a large class of graphs that contains chordal graphs [7].

In Sections 3.6.1 and 3.6.2 we have indicated how linear programming can be a useful technique in dealing with some fractional parameters. Recall, however, that in dealing with irredundance parameters, (Theorem 3.18) such methods were not used. It remains to be seen what additional light LP theory may shed in dealing with fractional irredundance.

3.6.3   Computational aspects

The parameters γ_f and γ_f^0 can be computed efficiently for any graph using linear programming, since a linear program can be solved in polynomial time. However, such is not the case for the parameters Γ_f and Γ_f^0. Even though an LP will find γ_f for any graph, more efficient algorithms exist for finding it for restricted families of graphs. Farber [23] and Chang and Nemhauser [6] describe procedures for obtaining γ_f for strongly chordal graphs.

Concerning the algorithmic complexity of the upper parameters, let us consider the following decision problems.

> **Upper Fractional Domination** (UFD):
> Instance. A graph G and a rational number q.
> Question. Is $\Gamma_f(G) \geq q$?
> **Upper Fractional Total Domination** (UFTD):
> Instance. A graph G and a rational number q.
> Question. Is $\Gamma_f^0(G) \geq q$?

Theorem 3.64 [8] *UFD is NP-complete for arbitrary graphs.*

Theorem 3.65 [30] *UFTD is NP-complete for bipartite graphs.*

Cheston, Fricke, Hedetniemi and Jacobs [8] and Fricke, Hare, Jacobs and Majumdar [30] describe linear time algorithms to find Γ_f and Γ_f^0, respectively, for trees.

3.7 Open Problems

In this concluding section we state some open problems which include several mentioned in previous sections.

1. It is known that for certain classes of graphs, for example, chordal graphs, bipartite graphs, strongly perfect graphs, $\beta_0 = \Gamma = \Gamma_f = IR = IR_f$. Find some other classes of graphs for which these are equal.

2. Can we define the concept of a fractional independent function as a function $g: V \rightarrow [0. 1]$ in such a way that,

(i) the characteristic function of every independent set of vertices is an independent function,

(ii) every maximal independent set of vertices corresponds to a maximal independent function,

(iii) every maximal independent function is also a minimal dominating function?

3. It is known for trees, $\gamma_f = \gamma$, $\gamma_f^0 = \gamma_t$ and $\beta_0 = \Gamma = \Gamma_f = IR = IR_f$.

Which other fractional parameter pairs are equal for trees? What are other classes of graphs for which these are equal?

4. In Section 3.3 several k-parameters, namely, k-domination, k-packings, and k-irredundance have been studied, with the hope that they will help us to better understand their associated fractional parameters. The idea of k-parameters involving packings and coverings is just beginning to surface.

There are several open problems in this area.

Is it true that $\Gamma_k(G) < \Gamma_{k+1}(G)$ for every graph G?

It is known that $\gamma_f(G) = \min \{\gamma_k(G)/k \colon k \text{ is a positive integer}\}$ and $\Gamma_f(G) = \max \{\Gamma_k(G)/k \colon k \text{ is a positive integer}\}$ for every graph G, but no upper bound has been found on the value of k for which the minimum and maximum values are assumed. Find such bounds.

Bibliography

[1] B. D. Acharya and H. B. Walikar, Domination critical graphs. *Nat. Acad. Sci. Lett.* 2 (1979) 70-72.

[2] R. Aharoni, Fractional matchings and covers in infinite hypergraphs. *Combinatorica* 5(3) (1985) 181-184.

[3] C. Berge, Balanced Matrices. *Math. Programming* 2 (1972) 19-31.

[4] C. Berge, Packing problems in hypergraph theory : a survey. *Ann. Discrete Math.* 4 (1979) 3-37.

[5] A. E. Brouwer, P. Duchet and A. Schrijver, Graphs whose neighborhoods have no special cycles. *Discrete Math.* 47 (1983) 177-182.

[6] G. J. Chang and G. L. Nemhauser, The R-domination and R-stability on sun-free chordal graphs. *SIAM J. Alg. Discrete Math.* 5(3) (1984) 332-345.

[7] G. A. Cheston and G. H. Fricke, Classes of graphs for which upper fractional domination equals independence, upper domination and upper irredundance. *Discrete Appl. Math.* 55 (1994) 244-258.

[8] G. Cheston, G. Fricke, S. T. Hedetniemi and D. P. Jacobs, On the computational complexity of upper fractional domination. *Discrete Appl. Math.* 27 (1990) 195-207.

[9] F. R. K. Chung, Z. Furedi, M. R. Garey and R. L. Graham, On the fractional covering number of hypergraphs. *SIAM J. Discrete Math.* 1 (1988) 45-49.

[10] E. J. Cockayne and S. T. Hedetniemi, Towards a theory of domination in Graphs. *Networks* 7 (1977) 247-261.

[11] E. J. Cockayne, S. T. Hedetniemi and D. Miller, Properties of hereditary hypergraphs and middle graphs. *Canad. Math. Bull.* 21(4) (1978) 461-468.

[12] E. J. Cockayne, G. MacGillivray, C. M. Mynhardt, Convexity of minimal dominating functions of trees - II. *Discrete Math.* 125 (1994) 137-146.

[13] E. J. Cockayne and C. M. Mynhardt, A characterization of universal minimal total dominating functions of trees. To appear in *Discrete Math.*

[14] E. J. Cockayne and C. M. Mynhardt, Convexity of minimal dominating function trees - a survey. *Quaestiones Math.* 16(3) (1993) 301-317.

[15] E. J. Cockayne and C. M. Mynhardt, Minimality and convexity of domination and related functions in graphs: a unifying theory. To appear in *Utilitas Math.*

[16] E. J. Cockayne, C. M. Mynhardt, B. Yu, Universal minimal total dominating functions in graphs. *Networks* 24 (1994) 83-90.

[17] G. S. Domke, *Variations of colorings, coverings and packings of graphs.* Ph.D. Dissertation, Clemson University (1988).

[18] G. S. Domke, S. T. Hedetniemi, and R. C. Laskar, Fractional packings, coverings and irredundance in graphs. *Congr. Numer.* 66 (1988) 227-238.

[19] G. S. Domke, S. T. Hedetniemi, R. C. Laskar and R. Allan, Generalized packings and coverings of graphs. *Congr. Numer.* 62 (1988) 259-270.

[20] G. Domke, S. T. Hedetniemi, R. C. Laskar, and G. Fricke, Relationships between integer and fractional parameters of graphs. *Graph Theory, Combinatorics, and Applications*, John Wiley & Sons, Inc. 2 (1991) 371-387.

[21] G. S. Domke and R. Laskar, Fractional-bondage and fractional-reinforcement numbers of graphs. To appear in *Discrete Math.*

[22] M. Farber, Characterization of strongly chordal graphs. *Discrete Math.* 43 (1983) 173-189.

[23] M. Farber, Domination, independent domination and duality in strongly chordal graphs. *Discrete Appl. Math.* 7 (1986) 115-130.

[24] J. F. Fink and M. S. Jacobson, *n*-domination in graphs. *Graph Theory with Application to Algorithms and Computer Science.* Wiley-Intersci. Publishers, Wiley, New York (1985) 283-300.

[25] J. F. Fink, M. S. Jacobson, L. F. Kinch and J. Roberts, The bondage number of a graph. *Discrete Math.* 86 (1990) 47-57.

[26] D. C. Fisher, Domination, fractional domination, 2-packing and graph products. *SIAM Discrete Math.* 7(3) (1994) 493-498.

[27] D. C. Fisher, J. Ryan, G. S. Domke and A. Majumdar, Fractional domination of strong direct products. *Discrete Appl. Math.* 50 (1994) 89-91.

[28] G. Fricke, On the equivalence of upper irredundance and fractional upper irredundance numbers of graphs. Unpublished manuscript (1989).

[29] G. Fricke, Upper domination on double cone graphs. *Congr. Numer.* 72 (1990) 199-207.

[30] G. Fricke, E. Hare, D. Jacobs, and A. Majumdar, On integral and fractional total domination. *Congr. Numer.* 77 (1990) 87-95.

[31] D. R. Fulkerson, A. J. Hoffman, R. Oppenheim, On balanced matrices. *Math. Programming Study* 1 (1974) 120-132.

[32] J. Ghoshal, R. Laskar, D. Pillone and C. Wallis, Strong bondage and strong reinforcement numbers of graphs. *Congr. Numer.* 108 (1995) 33-42.

[33] M. C. Golumbic, *Algorithmic Graph Theory and Perfect Graphs.* Academic Press, New York (1989).

[34] D. Grinstead and P. J. Slater, Fractional domination and fractional packings in graphs. *Congr. Numer.* 71 (1990) 153-172.

[35] E. O. Hare, Fibonacci numbers and fractional domination of $P_m \times P_n$. *Fibonacci Quart.* 32 (1994) 69-73.

[36] E. O. Hare, K-weight domination and fractional domination of $P_m \times P_n$. *Congr. Numer.* 78 (1990) 71-80.

[37] S. M. Hedetniemi, S. T. Hedetniemi and T. V. Wimer, Linear time resource allocation algorithms for trees. Technical Report URI-014, Department of Mathematics, Clemson University (1987).

[38] A. J. Hoffman, A. W. Kolen and M. Sakarovitch, Totally balanced and greedy matrices. *SIAM J. Alg. Discrete Math.* G(4) (1985) 721-730.

[39] J. Kok and C. M. Mynhardt, Reinforcement in graphs. *Congr. Numer.* 79 (1990) 225-231.

[40] M. Larsen, J. Propp and D. Ullman, The fractional chromatic number of Mycielski's graphs. *J. Graph Theory* 19 (1995) 411-416.

[41] R. Laskar, A. Majumdar, G. S. Domke and G. H. Fricke, A fractional view of graph theory. *Sankhyā* 54 (1992) 265-279.

[42] L. Lovasz, Combinatorial problems and exercises. Budapest, 528 (1979).

[43] A. Majumdar, *Neighborhood hypergraphs: a framework for covering and packing parameters in graphs.* Ph.D. Dissertation, Clemson University (1992).

[44] A. Majumdar, private communication (1988).

[45] A. Meir and J. W. Moon, Relations between packing and covering numbers of a tree. *Pacific J. Math.* 61(1) (1975) 225-233.

[46] W. R. Pulleyblank, Fractional matchings and the Edmonds-Gallai theorem. *Discrete Math.* 16 (1987) 51-58.

[47] L. S. Stewart and E. O. Hare, Fractional domination of $P_m \times P_n$. *Congr. Numer.* 91 (1992) 35-42.

[48] D. P. Sumner and P. Blitch, Domination critical graphs. *J. Combin. Theory B* 34 (1983) 65-76.

Chapter 4

Majority Domination and Its Generalizations

Johannes H. Hattingh
Department of Mathematics
Rand Afrikaans University
P. O. Box 524
Auckland Park
2006
Johannesburg, South Africa

4.1 Introduction

For any real valued function $f : V \to \mathbf{R}$ and $S \subseteq V$, let $f(S) = \sum_{u \in S} f(u)$. The *weight* of f is defined as $f(V)$. We will also denote $f(N[v])$ by $f[v]$, where $v \in V$. A *minus dominating function* is defined in [6] as a function $f : V \to \{-1, 0, 1\}$ such that $f[v] \geq 1$ for every $v \in V$. The *minus domination number* of a graph G is $\gamma^-(G) = \min\{f(V) \mid f$ is a minus dominating function on $G\}$. A *signed dominating function* is defined in [7] as a function $f : V \to \{-1, 1\}$ such that $f[v] \geq 1$ for every $v \in V$. The *signed domination number* of a graph G is $\gamma_s(G) = \min\{f(V) \mid f$ is a signed dominating function on $G\}$. A *majority dominating function* is defined in [3] as a function $f : V \to \{-1, 1\}$ such that $f[v] \geq 1$ for at least half the vertices $v \in V$. The *majority domination number* of a graph G is $\gamma_{maj}(G) = \min\{f(V) \mid f$ is a majority dominating function of $G\}$. Let k be a positive integer such that $1 \leq k \leq |V|$. A *signed k-subdominating function* (kSF) for G is defined in [4] as a function $f : V \to \{-1, 1\}$ such that $f[v] \geq 1$ for at least k vertices of G. The *signed k-subdomination number* of a graph G is $\gamma_{ks}^{-11}(G) = \min\{f(V) \mid f$ is a signed kSF of $G\}$. In the special cases where $k = |V|$ and $k = \lceil \frac{|V|}{2} \rceil$, $\gamma_{ks}^{-11}(G)$ is, respectively, the signed domination number and the majority domination number.

A *minus k-subdominating function (kSF)* for G is defined in [2] as a function $f : V \to \{-1, 0, 1\}$ such that $f[v] \geq 1$ for at least k vertices of G. The *minus k-subdomination number* of a graph G, denoted by $\gamma_{ks}^{-101}(G)$, is equal to $\min\{f(V) \mid f \text{ is a minus } kSF \text{ of } G\}$. The kSF f is called *minimal* if no $g < f$ is a kSF.

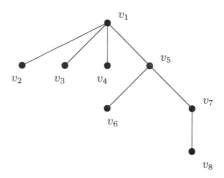

Figure 4.1: The tree T.

To demonstrate the notion of minus k-subdomination in graphs we consider the tree T of Figure 4.1. We now define a function $f : V(T) \to \{-1, 0, 1\}$ by

$$f(v) = \begin{cases} 1 & \text{if } v = v_1 \\ 0 & \text{if } v \in \{v_2, v_3, v_4\} \\ -1 & \text{otherwise.} \end{cases}$$

Then $f[v] \geq 1$ for all $v \in \{v_2, v_3, v_4\}$. Hence, f is a minus 3-subdominating function of T and $\gamma_{3s}^{-101}(T) \leq -3$. In fact, $\gamma_{3s}^{-101}(T) = -3$.

The motivation for studying these variations of the domination number is rich and varied from a modelling perspective. By assigning the values -1 or $+1$ to the vertices of a graph, we can model positive and negative spins on an electron; positive and negative polarities; positive and negative electrical charges or any number of real-word situations characterized by polarities or opposites. By assigning the values -1, 0 or $+1$ to the vertices of a graph, we can model negative or neutral responses of preferences in such things as political voting or social behaviors. By examining these parameters, we study situations in which, in spite of the presence of negative vertices, the closed neighborhoods of at least k of the vertices are required to maintain a positive sum.

A *remote vertex* v of a graph G is a vertex which is adjacent to an endvertex of G. Let L denote the set of endvertices of T. For each $v \in V$, define $L(v) = N(v) \cap L$ and $\ell(v) = |L(v)|$.

For a rooted tree T and any vertex u of T, let $T(u)$ denote the subtree of T induced by u and its descendants. A *full t-ary tree of height h* is a rooted tree such that each vertex which is not a leaf has exactly t children and all leaves are at distance h from the root.

In the next section, we survey the results concerning majority domination in graphs. Then, in the following section, results concerning a generalization of majority domination, namely signed k-subdomination, are surveyed. Lastly, results concerning a generalization of minus domination, namely minus k-subdomination, are surveyed.

4.2   Majority Domination in Classes of Graphs

The majority domination numbers of the following classes of graphs have been determined. Broere, Hattingh, Henning and McRae [3] determined the following.

Theorem 4.1 [3] *For* $n \geq 1$,

$$\gamma_{maj}(K_n) = \begin{cases} 1 \ if \ n \ is \ odd \\ 2 \ if \ n \ is \ even \end{cases} .$$

Theorem 4.2 [3] *For* $n \geq 2$,

$$\gamma_{maj}(K_{1,n-1}) = \begin{cases} 1 \ if \ n \ is \ odd \\ 2 \ if \ n \ is \ even \end{cases} .$$

Theorem 4.3 [3] *For* $t \geq s \geq 2$,

$$\gamma_{maj}(K_{s,t}) = \begin{cases} 2 - t \ if \ s \ is \ even \\ 3 - t \ if \ s \ is \ odd \end{cases} .$$

Theorem 4.4 [3] *For* $t \geq s \geq 1$,

$$\gamma_{maj}(K_s \cup K_t) = \begin{cases} 1 - s \ if \ t \ is \ odd \\ 2 - s \ if \ t \ is \ even \end{cases} .$$

The majority domination number has also been determined for paths and cycles. This will be addressed in Section 4.3, where we look at the first generalization of majority domination in graphs.

Lastly, for t an integer, let $p(t, \gamma_{maj})$ be the smallest order of a connected graph with majority domination number equal to t.

Theorem 4.5 [3] *For* $t \geq 0$ *an integer,* $p(-t, \gamma_{maj}) = t + 4$.

An extremal graph here is $K_{2,t+2}$, which is a connected graph of order $t+4$ with majority domination number equal to $-t$.

4.3   Signed k-subdomination in Graphs

In this section, we survey some recent results concerning signed k-subdomination in graphs.

Let f be a signed kSF of $G = (V, E)$. We say $v \in V$ is *covered by* f if $f[v] \geq 1$ and denote the set of vertices covered by f, by C_f. Let $P_f = \{v \in V \mid f(v) = 1\}$ and $B_f = \{v \in V \mid f[v] \in \{1, 2\}\}$. For $A, B \subseteq V$, we say A *dominates* B, denoted by $A \succ B$, if for each $b \in B, N[b] \cap A \neq \emptyset$.

Theorem 4.6 (Cockayne and Mynhardt [4]) *A signed kSF f is minimal if and only if for each k-subset K of C_f, $K \cap B_f \succ P_f$.*

Let $\gamma(n, k)$ be the minimum value of γ_{ks}^{-11} taken over all trees of order n ($n \geq k$) and $\mathcal{S}(n, k)$ be the set of such trees T with $\gamma_{ks}^{-11}(T) = \gamma(n, k)$. Further, let $\sigma(T)$ be the degree sum of all vertices of T with degree at least three and define $\mathcal{T}(n, k) = \{T \in \mathcal{S}(n, k) \mid \sigma(T) \text{ is minimum}\}$.

Theorem 4.7 [4] *For any n, $\mathcal{S}(n, k) = \{P_n\}$.*

Theorem 4.8 [4] *For $n \geq 2$ and $1 \leq k \leq n$,*

$$\gamma_{ks}^{-11}(P_n) = 2\lfloor (2k + 4)/3 \rfloor - n.$$

Note that if $k = \lceil \frac{n}{2} \rceil$, then we obtain the following result.

Theorem 4.9 [4] *For $n \geq 3$,*

$$\gamma_{maj}(P_n) = \begin{cases} -2\lfloor \frac{n-4}{6} \rfloor & \text{for } n \text{ even} \\ 1 - 2\lfloor \frac{n-3}{6} \rfloor & \text{for } n \text{ odd} \end{cases}.$$

Using Theorems 4.7 and 4.8, Cockayne and Mynhardt established the following result.

Theorem 4.10 [4] *If T is a tree of order $n \geq 2$ and k is an integer such that $1 \leq k \leq n$, then*

$$\gamma_{ks}^{-11}(T) \geq 2\lfloor (2k + 4)/3 \rfloor - n$$

with equality for $T = P_n$.

This result sheds new light on the following result.

Theorem 4.11 (Dunbar, Hedetniemi, Henning and Slater [7]) *Let T be a tree of order $n \geq 2$. Then $\gamma_s(T) \geq \frac{n+4}{3}$ with equality if and only if T is a path on $3j + 2$ vertices, for $j \geq 0$.*

Let $n \geq 2$ be an integer and let k be an integer such that $1 \leq k \leq n$. Trees T of order n for which $\gamma_{ks}^{-11}(T) = 2\lfloor(2k+4)/3\rfloor - n$ were recently characterised by Hattingh and Ungerer. The statement of this result is rather intricate and the reader is therefore referred to [13] for the details.

The *comet* $C_{s,t}$, where s and t are positive integers, denotes the tree obtained by identifying the center of the star $K_{1,s}$ with an end vertex of P_t, the path of order t. So $C_{s,1} \cong K_{1,s}$ and $C_{1,p-1} \cong P_p$. Beineke and Henning (see [1]) computed $\gamma_{ks}^{-11}(C_{s,t})$ for $k = s + t$ and for $k = \lceil\frac{s+t}{2}\rceil + 1$. Hattingh and Ungerer extended their result as follows.

Theorem 4.12 [13] *Let n, s and t be positive integers such that $n = s + t$ and let $G = C_{s,t}$. If $s, t \geq 2$, then*

$$\gamma_{ks}^{-11}(G) = \begin{cases} 2\lfloor(2k+4)/3\rfloor - n & \text{if } k \leq t - 1 \\ 2(k - \lceil\frac{t}{3}\rceil + 2) - n & \text{if } t \leq k \text{ and} \\ & \quad (k \leq t + \lfloor\frac{s}{2}\rfloor - 2, \ t \equiv 0 \ (mod \ 3) \text{ or} \\ & \quad k \leq t + \lfloor\frac{s}{2}\rfloor, \ t \equiv 1 \ (mod \ 3) \text{ or} \\ & \quad k \leq t + \lfloor\frac{s}{2}\rfloor - 1, \ t \equiv 2 \ (mod \ 3)) \\ 2(k - \lceil\frac{t}{3}\rceil + 1) - n & \text{otherwise.} \end{cases}$$

The value of $\gamma_{ks}^{-11}(C_n)$ is calculated in [10].

Theorem 4.13 (Hattingh, Henning and Ungerer [10]) *If $n \geq 3$ and $1 \leq k \leq n - 1$, then*

$$\gamma_{ks}^{-11}(C_n) = \begin{cases} \frac{n-2}{3} & \text{if } k = n - 1 \text{ and } k \equiv 1 \ (mod \ 3) \\ 2\lfloor\frac{2k+4}{3}\rfloor - n & \text{otherwise.} \end{cases}$$

This result generalizes the following.

Theorem 4.14 [3] *If $n \geq 3$, then*

$$\gamma_{maj}(C_n) = \gamma_{maj}(P_n).$$

Theorem 4.15 [10] *If $n \geq 3$ and $1 \leq k \leq n$, then for every r-regular $(r \geq 2)$ graph G of order n,*

$$\gamma_{ks}^{-11}(G) \geq \begin{cases} k\frac{r+3}{r+1} - n & r \text{ odd} \\ \\ k\frac{r+2}{r+1} - n & \text{for } r \text{ even,} \end{cases}$$

and these bounds are best possible.

The following result combines results due to Dunbar, Henning, Hedetniemi and Slater (r even) (see [7]) and Henning and Slater (r odd) (see [16]).

Theorem 4.16 *For every r-regular ($r \geq 2$) graph G of order n,*

$$\gamma_s(G) \geq \begin{cases} \frac{2n}{r+1} & r \ odd \\\\ \frac{n}{r+1} & r \ even. \end{cases}$$

Theorem 4.17 (Zelinka [17]) *For every cubic graph G of order n,* $\gamma_{maj}(G) \geq -\frac{n}{4}$.

Since $\gamma_{maj}(2K_4) = -2 = -\frac{8}{4}$, this bound is best possible.

Theorem 4.18 (Henning [15]) *For every r-regular ($r \geq 2$) graph $G = (V, E)$ of order n,*

$$\gamma_{maj}(G) \geq \begin{cases} \left(\frac{1-r}{2(r+1)}\right) n & r \ odd \\\\ \left(\frac{-r}{2(r+1)}\right) n & r \ even, \end{cases}$$

and these bounds are best possible.

Note that, if $k = n$ in the statement of Theorem 4.15, then we obtain the result of Theorem 4.16, and, if $k = \lceil \frac{1}{2}n \rceil$, then we obtain the result of Theorem 4.18.

Theorem 4.19 (Alon [3]) *For any connected graph G of order n,*

$$\gamma_{maj}(G) \leq \begin{cases} 1 \ if \ n \ is \ odd \\\\ 2 \ if \ n \ is \ even \end{cases}.$$

Proof. Let $G = (V, E)$ be a connected graph of order n. If $n = 2\ell + 1$, then we partition V into sets V_1 and V_2 with $|V_1| = \ell$ and $|V_2| = \ell + 1$ such that the sum of the number of edges induced by V_1 and V_2 is as large as possible. Then each vertex v in V_2 is adjacent to at least as many vertices in V_2 as in V_1, otherwise we may remove v from the set V_2 and add it to the set V_1 to produce a new partition of V in which the sum of the number of edges induced by the resulting partite sets exceeds that of the original partition of V. Hence assigning to each vertex of V_2 the value 1 and to each vertex of V_1 the value -1, we produce a majority dominating function of weight 1. Hence, if G has odd order, then $\gamma_{maj}(G) \leq 1$. On the other hand, if n is even, then let v be any vertex in G and consider the graph $G - v$. As above, we may produce a majority dominating function of $G - v$ of weight 1. We now extend this function to a majority dominating function of G of weight 2 by assigning the vertex v the value 1. Hence, if G has even order, then $\gamma_{maj}(G) \leq 2$. $\square$

Obviously, if f is a majority dominating function of $G = (V, E)$, then f is a signed kSF for each $k \leq \lceil |V|/2 \rceil$. Hence we have the following corollary.

Corollary 4.20 [4] *For any connected graph G of order n and integer $k \leq \lceil \frac{1}{2}n \rceil$,*

$$\gamma_{ks}^{-11}(G) \leq \begin{cases} 1 \text{ if } n \text{ is odd} \\ 2 \text{ if } n \text{ is even} \end{cases}.$$

That this bound is sharp can be seen by noting that $\gamma_{ks}^{-11}(K_{2t+1}) = 1$ and $\gamma_{ks}^{-11}(K_{2t}) = 2$ for each $k \leq t + 1$.

In [4], this bound is improved for trees and extended to an upper bound for γ_{ks}^{-11} for all $k \in \{1, \ldots, n\}$. In order to prove this result, we need the following definitions. A remote vertex x is *very remote* if x is adjacent to at most one vertex of T that is not a leaf. Note that each tree has at least one very remote vertex. Let $r(T)$ denote the radius and $C(T)$ the center of T. Let $z \in C(T)$ and consider any leaf y at distance $r(T)$ from z. Let x be the vertex adjacent to y. Then x is a very remote vertex of T. A tree T' is a *full subtree* of T if $T' = T$ or T' is a component of $T - e$ for some edge e of T. In the latter case, if $e = uv$ where $u \in V(T')$, we say that T' is *attached at u*. If $T' \cong K_{1,m}$ is a full subtree of T, then T' is called a *full substar* of T. Note that if T' is a full subtree attached at a very remote vertex x of T, then T' is a full substar of T with center x. We write $L(T), L_T(v)$ and $\ell_T(v)$ to emphasize that T is the tree under consideration and if $T = S_i$ (say), we write $\ell_i(v)$ for $\ell_T(v)$.

Theorem 4.21 [4] *For any tree T of order n and integer $k \in \{1, \ldots, n\}$, $\gamma_{ks}^{-11}(T) \leq 2(k+1) - n$.*

Proof. The result clearly holds if $T = K_2$ or if $k = 1$; thus we assume that $k \geq 2$ and $n \geq 3$. Set $S_0 = T$ and $s_0 = k$. We construct a sequence $T_1, \ldots, T_n$ of disjoint subtrees of T as follows: If S_0 contains a full substar G_1 with center v_1 such that $s_0 \leq \ell_0(v_1)$, let T_1 be a (nontrivial) full subtree of S_0 induced by v_1 and any s_0 leaves of G_1 and set $s_1 = -1$. Otherwise, let T_1 be a (nontrivial) full subtree of S_0 of order $k_1 \leq s_0$ attached at v_1 (if $T_1 \neq S_0$) and define $s_1 = s_0 - k_1$. Continuing in this way, if $s_i > 0$, define $S_i = S_{i-1} - T_i$. If S_i contains a full substar of G_{i+1} with center v_{i+1}, where $s_i \leq \ell_i(v_{i+1})$, let T_{i+1} be the subtree of S_i induced by v_{i+1} and any s_i leaves of G_{i+1} and set $S_{i+1} = -1$. Otherwise, let T_{i+1} be a full subtree of S_i of order $k_{i+1} \leq s_i$ attached at v_{i+1} (if $T_{i+1} \neq S_i$) and set $s_{i+1} = s_i - k_{i+1}$. We thus obtain a finite sequence of disjoint subtrees $T_1, \ldots, T_r$ of T and a sequence of integers $s_0 > s_1 > \ldots > s_r$, where $s_r \in \{0, -1\}$.

Let F be the (possibly disconnected) subgraph of T induced by $\cup_{i=1}^{r} V(T_i)$. Note that $|V(F)| = k + 1$ if $s_r = -1$ and $|V(F)| = k$ otherwise. Define $f : V(T) \to \{-1, 1\}$ by $f(x) = 1$ if $x \in V(F)$ and $f(x) = -1$ otherwise. Note that for each $i = 1, \ldots, r$, v_i is the only vertex of T_i that is possibly adjacent to a vertex of $S_r = T - F$ and since T_i is full (except possibly T_r if $s_r = -1$), v_i (except possibly v_r) is adjacent to at most one vertex of S_r. Moreover, if $v_i, i = 1, \ldots, r$, is adjacent to a vertex of S_r, then v_i is not a leaf of T (since T_i is nontrivial). Hence f covers each vertex of F except possibly v_r if $s_r = -1$. In either case, f covers at least k vertices of T and $f(V(T)) \leq 2(k+1) - n$. $\square$

That this bound is exact for trees of order n when $k \leq \frac{1}{2}n$ follows easily since $\gamma_{ks}^{-11}(K_{1,n-1}) = 2(k+1) - n$ if $k \leq \frac{1}{2}n$. However, Cockayne and Mynhardt have not been able to find a tree T (or any other connected graph) of order n for which $\gamma_{ks}^{-11}(T) = 2(k+1) - n$ if $k > \frac{1}{2}n$. Hence they formulate the following conjectures.

Conjecture 4.22 [4] *For any tree T of order n and any k with $\frac{1}{2}n < k \leq n, \gamma_{ks}^{-11}(T) \leq 2k - n$.*

Conjecture 4.23 [4] *For any connected graph G of order n and any k with $\frac{1}{2}n < k \leq n, \gamma_{ks}^{-11}(G) \leq 2k - n$.*

If either of these conjectures is false, there still remains the problem of determining the smallest integer $p = p(n)$ such that $\gamma_{ks}^{-11}(G) \leq 2k - n$ for all graphs G (trees) of order n and all $k \geq p$.

The remainder of Cockayne and Mynhardt's paper (see [4]) is devoted to determining conditions on k such that $\gamma_{ks}^{-11}(T) \leq 2k - n$ for certain classes of trees of order n.

Theorem 4.24 [4] *Let T be a tree of order n rooted at v, where $\deg(v) = s$ and $\ell(v) = t$; say $N(v) = \{w_1, \ldots, w_t, u_1, \ldots, u_{s-t}\}$ where $L(v) = \{w_1, \ldots, w_t\}$ and $|V(T(u_1))| \leq \ldots \leq |V(T(u_{s-t}))|$. If $r = \lceil \frac{1}{2}(s+2) \rceil \leq s - t$ and $n \geq k \geq |V(T(u_1))| + \ldots + |V(T(u_r))|$, then $\gamma_{ks}^{-11}(T) \leq 2k - n$.*

For any vertex v of T, define $\eta(v)$ by

$$\eta(v) = \begin{cases} \frac{d+2}{2d+1-\ell(v)} & \text{if } \deg(v) = 2d + 1 \text{ for some integer } d \geq 1 \\ \frac{d+1}{2d-\ell(v)} & \text{if } \deg(v) = 2d \text{ for some integer } d \geq 1. \end{cases}$$

Corollary 4.25 [4] *If T is a tree of order n such that $\eta(v) \leq 1$ and $\eta(v)(n - \ell(v) - 1) \leq k \leq n$ for some vertex v of T, then $\gamma_{ks}^{-11}(T) \leq 2k - n$.*

Theorem 4.26 [4] *For any full t-ary tree T of order n, $\gamma_{ks}^{-11}(T) \leq 2k - n$ whenever $2\lceil \frac{1}{2}(t+3) \rceil \leq k \leq n$.*

Lastly, we consider the decision problem.

PARTIAL SIGNED DOMINATING FUNCTION (PSDF)

INSTANCE: A graph G, positive rational number $r \leq 1$ (in its simplest form) and an integer ℓ.

QUESTION: Is there a function $f : V(G) \to \{-1, 1\}$ of weight ℓ or less for G such that $|C_f| \geq r|V(G)|$?

Hattingh, Henning and Ungerer (see [10]) showed that **PSDF** is NP-complete by describing a polynomial transformation from the following problem (see [8]).

PLANAR 4-REGULAR DOMINATING SET
INSTANCE: A planar 4-regular graph $G = (V, E)$ and a positive integer $k \leq \frac{|V|}{2}$.

QUESTION: Is there a dominating set of cardinality k or less for G?

If $r = 1$, then **PSDF** is the NP-complete problem **SIGNED DOMINATION** (see [9]). Hence, we also assume that $0 < r < 1$.

Theorem 4.27 [10] *The decision problem* **PSDF** *is NP-complete.*

This result generalizes the corresponding result of [3].

4.4   Minus k-subdomination in Graphs

In this section, we survey some recent results concerning minus k-subdomination in graphs.

Let f be a minus kSF for the graph $G = (V, E)$. We use three sets for such an f:

$$
\begin{aligned}
B_f &= \{v \in V | f[v] = 1\}, \\
P_f &= \{v \in V | f(v) \geq 0\} \\
\text{and } C_f &= \{v \in V | f[v] \geq 1\}.
\end{aligned}
$$

As before, a vertex $v \in C_f$ is *covered* by f; all other vertices are *uncovered* by f. Note that $B_f \subseteq C_f$.

Theorem 4.28 (Broere, Dunbar and Hattingh [2]) *A minus kSF f is minimal if and only if for each k-subset K of C_f we have $B_f \cap K \succ P_f$.*

Theorem 4.29 [2] *If $n \geq 2$ and $1 \leq k \leq n-1$, then $\gamma_{ks}^{-101}(P_n) = \lceil \frac{k}{3} \rceil + k - n + 1$.*

The following result is proved in [6] by Dunbar, Hedetniemi, Henning and McRae.

Theorem 4.30 [6] *For the path P_n, $\gamma_{ns}^{-101}(P_n) = \lceil \frac{n}{3} \rceil$.*

Hattingh and Ungerer ([12]) established the following result.

Theorem 4.31 [12] *If T is a tree of order $n \geq 2$ and k is an integer such that $1 \leq k \leq n - 1$, then*

$$\gamma_{ks}^{-101}(T) \geq k - n + 2.$$

Moreover, this bound is best possible.

Proof. The proof is by induction on n. If $n = 2$, then $T = P_2$ and $k = 1$. Then, by Theorem 4.29, $\gamma_{ks}^{-101}(T) = \lceil \frac{k}{3} \rceil + k - n + 1 = k - n + 2$. Let $n \geq 3$ and assume that for all trees T' of order n' ($\leq n - 1$) that $\gamma_{\ell s}^{-101}(T') \geq \ell - n' + 1$, where ℓ is an integer such that $1 \leq \ell \leq n' - 1$. Let T be a tree of order $n \geq 3$ and let k be an integer such that $1 \leq k \leq n - 1$. Also, let f be a kSF of T with weight $\gamma_{ks}^{-101}(T)$.

We distinguish three cases:

Case 1 Suppose $2 \leq k \leq n - 2$.

Case 1.1 There exists an endvertex $v \in V(T)$ such that $v \notin C_f$.

Let $T' = T - v$ and let g be the restriction of f to $V(T')$. If $f(v) \in \{-1, 0\}$, then, clearly, g is a kSF for T'. Since $1 \leq k \leq (n-1) - 1$, we have, by the induction hypothesis, that $\gamma_{ks}^{-101}(T') \geq k - (n-1) + 2 = k - n + 3$. Hence $\gamma_{ks}^{-101}(T) = f(V(T)) = g(V(T')) + f(v) \geq \gamma_{ks}^{-101}(T') + f(v) \geq k - n + 3 - 1 = k - n + 2$. If, on the other hand, $f(v) = 1$, then, since $v \notin C_f$, we must have that $f(w) = -1$, where w is the vertex adjacent to v. Since $f(v) = 1$, Theorem 4.28 implies that $f[w] = 1$, whence $g[w] = 0$. It follows, therefore, that g is a $(k-1)SF$ of T'. Since $1 \leq k - 1 \leq (n-1) - 1$, we have, by the induction hypothesis, that $\gamma_{ks}^{-101}(T') \geq k - 1 - (n-1) + 2 = k - n + 2$. Hence $\gamma_{ks}^{-101}(T) = f(V(T)) = g(V(T')) + f(v) \geq \gamma_{ks}^{-101}(T') + f(v) \geq k - n + 2 + 1 \geq k - n + 2$.

Case 1.2 All endvertices of T are covered by f.

Note that, if v is an endvertex, then $f(v) \in \{0, 1\}$. If there is an endvertex v such that $f(v) = 0$, let $T' = T - v$ and let g be the restriction of f to $V(T')$. Note that g is a $(k-1)SF$ of T' and again $\gamma_{ks}^{-101}(T) \geq k - n + 2$. We therefore assume that every endvertex is assigned the value 1 under f. Let v be an arbitrary endvertex and let w be the vertex adjacent to v. Clearly, $f(w) \in \{0, 1\}$.

If $f(w) = 0$, then we define a function $h : V(T) \rightarrow \{-1, 0, 1\}$ as follows: $h(w) = 1$, $h(v) = 0$ and $h(x) = f(x)$ for all $x \in V(T) - \{v, w\}$. Note that h is a kSF of T of weight equal to that of f. As before, $\gamma_{ks}^{-101}(T) \geq k - n + 2$.

We therefore assume that all endvertices and vertices adjacent to endvertices are assigned the value 1 under f. Root T at r and choose an endvertex v such that $d(r, v)$ is a maximum. Let w be the parent of v and let u be the parent of w. Since $f(v) \geq 1$ and $f[v] = 2$, Theorem 4.28 implies that $f[w] = 1$. Note that if $t \in N[w] - \{u, v, w\}$, then t must be a child of w. If $deg_T(t) \geq 2$, then there must be an endvertex x of T such that $d(r,x) > d(r,v)$, which contradicts our choice of v. Hence, t must be an endvertex of T, so that $f[w] \geq f(w) + f(t) + f(v) + f(u) \geq 1 + 1 + 1 - 1 = 2$, which is a contradiction. Therefore, $N[w] = \{u, v, w\}$, so that $f(u) = f[w] - f(v) - f(w) = 1 - 1 - 1 = -1$. Define a function $h : V(T) \rightarrow \{-1, 0, 1\}$ as follows: $h(u) = h(v) = 0$ and $h(x) = f(x)$ for all $x \in V(T) - \{u, v\}$. Note that h is a kSF of T of weight equal to that of f. As before, $\gamma_{ks}^{-101}(T) \geq k - n + 2$.

Case 2 Suppose $k = 1$.

Since T has at least two endvertices, there must be an endvertex v such that $v \notin C_f$. Let $T' = T - v$ and let g be the restriction of f to $V(T')$. If $f(v) \in \{-1, 0\}$, then g is a kSF of T'. Also, the induction hypothesis implies that $\gamma_{ks}^{-101}(T') \geq 1 - (n - 1) + 2 = 4 - n$. Hence, $\gamma_{ks}^{-101}(T) = g(V(T')) + f(v) \geq (4 - n) - 1 = 1 - n + 2$. We now assume that $f(v) = 1$. Let w be the vertex adjacent to v. Since $v \notin C_f$, it follows that $f(w) = -1$. Define a function $h : V(T) \rightarrow \{-1, 0, 1\}$ as follows: $h(w) = h(v) = 0$ and $h(x) = f(x)$ for all $x \in V(T) - \{v, w\}$. Note that h is a kSF of T of weight equal to that of f. As before, $\gamma_{ks}^{-101}(T) \geq k - n + 2$.

Case 3 Suppose $2 \leq k = n - 1$.

Since T has at least two endvertices and all but possibly one vertex are covered, there is an endvertex which is covered. Let v be any such an endvertex. Let w be the vertex adjacent to v. Note that $f(v) \in \{0, 1\}$, while $f(w) \in \{0, 1\}$. Also, the case $f(v) = f(w) = 0$ is impossible.

If $f(v) = 0$ and $f(w) = 1$, then g, the restriction of f to the vertex set of $T' = T - v$, is a $(k - 1)SF$ for T'. Also, by the induction hypothesis, we have that $\gamma_{ks}^{-101}(T') \geq (k - 1) - (n - 1) + 2 = k - n + 2$. Hence, $\gamma_{ks}^{-101}(T) = g(V(T')) + f(v) \geq \gamma_{ks}^{-101} \geq k - n + 2$. If $f(v) = 1$ and $f(w) = 0$, then the function $h : V(T) \rightarrow \{-1, 0, 1\}$ defined by $h(v) = 0$, $h(w) = 1$ and $h(x) = f(x)$ for all $x \in V(T) - \{v, w\}$, is a kSF of T of weight equal to that of f. As before, $\gamma_{ks}^{-101}(T) \geq k - n + 2$.

Hence, if v is an endvertex such that $v \in C_f$ and $w \in N(v)$, then $f(v) = f(w) = 1$. Root the tree at r and let v be a covered endvertex such that $d(v, r)$ is a maximum. Let $w \in N(v)$. By Theorem 4.28, it follows that $f[w] = 1$.

Suppose w has children v' and v'', both distinct from v, such that $f(v') \in \{-1, 0\}$ and $f(v'') \in \{-1, 0\}$. If both v' and v'' are endvertices of T, then $f(v') = f(v'') = -1$ (for if $f(v') = 0$ or $f(v'') = 0$, then $v' \in C_f$ or $v'' \in C_f$, which is contrary to our assumption). Then $v' \notin C_f$ and $v'' \notin C_f$, which contradicts the fact that $|C_f| \geq n - 1$. Suppose, without loss of generality, that $deg(v') \geq 2$. Let w' be an endvertex of the maximal tree rooted at v'. Note that, by our choice of v, $w' \notin C_f$ and $\deg(v'') = 1$. Also, since $v'' \in C_f$, we must have that $f(v'') = 0$, contrary to our assumption. Hence, if $f(v') \in \{-1, 0\}$, we must have that $f(v'') = 1$.

Let $v_1, \ldots, v_d, d \geq 3$, denote the children of w distinct from v and let u be the parent of w. Since $f[w] = 1$, not all the vertices $v_1, \ldots, v_d$ are assigned the value 1 under f. Hence, some vertex v_1, say, is assigned either -1 or 0. Then $f(v_2) = \ldots = f(v_d) = 1$, so that $f(u) + f(v_1) = f[w] - (f(v) + f(w) + f(v_2) + \ldots + f(v_d)) = 1 - (2 + (d-1)) = -d \leq -2$, which implies that $f(u) = f(v_1) = -1$. Note that $d = 2$. Define the function $h : V(T) \to \{-1, 0, 1\}$ by $h(v) = h(u) = 0$ and $h(x) = f(x)$ for all $x \in V(T) - \{u, v\}$. Then h is a kSF of T of weight equal to that of f. As before, $\gamma_{ks}^{-101}(T) \geq k - n + 2$.

Now suppose v and v_1 are the only children of w. If $f(v_1) = 1$, then $f[w] = 3 + f(u) \geq 2$, which is a contradiction. Since $d(v, r)$ is maximum, v_1 must be an endvertex of T. Thus, if $f(v_1) = 0$, then $v_1 \in C_f$, which contradicts our earlier assumption (any endvertex that belongs to C_f is assigned the value 1 under f). Hence $f(v_1) = -1$ and $f(u) = 0$. Define the function $h : V(T) \to \{-1, 0, 1\}$ by $h(v) = h(v_1) = 0$ and $h(x) = f(x)$ for all $x \in V(T) - \{v, v_1\}$. Then h is a kSF of T of weight equal to that of f. As before, $\gamma_{ks}^{-101}(T) \geq k - n + 2$.

If v is the only child of w, then $f(u) = -1$. Define the function $h : V(T) \to \{-1, 0, 1\}$ by $h(v) = h(u) = 0$ and $h(x) = f(x)$ for all $x \in V(T) - \{u, v\}$. Then h is a kSF of T of weight equal to that of f. As before, $\gamma_{ks}^{-101}(T) \geq k - n + 2$.

This completes the proof of our lower bound.

We now show that this bound is best possible. Let $n \geq 2$ be an integer, let $T = K_{1,n-1}$, let v be the central vertex of T and let $U \subseteq V(T) - \{v\}$ such that $|U| = k$. Define $f : V(T) \to \{-1, 0, 1\}$ by $f(v) = 1$, $f(u) = 0$ for all $u \in U$ and $f(u) = -1$ otherwise. Note that $S \subseteq C_f$, so that f is a kSF for T of weight $1 - (n - k - 1) = k - n + 2$. Hence $\gamma_{ks}^{-101}(T) \leq f(V(T)) = k - n + 2$. Our lower bound implies that $\gamma_{ks}^{-101}(T) = k - n + 2$. $\square$

However, trees which achieve the lower bound were not characterised in [12]. The following result solves this problem.

Theorem 4.32 (Hattingh, McRae and Ungerer [11]) *Let $n \geq 2$ and let $1 \leq k \leq n - 1$ be an integer. Then, for a tree T of order n, $\gamma_{ks}^{-101}(T) = k - n + 2$ if and only if one of the following holds.*

(a) T has a vertex v adjacent to at least k endvertices.

(b) T has a vertex v with $\deg(v) = k$ and at least $k - 1$ neighbors of v are endvertices.

(c) T has two adjacent vertices u and v with $\deg(u) + \deg(v) = k + 2$ such that u and v together are adjacent to at least $k - 2$ endvertices.

(d) T has a vertex w of degree three and two of the neighbors of w together are adjacent to exactly $k - 3$ other vertices, all of which are endvertices.

This result supplements the following result of Dunbar, Hedetniemi, Henning and McRae (see [6]).

Theorem 4.33 [6] *If T is a tree, then $\gamma_{ns}^{-101}(T) \geq 1$. Furthermore, equality holds if and only if T is a star.*

The value of $\gamma_{ks}^{-101}(G)$, where G is a comet, is calculated by Hattingh and Ungerer in [14].

Theorem 4.34 [14] *Let n, s and t be positive integers such that $n = s + t$, let k be an integer such that $1 \leq k \leq n - 1$ and let $G = C_{s,t}$. If $t \geq 2$ and $s \geq 2$, then*

$$\gamma_{ks}^{-101}(G) = \begin{cases} k - n + 2 & \text{if } 1 \leq k \leq s \\ \lceil \frac{k-s+1}{3} \rceil + k - n + 1 & \text{if } s + 1 \leq k \leq n. \end{cases}$$

Note that $\gamma^-(G_{s,t}) = \lceil \frac{t+1}{3} \rceil$, where s and t are positive integers. The value of $\gamma_{ks}^{-101}(C_n)$ is calculated in [12].

Theorem 4.35 [12] *If $n \geq 3$ and $1 \leq k \leq n - 1$, then*

$$\gamma_{ks}^{-101}(C_n) = \begin{cases} \lceil \frac{n-2}{3} \rceil & \text{if } k = n - 1 \text{ and } (k \equiv 0 \text{ or } k \equiv 1 \ (\text{mod } 3) \ ) \\ 2 \lfloor \frac{2k+4}{3} \rfloor - n & \text{otherwise.} \end{cases}$$

This result supplements the following result.

Theorem 4.36 [6] *If $n \geq 3$, then $\gamma_{ns}^{-101}(C_n) = \lceil \frac{n}{3} \rceil$.*

In [2], the least order of a connected graph G for which $\gamma_{ks}^{-101}(G) = -t$ is determined, where t is a positive integer t.

Theorem 4.37 [2] *Let t be a positive integer and let $G = (V, E)$ be a connected graph such that $\gamma_{ks}^{-101}(G) = -t$ with k an integer such that $1 \leq k \leq n = n(G)$. Then*
(a) if $k = 1$, then $n \geq t + 3$,
(b) if $2 \leq k \leq n - 2$, then $n \geq t + 4$,
(c) if $k = n - 1$, then $n \geq 2\ell + t$, where $\ell = \min\{\ell \in \mathbf{Z}^+ \mid t \leq (\ell - 1)^2 - \ell\}$,
(d) if $k = n$, then $n \geq 2\ell + t$, where $\ell = \min\{\ell \in \mathbf{Z}^+ \mid t \leq \frac{\ell^2 - 3\ell}{2}\}$.
All these bounds are best possible.

For each integer $q \geq 1$, let $I_q = \{q(q+1)/2, q(q+1)/2 + 1, \ldots, q(q+1)/2 + q\}$. Then the smallest integer in I_q is one larger than the largest integer in I_{q-1} (if $q \geq 2$), while the largest integer in I_q is one smaller than the smallest integer in I_{q+1}. Hence each positive integer is contained in a unique interval I_q for some $q \geq 1$.

Theorem 4.38 [6] *Let $q \geq 1$ be an integer and $t \in I_q$. Let G be a connected graph of order n with $\gamma_{ns}^{-101}(G) = -t$. Then $n \geq 2(q+3) + t$.*

The statement of Theorem 4.38 and statement (d) of Theorem 4.37 are equivalent, as may be seen from the following: If $q = \ell - 3$, then $t \leq \frac{\ell^2 - 3\ell}{2}$ if and only if $t \leq \frac{q(q+1)}{2} + q$.

For each integer $q \geq 1$, let $J_q = \{(q-1)q, (q-1)q+1, \ldots, (q-1)q + 2q - 1\}$. Then the smallest integer in J_q is one larger than the largest integer in J_{q-1} (if $q \geq 2$), while the largest integer in J_q is one smaller than the smallest integer in J_{q+1}. Hence, each positive integer is contained in a unique interval J_q for some $q \geq 1$. In this way we obtain an equivalent statement for statement (c) of Theorem 4.37 by letting $q = \ell - 2$.

Theorem 4.39 [2] *Let $q \geq 1$ be an integer and $t \in J_q$. Let G be a connected graph of order n with $\gamma_{(n-1)s}^{-101}(G) = -t$. Then $n \geq 2(q+2) + t$.*

Lastly, consider the following decision problem.

PARTIAL MINUS DOMINATING FUNCTION (PMDF)

INSTANCE: A graph G, positive rational number $r \leq 1$ (in its simplest form) and an integer ℓ.

QUESTION: Is there a function $f : V(G) \to \{-1, 0, 1\}$ of weight ℓ or less for G such that $|C_f| \geq r|V(G)|$?

Hattingh, McRae and Ungerer (see [11]) showed that **PMDF** is NP-complete by describing a polynomial transformation from the following NP-complete problem (see [8]):

EXACT COVER BY 3-SETS (X3C)

INSTANCE: A set $X = \{x_1, \ldots, x_{3q}\}$ and a set $\mathcal{C} = \{C_1, \ldots, C_t\}$ where $C_j \subseteq X$ and $|C_j| = 3$ for $j = 1, \ldots, t$.

QUESTION: Does $\mathcal{C}$ have a pairwise disjoint q-subset of $\mathcal{C}$ whose union is X (i.e. an exact cover)?

If $r = 1$, then **PMDF** is the NP-complete problem **MINUS DOMINATING FUNCTION** (see [5]). Hence, we also assume that $r < 1$.

Theorem 4.40 [11] **PMDF** *is NP-complete, even for bipartite graphs.*

4.5   Open Problems

We close with a list of open problems.

1. Settle Conjecture 4.22.

2. Settle Conjecture 4.23.

3. Let G be a connected graph of order n and let $k > \lceil \frac{n}{2} \rceil$. Find good upper bounds for $\gamma_{ks}^{-11}(G)$ and $\gamma_{ks}^{-101}(G)$.

4. Calculate γ_{ks}^{-11} and γ_{ks}^{-101} for other classes of trees or graphs.

5. Are there linear algorithms to compute $\gamma_{ks}^{-11}(T)$ and $\gamma_{ks}^{-101}(T)$ for an arbitrary tree T?

6. Characterize those graphs G (trees) for which $\gamma_{ks}^{-101}(G) = \gamma_{ks}^{-11}(G)$. Note that, for any graph G, $\gamma_{ks}^{-101}(G) \leq \gamma_{ks}^{-11}(G)$.

Bibliography

[1] L.W. Beineke and M.A. Henning, Opinion functions on trees. To appear in *Discrete Math.*

[2] I. Broere, J.E. Dunbar and J.H. Hattingh, Minus k-subdomination in graphs. To appear in *Ars Combin.*

[3] I. Broere, J.H. Hattingh, M.A. Henning and A.A. McRae, Majority domination in graphs. *Discrete Math.* 138 (1995) 125-135.

[4] E.J. Cockayne and C.M. Mynhardt, On a generalization of signed dominating functions of graphs. *Ars Combin.* 43 (1996) 235-245.

[5] J. E. Dunbar, W. Goddard, S.T. Hedetniemi, M. A. Henning, A.A. McRae, The algorithmic complexity of minus domination in graphs. *Discrete Appl. Math.* 68 (1996) 73-84.

[6] J.E. Dunbar, S.T. Hedetniemi, M.A. Henning and A.A. McRae, Minus domination in graphs. Submitted for publication.

[7] J.E. Dunbar, S.T. Hedetniemi, M.A. Henning and P.J. Slater, Signed domination in graphs. *Graph Theory, Combinatorics and Applications*, Eds. Y. Alavi and A. Schwenk, John Wiley and Sons, Inc. 1 (1995) 311-322.

[8] M.R. Garey and D.S. Johnson, *Computers and Intractability: A Guide to the Theory of NP-Completeness.* Freeman, New York (1979).

[9] J.H. Hattingh, M.A. Henning and P.J. Slater, The algorithmic complexity of signed domination in graphs. *Australas. J. Combin.* 12 (1995) 101-112.

[10] J.H. Hattingh, M.A. Henning and E. Ungerer, Partial signed domination in graphs. To appear in *Ars Combin.*

[11] J.H. Hattingh, A.A. McRae and E. Ungerer, Minus k-subdomination in graphs III. Submitted for publication.

[12] J.H. Hattingh and E. Ungerer, Minus k-subdomination in graphs II. To appear in *Discrete Math.*

[13] J.H. Hattingh and E. Ungerer, On the signed k-subdomination number of trees. Submitted for publication.

[14] J.H. Hattingh and E. Ungerer, The signed and minus k-subdomination numbers of comets. To appear in *Discrete Math.*

[15] M.A. Henning, Domination in regular graphs. *Ars Combin.* 43 (1996) 263-271.

[16] M.A. Henning and P.J. Slater, Inequalities relating domination parameters in cubic graphs. *Discrete Math.* 158 (1996) 87-98.

[17] B. Zelinka, Some remarks on domination in cubic graphs. *Discrete Math.* 158 (1996) 249-255.

Chapter 5

Convexity of Extremal Domination-Related Functions of Graphs

E. J. Cockayne
Department of Mathematics and Statistics
University of Victoria
P.O. Box 3045
Victoria, BC V8W 3P4
Canada

C. M. Mynhardt
Department of Mathematics
University of South Africa
P.O. Box 312
Pretoria 0003
South Africa

Abstract. For various properties P related to domination and a given graph $G = (V, E)$, the class of vertex subsets of G having property P have been embedded in some set F_P of real valued functions defined on V. For example, the dominating sets of G have been embedded in the set of dominating functions of G. This work surveys results about the convexity of the set of extremal elements of F_P when P is domination, total domination and packing. In each case most of the results concern the existence of universal extremal elements of F_P. Some of these results are special cases of more general theorems in a unifying theory, that of η functions.

5.1 Introduction

A considerable part of graph theory is concerned with the study of subsets of the vertices or edges of graphs which have certain properties. For example, independent sets, dominating sets, total dominating sets, packing sets, vertex covers and irredundant sets are well-known vertex subsets, while matchings and edge covers are familiar edge subsets of a graph.

A recent trend has been to generalise these various concepts by defining a set of functions, from V (or E) to a subset of the reals, which contains the characteristic functions of all the subsets of V (or E) of the type in question. Perhaps the most studied case is that of *dominating functions, i.e.,* functions

$f : V \rightarrow [0,1]$ such that for each $v \in V$, $f[v] = \sum_{u \in N[v]} f(u) \geq 1$, where $N[v]$ denotes the *closed neighbourhood* of v, that is, the vertex v together with all its neighbours. We use the abbreviation DF for dominating function. The set of dominating functions of G contains the characteristic function of each dominating set of G. The dominating functions may also be defined as the feasible solutions of the linear programming relaxation of the minimum dominating set integer programme.

The study of the more general $[0,1]$-functions has been called fractional theory; for example, fractional domination has been studied in [7, 9–12, 19, 22, 28], fractional total domination in [17, 18, 21, 33, 34], fractional packing in [6, 22], fractional covering in [5] and fractional matching in [29]. The reader is also referred to [26] for further work in fractional graph theory. Note that the theory of fractional domination is also studied in Chapters 1, 2 and 3 of this book.

For functions f, g from $V \rightarrow [0,1]$ we write $f \leq g$ if for each $v \in V$, $f(v) \leq g(v)$. Further, we write $f < g$ if $f \leq g$ and for some $v \in V$, $f(v) < g(v)$. A DF g of G is *minimal* (*i.e.*, g is an MDF) if for all functions $f : V \rightarrow [0,1]$ such that $f < g$, f is not dominating. The *aggregate* (originally called *value*) $ag(f)$ of a function $f : V \rightarrow [0,1]$ is defined by $ag(f) = \sum_{u \in V} f(u)$. These definitions are given in [7].

The particular aspect of fractional theory surveyed in this chapter was originally motivated by the following question which was raised by Hedetniemi [27] and is answered in Section 5.5:

(Q1) Suppose f, g are MDFs of G and $ag(f) < x < ag(g)$. Does there exist an MDF of G with aggregate x?

For real-valued functions f, g defined on V and $\lambda \in (0,1)$, the *convex combination* h_λ of f and g is defined by

$$h_\lambda(v) = \lambda f(v) + (1 - \lambda) g(v) \quad \text{for each} \quad v \in V. \tag{1}$$

It was noticed that in some cases convex combinations answer the question Q1. For MDFs f, g of G it is easy to show that h_λ is a DF of G with $ag(h_\lambda) = x$ for suitable choice of λ. Thus, provided h_λ is an MDF, we have an affirmative answer for Q1. However, as we will observe in Section 5.2, convex combinations of MDFs are DFs but not necessarily minimal, hence the motivation for studying convexity properties of the set of MDFs of G.

In Section 5.2 we discuss results on the convexity of MDFs and of a second class of functions which also generalize the concept of a dominating set. Sections 5.3 and 5.4 deal with analogous theories for minimal total dominating functions and maximal packing functions of graphs. In each of these sections the main results are concerned with the existence of *universal* extremal functions in graphs. For example, a universal MDF g has the property that for *any* other MDF f, h_λ (defined by (1)) is also an MDF.

In Section 5.5, we consider the theory of η-functions of graphs. It will be demonstrated that some of the results surveyed in Sections 5.2 and 5.3 are special cases of more general theorems in this unifying theory. One result here is an affirmative answer to a class of interpolation questions which includes Q1.

Finally, a variety of unsolved problems and directions for further research are indicated in Section 5.6.

5.2   Domination

In Section 5.2.1 we discuss the convexity of MDFs which were defined in the introduction. Then in Sections 5.2.2 and 5.2.3, theorems concerning universal MDFs in general graphs and trees are surveyed. Finally, in Section 5.2.5 a second type of vertex function (namely a 0-dominating function) which also generalises the concept of a dominating set of a graph is defined and the convexity of minimal 0-dominating functions is also discussed.

5.2.1   Minimal dominating functions

The minimality of DFs was defined in the introduction. For a DF f of G, we define the *boundary B_f* of f by

$$B_f = \left\{ v \mid \sum_{u \in N[v]} f(u) = 1 \right\}$$

and the *positive set P_f* of f by

$$P_f = \{ v \mid f(v) > 0 \} .$$

A subset $V_1 \subseteq V$ is said to *dominate* $V_2 \subseteq V$ if each $v \in V_2 - V_1$ is adjacent to some vertex in V_1.

Proposition 5.1 [5] *A DF f of G is an MDF if and only B_f dominates P_f.*

Proof. Let f be an MDF of G with $v \in P_f$ and suppose, to the contrary, that B_f does not dominate v. Then $N[v] \cap B_f = \emptyset$ which implies that $f[u] > 1$ for each $u \in N[v]$. Choose $\epsilon > 0$ in such a way that $\epsilon \leq \min_{u \in N[v]} \{ f[u] - 1 \}$, and define $g : V \to [0, 1]$ by

$$g(v) = f(v) - \epsilon$$

and

$$g(x) = f(x) \quad \text{for} \quad x \in V - \{v\}.$$

For each $u \in N[v]$, $g[u] = f[u] - \epsilon \geq 1$ and for $x \in V - N[v]$, $g[x] = f[x] \geq 1$; hence g is a DF. But $g < f$, contradicting the minimality of f.

Conversely, suppose B_f dominates P_f and let $g : V \to [0,1]$ satisfy $g < f$; say $g(v) < f(v)$ for some $v \in V$. Then $v \in P_f$ and so, by assumption, $u \in B_f$ for some $u \in N[v]$. Now

$$
\begin{aligned}
g[u] &= \sum_{w \in N[u]} g(w) \\
&= g(v) + \sum_{w \in N[u] - \{v\}} g(w) \\
&< f(v) + \sum_{w \in N[u] - \{v\}} f(w) \\
&= f[u] \\
&= 1.
\end{aligned}
$$

Hence g is not a DF and it follows that f is an MDF. $\square$

As observed above, convex combinations of MDFs are DFs but perhaps not minimal. Proposition 5.1 is used to illustrate this fact in the following example.

Example 1. Consider the tree with

$$V = \{1, 2, 3, 4, 5, 6\} \quad \text{and} \quad E = \{13, 23, 34, 45, 56\}.$$

Let

$$(f(1), \cdots, f(6)) = (0, 0, 1, 0, 1, 0) \quad \text{and} \quad (g(1), \cdots, g(6)) = \left(\tfrac{1}{2}, \tfrac{1}{2}, \tfrac{1}{2}, \tfrac{1}{2}, 0, 1\right).$$

It is easy to verify that f, g are MDFs (*i.e.*, each is a DF whose boundary dominates its positive set). However, the convex combination h_λ of f and g, where $\lambda = \tfrac{1}{4}$, is given by

$$\left(h_{\frac{1}{4}}(1), \cdots, h_{\frac{1}{4}}(6)\right) = \left(\tfrac{3}{8}, \tfrac{3}{8}, \tfrac{5}{8}, \tfrac{3}{8}, \tfrac{1}{4}, \tfrac{3}{4}\right).$$

This is a DF whose boundary $\{1, 2, 6\}$ does not dominate its positive set $\{1, \cdots, 6\}$. Therefore $h_{\frac{1}{4}}$ is not an MDF.

The following result of [7] shows that the minimality of h_λ for MDFs f and g is an "all or nothing" situation, *i.e.*, either all convex combinations of f and g are themselves MDFs or none of them are MDFs.

Theorem 5.2 [7] *If f, g are MDFs of G, then h_λ ($\lambda \in (0,1)$) is also an MDF of G if and only if $B_f \cap B_g$ dominates $P_f \cup P_g$.*

Proof. We prove that $B_{h_\lambda} = B_f \cap B_g$ and $P_{h_\lambda} = P_f \cup P_g$. The result is then immediate from Proposition 5.1. If $v \notin P_f \cup P_g$, then $f(v) = g(v) = h_\lambda(v) = 0$. If, say, $v \in P_f$, then $h_\lambda(v) \geq \lambda f(v) > 0$. Thus $P_{h_\lambda} = P_f \cup P_g$.

Suppose $v \in B_f \cap B_g$. Then

$$h_\lambda[v] = \lambda f[v] + (1 - \lambda) g[v] = \lambda + (1 - \lambda) = 1.$$

A similar calculation shows $h_\lambda[v] > 1$ for $v \notin B_f \cap B_g$ and hence $B_{h_\lambda} = B_f \cap B_g$.
□

We are thus led to study the binary relation $\mathcal{R}$ on the set $\mathcal{F}$ of all MDFs of G, defined by $f\mathcal{R}g$ if and only if h_λ is an MDF for all $\lambda \in (0,1)$, *i.e.*, (using Theorem 5.2) $f\mathcal{R}g$ if and only if $B_f \cap B_g$ dominates $P_f \cup P_g$.

A *universal* MDF of G, defined in [7], is an MDF of G which is related by $\mathcal{R}$ to *every* other MDF of G. In the terminology of the literature on convex sets (see [32, pp. 4–5]), g is a universal MDF if and only if $\mathcal{F}$ is star-shaped relative to g. The set of universal MDFs is the kernel of the set $\mathcal{F}$. It follows that the set of universal MDFs of a graph is a convex set.

We remark that not all graphs have universal MDFs. Examples will be given below. If G does have a universal MDF h, question Q1 has an affirmative answer. With suitable choices of λ, convex combinations of h and f, if $x < ag(h)$, or convex combinations of h and g, if $x > ag(h)$, yield MDFs of aggregate x for any real x satisfying $ag(f) < x < ag(g)$.

5.2.2   Universal MDFs in graphs

The following simple criteria (easily deduced from Theorem 5.2 and in the case of (ii) from the obvious fact that each vertex of G is contained in a maximal independent and hence minimal dominating set of G) may be used to establish the existence of universal MDFs in various classes of graphs.

Proposition 5.3 [7]
(i)   *If the MDF g satisfies $B_g = V$ and, for all $f \in \mathcal{F}$, B_f dominates V, then g is a universal MDF.*
(ii)   *If g is a universal MDF, then B_g dominates V.*

Theorem 5.4 [7]   *The path P_n $(n \geq 1)$, the cycle C_n $(n \geq 3)$, the complete bipartite graph $K_{r,s}$ $(r,s \geq 1)$, the n-vertex wheel W_n $(n \geq 4)$ and the complete graph K_n $(n \geq 1)$ all have universal MDFs.*

Examples of universal MDFs for the graphs mentioned in Theorem 5.4 are as follows: For the path P_n, assign to consecutive vertices in the path suitable consecutive elements from the sequence $100100100\cdots$, depending on the congruency of n modulo 3. For C_n, assign $\frac{1}{3}$ to each vertex. Suppose $K_{r,s}$ has bipartition (A, B) with $|A| = r$ and $|B| = s$, and assign $(s-1)/(rs-1)$ to each vertex in A and $(r-1)/(rs-1)$ to each vertex in B. For W_n, let u be the vertex of degree $n-1$ and for K_n, let u be any vertex; assign the value 1 to u and 0 to all other vertices.

The next result demonstrates the existence of universal MDFs whose boundaries do not contain all vertices (*i.e.,* the conditions of Proposition 5.3(i) are not necessary). It also provides examples of graphs, all of whose MDFs are universal, *i.e.,* $\mathcal{R} = \mathcal{F} \times \mathcal{F}$.

Let u be a vertex of the graph H. By a *complete addition to H at u*, we mean the identification of u and a vertex of some complete graph K_t ($t \geq 2$). (For example, for t fixed, the complete addition of K_t to H at each vertex u of H is also known as the *corona* $H \circ K_{t-1}$ of H and K_{t-1}.)

Theorem 5.5 [7] *Let H be any graph. Form G from H by making one or more complete additions to H at u for each vertex u of H. Then each MDF of G is universal.*

In order to state the next result which is a characterisation of universal MDFs of graphs, we need some further definitions. A vertex $v \in V$ is said to *absorb* the vertex $u \neq v$, and u is said to *be absorbed by v*, if $N[u] \subset N[v]$, where $\subset$ denotes strict inclusion. In this case, v is also called an *absorbing vertex*, and u an *absorbed vertex*. Let

$$A = \{v \in V(G): \quad v \text{ is an absorbing vertex}\}$$

and

$$\Omega = \{u \in V(G): \quad u \text{ is an absorbed vertex}\}.$$

We now define a vertex w of a graph G to be f-*sharp*, where f is an MDF of G, if $B_f \cap N[w] \subseteq A$, and w is said to be *sharp* if w is f-sharp for some MDF f of G.

Example 2. Let G be the graph which consists of the 6-cycle with vertex sequence $1, 2, \cdots, 6$, together with the edges 13 and 15. Then $A = \{1, 3, 5\}$ and $\Omega = \{2, 6\}$. The function $f : V \to [0, 1]$ defined by $f(3) = f(5) = 1$ and $f(v) = 0$ otherwise, satisfies $B_f = \{2, 3, 5, 6\}$ and $P_f = \{3, 5\}$, thus f is an MDF of G (Proposition 5.1). Further, $B_f \cap N[4] = \{3, 5\} \subseteq A$ so that 4 is f-sharp.

We now state a characterisation of universal MDFs of graphs.

Theorem 5.6 [12] *The MDF g of $G = (V, E)$ is a universal MDF if and only if*

(i) $V - A \subseteq B_g$ *and*

(ii) $g(w) = 0$ *for each sharp vertex w of G.*

The proof of this theorem uses properties of absorbing and sharp vertices established in [12] and also the fact that any graph has an MDF f satisfying $P_f = V$ (proved in [7]).

Corollary 5.7 [12] *If a graph G has a vertex v such that each $u \in N[v]$ is sharp, then G has no universal MDF.*

The following example uses Corollary 5.7 to exhibit a graph with no universal MDF.

Example 3.   Let G be the circulant formed by adding edges $\{i, i+5\}$ for $i = 1, \cdots, 5$ to the cycle with vertex sequence $1, \cdots, 10$. Then, for example, the function f which is 1 on $\{1, 3, 6, 8\}$ and 0 elsewhere is an MDF with $B_f = \{4, 5, 9, 10\}$. We observe that $B_f \cap N[2] = \emptyset \subseteq A$ and hence vertex 2 is sharp. Since G is vertex transitive (*i.e.*, for each pair u, v of vertices of G, there exists an automorphism ϕ of G such that $\phi(u) = v$), all vertices are sharp and by Corollary 5.7 G has no universal MDF.

Further deductions from Theorem 5.6 lead to existence theorems for universal MDFs in regular and vertex-transitive graphs.

Theorem 5.8 [7]   *If G has a vertex v such that for each $u \in N[v]$ and some MDF f of G,*

$$B_f \cap N[u] = \emptyset, \tag{2}$$

then G has no universal MDF.

A vertex satisfying (2) is called f-*loose* (see [13]) and is a special case of an f-sharp vertex. Thus Theorem 5.8 is merely a special case of Corollary 5.7.

Corollary 5.9 [7]   *If G is vertex transitive, then G has a universal* MDF *if and only if for each* MDF f, B_f *dominates* V.

An MDF f of a graph $G = (V, E)$ is an *efficient* MDF if $B_f = V$. If f is an integer-valued efficient MDF of G, then the set $\{v \in V : f(v) = 1\}$ is said to be an *efficient dominating set* of G (*cf.* [1,2]). The following deduction from Theorem 5.6 can be used to obtain classes of graphs which have universal MDFs.

Theorem 5.10 [12]   *Let G be a graph without sharp or absorbing vertices. A function $f : V(G) \to [0, 1]$ is a universal* MDF *if and only if it is an efficient* MDF.

It is easy to see that regular graphs have no absorbing vertices. Also, if G is r-regular, then $f : V \to [0, 1]$ defined by $f(v) = 1/(r+1)$ for each $v \in V$ is an efficient MDF of G.

Corollary 5.11 [12]   *Any regular graph without sharp vertices has a universal* MDF.

Examples of regular graphs without sharp vertices include the cycle C_n ($n \geq 3$), the complete graph K_n and the complete bipartite graph $K_{r,r}$. Theorem 5.10 has also been used to prove the following result.

Theorem 5.12 [12]   *The complete r-partite graph $G = K_{m_1, m_2, \cdots, m_r}$ has a universal* MDF.

The last theorem in this section and two corollaries assert non-existence of universal MDFs in large classes of graphs.

Theorem 5.13 [12] *If G is any graph with a vertex v such that*

  (i)   deg$(u) \geq 3$ *for each u with* $d(u,v) \leq 2$,

  (ii) deg$(u) \geq 2$ *for each u with* $d(u,v) \leq 4$, *and*

  (iii) *for each* $u \in N[v]$, *the subgraph of G induced by the vertices at distance at most four from u, is a tree,*

then G has no universal MDF.

Corollary 5.14 [12] *If the graph G has a vertex v such that*

  (i)   deg$(u) \geq 3$ *for each u with* $d(u,v) \leq 2$,

  (ii) deg$(u) \geq 2$ *for each u with* $d(u,v) \leq 4$, *and*

  (iii) *v lies on no cycle of length less than 10 and for each u with* $d(u,v) = k$, *where* $1 \leq k \leq 4$, *u lies on no cycle of length less than* $12 - 2k$,

then G has no universal MDF.

Corollary 5.15 [12] *If G has minimum degree at least three and girth (i.e., the length of a shortest cycle) at least ten, then G has no universal* MDF.

5.2.3   Universal MDFs in trees

For any tree T, let L be set of vertices of degree one and R be the set of all neighbours of vertices of L. Elements of L, R are called *leaves* and *remote vertices*, respectively. It is clear that for $T \ncong K_2$, $A = R$ and $\Omega = L$. Thus for an MDF f of T, v is an f-sharp vertex (defined in Section 5.2.2) if and only if $B_f \cap N[v] \subseteq R$. Such vertices are called f-*cool* and were defined [9] before the more general concept of sharp was considered. Hence in the literature of MDFs for trees [9,10,11,13] the terms f-cool and cool are used. We emphasize that for trees, f-sharp and sharp have meanings identical to f-cool and cool. As suggested by Theorem 5.6 the location of the cool vertices of trees influences the existence of universal MDFs (although, as yet, it is not known precisely how!).

Theorem 5.16 [9] *Vertex v is a cool vertex of a tree if and only if*

  (i)   deg$(u) \geq 3$ *for each* $u \in N(v) - R$

        and

  (ii) $N(v)$ *contains at least two vertices, each of which is adjacent to at least two vertices of* $V - R$.

The papers [9,13] contain more results concerning cool vertices of trees. We now list theorems which assert the existence of 0-1 universal MDFs, (*i.e.*, all values of the function are 0 or 1) in various types of trees. The first is a simple consequence of Theorem 5.5.

Corollary 5.17 [7] *Any MDF of a tree, whose leaves dominate the tree, is universal.*

Theorem 5.18 [9] *If T may be rooted at v such that for all* $u \in V - (L \cup R)$, *u has a non-cool child, then T has a 0-1 universal MDF.*

Corollary 5.19 [9] *In a tree* T, *if each* $u \in V - (L \cup R)$ *has at least two non-cool neighbours, then* T *has a* 0-1 *universal* MDF.

A *generalised star* is a tree formed by taking k disjoint paths with vertex sequences $v_{i1}, v_{i2}, \cdots, v_{it(i)}$, $i = 1, \cdots, k$ and then identifying the endvertices $v_{11}, v_{21}, \cdots, v_{k1}$. A *caterpillar* is a tree consisting of a path with vertex sequence $v_1, \cdots, v_t$ and each v_i is adjacent to a set L_i (possibly empty) of leaves.

Theorem 5.20 [9] *If the tree* T *is a generalised star or a caterpillar, then* T *has a* 0-1 *universal* MDF.

Theorem 5.21 [9] *If* T *is a tree with* $\mathrm{diam}(T) \leq 7$, *then* T *has a* 0-1 *universal* MDF.

Theorem 5.22 [9] *If* $\mathrm{diam}(T) = 8$ *and the central vertex of* T *is not cool, then* T *has a* 0-1 *universal* MDF.

The next result shows that if a tree has any universal MDF, then it has a 0-1 universal MDF. This is not true for graphs in general. For example, the graph C_5 has a universal MDF (Theorem 5.4). However, C_5 has no efficient dominating set and no sharp or absorbing vertex and hence no 0-1 universal MDF, by Theorem 5.10.

Theorem 5.23 [11] *If* T *has a universal* MDF, *then* T *has a* 0-1 *universal* MDF.

We now present two results concerning non-existence of universals. In view of Theorem 5.21 and Corollary 5.19, trees without universal MDFs must have at least two vertices with at most one non-cool neighbour and diameter at least eight.

Theorem 5.24 [9] *Let* x *and* y *be distinct vertices of a tree* T *satisfying*
(i) $N[x] \cap N[y] = \{v\}$ *where* $v \neq x, y$
     and
(ii) *each* $u \in N(x) \cup N(y)$ *is cool.*
Then T *has no universal* MDF.

The tree of Figure 5.1 is the smallest tree known to us, which has no universal MDF (by Theorem 5.24). The vertices depicted by large dots are cool.

The last theorem of this section exhibits an infinite class of trees which have no universal MDF. Let $\mathcal{T}$ be the set of all trees T with the following properties:

1. $\{v\}$, $\{a_1, \cdots, a_r\}$, $\{b_1, \cdots, b_s\}$, $\{c_1, \cdots, c_t\}$, A, B, C is a partition of the vertex set V of T, which we assume is rooted at v.

2. The sequences $(v, a_1, \cdots, a_r)$, $(v, b_1, \cdots, b_s)$ and $(v, c_1, \cdots, c_t)$ are the vertex sequences of paths in T.

3. The sets of descendants of a_r, b_s, c_t are respectively A, B, C.

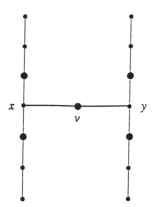

Figure 5.1: A tree with no universal MDF.

4. The sets of children $W(a_r)$, $W(b_s)$ and $W(c_t)$ contain only cool vertices of T.

5. Two of r, s, t are congruent to 1 (mod 3) and the third is congruent to 2 (mod 3).

Theorem 5.25 [13] *If $T \in \mathcal{T}$, then T has no universal MDF.*

In Figure 5.2 we exhibit a tree $T \in \mathcal{T}$ as an example of Theorem 5.25.

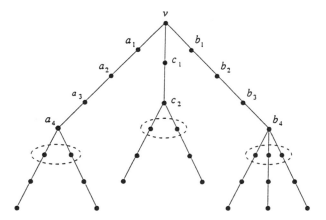

Figure 5.2: A tree of $\mathcal{T}$ with no universal MDF.

This concludes the results on existence and non-existence of universal MDFs of trees.

In [10] a linear time algorithm is presented which finds a 0-1 universal MDF for any tree (if it exists).

5.2.4   0-Dominating functions

In [23,24], a second type of function which also generalizes the concept of a dominating set, is defined and studied.

A *zero-dominating* or *0-dominating function* (0DF) of a graph G is a function $f : V \to [0,1]$ such that $f[v] = \sum_{x \in N[v]} f(x) \geq 1$ for all $v \in V$ for which $f(v) = 0$. A 0DF f of G is a *minimal 0-dominating function* (M0DF) if, for any $g : V \to [0,1]$ such that $g < f$, g is not a 0DF.

It is clear from the definition that any DF is also a 0DF. However, if v_1, v_2, v_3 is the vertex sequence of the path P_3 and $f(v_1), f(v_2), f(v_3) = \frac{1}{2}, 0, \frac{1}{2}$, respectively, then f is an M0DF of P_3 which is not a DF.

Although the set of all 0DFs of a graph is a convex set, convex combinations of M0DFs are not necessarily also minimal. An example is depicted in Figure 5.3 in which f, g are two M0DFs of K_3, but the convex combination $h_{\frac{1}{2}}$ is not minimal.

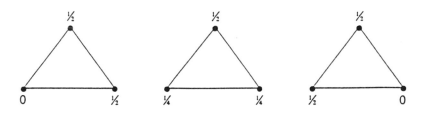

Figure 5.3: The convex combination $h_{\frac{1}{2}}$.

Thus one can study the convexity of the set of M0DFs of a graph and define a universal M0DF g to be a M0DF whose convex combinations with any other M0DF are minimal. Surprisingly, so far no graph with a universal M0DF has been found. Various results which will be useful in resolving this problem are proved in [21] along with proofs of non-existence of universal M0DFs in various classes of graphs.

Theorem 5.26 [23] *The following classes of graphs have no universal M0DF: trees, vertex-transitive graphs, edge-transitive graphs, complete n-partite graphs, graphs with a vertex v satisfying $N[v] = V$.*

5.3   Total Dominating Functions

In this section we survey the theory, analogous to that in Sections 5.2.1, 5.2.2 and 5.2.3, for total dominating functions which arise when we simply change "closed" neighbourhood in the definition to "open". A *total dominating function* (TDF) of G is a function $f : V \to [0,1]$ such that $\sum_{u \in N(v)} f(u) \geq 1$ for each $v \in V$. In order to ensure the existence of a TDF, we restrict our attention to graphs without isolated vertices. The integer-valued TDFs are precisely the characteristic functions of *total dominating sets*, (*i.e.*, sets X of vertices such that any $v \in V$ is adjacent to at least one $x \in X$). Total dominating functions (TDFs) have been studied in [14, 17, 18, 21, 33, 34]. A TDF f is minimal, (*i.e.*, is an MTDF) if no function $g < f$ is total dominating. We are concerned with *universal* MTDFs, *i.e.*, MTDFs g such that convex combinations of g and any other MTDF are also MTDFs.

For the TDF f we define TB_f, the *total boundary* of f, by

$$TB_f = \left\{ v \mid \sum_{u \in N(v)} f(u) = 1 \right\}$$

and P_f, the positive set of f, by

$$P_f = \{v \mid f(v) > 0\}.$$

For subsets A, B of V we say that A *totally dominates* B if $N(v) \cap A \neq \emptyset$ for each $v \in B$. We use the term 'total boundary' here to avoid confusion with 'boundary' used in Section 5.2. In the literature where no confusion could arise, total boundary was simply called boundary. The following two results are analogous to Proposition 5.1 and Theorem 5.2 respectively.

Proposition 5.27 [18]   *The TDF f is minimal if and only if TB_f totally dominates P_f.*

Theorem 5.28 [18]   *Let f, g be MTDFs. The convex combination $h_\lambda = \lambda f + (1 - \lambda)g$ is also an MTDF if and only if $TB_f \cup TB_g$ totally dominates $P_f \cup P_g$.*

5.3.1   Universal MTDFs in graphs

The existence of universal MTDFs in some simple classes of graphs can be proved by using the following proposition.

Proposition 5.29 [18]   *If the MTDF g satisfies $TB_g = V$ and for all MTDFs f, TB_f totally dominates V, then g is a universal MTDF.*

Theorem 5.30 [18]   *The cycle C_n ($n \geq 3$), the complete bipartite graph $K_{r,s}$, the wheel W_n ($n \geq 4$) and the complete graph K_n all have universal MTDFs.*

The next result may be used to prove non-existence of universal MTDFs.

Proposition 5.31 [18] *Let $v \in V$ be such that for each $u \in N(v)$ there exists an MTDF f_u where TB_{f_u} does not totally dominate $\{u\}$. Then G has no universal MTDF.*

Corollary 5.32 [18] *If G is vertex-transitive, then G has a universal MTDF if and only if for each MTDF f, TB_f totally dominates V.*

Example 4. In Figure 5.4 we exhibit an MTDF f_1 of a graph G. The vertices depicted by solid squares form the total boundary TB_{f_1}. Note that TB_{f_1} does not totally dominate $\{u_1\}$, where $u_1 \in N(v)$. By symmetry, for $i = 2, 3$, f_i may be defined so that TB_{f_i} does not totally dominate $\{u_i\}$. By Proposition 5.31, G has no universal MTDF.

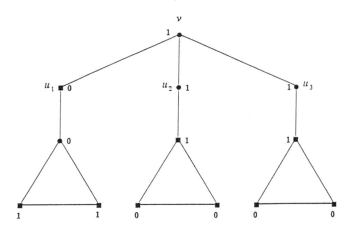

Figure 5.4: Graph G illustrating Proposition 5.31.

In Section 5.2, Theorem 5.6 gave a characterisation of those MDFs which are universal. We do not yet have a similar result for MTDFs but good progress towards this has been made in [34].

A vertex v is called *totally absorbing* if there exists a vertex u such that $N(u) \subset N(v)$ (proper containment). In this section A will denote the set of totally absorbing vertices. (We have changed the overworked term "dominating" used in [34].) Let f be an MTDF of G. Vertex v is called *low* if for some MTDF f, $TB_f \cap N(v) \subseteq A$.

We observe that totally absorbing and low vertices are the total analogue of the absorbing and sharp vertices considered in Section 5.2. Let W be the subset of vertices v such that for each $u \in N(v)$, there exists $t \in N(u) - v$ with $t \in TB_f$ for every MTDF f.

Theorem 5.33 [34]

(i)   *If the* MTDF *g satisfies*
    (a) $V - (A \cup W) \subseteq TB_g$ *and*
    (b) $g(v) = 0$ *for each low vertex* v, *then* g *is a universal* MTDF.

(ii) *If* g *is a universal* MTDF, *then* $g(v) = 0$ *for each low vertex* v.

Corollary 5.34 [34] *If a regular graph has no low vertex, then it has a universal* MTDF.

Corollary 5.35 [34] *If a graph G has a vertex v, all of whose neighbours are low, then G has no universal* MTDF.

Example 5 [34] Using Proposition 5.27, it is easy to check that g in Figure 5.5 is an MTDF. Note that TB_g consists of the solid square vertices and that x is g-low since $TB_g \cap N(x) = \emptyset$. By symmetry y and z are also low vertices. Hence all neighbours of v are low vertices. By Corollary 5.35 this graph has no universal MTDF.

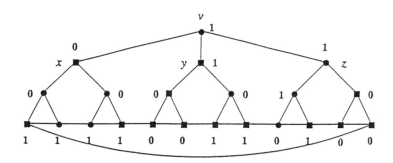

Figure 5.5: A regular graph without a universal MTDF.

A vertex splitting operation is introduced in [34]. The operation is used to show that several further classes of graphs have universal MTDFs. Let $v \in V(G)$ be any vertex and $w \notin V(G)$. Then v is *split* when we add w and the edges $\{wx \mid x \in N(v)\}$ to G. Note that $N(w) = N(v)$. The inverse of the splitting operation is to delete a vertex v from G if there exists another vertex w such that $N(v) = N(w)$. Let $Q(G)$ denote the class of graphs obtained by starting from G and successively performing the splitting operation or inverse splitting operation.

Theorem 5.36 [34] *Graph G has a* (0-1) *universal* MTDF *if and only if every graph in $Q(G)$ has a* (0-1) *universal* MTDF.

Corollary 5.37 [34] *Each graph in $Q(C_n)$, $Q(K_n)$ and $Q(W_n)$ has a universal* MTDF.

Corollary 5.38 [34] *The complete r-partite graph $K_{m_1,m_2,\cdots,m_r}$ has a universal MTDF.*

5.3.2   Universal MTDFs in trees

The sufficient condition for an MTDF of a graph to be universal, *i.e.*, Theorem 5.33(i), is also necessary if the graph is a tree. Hence the following characterisation of universal MTDFs of trees is obtained.

Theorem 5.39 [14] *Let g be an MTDF of the tree T. Then g is a universal MTDF of T if and only if*
  (a) $V - (A \cup W) \subseteq TB_g$
       and
  (b) $g(v) = 0$ *for each low vertex v of T.*

The reader might observe that the statement of this theorem is different in [14, Theorem 9]. Specifically the totally absorbing vertices A and the low vertices of our Theorem 5.39 are replaced in [14] by *short* vertices S and *hot* vertices which are defined in [14]. However, it is easily seen that for trees $A = S$ and hot and low vertices are precisely the same, so that in fact there is no discrepancy and the statements are equivalent.

In order to use Theorem 5.39 to show existence or non-existence of universal MTDFs in trees, we need a practical characterisation of low vertices of trees.

Theorem 5.40 [17] *Vertex v of a tree is low (equivalently hot) if and only if for each $u \in N(v) - A$, there are at least two vertices $x \in N(u) - \{v\}$ with $N(x) - \{u\} \not\subseteq A$.*

Corollary 5.41 [17] *If vertex v of a tree is low, then $\deg(u) \geq 3$ for each $u \in N(v) - A$.*

Let $\mathcal{T}_n$ denote the set of trees T rooted at vertex v, with the following properties. If H_k denotes the set of vertices at distance k from v (where $H_o = \{v\}$), then for $k = 0, \cdots, n - 1$, each $u \in H_k$ has at least two children and $H_n = L$.

Theorem 5.42 [17]
  (i)   *If $n = 5, 7,$ or 9, $T \in \mathcal{T}_n$ and $\deg(v) \geq 3$, then T has no universal MTDF.*
  (ii) *If $n = 8$ or $n \geq 10$ and $T \in \mathcal{T}_n$, then T has no universal MTDF.*

Let $\mathcal{C}$ be the set of caterpillars.

Theorem 5.43 [17] *Each $T \in \mathcal{C}$ has a 0-1 universal MTDF.*

Referring to Theorem 5.36, we deduce the following.

Corollary 5.44 [34] *Each graph in $Q(T)$ where $T \in \mathcal{C}$, has a universal MTDF.*

Finally, we state the total analogue of Theorem 5.23.

Theorem 5.45 [31] *If a tree T has a universal MTDF, then T has a 0-1 universal MTDF.*

5.4   Packing Functions

A packing function of a graph is defined by simply reversing the direction of the inequality in the definition of a dominating function. Formally, a *packing function* (PF) of G is a function $f : V \to [0,1]$ such that for each $v \in V$, $\sum_{u \in N[v]} f(u) \leq 1$ and a PF f is maximal, (*i.e.,* is an MPF) if for all $g > f$, g is not a PF. We note that the integer-valued (M)PFs are precisely the characteristic functions of (maximal) 2-packings, *i.e.,* (maximal) subsets X of V such that $d(x,y) \geq 3$ for all $x, y \in X$.

If f, g are PFs it is easy to show that for each $\lambda \in (0,1)$, the convex combination $h_\lambda = \lambda f + (1 - \lambda)g$ is also a PF. However, if f, g are MPFs, h_λ is not necessarily maximal. An example using MPFs of the path P_9 is given in [6]. In this section we consider the existence of a *universal* MPF in a graph, *i.e.,* an MPF f such that for every MPF g, h_λ is in fact maximal. We use the definitions of "boundary" and "dominates" given in Section 5.2.1, of A and Ω given in Section 5.2.2 and of L and R given in Section 5.2.3. We will also need the vertex subset Y defined by $Y = \{y \in V - R \mid R \text{ dominates } N[y]\}$.

No characterisation of the MPFs of a general graph which are universal has so far been obtained. A sequence of simple propositions [6] lead to the following sufficient conditions.

Theorem 5.46 [6] *If the MPF g satisfies $V - (\Omega \cup Y) \subseteq B_g$, then g is a universal MPF.*

Corollary 5.47 [6] *If G has an MPF g with $B_g = V$, then g is a universal MPF.*

We call MPFs g for which $B_g = V$, *efficient packing functions* (EPFs). Graphs with EPFs have been called "fractionally efficiently dominatable" [22] and include graphs with dominating sets whose characteristic functions have boundary V. In particular, graphs with domination number 1, regular graphs and complete multipartite graphs have EPFs and hence (by Corollary 5.47) have universal MPFs (see [12]). Further, let G be a bi-regular graph, *i.e.,* a bipartite graph with bipartition P, Q and $\deg(v) = p(q)$ for $v \in P(Q)$. Define g by:

$$
g(v) = \begin{cases} \dfrac{p-1}{pq-1} & \text{if } v \in Q. \\[2ex] \dfrac{q-1}{pq-1} & \text{if } v \in P. \end{cases}
$$

Then g satisfies $B_g = V$ and is a universal MPF.

The condition of Theorem 5.46 may also be used to establish the existence of universal MPFs in another class of graphs.

Theorem 5.48 [6] *If G is a graph all of whose blocks are complete graphs, then G has a 0-1 universal MPF.*

Although some graphs have universal MPFs but no 0-1 *universal* MPFs (an example is presented in [6]), this is not the case for trees.

Corollary 5.49 [6] *Every tree has a* 0-1 *universal* MPF.

It is interesting to note that not all trees have universal MDFs and universal MTDFs (see Sections 5.2 and 5.3). For trees, $\Omega = L \subseteq Y$ and the sufficient condition on Theorem 5.46 is also necessary. Hence we have the following characterisation of universal MPFs for trees.

Theorem 5.50 [6] *The* MPF *of a tree is universal if and only if* $V - Y \subseteq B_g$.

In order to find examples of graphs without universal MPFs, results concerning MPFs and the orbits of the automorphism group of G were obtained [6]. For example it was shown that if G has a universal MPF, then G also has a universal MPF which is constant on each orbit. Ideas of this type were also presented in [22]. This analysis (which is omitted for brevity) leads to the following infinite class of 2-orbit graphs which do not have universal MPFs.

Let $A = \{a_1, \cdots, a_n\}$, $B = \{b_1, \cdots, b_n\}$, $C = \{c_1, \cdots, c_n\}$ where $n \geq 3$ and $V(G_n) = A \cup B \cup C$. The edges of G_n are as follows:

$$\{a_i a_j \mid i \neq j\}, \textit{ i.e., the subgraph induced by } A \textit{ is } K_n,$$
$$\{b_i b_{i+1}, c_i c_{i+1} \mid i = 1, \cdots, n(\text{addition modulo } n)\},$$

i.e., the subgraphs induced by B, C are copies of the cycle C_n, and

$$\{a_i b_i, a_i c_i, a_i b_{i+1}, a_i c_{i-1} \mid i = 1, \cdots, n(\text{addition modulo } n)\}.$$

Theorem 5.51 [6] *For* $n \geq 5$, G_n *has no universal* MPF.

5.5 Minimal η-functions

The reader will observe similarities in the conditions for DFs and TDFs to be minimal (Proposition 5.1 and 5.27) and also in the results (Theorems 5.6 and 5.33) which give conditions for MDFs and MTDFs to be universal. These similarities suggest the existence of a more general type of function of which the DFs (Section 5.2) and TDFs (Section 5.3) are special cases and that certain results about DFs and TDFs are particular cases of more general theorems in a comprehensive theory which unifies these and other results. The question Q1 (as given in the introduction) is also a special case of a more general question that will be answered in this section.

The preceding paragraph provides the motivation for the following definition of η-functions of graphs given in [16].

For $G = (V, E)$, let $\eta = (V, \mathcal{X}, m, \alpha)$, where

- $\mathcal{X}$ is a family of $|V|$ not necessarily distinct subsets X_v of V indexed by the vertices $v \in V$ (usually, the sets X_v are certain types of neighbourhoods of the vertices v of G);

- m is a function from V to the non-negative reals $R^+ \cup \{0\}$;

- $\alpha \in \{0, 1\}$.

For a function f defined on V, let $f[v] = \sum_{x \in X_v} f(x)$. We say $f : V \to [0, 1]$ is an η-*function* if

$$f[v] \geq m(v) \tag{1}$$

for all $v \in V$ such that

$$f(v) \leq \alpha. \tag{2}$$

Note that if $\alpha = 1$, then an η-function exists if and only if $|X_v| \geq m(v)$ for all v and if $\alpha = 0$, then an η-function always exists, the constant function $f = 1$ being an example of such a function in both cases.

In [16] it is demonstrated that DFs, ODFs and TDFs are special cases of η-functions and that a variety of other kinds of well-studied vertex subsets may be generalised to functions which are also special η-functions (see Examples 1-4 of [16]). An η-function f is a *minimal η-function* (MηF) if no function $g < f$ is an η-function.

In order to give a simple and necessary and sufficient condition for an η-function to be minimal, we use the notation

- $M(v) = \{u \in V \mid v \in X_u\}$, for $v \in V$,

- $N(W) = \cup_{u \in W} X_u$, for $W \subseteq V$,

- $V_f = \{v \in V \mid f(v) \leq \alpha\}$, where f is an η-function,

- $B_f = \{v \in V_f \mid f[v] = m(v)\}$, called the η-*frontier* of f,

- $P_f = \{v \in V \mid f(v) > 0\}$, called the *positive set* of f.

Note that we have re-named B_f as η-frontier to replace the term boundary [16] which is already used in this chapter.

Proposition 5.52 [16]   *The η-function f is minimal if and only if*

$$P_f \subseteq \bigcup_{u \in B_f} X_u = N(B_f).$$

Observe that Propositions 5.1 and 5.27 are special cases of Proposition 5.52. Other results which are special cases of this result include [3, Theorem 1], [4, Proposition 1], [15, Theorem 1], [20, Proposition 2], [24, Proposition 1] and [25, Proposition 1].

The set of all η-functions is convex but, as we have already observed in the special cases (MDFs, MTDFs), convex combinations of MηFs are not necessarily minimal. The next result generalises Theorems 5.2 and 5.28.

Theorem 5.53 [16] *For any $\lambda \in (0,1)$, the convex combination h_λ of the $M\eta F$s f and g is an $M\eta F$ if and only if $P_f \cup P_g \subseteq N(B_f \cap B_g)$.*

Denote the set of all $M\eta F$s by $\mathcal{F}_\eta$ and define the relation $\mathcal{R}$ on $\mathcal{F}_\eta$ by $f\mathcal{R}g$ if and only if $P_f \cup P_g \subseteq N(B_f \cap B_g)$. A *universal* $M\eta F$ is an $M\eta F$ g satisfying $g\mathcal{R}f$ for all $f \in \mathcal{F}_\eta$. In the terminology of the literature on convex sets, g is a universal $M\eta F$ if and only if $\mathcal{F}_\eta$ is star-shaped relative to g. The set of universal $M\eta F$s is the kernel of the set $\mathcal{F}_\eta$ and it follows that the set of universal $M\eta F$s of a graph is a convex set. (See [32, pp. 4–5].)

In order to give the known conditions for an $M\eta F$ to be universal we require further definitions. A vertex $v \in V$ is an η-*absorbing* vertex if there exists $u \in V$ such that $X_v \neq X_u$ and $m(v) \leq m(u) - |X_u - X_v|$. We denote the set of η-absorbing vertices by $\mathcal{A}$. Observe that if m is constant, then v is an η-absorbing vertex if and only if $X_u \subset X_v$ (strict inclusion).

A vertex $w \in V$ is called η-*sharp* if for some $M\eta F$ f, $B_f \cap M(w) \subseteq \mathcal{A}$. Define the *core* of G by $\mathrm{cor}(G) = \{v \in V \mid v \in B_f \text{ for each } M\eta F \, f\}$ and let $\mathcal{W} = \{v \in V \mid \text{for each } x \in X_v \text{ there exists } u \neq v \text{ with } u \in M(x) \cap \mathrm{cor}(G)\}$. Further, let $D = \{v \in V \mid f(v) = 0 \text{ for each } M\eta F \, f\}$. We now state conditions for an $M\eta F$ to be a universal $M\eta F$.

Theorem 5.54 [16] *Let g be an $M\eta F$. (a) If (i) $V - (\mathcal{A} \cup \mathcal{W}) \subseteq B_g$ and (ii) $g(w) = 0$ for each η-sharp vertex $w \in V$, then g is a universal $M\eta F$. (b) If g is a universal $M\eta F$, then (ii) $g(w) = 0$ for each η-sharp element w and (iii) $V - D \subseteq N(B_g)$.*

In the special case when the η-functions are the TDFs of G, Theorem 5.54 specialises to Theorem 5.33. In this case the sets of η-absorbing and η-sharp vertices are the totally absorbing and low vertices of Section 5.3.

Note that we do not have a characterisation of universal $M\eta F$s (*i.e.*, the necessary conditions of Theorem 5.54 differ from the sufficient conditions). Under certain more restrictive conditions on η, a characterisation can be obtained ([16, Theorem 13]) and Theorem 5.6 is a special case of this characterisation theorem.

In [8] results concerning aggregates of η-functions are obtained. The *aggregate* of an η-function is defined by $\mathrm{ag}(f) = \sum_{v \in V} f(v)$. Let

$$I = \{x \mid x \text{ is real and } x = \mathrm{ag}(f) \text{ for some } M\eta F \, f\}.$$

Theorem 5.55 [8] *Max $I = \sup I$ and $\min I = \inf I$, i.e., there exist $M\eta F$s f and g such that*

$$\mathrm{ag}(f) = \inf I \quad and \quad \mathrm{ag}(g) = \sup I.$$

This result shows, for example, that the upper and lower fractional domination numbers of a graph, (*i.e.*, the maximum and minimum aggregates of MDFs, see *e.g.* [28]) actually exist.

Finally, we state an interpolation theorem which includes among its special cases, the answer to question Q1 considered in the introduction.

Theorem 5.56 [8]   *Let f, g be MηFs and* $\operatorname{ag}(f) < x < \operatorname{ag}(g)$. *There exists an* MηF *h with* $\operatorname{ag}(h) = x$.

5.6   Open Problems

The theories surveyed in this chapter are in their infancy and there is much scope for further research. We list some specific areas where further research might well produce new and interesting results.

1. Obtain more (non-) existence theorems for universal MDFs, MTDFs, MPFs of graphs and special classes of graphs, (*e.g.* trees).

2. Characterize the trees which have universal MDFs.

3. Does there exist a graph with a universal M0DF?

4. Characterize the MTDFs of graphs which are universal.

5. Characterize low vertices and sharp vertices of graphs (both have been characterized for trees).

6. Is there a good algorithm for deciding whether a graph has a universal MDF (MTDF)?

7. It was shown in [11] that the relation $\mathcal{R}$ on the set of MDFs (defined in Section 5.2.1) may be studied with the aid of a finite graph called the *convexity graph of the* MDFs. A few properties of convexity graphs of MDFs of trees have been obtained [11]. This may well be a fruitful area for further research.

8. Universal MηFs have only been characterized when certain extra conditions are imposed on η. An important problem is to characterize them for general η. (4 is a special case of this problem.)

9. We have so far only considered functional generalisations of vertex subsets. Edge subsets may also be embedded into sets of functions (see the introduction) and analogous theories of convexity, "η-type" edge-functions etc. could also be developed.

Acknowledgements

Financial support from the University of South Africa, the South African Foundation of Research Development and the Canadian Natural Sciences and Engineering Research Council is gratefully acknowledged.

This chapter was completed while E.J. Cockayne held a visiting position in the Department of Mathematics, Applied Mathematics and Astronomy at UNISA in 1995.

Bibliography

[1] D. W. Bange, A. E. Barkauskas and P. J. Slater, Disjoint dominating sets in trees. Sandia Laboratories Report, SAND 78-1087J, (1978).

[2] D. W. Bange, A. E. Barkauskas and P. J. Slater, Efficient dominating sets in graphs. *Applications of Discrete Mathematics SIAM*, Eds. R. D. Ringeisen and F. S. Roberts, Philadelphia, PA, (1988) 189–199.

[3] I. Broere, J. E. Dunbar and J. H. Hattingh, Minus k-subdomination in graphs. Preprint.

[4] I. Broere, J. H. Hattingh, M. A. Henning and A. A. McRae, Majority domination in graphs. *Discrete Math.* 138 (1995) 125–135.

[5] F. R. K. Chung, Z. Füredi, M. R. Garey and R. L. Graham, On the fractional covering number of hypergraphs. *SIAM J. Discrete Math.* 1 (1988) 45–49.

[6] E. J. Cockayne, O. Favaron and C. M. Mynhardt, Universal maximal packing functions of graphs. *Discrete Math.* 159 (1996) 57–68.

[7] E. J. Cockayne, G. Fricke, S. T. Hedetniemi and C. M. Mynhardt, Properties of minimal dominating functions of graphs. *Ars Combin.* 41 (1995) 107–115.

[8] E. J. Cockayne, G. Fricke and C. M. Mynhardt, Extremum aggregates of hypergraph functions: Existence and interpolation theorems. Submitted for publication.

[9] E. J. Cockayne, G. MacGillivray and C. M. Mynhardt, Convexity of minimal dominating functions of trees. *Utilitas Math.* 48 (1995) 129–144.

[10] E. J. Cockayne, G. MacGillivray and C. M. Mynhardt, A linear algorithm for universal minimal dominating functions in trees. *J. Combin. Math. Combin. Comput.* 10 (1991) 23–31.

[11] E. J. Cockayne, G. MacGillivray and C. M. Mynhardt, Convexity of minimal dominating functions of trees II. *Discrete Math.* 125 (1994) 137–146.

[12] E. J. Cockayne, G. MacGillivray and C. M. Mynhardt, Convexity of minimal dominating functions and universals of graphs. *Bull. Inst. Combin. Appl.* 5 (1992) 37–48.

[13] E. J. Cockayne and C. M. Mynhardt, Convexity of minimal dominating functions of trees: A survey. *Quaestiones Math.* 16 (1993) 301–317.

[14] E. J. Cockayne and C. M. Mynhardt, A characterisation of universal minimal total dominating functions in trees. *Discrete Math.* 141 (1995) 107–115.

[15] E. J. Cockayne and C. M. Mynhardt, On a generalisation of signed dominating functions of graphs. *Ars Combin.* 43 (1996) 235–245.

[16] E. J. Cockayne and C. M. Mynhardt, Minimality and convexity of dominating and related functions in graphs: A unifying theory. To appear in *Utilitas Math.*

[17] E. J. Cockayne, C. M. Mynhardt and B. Yu, Total dominating function in trees: minimality and convexity. *J. Graph Theory* 19 (1995) 83–92.

[18] E. J. Cockayne, C. M. Mynhardt and B. Yu, Universal minimal total dominating functions in graphs. *Networks* 24 (1994) 83–90.

[19] G. S. Domke, G. Fricke, S. T. Hedetniemi and R. C. Laskar, Relationships between integer and fractional parameters in graphs. *Graph Theory, Combinatorics and Applications*, Eds. Y. Alavi, G. Chartrand, O. Oellermann and A. Schwenk, John Wiley and Sons, Inc. 1 (1991) 371–387.

[20] J. E. Dunbar, S. T. Hedetniemi, M. A. Henning and P. J. Slater, Signed domination in graphs. *Graph Theory, Combinatorics and Applications*, Eds. Y. Alavi and A. Schwenk, John Wiley and Sons, Inc. (1995) 311–322.

[21] G. Fricke, E. O. Hare, D. P. Jacobs and A. Majumdar, On integral and fractional total domination. *Congr. Numer.* 77 (1990) 87–95.

[22] D. Grinstead and P. J. Slater, Fractional domination and fractional packing in graphs. *Congr. Numer.* 71 (1990) 153–172.

[23] P. J. P. Grobler, *Functional generalisations of dominating sets of graphs.* Masters Thesis, University of South Africa (1994).

[24] P. J. P. Grobler and C. M. Mynhardt, Extremum aggregates of minimal 0-dominating functions of graphs. *Quaestiones Math.* 19 (1996).

[25] J. H. Hattingh, M. A. Henning and P. J. Slater, Three-valued k-neighbourhood domination in graphs. *Australasian J. Comb.* 9 (1994) 233–242.

[26] S. T. Hedetniemi, Bibliography on fractional parameters of graphs. Clemson University, Clemson, SC (1991).

[27] S. T. Hedetniemi, private communication (1990).

[28] R. Laskar, A. Majumdar, G. Domke and G. Fricke, A fractional view of graph theory. *Sankhyā* 54 (1992) 265–279.

[29] L. Lovasz and M. D. Plummer, *Matching Theory. (Annals of Discrete Math.* 29.) North-Holland, Amsterdam, 1986.

[30] D. F. Rall, A fractional version of domatic number. *Congr. Numer.* 74 (1990) 100–106.

[31] A. Stacey, On the existence of 0-1 universal minimal total dominating functions of trees. Submitted for publication.

[32] F. A. Valentine, *Convex sets*. McGraw-Hill, New York, 1964.

[33] B. Yu, *Convexity of minimal total dominating functions in graphs*. Masters Thesis, University of Victoria (1992).

[34] B. Yu, Convexity of minimal total dominating functions in graphs. Proceedings of the First Ann. Internat. Conf. on Computing and Combinatorics, D. Z. Du and M. Li (Eds.), *Lecture notes in Computer Science*, Springer, 959 (1995) 357–365.

Chapter 6

Combinatorial Problems on Chessboards: II

Sandra M. Hedetniemi
Stephen T. Hedetniemi
Dept. of Computer Science
Clemson University
Clemson, SC   29634-1906 USA

Robert Reynolds
Data General Corp.
Research Triangle Park, NC
27709 USA

6.1   Introduction

In 1993, Fricke, Hedetniemi, Hedetniemi, McRae, Wallis, Jacobson, Martin and Weakley [29] pooled their results and collectively wrote a survey of progress that they and others had made on solving some 36 basic domination-related problems on chessboards. They also stated approximately 50 open problems. We will refer to this paper hereafter as Survey I.

The purpose of this chapter is to present a second survey of domination related problems on chessboards, including several improvements to results mentioned in Survey I, a variety of illustrations not given in Survey I, and discussions of several topics not included in Survey I.

The 36 basic problems discussed in Survey I are to determine (i) the irredundance number, $ir(G)$, (ii) the domination number, $\gamma(G)$, (iii) the independent domination number, $i(G)$, (iv) the independence number, $\beta(G)$, (v) the upper domination number, $\Gamma(G)$, and (vi) the upper irredundance number, $IR(G)$, for each of six types of chessboard graphs: (a) the Queens graph, Q_n, (b) the Kings graph, K_n, (c) the Rooks graph, R_n, (d) the Bishops graph, B_n, (e) the Knights graph, N_n, and (f) the Grid graph, G_n.

Each of these classes of graphs, Q_n, K_n, R_n, B_n, N_n and G_n, has a set V of n^2 vertices (in 1-1 correspondence with the n^2 squares of an order-n chessboard), where two vertices (squares) are adjacent if and only if a piece of a given type, when placed on one of the squares can reach (or attack) the other square in one move.

Recall the following moves of the chess pieces:

(a) a queen can move any number of squares horizontally, vertically or diagonally;

(b) a king can move one square in any direction (horizontally, vertically or diagonally);

(c) a rook can move any number of squares horizontally or vertically;

(d) a bishop can move any number of squares diagonally;

(e) a knight at square (i, j) can move to any square on the board with indices $(i \pm 2, j \pm 1)$ or $(i \pm 1, j \pm 2)$;

(f) a vertex in a grid graph, at square (i, j), is adjacent to vertices $(i - 1, j)$, $(i, j + 1)$, $(i + 1, j)$ and $(i, j - 1)$, which are located one square to the north, east, south or west of (i, j), respectively.

It can be noted that: the Rooks graph R_n is the cartesian product of the complete graph on n vertices with itself; the Rooks graph R_n, Bishops graph B_n and Kings graph K_n are subgraphs of the Queens graph Q_n; the Bishops graph B_n consists of two disjoint subgraphs; and the Knights graph N_n and the Grid graph G_n are both bipartite.

It is also clear that if $G \subset H$ means that G is a proper, spanning subgraph of H, then:

$$\begin{array}{lll} G_n & \subset & R_n & \subset & Q_n, \\ B_n & \subset & Q_n, & & \text{and} \\ G_n & \subset & K_n & \subset & Q_n. \end{array}$$

The following figure presents the state of knowledge of these 36 basic problems as presented in Survey I.

	ir	γ	i	β	Γ	IR
Queens Q_n	?	A	B	n	C	?
Kings K_n	?	D	D	E	?	F
Rooks R_n	n	n	n	n	n	$2n-4$
Bishops B_n	n	n	n	$2n-2$	$2n-2$	$4n-14$
Knights N_n	?	?	?	G	G	G
Grid G_n	?	H	?	G	G	G

A: Grinstead, Hahne and Van Stone [41]: $\gamma(Q_n) \le 14n/23 + O(1)$.
B: Grinstead, Hahne and Van Stone [41]: $i(Q_n) \le 2n/3 + O(1)$.
C: Weakley [75]: $\Gamma(Q_n) \ge 2n - 5$.
D: Yaglom, Yaglom [78]: $\gamma(K_n) = i(K_n) = (\lfloor (n+2)/3 \rfloor)^2$.
E: Yaglom, Yaglom [78]: $\beta(K_n) = (\lfloor (n+1)/2 \rfloor)^2$.
F: Fricke and Pritikin [30]: $(n-1)^2/3 \le IR(K_n) \le n^2/3$.
G: Cockayne, Favaron, Payan and Thomason [16]:

$$
\begin{aligned}
\beta(N_n) = \Gamma(N_n) = IR(N_n) &= n^2/2, && \text{for } n \text{ even;} \\
&= (n^2+1)/2, && \text{for } n \text{ odd;} \\
\beta(G_n) = \Gamma(G_n) = IR(G_n) &= n^2/2, && \text{for } n \text{ even;} \\
&= (n^2+1)/2, && \text{for } n \text{ odd.}
\end{aligned}
$$

H: Cockayne, Hare, Hedetniemi and Wimer [19]:

$$
(n^2 + n - 3)/5 \le \gamma(G_n) \le (n^2 + 4n - \alpha)/5,
$$
where $\alpha = 16, 17, 20, 20$ or 17
if $n = 5k - 2, 5k - 1, 5k, 5k + 1$ or $5k + 2$, resp., for $k \ge 2$.

Figure 6.1: Best known results of 36 chessboard problems as of 1993.

We should also note that the values $\gamma(R_n) = i(R_n) = \beta(R_n) = n, \gamma(B_n) = i(B_n) = n$ and $\beta(B_n) = 2n - 2$ are due to Yaglom and Yaglom [78]; while the value $\beta(Q_n) = n$ is due to Ahrens [2]. All other values in Figure 6.1 are due to one or more of the authors of Survey I.

6.2 Updates to Survey I

Since the publication of Survey I there have been several notable improvements to the solutions given in Figure 6.1. First, Burger, Mynhardt and Cockayne [9] have several new results on the domination number of the Queens graph, Q_n. Using a creative combination of mathematical analysis and computer programming they were able to determine the exact value of $\gamma(Q_{4k+1})$ for $k = 9, 12, 13$ and 15. Using the lower bound of Weakley [74], that $\gamma(Q_{4k+1}) \geq 2k + 1$, they were able to construct dominating sets of $2k + 1$ queens for $k = 9, 12, 13$ and 15. Recently Gibbons and Webb [36] have extended these results even further by determining that $\gamma(Q_{14}) = 8$, $\gamma(Q_{29}) = 17$, $\gamma(Q_{41}) = 21$, $\gamma(Q_{45}) = 23$ and $\gamma(Q_{57}) = 29$. They also determined that $i(Q_{14}) = 8$, $i(Q_{15}) = 9$, and $i(Q_{16}) = 9$ [36]. The current state of knowledge of $\gamma(Q_n)$ and $i(Q_n)$, as given by Weakley [74], Burger, Mynhardt and Cockayne [9] and Gibbons and Webb [36], is shown in Figure 6.2. We should also note that P. H. Spencer's theorem that $\gamma(Q_n) \geq (n-1)/2$ is the basis for much of the recent work on $\gamma(Q_n)$ (cf. [74]).

n	1	2	3	4	5	6	7	8	9	10	11	12	13	14	15
$\gamma(Q_n)$	1	1	1	2	3	3	4	5	5	5	5	6	7	8	≤ 9
$i(Q_n)$	1	1	1	3	3	4	4	5	5	5	5	7	7	8	9

n	16	17	18	19	20	21	22	23	24	25	29	33	37
$\gamma(Q_n)$	≤ 9	9	9	≤ 10	≤ 11	11	≤ 12	≤ 13	≤ 13	13	17	17	19
$i(Q_n)$	9	9	≤ 10	≤ 11	≤ 11	11	≤ 13	≤ 13	≤ 13	13	≤ 17	17	

n	41	45	49	53	57	61	65
$\gamma(Q_n)$	21	23	25	27	29	31	≤ 35

Figure 6.2: Values of $\gamma(Q_n)$ and $i(Q_n)$.

We should also note that the current best known upper bounds for $\gamma(Q_n)$ and $i(Q_n)$ are the following by Burger, Cockayne and Mynhardt [8] and Eisenstein, Grinstead, Hahne and Van Stone [22], respectively:

$$\gamma(Q_n) \leq (31n/54) + O(1) \quad \text{and} \quad i(Q_n) \leq (7n/12) + O(1)$$

Although no good upper bound (other than $IR(Q_n)$) is known for $\Gamma(Q_n)$, the best lower bound for $\Gamma(Q_n)$ is due to Burger, Cockayne and Mynhardt [8]:

$$\Gamma(Q_n) \geq (5n/2) - O(1)$$

The best known upper bound for $IR(Q_n)$ is also found in [8]:

$$IR(Q_n) \leq \left\lfloor 6n + 6 - 8\sqrt{n + 1 + \sqrt{n}} \right\rfloor.$$

Recently Fricke and Pritikin [30] and Favaron and Puech [26] have provided new results (some identical) concerning irredundance and domination in Kings graphs. Fricke and Pritikin have provided the following lower and upper bounds for $IR(K_n), ir(K_n)$ and $\Gamma(K_n)$:

$$
\begin{array}{rcll}
(n-1)^2/3 & \leq & IR(K_n) \leq & n^2/3, \quad \text{for } n \geq 6; \\
n^2/9 & \leq & ir(K_n) \leq & (\lfloor (n+2)/3 \rfloor)^2, \quad \text{and} \\
& & ir(K_n) = & n^2/9, \quad \text{if } n = 0 \pmod 3; \\
& & ir(K_n) \leq & (\lfloor (n+2)/3 \rfloor)^2 - 1, \quad \text{if } n = 4 \pmod 6; \\
\lceil (n^2 - 4n + 9)/3 \rceil & \leq & \Gamma(K_n) \leq & n^2/3, \quad \text{for } n \geq 6.
\end{array}
$$

They also determined, by mathematical analysis, the first five values of $IR(K_n)$ and the first nine values of $ir(K_n)$ (given later).

Favaron and Puech:

1. also determined the first few values of $ir(K_n), \Gamma(K_n)$ and $IR(K_n)$,

2. also obtained the lower bound of $n^2/9$ for $ir(K_n)$,

3. showed that $ir(K_n) = \gamma(K_n) = i(K_n) = n^2/9$, for every $n = 0 \pmod 3$,

4. showed that $\Gamma(K_n) \geq \lceil (n-2)^2/3 \rceil + 3$,

5. showed that as $n \to \infty$, both ratios $ir(K_n)/i(K_n)$ and $\Gamma(K_n)/IR(K_n) \to 1$, and

6. showed that for every positive integer t such that $i(K_n) \leq t \leq \beta(K_n)$, there exists in K_n a maximal independent set of t kings.

Several authors have made improvements in determining the domination numbers of $m \times n$ grid graphs, including Fisher [28], Clark, Colbourn and Johnson [15], Chang [12], Chang and Clark [13], Chang, Clark, and Hare [14] and Ma and Lam [54]. A brief summary of the current state of knowledge is contained in the following list:

$$
\begin{array}{rll}
[50] \quad \gamma(G_{1,n}) & = & \lceil n/3 \rceil; \\
\gamma(G_{2,n}) & = & \lceil (n+1)/2 \rceil; \\
\gamma(G_{3,n}) & = & \lceil (3n+1)/4 \rceil; \\
\gamma(G_{4,n}) & = & n+1, \quad \text{if } n = 1,2,3,5,6,9; \\
& = & n \quad \text{otherwise}; \\
[13] \quad \gamma(G_{5,n}) & = & \lceil (6n+4)/5 \rceil - 1, \quad \text{if } n = 2,3,7; \\
& = & \lceil (6n+4)/5 \rceil \quad \text{otherwise}; \\
\gamma(G_{6,n}) & = & \lceil (10n+4)/7 \rceil + 1, \quad \text{if } n = 3 \pmod 7, n \neq 3; \\
& = & \lceil (10n+4)/7 \rceil \quad \text{otherwise};
\end{array}
$$

$$
\begin{aligned}
[54] \quad \gamma(G_{8,8}) &= 16; \\
\gamma(G_{8,9}) &\geq 17; \\
\gamma(G_{9,9}) &\geq 19; \\
[28] \quad \gamma(G_{7,n}) &= \lceil (5n+1)/3 \rceil + 1, && \text{if } n = 1; \\
&= \lceil (5n+1)/3 \rceil && \text{otherwise}; \\
\gamma(G_{8,n}) &= \lceil (15n+7)/8 \rceil - 1, && \text{if } n = 4, 6; \\
&= \lceil (15n+7)/8 \rceil && \text{otherwise}; \\
\gamma(G_{9,n}) &= \lceil (23n+10)/11 \rceil - 1, && \text{if } n = 2, 3; \\
&= \lceil (23n+10)/11 \rceil && \text{otherwise}; \\
\gamma(G_{10,n}) &= \lceil (30n+15)/13 \rceil - 1, && \text{if } n = 3, 4, 7, 13, 16 \text{ or} \\
&&& 10 \pmod{13}; \\
&= 15 && \text{if } n = 6; \\
&= \lceil (30n+15)/13 \rceil && \text{otherwise}; \\
\gamma(G_{11,n}) &= \lceil (38n+22)/15 \rceil - 1, && \text{if } n = 2, 3, 4, 5, 7, 9, 11, 18, \\
&&& 20, 22, 33; \\
&= \lceil (38n+22)/15 \rceil && \text{otherwise}; \\
\gamma(G_{12,n}) &= \lceil (80n+38)/29 \rceil - 1, && \text{if } n = 1, 4, 9; \\
&= \lceil (80n+38)/29 \rceil && \text{otherwise}; \\
\gamma(G_{13,n}) &= \lceil (98n+54)/33 \rceil - 1, && \text{if } n = 13, 16, 18, 19 \pmod{33} \text{ or} \\
&&& n = 2, 3, 4, 7, 10; \\
&= \lceil (98n+54)/33 \rceil && \text{otherwise}; \\
\gamma(G_{14,n}) &= \lceil (35n+20)/11 \rceil - 1, && \text{if } n = 2, 3, 4 \text{ or} \\
&&& n = 7 \pmod{11} \\
&&& \text{except } n = 40; \\
&= \lceil (35n+20)/11 \rceil && \text{otherwise}; \\
\gamma(G_{15,n}) &= \lceil (44n+28)/13 \rceil - 1, && \text{if } n = 1, 2, 3, 4, 6, 8 \text{ or} \\
&&& n = 5 \pmod{13} \\
&&& \text{except } n = 18, 44; \\
&= \lceil (44n+28)/13 \rceil && \text{otherwise}; \\
\gamma(G_{16,n}) &= \lceil (18n+12)/5 \rceil - 1, && \text{if } n = 2, 3, 4, 5, 7, 8, 10, 13; \\
&= \lceil (18n+12)/5 \rceil && \text{otherwise}; \\
[12] \quad \gamma(G_{m,n}) &\leq \lfloor (m+2)(n+2)/5 \rfloor - 4 && \text{if } m, n \geq 8.
\end{aligned}
$$

Based on these results, Fisher has made the following conjecture:

Conjecture 6.1 [28] $\gamma(G_{m,n}) = \lfloor (m+2)(n+2)/5 \rfloor - 4.$

6.3 First 10 Values of Chessboard Domination Parameters

In Survey I it was pointed out that the first 10 or so values of many of the domination-related parameters of the various chessboard graphs had not been determined. In this section we would like to present the first values of these parameters, some of which have recently been determined by exhaustive computer search. We will also present a selection of optimal configurations in order to give the reader some insights into solutions of these problems.

Queens graph Q_n

n	1	2	3	4	5	6	7	8	9	10	
$ir(Q_n)$	1	1	1	2							
$\gamma(Q_n)$	1	1	1	2	3	3	4	5	5	5	computer search
$i(Q_n)$	1	1	1	3	3	4	4	5	5	5	computer search
$\beta(Q_n)$	1	1	2	4	5	6	7	8	9	10	formula: n
$\Gamma(Q_n)$	1	1	2	4	5	7	9				computer search
$IR(Q_n)$	1	1	2	4	5	7	9	11			computer search

Several solutions for $IR(Q_n)$ (private neighbors indicated) follow.

$IR(Q_5) = 5$				
$Q1$	$Q2$	$Q3$	$-$	$Q4$
$-$	$-$	$-$	$-$	4
$-$	2	$-$	$-$	$-$
$-$	$-$	3	1	$-$
$Q5$	$-$	$-$	5	$-$

$IR(Q_6) = 7$					
$Q1$	$Q2$	$Q3$	$-$	$-$	$-$
$-$	$Q4$	$-$	$-$	4	$-$
$-$	$-$	$Q5$	$-$	$-$	5
1	$-$	$-$	$-$	2	3
$-$	$-$	$Q6$	6	$-$	$-$
$-$	$Q7$	$-$	$-$	7	$-$

$IR(Q_7) = 9$						
Q1	Q2	Q3	–	–	–	–
–	–	–	3	7	8	9
Q4	Q5	–	–	–	–	–
Q6	–	–	–	–	–	6
–	Q7	–	–	–	–	–
–	–	–	4	5	1	2
Q8	Q9	–	–	–	–	–

$IR(Q_8) = 11$							
Q1	Q2	Q3	–	–	–	–	–
–	–	–	–	7	9	10	11
–	Q4	–	–	–	–	–	4
Q5	Q6	Q7	–	–	–	–	–
Q8	–	–	–	–	–	–	8
–	–	–	–	–	1	2	3
Q9	Q10	Q11	–	–	–	–	–
–	–	–	–	5	6	–	–

Kings graph K_n

n	1	2	3	4	5	6	7	8	9	10	
$ir(K_n)$	1	1	1	3	4	4	8	9	9		proof [30][26]
$\gamma(K_n)$	1	1	1	4	4	4	9	9	9	16	formula: $\lfloor n+2)/3 \rfloor^2$
$i(K_n)$	1	1	1	4	4	4	9	9	9	16	formula: $\lfloor n+2)/3 \rfloor^2$
$\beta(K_n)$	1	1	4	4	9	9	16	16	25	25	formula: $\lfloor n+1)/2 \rfloor^2$
$\Gamma(K_n)$	1	1	4	4	9	9	16				proof [30][26]
$IR(K_n)$	1	1	4	4	9	9	16				proof [30][26]

Rooks graph R_n

n	1	2	3	4	5	6	7	8	9	10	
$ir(R_n)$	1	2	3	4	5	6	7	8	9	10	formula: n
$\gamma(R_n)$	1	2	3	4	5	6	7	8	9	10	formula: n
$i(R_n)$	1	2	3	4	5	6	7	8	9	10	formula: n
$\beta(R_n)$	1	2	3	4	5	6	7	8	9	10	formula: n
$\Gamma(R_n)$	1	2	3	4	5	6	7	8	9	10	formula: n
$IR(R_n)$	1	2	3	4	6	8	10	12	14	16	formula: $2n-4$

Bishops graph B_n

n	1	2	3	4	5	6	7	8	9	10	
$ir(B_n)$	1	2	3	4	5	6	7	8	9	10	formula: n
$\gamma(B_n)$	1	2	3	4	5	6	7	8	9	10	formula: n
$i(B_n)$	1	2	3	4	5	6	7	8	9	10	formula: n
$\beta(B_n)$	1	2	4	6	8	10	12	14	16	18	formula: $2n - 2$
$\Gamma(B_n)$	1	2	4	6	8	10	12	14	16	18	formula: $2n - 2$
$IR(B_n)$	1	2	4	6	8	10	14	18	22	26	formula: $4n - 14$

Knights graph N_n

n	1	2	3	4	5	6	7	8	9	10	
$ir(N_n)$	1	4									
$\gamma(N_n)$	1	4	4	4	5	8	10	12	14	16	algorithm known [43]
$i(N_n)$	1	4	4	4	5	8	13	14	14	16	computer search
$\beta(N_n)$	1	4	5	8	13	18	25	32	41	50	formula: $n^2/2$ or $(n^2 + 1)/2$
$\Gamma(N_n)$	1	4	5	8	13	18	25	32	41	50	formula: $n^2/2$ or $(n^2 + 1)/2$
$IR(N_n)$	1	4	5	8	13	18	25	32	41	50	formula: $n^2/2$ or $(n^2 + 1)/2$

Several solutions for $i(N_n)$:

$i(N_3) = 4$

$i(N_4) = 4$

$i(N_5) = 5$

$i(N_6) = 8$					
–	–	–	–	–	–
–	–	–	–	–	–
N	N	–	–	N	N
N	N	–	–	N	N
–	–	–	–	–	–
–	–	–	–	–	–

$i(N_7) = 13$						
–	–	–	–	–	–	–
–	–	–	–	–	–	–
N	N	N	N	N	N	N
–	–	–	–	–	–	–
–	–	–	–	–	–	–
–	–	–	N	–	–	–
N	–	N	N	N	–	N

Grid graph G_n

n	1	2	3	4	5	6	7	8	9	10	
$ir(G_n)$	1	2	3								
$\gamma(G_n)$	1	2	3	4	7	10	12	16	20	24	algorithm known [44]
$i(G_n)$	1	2	3								
$\beta(G_n)$	1	4	5	8	13	18	25	32	41	50	formula: $n^2/2$ or $(n^2+1)/2$
$\Gamma(G_n)$	1	4	5	8	13	18	25	32	41	50	formula: $n^2/2$ or $(n^2+1)/2$
$IR(G_n)$	1	4	5	8	13	18	25	32	41	50	formula: $n^2/2$ or $(n^2+1)/2$

6.4   The N-queens Problem

With approximately 100 literature citations, the N-queens problem is one of the most studied of all chessboard problems. Originally stated by a German chess player, Max Bezzel, in the September 1848 issue of the chess newspaper Berliner Schachzeitung [6], the problem is to place n queens on an order n board so that no two queens attack each other, i.e., no two queens lie on a common row, common column or common diagonal. Campbell [10] provides an interesting discussion of the early history of this problem, during the period 1848-1899.

Although the famous mathematician Gauss worked on the number of different solutions to the problem on the order-8 board, it was Dr. Franz Nauck [61, 60] who first pointed out in 1850, apparently without proof, that there are 92 solutions, which fall into 12 classes: 8 solutions can be generated by rotations and reflections of each of 11 basic solutions, while a twelfth basic solution generates four others ($11 \times 8 + 1 \times 4 = 92$).

Gauss, however, is given credit [10] for giving the 8-queens problem an arithmetic reformulation, as follows: arrange the integers 1 through 8 in a sequence in such a way that adding 1 to the first integer, 2 to the second integer, and so forth, adding 8 to the eighth integer in the sequence, produces a set of eight distinct sums, while for the same sequence, adding 8 to the first integer, 7 to the second integer, and so forth, and 1 to the eighth integer, also produces a set of eight distinct sums. The two sets of eight distinct sums may have values in common. With this formulation, the basic 12 solutions can be given in lexicographic order as follows:

1. 2 7 3 6 8 5 1 4
2. 3 8 4 7 1 6 2 5
3. 4 1 5 8 2 7 3 6
4. 4 2 7 3 6 8 1 5
5. 4 2 7 3 6 8 5 1
6. 4 8 5 3 1 7 2 6
7. 5 1 8 4 2 7 3 6
8. 5 3 8 4 7 1 6 2
9. 5 7 2 4 8 1 3 6
10. 6 4 1 5 8 2 7 3
11. 6 4 7 1 8 2 5 3
12. 8 3 1 6 2 5 7 4

Consider, for example, the first solution above and the two sets of distinct sums that it generates:

$$
\begin{array}{c}
1\ 2\ 3\quad 4\quad 5\quad 6\ 7\quad 8 \\
+\ 2\ 7\ 3\quad 6\quad 8\quad 5\ 1\quad 4 \\
\hline
3\ 9\ 6\ 10\ 13\ 11\ 8\ 12
\end{array}
\qquad
\begin{array}{c}
8\quad 7\ 6\quad 5\quad 4\ 3\ 2\ 1 \\
+\ 2\quad 7\ 3\quad 6\quad 6\ 5\ 1\ 4 \\
\hline
10\ 14\ 9\ 11\ 12\ 8\ 3\ 5
\end{array}
$$

Figure 6.3 shows the configuration of 8 queens corresponding to the sequence $2, 7, 3, 6, 8, 5, 1, 4$.

	1	2	3	4	5	6	7	8
1	−	−	−	−	−	−	Q	−
2	Q	−	−	−	−	−	−	−
3	−	−	Q	−	−	−	−	−
4	−	−	−	−	−	−	−	Q
5	−	−	−	−	−	Q	−	−
6	−	−	−	Q	−	−	−	−
7	−	Q	−	−	−	−	−	−
8	−	−	−	−	Q	−	−	−

Figure 6.3: Solution {2,7,3,6,8,5,1,4}.

The requirement that the integers in the first set of sums be distinct assures that no two queens lie on a common upward diagonal (lower left to upper right), while the requirement that the integers in the second set of sums be distinct assures that no two queens lie on a common downward diagonal.

In 1880, Lucas [57, 58] noted that seven of the 12 basic solutions can be obtained from each other by rearranging rows or columns, while in 1973, Hollander [49] pointed out that these seven solutions are related as overlapping areas of tessellations of one of the solutions, (e.g., the solution 2,7,3,6,8,5,1,4). In Figure 6.4 this solution is repeated nine times in a three-by-three array. In this iterated array, solutions 2, 3, 5, 8, 10 and 12 can also be found (their upper left corner is indicated by the squares so numbered).

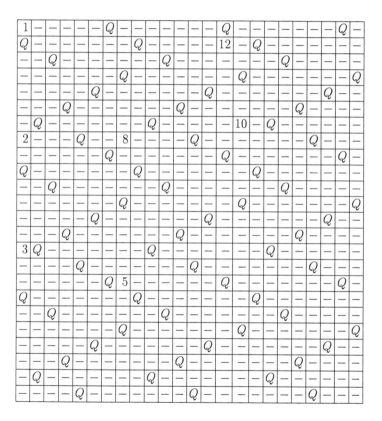

Figure 6.4: A repeated pattern containing eight solutions.

In 1938, J. Ginsburg [37] pointed out that the N-queens problem has another arithmetic formulation: write the integers 9 to 16 on a line, under which you are to arrange the integers from 1 to 8 in such a way that different values result when each pair is subjected to addition and subtraction. For example, using solution 1 above, we get:

$$
\begin{array}{r}
\; 9 \quad 10 \quad 11 \quad 12 \quad 13 \quad 14 \quad 15 \quad 16 \\
+ \;\; 2 \quad\; 7 \quad\; 3 \quad\; 6 \quad\; 8 \quad\; 5 \quad\; 1 \quad\; 4 \\
\hline
11 \quad 17 \quad 14 \quad 18 \quad 21 \quad 19 \quad 16 \quad 20
\end{array}
$$

$$
\begin{array}{r}
\; 9 \quad 10 \quad 11 \quad 12 \quad 13 \quad 14 \quad 15 \quad 16 \\
- \;\; 2 \quad\; 7 \quad\; 3 \quad\; 6 \quad\; 8 \quad\; 5 \quad\; 1 \quad\; 4 \\
\hline
7 \quad\; 3 \quad\; 8 \quad\; 6 \quad\; 5 \quad\; 9 \quad 14 \quad 12
\end{array}
$$

A close relation between solutions to the N-queens problem and the construction of magic squares has recently been observed by Demirors, Rafraf and Tanik [21].

In his book, Mathematical Recreations [53], Kraitchik classifies solutions to the N-queens problem according to their symmetries, as follows:

Class O solutions are those which are invariant only under the identity transformation;

Class C solutions are those which are invariant only under the identity and rotations through two right angles; and

Class Q solutions are those which are invariant only under the identity and rotations through one right angle.

Kraitchik then lists the number of solutions to the N-queens problem for various values of n:

n	$O(\times 8)$	$C(\times 4)$	$Q(\times 2)$	Total
1				
2				
3				
4			1	2
5	1		1	10
6		1		4
7	4	2		40
8	11	1		92
9	42	4		352
10	89	3		724
11	329	12		2,680
12	1,765	18	1	14,200

Ahrens [2] in 1910 was apparently the first person to prove that the N-queens problem has a solution for every positive integer $n \geq 4$. This was independently rediscovered by the Russian Yaglom brothers in 1964 [78], who found essentially the same proof. Other independent proofs of this result have been given by Hoffman, Loessi and Moore in 1969 [48], Falkowski and Schmitz in 1986 [25], Reichling in 1987 [63] and Erbas, Tanik and Aliyazicioglu in 1992 [24].

Ahrens' solutions for arbitrary values of n involve one pattern which for even N generates solutions for $N = 6a$ and $6a + 4$, and a second pattern which generates solutions for $N = 6a + 2$. Using the fact that neither of these patterns places a queen on the main diagonal, Ahrens then developed a third pattern, which places a queen on the main diagonal for odd values of N, using an even pattern for $N - 1$.

The solutions presented by Yaglom and Yaglom [78], Hoffman, Loessi and Moore [48] and Reichling [63] are similar. Falkowski and Schmitz [25] present five solution patterns, while Erbas, Tanik and Aliyazicioglu [24] use linear congruence relations, which provide increasing numbers of solutions as N increases.

Modern work on the N-queens problem has had three aspects: developing efficient and fast algorithms for finding at least one solution for every value of N (e.g., Bernhardsson [5]); finding families of solutions for certain values of N (e.g., Larson [55]); and finding all solutions in lexicographic order, for a given value of N (e.g., Rohl [67] and Hitotumatsu and Noshita [47]).

Perhaps the most concentrated attention has been on different algorithmic design paradigms which can be used to find solutions for all values of N. Such design paradigms include backtrack programming (Rohl [68], Hitotumatsu and Noshita [47], Topor [71], Stone and Stone [70], Sosic and Gu [69], and Golomb and Baumert [38]); dynamic programming (Rivin and Zabih [65]); divide and conquer and constraint satisfaction (Abramson and Yung [1], Nadel [59]); general program development (Wirth [77], Naur [62]); and distributed and parallel computing (Finkel and Manber [27]).

In conclusion, it is interesting to note the wide variety of other areas of study in which the N-queens problem is discussed. Erbas, Tanik and Aliyazicioglu [24] list the following: permutation generation, integer programming, parallel memory storage schemes allowing conflict-free access to portions of two-dimensional arrays [23], VLSI testing, traffic control and deadlock prevention.

6.5   Queens on a Column and on a Diagonal

Cockayne and others [20], [17] have asked the questions: what is the minimum number of queens which placed on a single diagonal, or on a single column, will suffice to dominate every square on an order-n chessboard? Let us denote these two numbers by $\text{diag}(Q_n)$ and $\text{col}(Q_n)$. It follows by definition that: $\gamma(Q_n) \leq \text{col}(Q_n) \leq n$ and $\gamma(Q_n) \leq \text{diag}(Q_n) \leq n$. The following is a brief summary of what is known about these two upper bounds for $\gamma(Q_n)$.

Let $r_3(n)$ be the largest cardinality of a subset of $N = \{1, 2, ..., n\}$ which contains no 3-term arithmetic progression.

Cockayne and Hedetniemi [19]:

$$\text{diag}(Q_n) = n - r_3(\lfloor n/2 \rfloor)$$
$$\lim_{n \to \infty} (\text{diag}(Q_n)/n) = 1$$
for n sufficiently large, $\gamma(Q_n) < \text{diag}(Q_n)$.

Cockayne, Gamble and Shepherd [18]:

$$\text{col}(Q_n) = \min\lfloor A(n), B(n)\rfloor \ n \geq 2, \text{ where}$$
$$A(n) = n - r_3(\lceil n/3\rceil)$$
$$B(n) = n - \max\lfloor r_3(\lceil k/2\rceil) + r_3(\lceil l/2\rceil)\rfloor,$$
$$k + l = \lceil n/2\rceil, \ k, l \geq 0.$$
$$\text{for all } n \ \text{col}(Q_n) \leq \text{diag}(Q_n).$$

6.6 Non-attacking Queens

In his book published in 1892 [4], W. W. Rouse Ball attributed to Captain Turton the problem of placing eight queens on a chessboard "so as to command the fewest possible squares". Ball presented the possible solution given in Figure 6.5, in which the 11 numbered squares are not attacked by any of the eight queens.

$-$	Q	$-$	$-$	$-$	$-$	Q	$-$
Q	$-$	Q	$-$	$-$	$-$	Q	$-$
$-$	Q	$-$	$-$	$-$	$-$	$-$	$-$
$-$	$-$	$-$	$-$	$-$	5	$-$	8
$-$	$-$	$-$	$-$	3	$-$	$-$	9
$-$	$-$	$-$	1	$-$	6	$-$	10
Q	Q	$-$	$-$	$-$	$-$	$-$	$-$
$-$	$-$	$-$	2	4	7	$-$	11

Figure 6.5: Non-attacking Queens.

Ball then asked, "Is it possible to place eight queens so as to leave more than eleven cells out of check? I have never succeeded in doing so, nor in showing that it is impossible to do it."

Martin Gardner considered the same problem in his 1983 book [34], stating that "There are at least six basic ways to do it (the exact number is not known).... Eleven unchecked cells is undoubtedly maximum, although no proof is known to me." Gardner then considered the general problem of "placing n queens on an order-n board to leave a maximum number of unattacked vacant cells."

Let $Free(Q_n)$ equal the maximum number of unattacked squares on an order-n chessboard by a placement of n queens. Stating that "this problem has not, to my knowledge, been fully analyzed," Gardner presented the best known values of $Free(Q_n)$ given in Figure 6.6.

n :	1	2	3	4	5	6	7	8	9	10	11	12
$Free(Q_n)$:	0	0	0	1	3	5	7	11	16	22	27	36

Figure 6.6: Gardner's conjectured values of $Free(Q_n)$.

Gardner pointed out that the solution for $n = 5$ is unique, except for rotations and reflections, and presented the two "solutions" given in Figure 6.7.

Free(Q_5) = 3				
−	1	−	−	2
−	−	−	−	3
Q	−	−	−	−
Q	−	−	Q	−
−	−	Q	Q	−

Free(Q_{11}) = 11							
−	1	−	−	−	−	5	9
−	2	−	−	−	3	6	10
−	−	−	−	−	4	7	−
Q	−	Q	−	−	−	−	−
Q	−	−	−	−	−	−	−
Q	−	−	−	Q	−	−	
−	−	−	−	−	−	8	11
−	−	Q	Q	Q	−	−	−

Figure 6.7: Gardner's solutions for $n = 5$ and $n = 11$.

In an addendum to the chapter in which the above appears, Gardner presents the following information:

"The problem of placing five queens on a 5×5 board so that three cells are not attacked has appeared in many places since I introduced it in my 1972 column (cf. [32]). It is usually given in the following form: Place five queens of one color and three of another color on an order-5 board so that no queen attacks a queen of a different color. I myself gave it in this form in a later (February, 1978) column (cf. [31])."

"I had in 1972 confined the task to n queens on a board also of order n. When I gave it again in 1978 for the order-5 board, a number of readers generalized it to k queens on an order-n board. The best results came from Hiroshi Okuno, of Tokyo, whose computer search provided valuable data for low values of n and k. In 1983, Ronald L. Graham and Fan K. Chung, both of Bell Laboratories, turned their attention to Okuno's data and made some truly remarkable discoveries about the general problem. They will be reported in a paper that may appear before this book."

According to Graham [40] this paper was never finished, but may soon appear and will provide the most recent work that has been done on the non-attacking queens problem. Velucchi [72] has provided the following best known values of $Free(Q_n)$, the number of distinct solutions known for each value of n, and the authors of these values of $Free(Q_n)$. According to Velucchi, for $n \leq 16$ the values of $Free(Q_n)$ and the number of distinct solutions have been determined exactly by computer search.

n	$Free(Q_n)$	#solutions known[72]	Author(s) of first solution [72]
1	0	0	Obvious
2	0	0	Obvious
3	0	0	Obvious
4	1	25	Stephen Ainley, 1977[3]
5	3	1	Ainley, 1977
6	5	3	Ainley, 1977
7	7	38	Ainley, 1977
8	11	7	Ball,1982
9	18	1	Diego Bracamonte, Argentina, 1993
10	22	1	Ainley, 1977
11	30	2	Rodolfo Kurchan, Argentina, 1993, and Velucchi, Italy, 1995
12	36	7	Ainley, 1977
13	47	1	Johan Claes, Belgium, and Velucchi, 1995
14	56	4	Ainley, 1977
15	72	3	Claes and Velucchi, 1995
16	82	1	Claes and Velucchi, 1995
17	97	1	Claes and Velucchi; Miyoshi Nagai, Japan, 1995
18	111	1	Ainley, 1977

Figure 6.8: Best known values of $Free(Q_n)$.

Although limitations of space will not permit us to present the best known solutions, we present several solutions constructed independently by the authors of this chapter.

$Free(Q_6) = 5$					
–	–	–	1	2	–
Q	Q	–	–	–	–
–	–	–	–	–	3
–	–	–	–	4	5
–	Q	Q	–	–	–
Q	Q	–	–	–	–

$Free(Q_7) = 7$						
–	–	–	–	1	2	–
–	–	–	3	4	–	–
Q	Q	–	–	–	–	–
Q	–	–	–	–	–	–
–	–	–	–	5	6	7
–	Q	Q	–	–	–	–
Q	Q	–	–	–	–	–

Free(Q_{12}) = 36											
−	−	−	−	1	6	10	13	15	−	−	−
−	−	−	−	2	7	11	14	−	−	−	−
−	−	−	−	3	8	12	−	−	−	−	31
−	−	−	−	4	9	−	−	−	−	26	32
−	−	−	−	5	−	−	−	−	22	27	33
−	−	−	−	−	−	−	−	19	23	28	34
−	−	−	−	−	−	−	17	20	24	29	35
−	−	−	−	−	−	16	18	21	25	30	36
−	Q	Q	Q	−	−	−	−	−	−	−	−
Q	Q	Q	Q	−	−	−	−	−	−	−	−
Q	Q	Q	−	−	−	−	−	−	−	−	−
Q	Q	−	−	−	−	−	−	−	−	−	−

It is interesting to note that $Free(Q_7) = 7$. This means that given a placement of seven queens which leaves seven squares unattacked, the unattacked squares also form a solution. Thus, for $N = 7$, solutions come in pairs. Velucchi [72] points out that this phenomenon occurs for other boards and other numbers of queens placed on a board. Let $Free(k,n)$ be the maximum number of free queens possible by placing k queens on an order-n board. Velucchi gives the following:

$$Free(4,5) = 4;$$
$$Free(7,7) = 7;$$
$$Free(12,9) = 12;$$
$$Free(21,12) = 21;$$
$$Free(37,16) = 37;$$
$$Free(42,17) = 42;$$
$$Free(64,21) = 64.$$

Notice also that Ball's solution (Figure 6.5) can be obtained from Gardner's solution (Figure 6.7) by placing a new set of queens on Gardner's unattacked squares numbered 1, 2, 3, 5, 7, 8, 10 and 11.

We know of no work on the non-attacking queens problem other than that reported by Ball, Gardner and Velucchi. For example, the following are open problems:

1. What are good lower and upper bounds for $Free(Q_n)$? Velucchi [72] reports the following values for $Free(Q_{k,n})$, i.e., the maximum number of unattacked squares in a placement of k queens on an order-n board:

$$Free(Q_{1,n}) = n^2 - 3n + 2;$$
$$Free(Q_{2,n}) = n^2 - 5n + 7;$$

and conjectures:

$$Free(Q_{3,n}) = n^2 - 6n + 9;$$
$$Free(Q_{4,n}) = n^2 - 6n + 8.$$

2. Is $Free(Q_n) < Free(Q_{n+1})$?

We also know of no work that has been done on the concept of non-dominating sets in graphs or on non-attacking sets of other chessboard pieces, although preliminary results on non-attacking chess pieces have been reported by Velucchi [72] and Reynolds[64]. We illustrate some of these results with computer generated optimal non-attacking pieces on 8×8 Kings, Bishops, Knights, Rooks and Grid graphs [64]. In these figures, a '0' represents an unattacked square, a '1' represents an attacked square and a capital letter represents a placed piece.

$Free(G_8) = 51$							
G	G	G	G	1	0	0	0
G	G	G	1	0	0	0	0
G	1	1	0	0	0	0	0
1	0	0	0	0	0	0	0
0	0	0	0	0	0	0	0
0	0	0	0	0	0	0	0
0	0	0	0	0	0	0	0
0	0	0	0	0	0	0	0

$Free(R_8) = 25$							
R	R	R	1	1	1	1	1
R	R	R	1	1	1	1	1
R	R	1	1	1	1	1	1
1	1	1	0	0	0	0	0
1	1	1	0	0	0	0	0
1	1	1	0	0	0	0	0
1	1	1	0	0	0	0	0
1	1	1	0	0	0	0	0
1	1	1	0	0	0	0	0

$Free(N_8) = 44$							
N	N	1	0	0	0	0	0
0	0	1	1	0	0	0	0
N	N	1	0	0	0	0	0
0	0	1	1	0	0	0	0
N	N	1	0	0	0	0	0
0	0	1	1	0	0	0	0
N	N	1	0	0	0	0	0
0	0	1	1	0	0	0	0

$Free(K_8) = 49$							
K	K	K	K	1	0	0	0
K	K	K	K	1	0	0	0
1	1	1	1	1	0	0	0
0	0	0	0	0	0	0	0
0	0	0	0	0	0	0	0
0	0	0	0	0	0	0	0
0	0	0	0	0	0	0	0
0	0	0	0	0	0	0	0

$Free(B_8) = 46$							
0	B	0	1	0	0	0	0
B	0	B	0	0	0	0	0
0	B	0	1	0	0	0	0
1	0	1	0	1	0	0	0
0	0	0	1	0	1	0	1
0	0	0	0	1	0	B	0
0	0	0	0	0	B	0	B
0	0	0	0	1	0	B	0

Velucchi reports that $Free(R_n) = (n-a)(n-b) + (n-a)b + (n-b)a$, where $a = \lceil\sqrt{n}\rceil$ and $b = \lceil\frac{n}{a}\rceil$.

Reynolds [64] has used computer search methods to determine the exact values of $Free(N_n)$, $Free(B_n)$, $Free(K_n)$ and $Free(G_n)$, for the knights, bishops, kings and grid graphs, as follows:

Free	1	2	3	4	5	6	7	8
Rook	0	0	1	4	6	12	16	25
Knight	0	2	4	8	14	23	33	44
Bishop	0	2	5	10	16	24	33	46
King	0	0	3	8	15	24	35	49
Grid	0	0	3	8	16	26	37	51

6.7   Domination on Variant Chessboards

Wallis [73] and Laskar and Wallis [56] have studied domination, independence and irredundance parameters for what they call triangulated chessboard graphs. In chessboard graphs the vertex set consists of n^2 ordered pairs of positive integers $\{(i,j) : 1 \le i \le n, 1 \le j \le n\}$. The vertices of a triangulated chessboard consist of the $n(n-1)/2$ ordered pairs of distinct integers $\{(i,j) : 1 \le i < j \le n\}$. It is necessary to redefine adjacencies for each of the chess pieces in triangulated chessboards.

For example, two triangulated rooks are adjacent if and only if their corresponding ordered pairs have a symbol in common. In Figure 6.9 the triangulated rooks graph TR_5 consists of the squares above the main diagonal, where the rook R at square $(2,4)$ is adjacent to squares marked '@' located at positions $(1,2)$, $(1,4)$, $(2,3)$, $(2,5)$, $(3,4)$ and $(4,5)$.

	1	2	3	4	5
1	x	@	–	@	–
2	x	x	@	R	@
3	x	x	x	@	–
4	x	x	x	x	@
5	x	x	x	x	x

Figure 6.9: A triangulated Rooks graph.

Considerable work has been done on Rooks graphs in higher dimensions (cf. Golomb and Posner [39], Rodemich [66], Johnson [51], Blokhuis and Lam [7] and Carnielli [11], to mention a few). For example, the result that $\gamma(R_n^3) = \lceil n^2/2\rceil$ was first proved no later than 1970 (cf. [66]).

More recently Wallis [73] and Koehler [52] have obtained results on 3- and k-dimensional chessboards. Koehler has also shown that $\gamma(R_n^3) = i(R_n^3) = \lceil n^2/2\rceil$.

Figure 6.10 provides an illustration of the minimum domination pattern in the 3-dimensional Rooks graphs, R_6^3, which consists of six copies of the 6×6 Rooks graphs R_6.

Level 1						Level 2						Level 3					
R	–	–	–	–	–	–	R	–	–	–	–	–	–	R	–	–	–
–	R	–	–	–	–	–	–	R	–	–	–	R	–	–	–	–	–
–	–	R	–	–	–	R	–	–	–	–	–	–	R	–	–	–	–
–	–	–	–	–	–	–	–	–	–	–	–	–	–	–	–	–	–
–	–	–	–	–	–	–	–	–	–	–	–	–	–	–	–	–	–
–	–	–	–	–	–	–	–	–	–	–	–	–	–	–	–	–	–

Level 1						Level 2						Level 6					
–	–	–	–	–	–	–	–	–	–	–	–	–	–	–	–	–	–
–	–	–	–	–	–	–	–	–	–	–	–	–	–	–	–	–	–
–	–	–	–	–	–	–	–	–	–	–	–	–	–	–	–	–	–
–	–	–	R	–	–	–	–	–	–	R	–	–	–	–	–	–	R
–	–	–	–	R	–	–	–	–	–	–	R	–	–	–	R	–	–
–	–	–	–	–	R	–	–	–	R	–	–	–	–	–	–	R	–

Figure 6.10: A minimum dominating set in R_6^3.

We list results involving R_n^k:

Wallis [73]: $\beta(R_n^k) = n^{k-1}$.

Koehler [52]: $\Gamma(R_n^3) \geq 3(n-2)^2 + 4$, for $n > 3$.

Conjecture (Koehler): $ir(R_n^3) = \gamma(R_n^3)$.

Koehler also obtained several values, by exhaustive computer search for the domination number of the 3-dimensional Bishops graph, $\gamma(B_n^3)$:

$$\gamma(B_1^3) = 1; \quad \gamma(B_2^3) = 2; \quad \gamma(B_3^3) = 3; \quad \gamma(B_4^3) = 8;$$
$$\gamma(B_5^3) \leq 13; \quad \gamma(B_6^3) \leq 18; \quad \beta(B_7^3) \leq 27.$$

A minimum dominating set of bishops in B_4^3 is given in Figure 6.11, where a square marked i is dominated by the bishop B_i.

Level 1				Level 2				Level 3				Level 4			
7	5	1	2	6	7	5	4	1	6	7	$B5$	2	3	$B6$	$B7$
8	1	2	3	1	8	4	5	2	3	5	7	4	6	7	$B8$
$B1$	2	3	5	2	1	6	7	4	5	1	8	6	7	8	1
$B2$	$B3$	6	7	$B4$	2	3	8	5	4	2	3	7	8	4	2

Figure 6.11: A minimum dominating set of bishops in B_4^3.

6.8   Total Domination Problems

As early as 1892, W. W. Rouse Ball [4] discussed the problem of placing a minimum number of chess pieces of a given kind in such a way that every square and every piece is attacked by another piece. In graph theoretic terminology, this is the problem of determining the total domination number $\gamma_t(G)$ of the various chessboard graphs. Ball obtained the following values for the standard order-8 board:

$\gamma_t(Q_8) = 5$; $\gamma_t(B_8) = 10$; $\gamma_t(N_8) = 14$; and $\gamma_t(R_8) = 8$.

It is immediately obvious that $\gamma_t(R_n) = n$, as any set of n rooks on the same row is a minimum cardinality total dominating set of R_n.

Cockayne, Gamble and Shepherd [18] determined that

$$\gamma_t(B_n) = 2\lfloor 2(n-1)/3 \rfloor \text{ for } n \geq 3.$$

Garnick and Nieuwejaar [35] have determined the following bounds for total domination numbers in Kings graphs, Knights graphs and Grid graphs, for more general $m \times n$ boards:

(i) $mn/7 \leq \gamma_t(K_{m,n}) \leq (mn + 2n + 89)/7$;

(ii)   for $m = n \pmod 2$, $mn/8 \leq \gamma_t(N_{m,n}) \leq (mn + 5m + 6n + 56)/8$, and
       for m odd, n even, $mn/8 \leq \gamma_t(N_{m,n}) \leq (mn + 5m + 5n + 43)/8$;

(iii)   for $m, n > 2$, $\gamma_t(G_{m,n}) \leq (mn + 2m + 2n + 5)/4$.

Garnick and Nieuwejaar also determined most values of $\gamma_t(K_n), \gamma_t(N_n)$ and $\gamma_t(G_n)$ for $2 \leq n \leq 12$ by computer search, and obtained reasonably good upper bounds on these numbers for $n \leq 30$.

We conclude this section by commenting that we know of no work on the total domination number of the Queens graphs nor any other work on the upper total domination numbers of chessboard graphs. It would also be interesting to see some results on the connected domination numbers $\gamma_c(G)$ of chessboard graphs, particularly on Queens graphs, in light of the simple inequality chain: $\gamma(G) \leq \gamma_t(G) \leq \gamma_c(G)$.

6.9   Algorithms Involving Weighted Rooks

Hedetniemi, Hedetniemi and Jacobson [46] have obtained a variety of algorithmic results concerning weighted Rooks graphs. Assume that the squares of an order-n chessboard have been weighted with non-negative integers. The following is a brief summary of algorithms that have been designed:

(i) an $O(n^3)$ weighted matching algorithm for finding a minimum or maximum weight of a maximal independent set of n rooks in R_n;

(ii) an $O(n^2)$ greedy algorithm for finding a minimum or maximum weight of a minimal dominating set of n rooks in R_n; this algorithm simply finds, for example, the smallest weight in each row of R_n, then finds the smallest weight in each column of R_n, and takes the smaller of the sum of the weights chosen in each row and the sum of the weights chosen in each column; the smaller sum is the minimum weight of a minimal dominating set of n rooks in R_n; the correctness of this algorithm is based on the observation that any minimal dominating set of rooks in R_n must either contain one rook in every row or one rook in every column;

(iii) an $O(n^{10})$ algorithm for finding a minimum or maximum weight of a maximal irredundant set of n rooks in R_n;

(iv) an $O(n^3)$ algorithm for finding a minimum or maximum weight of a maximal irredundant set of $2n - 4$ rooks in R_n;

(v) an $O(\log^3 n)$ randomized parallel algorithm for finding a minimum or maximum weight of an independent dominating set of rooks in R_n, using $O(n^{5.5})$ operations;

(vi) an $O(\log n)$ parallel algorithm for finding a minimum or maximum weight of a minimal dominating set of rooks in R_n, using $O(n^2)$ processors and $O(n^2)$ operations.

6.10   Other Results

Wilf [76] has studied the number of ways, $f_m(n)$, that a maximum number of nonattacking kings can be placed on a $2m \times 2n$ chessboard. The maximum number of nonattacking kings is mn, since each 2×2 array of squares can have at most one king. Wilf has shown that for each $m = 1, 2, 3, ...$, there are constants $c_m > 0$, d_m, and $0 \le \omega_m < m + 1$ such that $f_m(n) = (c_m n + d_m)(m + 1)^n + O(\omega_m^n)$, $(n \to \infty)$.

This problem has also been studied by Larsen [54], who shows that for the number $f(n)$ of configurations of n^2 mutually non-attacking kings on a $2n \times 2n$ board, the term $\log f(n)$ grows like $2n/\log n - 2n/\log 2$.

Fisher [28] has determined the 2-packing number of complete grid graphs of all sizes, thereby verifying a conjecture of Hare and Hare [42]. A 2-packing of a grid is a set of vertices, no two of which are within distance two of each other; $\rho_2(G)$ is the maximum cardinality of a 2-packing in a graph G. Let $\rho_2(G_{m,n})$ equal the 2-packing number of the $m \times n$ complete grid graph. Fisher's result is that $\rho_2(G_{m,n}) = \lceil mn/5 \rceil$ when $m, n \ge 8$ except when $(m, n) = (8, 10)$ or $(10, 8)$. Fisher also determined all other values of $\rho_2(G_{m,n})$.

A variant of dominating queens was introduced by Scott Kim and presented in a 1981 issue of Scientific American [33]: what is the greatest number of queens which can be placed on an order-n board under the condition that each queen attacks precisely k others, where $k = 1, 2, 3$, or 4? For $k = 0$ this is the N-queens problem. Let $Q_k(n)$ denote this maximum value. In this problem it is assumed that if three queens are placed on a row, then the middle queen attacks the other two, but the leftmost queen can only attack the middle queen. Thus, queens can "block" attacks by other queens in the same row, column or diagonal. The latest results on this problem have been presented by Hayes [45], who lists the following known results:

$$
\begin{aligned}
k = 2 \quad & n = 1 \quad && \text{no solution} \\
& n = 2 \quad && Q_2(2) = 3 \\
& n \geq 3 \quad && Q_2(n) = 2n - 2
\end{aligned}
$$

$$
\begin{aligned}
k = 3 \quad & n = 1 \quad && \text{no solution} \\
& n \geq 2 \quad && Q_3(n) \leq 2\lfloor (6n - 2)/5 \rfloor
\end{aligned}
$$

$$
\begin{aligned}
k = 4 \quad & n = 1, 2, 3 \quad && \text{no solution} \\
& n = 4, 5 \quad && Q_4(n) = 8, 11 \\
& n \geq 6 \quad && Q_4(n) = 3n - 3
\end{aligned}
$$

A solution for $Q_3(8) = 18$ is given in Figure 6.12.

Q	$-$	Q	$-$	$-$	Q	Q	Q
Q	$-$	$-$	$-$	$-$	$-$	$-$	$-$
$-$	Q	$-$	$-$	$-$	$-$	Q	$-$
$-$	$-$	$-$	$-$	Q	$-$	$-$	$-$
Q	$-$	$-$	$-$	$-$	$-$	$-$	Q
$-$	$-$	Q	$-$	$-$	$-$	$-$	$-$
$-$	$-$	$-$	$-$	$-$	$-$	$-$	Q
$-$	Q	Q	Q	$-$	Q	$-$	Q

Figure 6.12: $Q_3(8) = 18$.

It is interesting to observe that for $k \geq 5$, there are no solutions to this problem. In any solution in which, say, each queen attacks precisely 5 other queens, consider the location of the top-most and left-most queen. Since this queen does not attack any queen to its left (in its row) or above it, it can only attack a queen in four directions (east, southeast, south or southwest), i.e., it can attack at most four other queens.

This problem suggests a variant of the basic n-domination problem, in which a vertex in a dominating set S can dominate an adjacent vertex in $V - S$ only if it is not 'blocked' from doing so by the presence of another vertex in S. To the best of our knowledge this variant has not been studied in the domination literature.

6.11 Review of Open Problems

Survey I listed approximately 50 open problems concerning dominating, irredundant and independent sets of pieces on chessboards. Although nearly all of these remain open, we will include in this section some that we think are the most interesting and challenging.

Problems involving queens on chessboards will perhaps always be more interesting and difficult than corresponding problems for other chess pieces. Next most difficult would be problems involving knights. Problems involving rooks, bishops and kings are usually much simpler. Problems involving Grid graphs vary considerably in complexity.

Among the most interesting open problems are the following:

1. (a) What is the value of $\gamma(Q_n)$? This is perhaps the oldest and most difficult problem of them all, dating back to the 1850's. (b) Can you determine the exact value of $\gamma(Q_n)$ for any new values of n? (c) Can you establish a good lower bound for $\gamma(Q_n)$?

2. Is the function $\gamma(Q_n)$ monotonic non-decreasing? That is, is $\gamma(Q_n) \leq \gamma(Q_{n+1})$? This would appear to be obviously true, but no proof has yet been found. For a solution to this problem, a \$100 prize has been offered by S. T. Hedetniemi.

3. Is $\gamma(Q_{4k+1}) = 2k + 1$ for all $k \geq 1$? This is true for $k \leq 15$.

4. For sufficiently large values of n, is $\gamma(Q_n) = i(Q_n)$?

5. Do maximal irredundant sets of queens, minimal dominating sets of queens and maximal independent sets of queens interpolate? For example, if one can find minimal dominating sets of queens of sizes i and k, can one also find minimal dominating sets of size j, for every j, $i < j < k$?

6. Is the value $\Gamma(Q_n)$ monotonic non-decreasing?

7. Can you produce a good upper bound for $IR(Q_n)$? For example, it seems very likely that $IR(Q_n) \leq 5n$ and possibly even $IR(Q_n) \leq 4n$.

8. Can you provide a good upper bound for $\Gamma(Q_n)$?

9. In general, we know very little at all about $ir(Q_n), ir(K_n), ir(N_n)$ and $ir(G_n)$. In particular, we know of no good bounds, upper or lower, for these numbers, we do not know if maximal irredundant sets on these chessboards interpolate and we have yet to establish the expected result that these irredundance functions are monotonic non-decreasing.

10. We know very little about irredundant sets, dominating sets and independent dominating sets in Knights graphs. In particular no good upper bounds are known for $ir(N_n), \gamma(N_n)$ or $i(N_n)$; it has not been shown that these three functions are monotonic non-decreasing; and it has not been shown that maximal irredundant sets, minimal dominating sets and independent dominating sets of knights interpolate.

11. Is there a formula for $\gamma(G_{m,n})$? Or perhaps a finite number of formulas?

Bibliography

[1] B. Abramson and M. Yung, Divide and conquer under global constraints: a solution to the N-queens problem. *J. Parallel Distrib. Comput.* 6 (1989) 649–662.

[2] W. Ahrens, Mathematische unterhaltungen und spiele. *Leipzig-Berlin* (1910).

[3] S. Ainley, *Mathematical Puzzles.* G. Bell and Sons, Ltd., UK (1977).

[4] W. W. Rouse Ball, *Mathematical Recreation and Problems of Past and Present Times.* MacMillan, London (1892).

[5] B. Bernhardsson, Explicit solutions to the N-queens problem for all N. *SIGART Bull.* 2(2) (1991) 7.

[6] M. Bezzel, Schachfreund. *Berliner Schachzeitung* 3 (September 1848) 363.

[7] A. Blokhuis and C. W. H. La, More coverings by rook domains. *J. Combin. Theory Ser. A* 36 (1984) 240–244.

[8] A. P. Burger, E. J. Cockayne, and C. M. Mynhardt, Domination and irredundance in the queen's graph. Preprint (1994).

[9] A. P. Burger, C. M. Mynhardt, and E. J. Cockayne, Domination numbers for the queen's graph. *Bull. Inst. Combin. Appl.* 10 (1994) 73–82.

[10] P. J. Campbell, Gauss and the eight queens problem: a study in miniature of the propagation of historical error. *Historia Math.* 4 (1977) 397–404.

[11] W. A. Carnielli, On covering and coloring problems for rook domains. *Discrete Math.* 57 (1985) 9–16.

[12] T. Y. Chang, *Domination numbers of grid graphs.* Ph.D. Dissertation, University of South Florida (1992).

[13] T. Y. Chang and W. E. Clark, The domination numbers of the $5 \times n$ and the $6 \times n$ grid graphs. To appear in *J. Graph Theory.*

[14] T. Y. Chang, W. E. Clark, and E. O. Hare, Domination numbers of complete grid graphs, i. *Ars Combin.* 38 (1994) 97–111.

[15] B. N. Clark, C. J. Colbourn, and D. S. Johnson, Unit disk graphs. *Discrete Math.* 86 (1990) 145–164.

[16] E. J. Cockayne, O. Favaron, C. Payan, and A. G. Thomason, Contributions to the theory of domination, independence and irredundance in graphs. *Discrete Math.* 53 (1981) 249–258.

[17] E. J. Cockayne, B. Gamble, and B. Shepherd, Domination of chessboards by queens in a column. *Ars Combin.* 19 (1985) 105–118.

[18] E. J. Cockayne, B. Gamble, and B. Shepherd, Domination parameters for the bishops graph. *Discrete Math.* 58 (1986) 221–227.

[19] E. J. Cockayne, E. O. Hare, S. T. Hedetniemi, and T. V. Wimer, Bounds for the domination of grid graphs. *Congr. Numer.* 47 (1985) 217–228.

[20] E. J. Cockayne and S. T. Hedetniemi, On the diagonal queens domination problem. *J. Combin. Theory B* 42 (1986) 137–139.

[21] O. Demirors, N. Rafraf, and M. Tanik, Obtaining N-queens solutions from magic squares and constructing magic squares from N-queens solutions. *J. Recreational Math.* 24 (4) (1992) 272–280.

[22] M. Eisenstein, C. M. Grinstead, B. Hahne, and D. Van Stone, The queen domination problem. *Congr. Numer.* 91 (1992) 189–193.

[23] C. Erbas and M. M. Tanik, Storage schemes for parallel memory systems and the N-queens problem. In *Proc. 15th Ann. ASME ETCE Conference, Computer Applications Symp., Houston, TX* (Jan. 26-30, 1992).

[24] C. Erbas, M. M. Tanik, and Z. Aliyazicioglu, Linear congruence equations for the solutions of the N-queens problem. *Inform. Process. Lett.* 41 (1992) 301–306.

[25] B. J. Falkowski and L. Schmitz. A note on the queens problem. *Inform. Process. Lett.* 23 (1986) 39–46.

[26] O. Favaron and J. Puech, Domination and irredundance in the kings graph. Preprint (1994).

[27] R. Finkel and U. Manbe, Dib - a distributed implementation of backtracking. *ACM Trans. Programming Languages Systems* 9 (1987) 235–256.

[28] D. C. Fisher, The domination number of complete grid graphs. Preprint (1993).

[29] G. H. Fricke, S. M. Hedetniemi, S. T. Hedetniemi, A. A. McRae, C. K. Wallis, M. S. Jacobson, W. W. Martin, and W. D. Weakley, Combinatorial problems on chessboards: a brief survey. *Graph Theory, Combinatorics, and Applications* 1 (1995) 507–528.

[30] G. H. Fricke and D. Pritikin, Irredundance in kings graphs. Preprint (1994).

[31] M. Gardner, Mathematical games. *Scientific Amer.* (Feb. 1978) 20. (cf. also Scientific Amer., March 1978, p. 27).

[32] M. Gardner, Mathematical games. *Scientific Amer.* (May 1972) 114-115. (cf. also *Scientific Amer.*, June, 1972, p.117).

[33] M. Gardner, Mathematical games. *Scientific Amer.* (June 1981) 22–32.

[34] M. Gardner, Wheels, Life and Other Mathematical Amusement. Freeman (1983).

[35] D. Garnick and N. Nieuwejaar, Total domination of the $m \times n$ chessboard by kings, crosses and knights. Preprint (1992).

[36] P. B. Gibbons and J. A. Webb, Some new results for the queens domination problem. To appear in *Australian J. Combin.*

[37] J. Ginsburg, Gauss's arithmetization of the problem of 8 queens. *Scripta Math* 5(1) (1938) 63–66.

[38] S. W. Golomb and L. D. Baumert, Backtrack programming. *J. Assoc. Comput. Mach.* 12(4) (1965) 516–524.

[39] S. W. Golomb and E. C. Posner, Rook domains, Latin squares, affine planes, and error-distributing codes. *IEEE Trans. Inform. Theory* IT-10 (1964) 196–208.

[40] R. L. Graham, private communication (1995).

[41] C. M. Grinstead, B. Hahne, and D. Van Stone, On the queen domination problem. *Discrete Math.* 86 (1990) 21–26.

[42] E. O. Hare and W. R. Hare, k-packing of $P_m \times P_n$. *Congr. Numer.* 84 (1991) 33–39.

[43] E. O. Hare and S. T. Hedetniemi, A linear algorithm for computing the knight's domination number of a $k \times n$ chessboard. *Congr. Numer.* 59 (1987) 115–130.

[44] E. O. Hare, S. T. Hedetniemi, and W. R. Hare, Algorithms for computing the domination number of $k \times n$ complete grid graphs. *Congr. Numer.* 55 (1986) 81–92.

[45] P. Hayes, A problem of chess queens. *J. Recreational Math.* 24(4) (1992) 264–271.

[46] S. M. Hedetniemi, S. T. Hedetniemi, and M. S. Jacobson, private communication (1994).

[47] H. Hitotumatsu and K. Noshita, A technique for implementing backtrack algorithms and its applications. *Inform. Process. Lett.* 8(4) (1979) 174–175.

[48] E. J. Hoffman, J. C. Loessi, and R. C. Moore, Constructions for the solution of the m queens problem. *Math. Mag.* 42 (1969) 66–72.

[49] D. H. Hollander, An unexpected two-dimensional space-group containing seven of the twelve basic solutions to the eight queens problem. *J. Recreational Math.* 6(4) (1973) 287–291.

[50] M. S. Jacobson and L. F. Kinch, On the domination number of products of a graph *I*. *Ars Combin.* 10 (1983) 33–44.

[51] S. M. Johnson, A new lower bound for coverings by rook domains. *Utilitas Math.* 1 (1972) 121–140.

[52] E. G. Koehler, Domination problems in three-dimensional chessboard graphs. Masters Thesis, Clemson University (1994).

[53] M. Kraitchik, *Mathematical Recreations*. Norton, New York (1942).

[54] M. Larsen, The problem of kings. *Electron. J. Combin.* 2 (1995).

[55] L. C. Larson, A theorem about primes proved on a chessboard. *Math. Mag.* 50(2) (1977) 69–74.

[56] R. Laskar and C. Wallis, Domination parameters of graphs of three-class association schemes and variations of chessboard graphs. *Congr. Numer.* 100 (1994) 199–213.

[57] E. Lucas, La solution complete du probleme des huit reines. *Revue Scientifique* 9 (1880) 948–953.

[58] E. Lucas, Le probleme des huites reines au jeu des echecs. *Recreations Mathematiques, Paris* (1882) 57–86.

[59] B. A. Nadel, Representation selection for constraint satisfaction: a case study using the N-queens. *IEEE Expert* 5(3) (1990) 16–23.

[60] F. Nauck, Briefwechseln mit allen fur alle. *Illustrirte Zeitung* 15 (No. 377, Sept. 21, 1850) 182.

[61] F. Nauck, Schach: eine in das gebiet der mathematik fallende aufgabe von herrn dr. nauck in schleusingen. *Illustrirte Zeitung* 14 (No. 365, June 29, 1850) 416.

[62] P. Naur, An experiment on program development. *BIT* 12 (1971) 347–365.

[63] M. Reichling, A simplified solution for the N queens problem. *Inform. Process. Lett.* 25 (1987) 253–255.

[64] R. Reynolds, private communication (1995).

[65] I. Rivin and R. Zabih, A dynamic programming solution to the N-queens problem. *Inform. Process. Lett.* 41 (1992) 253–256.

[66] E. R. Rodemich, Coverings by rook domains. *J. Combin. Theory* 9 (1970) 117–128.

[67] J. S. Rohl, A faster lexicographical N queens algorithm. *Inform. Process. Lett.* 17 (1983) 231–233.

[68] J. S. Rohl, Generating permutations by choosing. *Comput. J.* 21(4) (1978) 302–305.

[69] R. Sosic and J. Gu, A polynomial time algorithm for the N-queens problem. *SIGART Bull.* 2(2) (1990) 7–11.

[70] H. S. Stone and J. M. Stone, Efficient search techniques - an empirical study of the N-queens problem. *IBM J. Res. Develop.* 31 (1987) 464–474.

[71] R. W. Topor, Fundamental solutions of the eight queens problem. *BIT* 22 (1982) 42–52.

[72] M. Velucchi, private communication (1995).

[73] C. K. Wallis, *Domination parameters of line graphs, of designs and variations of chessboard graphs.* Ph.D. Dissertation, Clemson University (1994).

[74] W. D. Weakley, Domination in the Queen's graph. *Graph Theory, Combinatorics, and Applications* 2 (1995) 1223–1232.

[75] W. D. Weakley, private communication (1991).

[76] H. S. Wilf, The problem of the kings. *Electron. J. Combin.* 2 (1995) 1–7.

[77] N. Wirth, *Systematic programming - an introduction.* Prentice-Hall (1973).

[78] A. M. Yaglom and I. M. Yaglom, Challenging mathematical problems with elementary solutions. *Volume 1: Combinatorial Analysis and Probability Theory* (1964).

Chapter 7

Domination in Cartesian Products: Vizing's Conjecture

Bert Hartnell
Saint Mary's University
Halifax, Nova Scotia
Canada B3H 3C3

Douglas F. Rall
Furman University
Greenville, SC 29613
U.S.A.

7.1 Introduction

Studies of how particular graphical parameters interact with graph products have been the impetus for several areas of research in graph theory. Notable examples of this are the Shannon capacity of a graph, which is a certain limiting value involving the vertex independence number of strong product powers of a graph, and Hedetniemi's coloring conjecture for the categorical product. One of the oldest unsolved problems in domination theory also comes from the area of graph products. Recently there has been renewed interest in a question first asked by V. G. Vizing [17] in 1963. Five years later he offered it as a conjecture [18], and in 1978 E. Cockayne included the conjecture as Problem 1 in his survey article [3].

Vizing's Conjecture: The domination number of the Cartesian product of any two graphs is at least as large as the product of their domination numbers.

In this chapter our main focus will be this conjecture; both progress and methods of attack on it. In addition we will briefly mention some recent work and open questions on other domination-type parameters and other graph products.

We will follow the approach of Imrich and Izbicki [11] and of Nowakowski and Rall [14]. By a *graph product*, denoted by $\otimes$, of two graphs G and H we mean a new graph, $G \otimes H$, in which the vertex set $V(G \otimes H)$ is the Cartesian product - as sets - of $V(G)$ and $V(H)$. That is, $V(G \otimes H) = V(G) \times V(H) = \{(x,y) | x \in V(G)$ and $y \in V(H)\}$. The edge set $E(G \otimes H)$ is completely determined by the adjacency, equality or non-adjacency of the vertices in the two *factors*, G and H. Of the resulting 256 products the most often studied are the following four: Cartesian, categorical, strong and lexicographic. These four graph products, as well as several others, have been studied to varying degrees. At times some confusion has arisen because of the different names used to refer to the same product. Notation used to indicate a particular graph product has also been inconsistent.

Two vertices (x_1, y_1) and (x_2, y_2) in $V(G) \times V(H)$ are adjacent in
(1) the *Cartesian product* $G \square H$ if $x_1 = x_2$ and $y_1 y_2 \in E(H)$, or if $y_1 = y_2$ and $x_1 x_2 \in E(G)$;
(2) the *categorical product* $G \times H$ if $x_1 x_2 \in E(G)$ and $y_1 y_2 \in E(H)$;
(3) the *strong product* $G \boxtimes H$ if $x_1 = x_2$ and $y_1 y_2 \in E(H)$, or if $y_1 = y_2$ and $x_1 x_2 \in E(G)$, or if $x_1 x_2 \in E(G)$ and $y_1 y_2 \in E(H)$;
(4) the *lexicographic product* $G \bullet H$ if $x_1 x_2 \in E(G)$, or if $x_1 = x_2$ and $y_1 y_2 \in E(H)$.

It is clear from the above definitions that for any pair of graphs, $G \boxtimes H$ is a spanning subgraph of $G \bullet H$ while each of $G \square H$ and $G \times H$ is a spanning subgraph of $G \boxtimes H$. The following result follows immediately from this observation. There are, of course, many other similar results for other domination-type parameters (upper and lower irredundance, total domination, independence and independent domination) as well as some of the related partition parameters (chromatic, achromatic, domatic and adomatic numbers). See [14].

Theorem 7.1 [14] *For any two graphs G and H,*

$$\gamma(G \bullet H) \leq \gamma(G \boxtimes H) \leq min\{\gamma(G \square H), \gamma(G \times H)\}.$$

The Cartesian, categorical, strong and lexicographic products are associative in the sense that for any graphs F, G and H, the function

$$f : (V(F) \times V(G)) \times V(H) \to V(F) \times (V(G) \times V(H))$$

defined by $f((x, y), z) = (x, (y, z))$ is an isomorphism for any of these four products. In addition, the first three are also commutative under the obvious isomorphism interpretation of this concept. Considerations such as associativity and commutativity are important if, for example, one investigates powers of a graph with respect to some product. In the Open Questions section we will include some questions and conjectures on some of the other three products above, but until then the only graph product we will consider is the Cartesian product. In fact, until then we concern ourselves mainly with Vizing's conjecture, which restated in terms of our notation is as follows.

Vizing's Conjecture: $\gamma(G \square H) \geq \gamma(G)\gamma(H)$ for every pair of graphs G and H.

We will say that a graph G *satisfies Vizing's conjecture* or that *Vizing's conjecture is true* for G if the above inequality holds for every graph H.

If A and B are nonempty subsets of $V(G)$, we will say A *dominates* B if every vertex of B is in $N[A]$, the closed neighborhood of A. When A dominates $V(G)$ and $|A| = \gamma(G)$ we call A a γ-*set* of G. A graph G is said to be *edge-maximal with respect to domination* if $\gamma(G + uv) < \gamma(G)$ for every pair of nonadjacent vertices u and v in G. A 2-*packing* of G is a subset I of $V(G)$ such that $N[x] \cap N[y] = \emptyset$ for any pair $x, y \in I$. The 2-*packing number* of G, $\rho_2(G)$, is the cardinality of a largest two-packing of G. We use $|G|$ to denote the order of G.

The context should make it clear what is meant, but if A and B are sets of vertices with no edge structure implied, then $A \times B$ will denote the Cartesian product of A and B as sets. For $B \subseteq V(H)$, G_B will denote $\langle V(G) \times B \rangle$, the subgraph of $G \square H$ induced by $V(G) \times B$. For $A \subseteq V(G)$, $H_A = \langle A \times V(H) \rangle$. When $B = \{y\}$ we simplify this to G_y and call G_y a G-*level* of $G \square H$. Similarly, $H_x = H_{\{x\}}$ is an H-level of $G \square H$. Two G-levels, G_y and G_z, are called *neighboring levels* if $yz \in E(H)$. The reason that G-levels and H-levels are important in what follows is that for $(x, y) \in V(G \square H)$ the closed neighborhood of (x, y) is a subset of $G_y \cup H_x$. Note also that $G_y \simeq G$ and $H_x \simeq H$.

7.2 Observations and Bounds

Many of the published results which support the truth of Vizing's conjecture are of the following general type. "If G is a graph satisfying property $\mathcal{P}$, then G satisfies Vizing's conjecture." Because of the nature of property $\mathcal{P}$, it is often the case that the proofs of these results are more easily obtained if one assumes that G has many edges. See, for example, [1], [9], and [10]. The following lemma shows that it is not always necessary to assume that all these edges are present.

Lemma 7.2 [9] *If G satisfies Vizing's conjecture and K is a spanning subgraph of G such that $\gamma(K) = \gamma(G)$, then K also satisfies Vizing's conjecture.*

Proof. The spanning subgraph K of G can be obtained from G by a finite sequence of edge removals, each of which leaves the domination number unchanged. Assume $e \in E(G)$ is such that $\gamma(G - e) = \gamma(G)$. The graph $(G - e) \square H$ is a spanning subgraph of $G \square H$ for every graph H. Therefore $\gamma((G - e) \square H) \geq \gamma(G \square H) \geq \gamma(G)\gamma(H) = \gamma(G - e)\gamma(H)$. That is, $G - e$ also satisfies Vizing's conjecture. □

Under certain circumstances it is also possible to conclude that Vizing's conjecture is true for a vertex-deleted subgraph of a graph satisfying the conjecture.

Theorem 7.3 [16] *If G is a graph which is known to satisfy Vizing's conjecture and x is a vertex of G such that $\gamma(G - x) < \gamma(G)$, then $G - x$ satisfies Vizing's conjecture.*

Proof. Let $K = G - x$ where G and x satisfy the hypotheses of the theorem. Then $\gamma(K) = \gamma(G) - 1$. Suppose there exists a graph H such that $\gamma(K \,\square\, H) < \gamma(K)\gamma(H)$. Let A be a γ-set for $K \,\square\, H$, and let B be a γ-set for H. Set $D = A \cup \{(x, b) | b \in B\}$. It is then clear that D dominates $G \,\square\, H$ and yet

$$|D| = |A| + |B| < \gamma(K)\gamma(H) + \gamma(H) = (\gamma(K) + 1)\gamma(H) = \gamma(G)\gamma(H).$$

But then $\gamma(G \,\square\, H) < \gamma(G)\gamma(H)$, which contradicts the assumption about G. Therefore, $G - x$ satisfies Vizing's conjecture. $\square$

An obvious question that can be raised here is the following:

QUESTION 1: Is the converse of Theorem 7.3 true? That is, suppose a graph K satisfies Vizing's conjecture. Let S be any set of vertices in K such that no member of S belongs to any γ-set of K and such that $\gamma(K - S) = \gamma(K)$. Form a graph G from K by adding a new vertex x and all edges of the form xs where $s \in S$. Must G satisfy Vizing's conjecture?

It is unlikely that this question has a negative answer. Indeed, a negative answer would immediately yield a counterexample to the conjecture, at which point there would probably be no interest in the question itself. However, as we will see in Section 7.4, a positive answer to the above would fit nicely into an attempt to prove the conjecture by a finite set of constructive operations.

A familiar graph which illustrates some of the properties of domination in a Cartesian product is the so-called "rooks graph." This graph of order 64 is obtained by representing each square on a standard chessboard by a vertex and making two vertices adjacent if and only if a rook placed on one of the corresponding squares can capture a piece on the other square (assuming, of course, no intervening square is occupied). This rooks graph is simply the Cartesian product $K_8 \,\square\, K_8$. If seven or fewer rooks are placed on the chessboard then there will be at least one horizontal row and at least one vertical column, neither of which contains a rook. A piece could be placed on the square at the intersection of this row and this column without being captured by the rooks on the board. Of course we can generalize this to a chessboard which is $m \times n$, and it is clear that at least $min\{m, n\}$ rooks would be required to dominate the board.

One way of picturing Cartesian product graphs is to imagine the vertices laid out in such a row and column fashion. Of course, unless the corresponding factors are complete graphs, the rows and columns would not induce cliques as in the rooks graph. Let $v_1, v_2, \ldots, v_n$ be the vertices of G and let $u_1, u_2, \ldots, u_m$ be the vertices of H. Assume D is any dominating set for $G \,\square\, H$. If $|D| < min\{n, m\}$, then there exists a G-level, G_{u_i}, and a H-level, H_{v_j}, such that $D \cap H_{v_j} = \emptyset = D \cap G_{u_i}$. But then the vertex $(v_j, u_i) \notin N[D]$, which contradicts the assumption about D. This establishes the following lower bound, which is due to El-Zahar and Pareek.

Lemma 7.4 [4] *For any pair of graphs G and H, $\gamma(G \,\square\, H) \geq min\{|G|, |H|\}$.*

The following somewhat unusual result illustrates one possible use of the above lemma. While it does not establish the conjecture for any graph, it does imply that for any given graph there will always be infinitely many other graphs so that the pair satisfies the inequality in the conjecture.

Corollary 7.5 [16] *Let H be an arbitrary graph. There exists a positive integer $r = r(H)$ such that if G is any graph with $\gamma(G) \leq r$ and $|G| \geq |H|$, then $\gamma(G \square H) \geq \gamma(G)\gamma(H)$.*

Proof. There is a constant c such $\gamma(H) = c\,|H|$ and $0 < c \leq \frac{1}{2}$. Let $r = \lfloor \frac{1}{c} \rfloor$. Now for any graph G with $|G| \geq |H|$ and $\gamma(G) \leq r$, it follows by Lemma 7.4 that

$$\gamma(G \square H) \geq min\{|G|, |H|\} = |H| = \frac{1}{c}\,(c\,|H|) \geq r\,\gamma(H) \geq \gamma(G)\gamma(H).$$

$\square$

By considering any dominating set for a Cartesian product, $G \square H$, and cleverly counting how it intersects the G-levels, Jacobson and Kinch [12] established the following lower bound and then used it to prove that $\gamma(C_t \square C_s) \geq \gamma(C_t)\gamma(C_s)$. In addition, they concluded that any cycle of length $3k$ satisfies Vizing's conjecture.

Theorem 7.6 [12] *For any graphs G and H, $\gamma(G \square H) \geq \frac{|H|}{\Delta(H)+1}\gamma(G)$.*

At the other extreme, a dominating set for $G \square H$ can be constructed by using a "copy" of a γ-set for H in each H-level or a "copy" of a γ-set for G in each G-level. This establishes the following upper bound due to Vizing.

Theorem 7.7 [17] *For any graphs G and H, $\gamma(G \square H) \leq min\{\gamma(G)|H|, |G|\gamma(H)\}$.*

Dominating sets in a graph are precisely those subsets of vertices which intersect each closed neighborhood in the graph. In this sense the entire study of domination in graphs reduces to an investigation of how the closed neighborhoods fit together. A natural place to start when trying to bound the domination number of a Cartesian product in terms of parameter values of the factors is thus to consider the closed neighborhoods of $G \square H$.

Lemma 7.8 [13] *If D is any dominating set for $G \square H$ and x is any vertex of G, then $|D \cap (N[x] \times V(H))| \geq \gamma(H)$.*

Proof. Let $N[x] = \{x, u_1, \ldots, u_k\}$ and let $p : N[x] \times V(H) \to H_x$ be the projection map defined by $p(u_i, h) = (x, h)$. Since D is a dominating set for $G \square H$, it follows that $p(D \cap (N[x] \times V(H)))$ dominates H_x. But $H_x \simeq H$, and so the result is established. $\square$

Consider the graph G in Figure 7.1. The subset of vertices $\{u, x, z\}$ is a γ-set for G. Note that within G we can find three disjoint closed neighborhoods, for example $N[a]$, $N[b]$, and $N[c]$. Let H be any graph and assume that D is a dominating set for $G \,\square\, H$. The closed neighborhood of any vertex in H_a is a subset of $H_{N[a]}$. Similar statements can be made about the closed neighborhoods of vertices in H_b and H_c. But $H_{N[a]}$, $H_{N[b]}$ and $H_{N[c]}$ are disjoint, and by the preceding lemma each must contain at least $\gamma(H)$ elements from D. Thus $|D| \geq 3\gamma(H)$, and so G satisfies Vizing's conjecture.

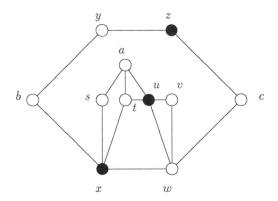

Figure 7.1: Graph G with $\rho_2(G) = \gamma(G)$.

In the above example, $\rho_2(G) = \gamma(G)$. Since any dominating set in an arbitrary graph must have a nonempty intersection with each closed neighborhood it follows that $\rho_2(G) \leq \gamma(G)$ for every G. Of course, an arbitrary graph requires more vertices to dominate it than can be included in a 2-packing. Jacobson and Kinch [13] observed that if $S = \{h_1, h_2, \ldots, h_r\}$ is a maximum 2-packing of H, then $\{V(G) \times N[h_1], \ldots, V(G) \times N[h_r]\}$ is a set of r pairwise disjoint subsets of $G \,\square\, H$. By combining this with the result in Lemma 7.8, they established the following lower bound.

Theorem 7.9 [13] *For any pair of graphs G and H,*

$$\gamma(G \,\square\, H) \geq max\{\gamma(G)\rho_2(H), \rho_2(G)\gamma(H)\}.$$

This bound can be improved as follows.

Theorem 7.10 [2] *For any pair of graphs G and H,*

$$\gamma(G \,\square\, H) \geq \gamma(G)\rho_2(H) + \rho_2(G)(\gamma(H) - \rho_2(H)).$$

7.3 Early Results

Nearly sixteen years elapsed between the time Vizing posed his question and the publication of the first paper announcing significant insight on what was then called the "external stability number" of a Cartesian product. In 1979 Barcalkin and German [1] proved that Vizing's conjecture is true for a rather large class of graphs including some well-known families. Perhaps their work on the conjecture was prompted by its inclusion on a list of unsolved problems in graph theory put forth by Vizing([18]) in 1968. Their main result, like most that follow it, involves imposing some type of partitioning condition on the vertex set. Since this general partitioning idea is central to much of what follows, we include our independently constructed proof from 1991 (see [9]) with slight changes in terminology.

It is always true that $\gamma(G) \leq \chi(\overline{G})$ since any proper coloring of $\overline{G}$ is a partition of $V(G)$ into cliques of G. A single vertex chosen from each clique forms a dominating set for G. Barcalkin and German considered those graphs in which $\gamma(G) = \chi(\overline{G})$. Suppose that G is a graph with $\gamma(G) = k$ which has the property that $V(G)$ can be partitioned into k subsets $C_1, C_2, \ldots, C_k$ so that each of the induced subgraphs $\langle C_i \rangle$ is a complete subgraph of G. The graph G is then said to be a *decomposable* graph. In addition, they define a collection of graphs called the *A-class* consisting of all graphs G' such that G' is a spanning subgraph of a decomposable graph G with $\gamma(G') = \gamma(G)$. Their far-reaching result establishes the conjecture for any graph belonging to the A-class.

Theorem 7.11 [1] *If K is a spanning subgraph of a decomposable graph G where $\gamma(K) = \gamma(G)$, then K satisfies Vizing's conjecture.*

Proof. Consider a decomposable graph G with $\gamma(G) = k$. Let $V(G) = C_1 \cup C_2 \cup \ldots \cup C_k$ be the partition where each C_i induces a complete subgraph of G. For an arbitrary graph H consider the resulting partition $(C_1 \times V(H)) \cup \ldots \cup (C_k \times V(H))$ of $V(G \square H)$. Let D be any γ-set for $G \square H$, and assume that $V(H) = \{u_1, u_2, \ldots, u_n\}$. For $1 \leq j \leq k$ set $D_j = D \cap (C_j \times V(H))$. Select a vertex $w_j \in C_j$, and let p_j be the natural projection map of $C_j \times V(H)$ onto H_{w_j}. That is, $p_j(x, u_i) = (w_j, u_i)$.

Consider the set of vertices in H_{w_j} dominated by $p_j(D_j)$. For each j we form a "missing level" list at clique C_j by letting $\mathcal{L}_j$ be the set of all vertices u_i such that (w_j, u_i) is not dominated by $p_j(D_j)$. If u_i belongs to some $\mathcal{L}_j$, then every vertex in $C_j \times \{u_i\}$ must be adjacent to a vertex in $D_t \cap G_{u_i}$, where C_t has a vertex adjacent to C_j. Note that if H_{w_j} is dominated by $p_j(D_j)$, then $\mathcal{L}_j$ is empty. However, if $|D_j| = \gamma(H) - m$, then $|\mathcal{L}_j| \geq m$ follows immediately from the fact that $\gamma(H_{w_j}) = \gamma(H)$.

Consider each $1 \leq i \leq n$, such that u_i belongs to the missing level list of at least one clique. Assume u_i belongs to the list of r such cliques and let $\mathcal{R}$ represent the set of these cliques. Consider the set $\mathcal{S}$ of s cliques from $\{C_1, C_2, \ldots, C_k\}$ which have at least one edge to at least one member of $\mathcal{R}$ and which have at least one element of D in G_{u_i}. If A is the subset of all vertices of D of the form (x, u_i) such that x belongs to one of the cliques in $\mathcal{S}$, then $|A| \geq r+s$. For if not, then $A' = A \cup \{(w_j, u_i)|C_j \notin \mathcal{R} \cup \mathcal{S}\}$ would be a dominating set of G_{u_i} of cardinality less than k. This is a contradiction since $G \simeq G_{u_i}$.

Hence $|D| \geq k\gamma(H)$, since for each j, either $|D_j| \geq \gamma(H)$ or else there are sufficient extra elements from D in neighboring cliques to achieve an average of $\gamma(H)$. That is, $\gamma(G \,\square\, H) \geq \gamma(G)\gamma(H)$. Now if K is a spanning subgraph of a decomposable graph G as in the statement of the theorem, then we can apply Lemma 7.2 to prove that K satisfies Vizing's conjecture as well. $\square$

Figure 7.2: Decomposable graph T' and a spanning tree T.

For the graphs in Figure 7.2, T is a spanning tree of T', which is decomposable, and both have domination number 3. In fact, the method of adding edges in this example works in general when the domination number and the 2-packing number are equal. Specifically, if $\rho_2(G) = \gamma(G) = k$ and $S = \{v_1, \ldots, v_k\}$ is a 2-packing of G, then edges can be added to G to make $N[v_1], \ldots, N[v_{k-1}]$ as well as $V(G) - (N[v_1] \cup \ldots \cup N[v_{k-1}])$ into cliques. The resulting graph G' is decomposable and still has k pairwise disjoint closed neighborhoods. This proves the second part of the following corollary of Barcalkin and German [1], which first established the truth of the conjecture for several large classes of graphs. This result was also obtained independently by Jacobson and Kinch [13] (see Theorem 7.9); Faudree, Schelp and Shreve [5]; Chen, Piotrowski and Shreve [2].

Corollary 7.12 [1] *If G is a graph such that $\gamma(G) = 2$ or $\rho_2(G) = \gamma(G)$, then G satisfies Vizing's conjecture. In particular, Vizing's conjecture is true for any tree T.*

As we noted earlier, $\gamma(G) \leq \chi(\overline{G})$ for any graph G. Rall obtained the following corollary of Theorem 7.11 and Corollary 7.12.

Corollary 7.13 [16] *If G is a graph whose complement $\overline{G}$ is 3-colorable, then G satisfies Vizing's conjecture.*

Proof. Let $F = \overline{G}$. If $\chi(F) = 1$, then G is complete and the result is immediate. If F is bipartite, then G clearly belongs to the A-class. Thus, assume that $\chi(F) = 3$. If $\gamma(G) = 3$, then G is decomposable and the result follows by Theorem 7.11. Otherwise, $\gamma(G) \leq 2$ and the conclusion follows from Corollary 7.12.   □

In particular, we can now see that any tree and the complement of any tree satisfies the conjecture. Using the chromatic number of the complement as a measure, the smallest open case is a graph G with domination number 3, whose complement is 4-chromatic.

When the 2-packing number of a graph is smaller than its domination number, it can be difficult to determine if Theorem 7.11 applies. As an easy case in point consider the 7-cycle in Figure 7.3(a). By simple counting it is clear that C_7 is not decomposable. However, C_7 does belong to the A-class.

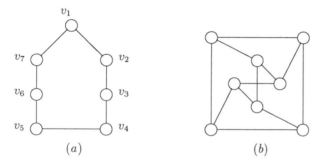

$$(a) \qquad\qquad\qquad\qquad (b)$$

Figure 7.3: Two graphs illustrating Theorem 7.11.

If edge $v_1 v_3$ is added to C_7 so that the vertex set partitions into 3 cliques, the resulting graph has domination number 2. However, adding edges $v_1 v_4$ and $v_1 v_5$ gives a decomposable graph with domination number 3, and so C_7 satisfies the hypothesis of Theorem 7.11. The graph H of Figure 7.3(b) is a graph which does not satisfy the hypothesis of Theorem 7.11. The interested reader is left the tedious exercise of showing that it is impossible to add edges to get a decomposable supergraph H' of H such that $\gamma(H') = 3$.

In a manner similar to that above with C_7, any cycle can be shown to be a spanning subgraph of a decomposable graph with the same domination number, and so we obtain the following corollary of Theorem 7.11. El-Zahar and Pareek, using induction on the order of the cycle and not the method of decomposable graphs, were actually the first to prove this result for cycles, and they independently showed the conjecture is true for graphs of domination number 2.

Corollary 7.14 [4] *For every* $k \geq 3$ *and every graph* H, $\gamma(C_k \square H) \geq \gamma(C_k)\gamma(H)$.

In 1990 Faudree, Schelp and Shreve imposed a partition condition to show that a large class of graphs satisfies Vizing's conjecture. In particular they make the following definition. A graph G satisfies *Condition CC* if there is a coloring, or partition, of $V(G)$ using $\gamma(G)$ colors such that any subset of $V(G)$ having order at most $\gamma(G) - 1$ fails to dominate some vertex in each color not represented in the subset. They prove that whenever G is a graph such that $\rho_2(G) = \gamma(G)$ then G satisfies CC and also prove the following theorem.

Theorem 7.15 [5] *If one of* G *or* H *satisfies CC, then* $\gamma(G \square H) \geq \gamma(G)\gamma(H)$.

Several years later Chen, Piotrowski and Shreve [2] pursued the following approach to the conjecture. Let $\mathcal{V} = \{V_1, \ldots, V_k\}$ be a partition of $V(G)$. The set V_i is said to be covered by $A \subseteq V(G)$ if $V_i \cap A \neq \emptyset$ or if $V_i \subseteq N(A)$. The partition $\mathcal{V}$ is said to be extracted if no subset A of $V(G)$ covers more than $|A|$ members of $\mathcal{V}$. The *extraction number* of G, denoted by $x(G)$, is the largest order of all extracted partitions of $V(G)$. They showed that the extraction number of a graph is never any smaller than the 2-packing number nor any larger than the domination number. The main result of [2] is the following. As we will see in Theorem 7.17, it is actually a corollary of Theorem 7.11.

Corollary 7.16 [2] *If* G *is a graph such that* $x(G) = \gamma(G)$, *then* G *satisfies Vizing's conjecture.*

It appears that the results of [1] were not widely known until 1991 (see [9]). Although the results of [5] and [2] were obtained independently of those of Barcalkin and German, the strength of the underlying clique partition idea in [1] can now be seen in the following result.

Theorem 7.17 *If* G *satisfies CC or if* $x(G) = \gamma(G)$, *then* G *is a spanning subgraph of a decomposable graph with the same domination number.*

Proof. Assume first that G satisfies CC. Let $k = \gamma(G)$ and let $\mathcal{V} = \{V_1, \ldots, V_k\}$ be a coloring of $V(G)$ guaranteed by CC. Add edges, if necessary, between any pair of nonadjacent vertices in the same color class (or member of the partition) to give a graph G'. If $B \subseteq V(G')$ has order at most $k - 1$, then some color is not represented in B. Assume $B \cap V_j = \emptyset$. There exists a vertex $x \in V_j$ such that $x \notin N_G[B]$. But since no edges were added between color classes, it follows that $x \notin N_{G'}[B]$, and so B does not dominate G'. Thus $\gamma(G') = k$ and G' is decomposable. Thus, G belongs to the A-class.

Now assume $x(G) = k = \gamma(G)$ and let $\mathcal{V} = \{V_1, \ldots, V_k\}$ be an extracted partition of $V(G)$. As in the previous case, add edges to form G'. Assume that $B \subseteq V(G')$ dominates G'. If $B \cap V_j = \emptyset$ for some j, then since B dominates G' it follows that $V_j \subseteq N_{G'}(B)$. In forming G' no edges were added between vertices from different members of $\mathcal{V}$, so B covers V_j in G as well. Thus B covers all k members of $\mathcal{V}$ in G, and so $k \leq |B|$. Therefore, in this case also we have shown that G is a spanning subgraph of a decomposable graph with the same domination number.   □

The collection of graphs satisfying CC is a proper subset of the A-class. It follows easily from the definitions that if G belongs to the A-class, then $x(G) = \gamma(G)$. That is, Chen, Piotrowski and Shreve [2] had rediscovered the result of Barcalkin and German of 1979.

7.4   Building Graphs Satisfying Vizing's Conjecture

Most of the positive results that have appeared on Vizing's conjecture consist of showing that if a graph satisfies some property then the conjecture is true for that graph. See the results of Sections 7.3 and 7.7. Another possible approach suggested by some is constructive in nature. This type of attack on the conjecture involves starting with some class, $\mathcal{C}$, of graphs for which the conjecture is known to be true. A collection of operations is then found such that each preserves membership in the class of graphs which satisfy the conjecture. The final - and probably most difficult step - is showing that every graph can be obtained from a seed graph in $\mathcal{C}$ by a finite sequence of these operations. Of course, no one has as yet succeeded with a program of this type. In this section we will present several operations which will allow us to "build" graphs satisfying the conjecture. Most of these results appear in [9].

To motivate one of the main ideas we begin with several examples. Consider the graph G_1 in Figure 7.4, and let H be any graph. Since $\rho_2(G_1) = 2 = \gamma(G_1)$, it follows from Corollary 7.12 that $\gamma(G_1 \square H) \geq 2\gamma(H)$. The graph $G_1 - z$ is isomorphic to C_5, and so it is also the case that $\gamma((G_1 - z) \square H) \geq 2\gamma(H)$. Suppose that we temporarily drop the requirement that the vertices of H_z be dominated. In particular, let $D \subseteq V(G_1 \square H)$ such that D dominates all vertices in $V(G_1 - z) \times V(H)$. That is, we are allowing members of $D \cap H_z$ to dominate their neighbors in H_w, but are not requiring H_z to be dominated.

If $p : H_z \to H_w$ is the projection map and we let

$$D' = [D \cap (V(G_1 - z) \times V(H))] \cup [p(D \cap H_z)],$$

then D' dominates $(G_1 - z) \square H$, and so it follows that $|D'| \geq 2\gamma(H)$. But $|D| \geq |D'|$. Thus, to dominate the smaller graph still requires $\gamma(G_1)\gamma(H)$ vertices, even if the dominating set is allowed to use vertices from the larger graph.

Contrast this with $G_2 \square C_4$, where the vertices of the 4-cycle are cyclically labeled v_1, v_2, v_3, v_4 and G_2 is the graph from Figure 7.4. If we let D be the sub-

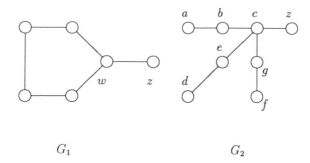

Figure 7.4: Graphs illustrating attachable sets.

set of vertices $\{(a, v_1), (b, v_3), (d, v_3), (e, v_1), (f, v_2), (g, v_4), (z, v_2)\}$, then D dom-inates the set $V(G_2 - z) \times V(C_4)$ and yet $|D| = 7 < \gamma(G_2)\gamma(C_4)$. With these two examples in mind we make the following definition.

Let $S \subseteq V(G)$ with $S \neq \emptyset$. The set S is called an *attachable set* in G if and only if it is the case that for every graph H and for every subset D of $V(G \square H)$ which dominates $V(G - S) \times V(H)$ it follows that $|D| \geq \gamma(G)\gamma(H)$. If $\{x\}$ is attachable in G we will sometimes call x an *attachable vertex*.

Thus, in particular, if G has an attachable set S then G must satisfy Vizing's conjecture, since any set of vertices D in $G \square H$ which dominates all of $G \square H$ certainly dominates $V(G - S) \times V(H)$ as well. It is an easy consequence of the definition that any nonempty subset of an attachable set in a graph is itself attachable in that graph. Note also that attachability is a property of subsets of vertices and not a property of a vertex itself. For example, in the path v_1, v_2, v_3, v_4, v_5 each vertex forms an attachable set of order one, but $\{v_1, v_4, v_5\}$ is not an attachable set. In the examples in Figure 7.4, $\{z\}$ is an attachable set of G_1 but is not an attachable set of G_2. The importance of this idea of attachable sets (and our choice of the name attachable) is its use in constructing graphs which satisfy Vizing's conjecture as illustrated in the following theorem.

Theorem 7.18 [9] *Let G_1 and G_2 be graphs with attachable sets S_1 and S_2, respectively. Let G be the graph obtained from the union of G_1 and G_2 by adding any subset of the set of edges $\{xy | x \in S_1 \text{ and } y \in S_2\}$. Then $S = S_1 \cup S_2$ is an attachable set in G.*

Proof. Let G be as in the statement of the theorem. Then it is clear that $\gamma(G) \leq \gamma(G_1) + \gamma(G_2)$. If H is any graph and D is any subset of $V(G \square H)$ which dominates every vertex in $V(G-S) \times V(H)$, then $D_i = D \cap V(G_i \square H)$ dominates every vertex in $V(G_i - S_i) \times V(H)$ for $i = 1, 2$. But since S_i is attachable in G_i it follows that

$$|D| = |D_1| + |D_2| \geq \gamma(G_1)\gamma(H) + \gamma(G_2)\gamma(H) \geq \gamma(G)\gamma(H).$$

Thus S is an attachable set in G.   □

It is not difficult to see that edges can be added between vertices of an attachable set in a graph G and the set is still attachable in the resulting graph. In some cases it is also possible to remove edges from a graph with an attachable set while maintaining the attachability of the set. The proofs of these two lemmas follow immediately from the definition of attachable set and are omitted.

Lemma 7.19 [16] *Let S be an attachable set in a graph G and let F be any subset of the set of edges $\{xy | x, y \in S\}$. The set S is an attachable set in the graph $K = G + F$.*

Lemma 7.20 [16] *Let S be an attachable set in a graph G and assume that F is a subset of $E(G)$ such that $\gamma(G - F) = \gamma(G)$. Then the set S is an attachable set in the graph $G - F$.*

We know of several ways to produce graphs with attachable sets. For example, if $x_1, x_2, \ldots, x_t$ is a collection of vertices, where x_i is adjacent to a set of vertices S_i, where S_i may induce any graph, then S_i is an attachable set in the closed neighborhood $N[x_i]$. Thus, by applying Theorem 7.18, one can build a graph with domination number t and having 2-packing number t as well. Of course, Corollary 7.12 guarantees that the resulting graph satisfies Vizing's conjecture. The additional fact about the attachable set is summarized in the following theorem.

Theorem 7.21 [9] *Let G be a graph such that $\rho_2(G) = \gamma(G)$. If $A \subseteq V(G)$ is any maximum 2-packing in G, then $N(A)$ is an attachable set in G.*

Other well known classes of graphs have attachable sets, as the next theorem shows.

Theorem 7.22 [9] *The cycle C_n has an attachable set if and only if $n \equiv 0, 2 \pmod 3$.*

Another way to construct graphs with attachable sets involves the operation of vertex cloning which we now define.

Let G be a graph and let $v \in V(G)$. Let G_1 be the graph with vertex set $V(G) \cup \{v'\}$ and edge set $E(G) \cup \{vv'\} \cup F$, where F is any subset of $\{wv' | w \in N(v)\}$. The graph G_1 is said to be obtained from G by a *cloning of type 1* and we refer to v' as being a *type 1 clone* of v in G_1. If $F \neq \emptyset$ and

$G_2 = G_1 - vv'$, then we say G_2 is obtained from G by a *cloning of type 2* and v' is then a *type 2 clone* of v in G_2.

Now we shall demonstrate how the cloning process fits into a possible constructive approach to settling the conjecture. The following lemma shows that any graph which satisfies the conjecture is an induced subgraph of many larger graphs which not only satisfy Vizing's conjecture but which also have attachable sets. The proof follows directly from the above definition and can be found in [9]. The weakness of the lemma is that it does not produce graphs of larger domination number which satisfy the conjecture.

Lemma 7.23 [9] *Let G be a graph that satisfies Vizing's conjecture. If v is a vertex of G that belongs to at least one minimum dominating set of G and v' is a type 1 clone of v in G_1, then $\{v'\}$ is an attachable set in G_1. If v' is a type 2 clone of v in G_2 and there exists an edge $v'w \in E(G_2)$, where w belongs to some minimum dominating set of G, then $\{v'\}$ is an attachable set in G_2. If, in addition, S is an attachable set in G, then $S \cup \{v'\}$ is attachable in the new graph.*

Corollary 7.24 [9] *The complete bipartite graph $K_{r,s}$ where $2 \leq r \leq s$ and $s \geq 3$, has an attachable set of vertices. In particular, $K_{r,s}$ satisfies Vizing's conjecture.*

Proof. Start with $K_{2,2} \simeq C_4$, which satisfies the conjecture. Perform a sequence of $s - 2$ clonings of type 2 to get $K_{2,s}$ and then a sequence of $r - 2$ clonings of type 2 to arrive at $K_{r,s}$ with the last new vertex forming an attachable set. In fact, the subset of $V(K_{r,s})$ consisting of the $r + s - 4$ new vertices added forms an attachable set in the resulting complete bipartite graph.   □

These ideas of cloning and attachable sets can be used to construct some graphs – often in several different ways – that satisfy Vizing's conjecture and have (several different) attachable sets. For example, consider the graph G in Figure 7.5(a). The graph G can be formed from the 4-cycle in (b) by a type 1 cloning of vertex u, so $\{v\}$ is an attachable set in G by 7.23; or by applying 7.18 to the two graphs in (c) where $\{u\}$ and $\{s,t\}$ are attachable, respectively, so $\{u,s,t\}$ is also an attachable set in G.

Graphs can be "put together" in various other ways producing graphs with attachable sets. For instance, consider any finite collection of cycles each of order congruent to 2 modulo 3 and form the graph in which one vertex from each is identified to form a new graph. See Figure 7.6 for an example. Here we can use the idea of 2-packing together with attachable sets to show the resulting graph has an attachable set. The set $\{x\}$ is attachable in one of the 5-cycles, and we can "pack"(i.e., use Lemma 7.8) at each of a,b and c to show that $\{x\}$ is attachable in G.

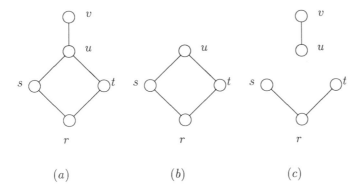

Figure 7.5: Type 1 cloning and an attachable set.

We conclude this section with two more ways to build larger graphs satisfying Vizing's conjecture. We have not been able to show that every graph can be constructed from some appropriate set of starting graphs by using the techniques of this section.

Theorem 7.25 *Let G_1 and G_2 be graphs with $u \in V(G_1)$ and $v \in V(G_2)$ such that the following conditions hold: (i) G_2 satisfies Vizing's conjecture, (ii) $\gamma(G_1 - u) \geq \gamma(G_1)$, and (iii) u belongs to some γ-set for G_1 and is attachable in G_1. Let $G = (V, E)$ be the graph with $V = V(G_1) \cup V(G_2) \cup \{x\}$ and $E = E(G_1) \cup E(G_2) \cup \{ux, xv\}$. Then $\{u, x\}$ is an attachable set of G.*

Proof. Let A be a γ-set for G_1 with $u \in A$, and let B be a γ-set for G_2. The set $A \cup B$ dominates G, and so $\gamma(G) \leq \gamma(G_1) + \gamma(G_2)$. Suppose $\gamma(G) < \gamma(G_1) + \gamma(G_2)$. Then if D is any minimum dominating set of G, it follows that $x \in D$ (for otherwise $D \cap V(G_i)$ dominates G_i for $i = 1, 2$ a contradiction).

Let $D_i = D \cap V(G_i)$. Since $|D| < \gamma(G_1) + \gamma(G_2)$, either $|D_1| < \gamma(G_1)$ or $|D_2| < \gamma(G_2)$. If $u \in D_1$ then D_1 dominates G_1. It follows that $|D_1| \geq \gamma(G_1)$ and so $|D_2| < \gamma(G_2)$. On the other hand, if $u \notin D_1$ then D_1 dominates $G_1 - u$ and then by (ii) it follows that $|D_1| \geq \gamma(G_1)$, which in turn implies that $|D_2| < \gamma(G_2)$. But then from the above it follows that $|D_2| \leq \gamma(G_2) - 2$. Let $B_2 = D_2 \cup \{v\}$. B_2 dominates G_2 and yet $|B_2| \leq \gamma(G_2) - 1$, a contradiction. Therefore, $\gamma(G) = \gamma(G_1) + \gamma(G_2)$.

Now let H be any graph and assume $A \subseteq V(G \square H)$, such that A dominates the vertices in $(V(G) - \{u, x\}) \times V(H)$. The subset $A_1 = A \cap V(G_1 \square H)$ dominates $(G_1 - \{u\}) \square H$, and since $\{u\}$ is attachable in G_1, it follows that $|A_1| \geq \gamma(G_1)\gamma(H)$.

Let $A_2 = (A \cap V(G_2 \square H)) \cup \{(v, h)|(x, h) \in A\}$. Then A_2 dominates $G_2 \square H$, and so $|A_2| \geq \gamma(G_2)\gamma(H)$ by (i). Therefore,

$$|A| \geq |A_1| + |A_2| \geq \gamma(G_1)\gamma(H) + \gamma(G_2\gamma(H) = \gamma(G)\gamma(H).$$

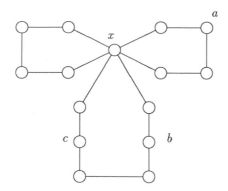

Figure 7.6: 2-Packing and an attachable set.

Thus $\{u, x\}$ is an attachable set in G.   □

The following theorem uses an extension of the idea of cloning to construct graphs with an attachable vertex. The proof is omitted but several examples of the construction are shown in Figure 7.7. In the two examples the vertices in A are marked with $*$. Note that the cloning of these vertices need not be complete, in the sense that the new vertex need not be made adjacent to every vertex of the open neighborhood.

Theorem 7.26 *Let G satisfy Vizing's conjecture and let $A \subseteq V(G)$ be a subset of some γ-set of G. Perform some type 1 cloning of the vertices in A to get a clone set S in the resulting graph G_1 such that $\gamma(G) = \gamma(G_1)$ and so that at least one vertex of S belongs to some γ-set of G_1. Form a new graph G_2 such that $V(G_2) = V(G_1) \cup \{v\}$ and $E(G_2) = E(G_1) \cup \{vs | s \in S\}$. Then $\{v\}$ is an attachable set in G_2.*

We conclude this section with two questions related to this particular constructive approach to settling Vizing's conjecture.

QUESTION 2: If G satisfies Vizing's conjecture and u is a vertex of G such that $\gamma(G - u) \geq \gamma(G)$, is $\{u\}$ an attachable set in G?

QUESTION 3: Are there well-known classes of graphs which we can "construct" using attachable sets, cloning and the results in this section? (For example, cycles of order $3k$ or $3k + 2$ and graphs with $\rho_2 = \gamma$ can be so constructed.)

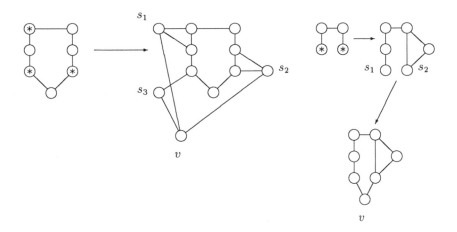

Figure 7.7: Constructions using Theorem 7.26.

7.5 Equality and Excess

One of the comments we have heard expressed by a number of people interested in Vizing's conjecture is that surely the conjecture must be correct since, for almost every pair of graphs G and H one selects, $\gamma(G \square H)$ is much larger than the product of $\gamma(G)$ and $\gamma(H)$. (Of course, since determining the domination number of a graph is a computationally hard problem, most of these examples are of small order or are members of some well-know family of graphs (see [12]).) Indeed, when we first became interested in the problem our experience was similar. It is somewhat surprising then to discover infinite families of graphs for which the conjectured bound is sharp.

In this section we will focus on graphs, G, for which there exists a graph H such that $\gamma(G \square H) = \gamma(G)\gamma(H)$. In most of these cases it turns out that once equality can be demonstrated using some pair of graphs, there emerge two infinite families $\mathcal{F}_1$ and $\mathcal{F}_2$ such that one of the following two statements is true.

$$\text{For every } G \in \mathcal{F}_1 \text{ and every } H \in \mathcal{F}_2, \ \gamma(G \square H) = \gamma(G)\gamma(H). \qquad (7.1)$$

$$\text{For every } G \in \mathcal{F}_1 \text{ there exists } H \in \mathcal{F}_2 \text{ such that } \ \gamma(G \square H) = \gamma(G)\gamma(H). \ (7.2)$$

In a paper that did not concern products directly, Payan and Xuong [15] provided the first examples of families of graphs satisfying (7.2) above. They defined a graph G of order n to be *domination balanced* if $\gamma(G)\gamma(\overline{G}) = n$. It is easy to verify that $\gamma(G \,\square\, \overline{G}) = n$ for any graph G of order n. Thus by letting $\mathcal{F}_1 = \mathcal{F}_2$ be the class of domination balanced graphs, we have an example of (7.2). See [15] for a characterization of the class of domination balanced graphs.

Fink, Jacobson, Kinch and Roberts [6] were the first to investigate the question of equality as in (7.1). A connected graph G is called a *generalized comb* (also called a corona) if each vertex of degree greater than one is adjacent to exactly one leaf. Let $\mathcal{C}$ denote the class of generalized combs. If $G \in \mathcal{C}$ then by the *interior* of G we mean the subgraph of $V(G)$ induced by the set of vertices of degree larger than one. This subgraph will necessarily be connected. Payan and Xuong [15] and Fink, Jacobson, Kinch and Roberts [6] independently showed that with the lone exception of the 4-cycle, the generalized combs are the only connected graphs which have domination number one-half their order. In addition, the second group of authors proved the following theorem and posed as a problem the characterization of those graphs with $\gamma(G \,\square\, H) = \gamma(G)\gamma(H)$.

Theorem 7.27 [6] *If both G and H are connected graphs of order at least four and have domination number one-half their order, then $\gamma(G \,\square\, H) = \gamma(G)\gamma(H)$.*

That is, they provided the first instance of (7.1) by letting $\mathcal{F}_1 = \mathcal{C} = \mathcal{F}_2$. In a related paper, Jacobson and Kinch [13] pursued this question when both factors were trees. They showed that one of the trees must be a generalized comb and deduced some structural properties which the other factor must possess. As a simplifying step in their attempt to characterize the trees in the other factor, they were able to reduce the study as follows.

Theorem 7.28 [6] *If H is a graph for which $\gamma(P_4 \,\square\, H) = 2\gamma(H)$, then for any generalized comb G of order at least four, $\gamma(G \,\square\, H) = \gamma(G)\gamma(H)$.*

Hartnell and Rall [9] gave five instances of families where equality is achieved as in (7.1) or (7.2). Several of them involve the class $\mathcal{C}$. The 4-cycle and graphs in $\mathcal{C}$ have minimum dominating sets with a property that turns out to be useful in constructing some of the infinite families satisfying (7.1) or (7.2). Consider the examples G and H of generalized combs in Figure 7.8.

Let H' be the interior of H. Then $\gamma(H') = 2$. If we let D_1 be any γ-set for H', say $D_1 = \{v_1, v_4\}$, then notice that $V(H) - N[D_1] = \{u_2, u_3, u_5, u_6\}$ and $D_1 \cup \{u_2, u_3, u_5, u_6\}$ is a γ-set of H. Let $D_2 = \{v_2, v_3, v_5, v_6\}$, let $D_1' = \{u_1, u_4\}$ and let $D_2' = \{u_2, u_3, u_5, u_6\}$. Choose any minimum dominating set A of the interior, G', of G, and let $B = V(G') - A$. For example, let $A = \{x_2, x_3\}$. Visualize $G \,\square\, H$ as a copy of H inside each vertex of G with appropriate additional edges. It follows from Lemma 7.8 that every dominating set of $G \,\square\, H$ must have at least $\gamma(H)$ vertices in common with each $H_{x_i} \cup H_{y_i}$. We can build a dominating set of this order by choosing a copy of D_1 from each H_{x_i} where $x_i \in A$, a copy of D_2

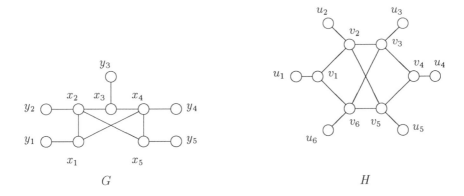

Figure 7.8: Two generalized combs.

from each H_{x_i}, where $x_i \in B$, a copy of D'_1 from each H_{y_i}, where $y_i \in V(H) - N[A]$ and finally a copy of D'_2 from each H_{y_i}, where $y_i \in V(H) - N[B]$.

The key property that makes this work is that each of the two graphs possesses a γ-set which partitions as $X_1 \cup X_2$ in such a way that if the closed neighborhood of X_1 is removed from the graph, only X_2 remains. A slightly stronger property is used by Hartnell and Rall [9] in several of their examples. We say a graph G has a *two-colored* γ-set D if D partitions as $D_1 \cup D_2$ such that $G - N[D_1] = D_2$ and $G - N[D_2] = D_1$. We let $\mathcal{A}$ denote the class of connected graphs having a two-colored γ-set, and let $\mathcal{B}$ be the subfamily of $\mathcal{A}$ consisting of all G such that each vertex of G belongs to a two-colored γ-set. The family $\mathcal{B}$ contains $K_{2r,2r} - rC_4$ for each $r \geq 2$, and so both of $\mathcal{A}$ and $\mathcal{B}$ are infinite families of graphs. In the list of examples of (7.1) or (7.2) below we also use the family $\mathcal{E}$ consisting of all connected graphs G such that for every vertex x of G, $\gamma(G - x) < \gamma(G)$. The family $\mathcal{E}$ is also infinite since, for example, it contains every graph of the form $K_r \square K_r$ for $r \geq 2$.

The following summary lists the known cases of equality as in (7.1) or (7.2). In most cases we do not provide specific examples but instead refer the reader to the appropriate reference.

I: An instance of (7.2) where $\mathcal{F}_1 = \mathcal{F}_2$ is the class of domination balanced graphs. See [15].

II: An instance of (7.2) where $\mathcal{F}_1 = \mathcal{C} = \mathcal{F}_2$. See [6] and [13].

III: An instance of (7.1) where $\mathcal{F}_1 = \mathcal{C}$ and $\mathcal{F}_2 = \mathcal{A}$. See [9].

IV: An instance of (7.2) where $\mathcal{F}_1 = \mathcal{E}$ and $\mathcal{F}_2$ is a family whose construction is given on page 92 of [9]. Note the large edge densities possible in these examples.

V: An instance of (7.2) where $\mathcal{F}_2 = \mathcal{C}$ and $\mathcal{F}_1 = \{K_{2,r} | r \geq 2\}$. Actually $\mathcal{F}_1$ is more general than this. See pages 93–94 of [9].

VI: An instance of (7.1) where $\mathcal{F}_1 = \mathcal{B}$ and $\mathcal{F}_2$ is an infinite family whose structure is given on page 95 of [9]. Note again the large edge densities of the graphs in $\mathcal{F}_2$.

VII: An instance of (7.2) where $\mathcal{F}_1 = \mathcal{B}$ and $\mathcal{F}_2$ is a family constructed using the notion of attachable sets in a collection of 5-cycles. See page 95 of [9] for an example.

It seems natural to then ask if for every G is there at least one graph H, of order at least two, such that $\gamma(G \square H) = \gamma(G)\gamma(H)$? The following result of Hartnell [8] answers this in the negative and in fact proves that some graphs G require a certain "excess"— over and above $\gamma(G)\gamma(H)$ — when dominating $G \square H$. The proof involves a careful counting of the intersection of any dominating set of $K_{1,k} \square H$ with the H-levels corresponding to leaves of $K_{1,k}$. For other related results on this idea of excess see [7], where Hartnell considers the family of graphs which he calls $[k, d]$-packable, which is a generalization of the class $\mathcal{C}$ of generalized combs.

Theorem 7.29 [7] *If k is a positive integer and H is any graph, then*

$$\gamma(K_{1,k} \square H) \geq \gamma(H) + k - 1.$$

If $\Delta(H) \leq k - 1$, then $\gamma(K_{1,k} \square H) = |H|$.

Corollary 7.30 [7] *Let G be the graph constructed by arbitrarily joining the centers of the stars $K_{1,n_1}, K_{1,n_2}, \ldots, K_{1,n_r}$ where $n_1 \leq n_2 \leq \ldots \leq n_r$. If H is any graph with $\Delta(H) \leq n_1 - 1$, then $\gamma(G \square H) = r|H|$.*

Hartnell [8] also showed that the 6-cycle does not belong to any family of graphs $\mathcal{F}_1$ or $\mathcal{F}_2$ satisfying (7.1) or (7.2) above. It is interesting to note that the existence of a graph H such that $\gamma(C_6 \square H) = \gamma(C_6)\gamma(H)$ would force H to have a "three-colored" γ-set. That is, H would have a minimum dominating set D which partitioned as $D_1 \cup D_2 \cup D_3$, such that if the closed neighborhood of any one of these parts of D is removed, only the union of the other two remain. It can be shown that no such H exists.

We conclude this section with several questions concerning equality or excess.

QUESTION 4: Is there a structural characterization of the graphs G such that there exists a graph H with $\gamma(G \square H) = \gamma(G)\gamma(H)$?

QUESTION 5: Does there exist a graph G for which there is a unique connected graph H of order at least two such that $\gamma(G \square H) = \gamma(G)\gamma(H)$? If so, can we give a characterization of them?

QUESTION 6: Can one characterize the graphs H for which $\gamma(C_4 \square H) = 2\gamma(H)$? More generally, for any fixed graph G, can we characterize the H for which $\gamma(G \square H) = \gamma(G)\gamma(H)$?

7.6 Recent Results

In this section we present some recent results of Hartnell and Rall [10] which
extend the collection of graphs for which Vizing's conjecture can be shown to
hold. The approach is similar to that in [1] in that the vertex set of a graph G
is assumed to possess a partition into $\gamma(G)$ subsets which induce certain types
of subgraphs. We will see that the main result generalizes Theorem 7.11 and
in fact shows that if the domination number and the 2-packing number of a
graph differ by no more than one, then the graph satisfies the conjecture. An
interesting feature of the argument used here is that it makes use of the trivial
fact that the domination number of a connected graph of order at least two is
no more then one-half its order.

Rather than include all the details of the proof we instead illustrate the
main idea on a small graph. Let G be the graph of Figure 7.9, and consider the
following partition of $V(G)$ into five subsets: $S_1 = \{v_1, v_3, v_4\}$, $S_2 = \{v_2, v_5, v_6\}$,
$B = \{v_9, v_{10}, v_{11}\}$, $SC = \{v_7, v_8, v_{12}\}$ and $C = \{v_{13}, v_{14}, v_{15}\}$.

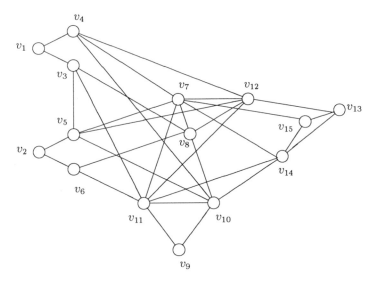

Figure 7.9: Graph G to demonstrate proof of Theorem 7.31.

Note that S_1 and S_2 induce stars and that B, C and SC induce cliques
in G. Both of S_1 and S_2 have a vertex (the center of the star) which has no
neighbors outside the star, and B also has such a vertex, v_9. However, every
vertex of both SC and C has at least one neighbor outside its respective clique,

but C has no neighbors in $S_1 \cup S_2$. It can be verified that $\gamma(G) = 5$, $\rho_2(G) = 4$, and that G is edge-maximal with respect to domination. The graph G is not decomposable and so G does not belong to the A-class of Barcalkin and German. Hence Theorem 7.11 is not applicable to G.

Let H be any graph and suppose that D is a dominating set for $G \,\square\, H$. Since $N[v_1] = S_1$, $N[v_2] = S_2$ and $N[v_9] = B$, it follows from Lemma 7.8 that D must have at least $\gamma(H)$ vertices in common with each of H_{S_1}, H_{S_2} and H_B. However, the same is not necessarily true for H_{SC} and H_C. We must show that any shortage in members of D in either of these is compensated by excess in $D \cap (H_{S_1} \cup H_{S_2} \cup H_B)$.

Project $D \cap H_{SC}$ onto $H_{v_{12}}$, project $D \cap H_C$ onto $H_{v_{13}}$, and call the resulting images X_1 and X_2, respectively. Form a missing level list $\mathcal{L}_1$ at SC by letting $\mathcal{L}_1$ be the set of all vertices $u \in V(H)$ such that (v_{12}, u) is not dominated by X_1. Similarly, let $\mathcal{L}_2$ be a missing level list at C. If D intersects each of H_{SC} and H_C in at least $\gamma(H)$ vertices then $|D| \geq \gamma(G)\gamma(H)$. But if, for example, $|D \cap H_{SC}| = \gamma(H) - r$, for some $1 \leq r \leq \gamma(H)$, then $|\mathcal{L}_1| \geq r$. A similar statement is true for H_C and $|\mathcal{L}_2|$.

Note that if $u \in \mathcal{L}_2$, then (v_{12}, u) and (v_7, u) must belong to D, since D dominates $C \times \{u\}$. Of course, it is also possible that (v_{10}, u) or (v_{11}, u) is in D, but this excess will not be needed to handle the missing level list $\mathcal{L}_2$. If a vertex w is in $\mathcal{L}_1$ but $SC \times \{w\}$ is dominated by $D \cap ((B \cup C) \times \{w\})$, then $(v_{10}, w) \in D$, and either $(v_{11}, w) \in D$ (so that $|D \cap ((B \cup SC) \times \{w\})| \geq 2$), or (v_{13}, w) as well as at least one of (v_{14}, w) and (v_{15}, w) is in D (so that $|D \cap ((B \cup C \cup SC) \times \{w\})| \geq 3$). In either case we again have shown that sufficient excess exists.

Thus we assume that $w \in \mathcal{L}_1$ but that $SC \times \{w\}$ is not completely dominated by the intersection of D with $(B \cup C) \times \{w\}$. Let $T_1 = S_1 - \{v_1\}$ and let $T_2 = S_2 - \{v_2\}$. Project $D \cap H_{T_1}$ onto H_{v_4} and project $D \cap H_{T_2}$ onto H_{v_6}. Call the images of these projections A_1 and A_2, respectively. Let F_1 (F_2) be the subgraph of H_{v_4} (H_{v_6}) induced by A_1 (A_2). These graphs F_1 and F_2 consist of components, some of which are isolated vertices and some of which are of order two or more.

One of the following two cases must hold:

Case 1: $w \in \mathcal{L}_1$ and for $i = 1, 2$, $|D \cap (T_i \times \{w\})| \leq 1$.

    Furthermore, if $|D \cap (T_i \times \{w\})| = 1$, then $(v_k, w) \notin D$ and for every neighboring level G_x to G_w, $(S_k \times \{x\}) \cap D = \emptyset$.

Case 2: $w \in \mathcal{L}_1$ and there is at least one j, $1 \leq j \leq 2$, such that either

    (a): $|D \cap (T_j \times \{w\})| = 2$, or

    (b): $|D \cap (T_j \times \{w\})| = 1$ and for some neighboring level G_x to G_w, $|D \cap (T_j \times \{x\})| \geq 1$ (so that (v_4, w) or (v_6, w) belongs to a component of order at least two in F_1 or F_2), or (c): $|D \cap (T_j \times \{w\})| = 1$, and either $(v_j, w) \in D$ or $(v_j, x) \in D$ for some $x \in N(w)$.

Consider Case 1 and assume, for example, that $|D \cap (T_1 \times \{w\})| = 1$ and $|D \cap (T_2 \times \{w\})| = 0$. If $(v_3, w) \in D$ and $(v_4, w) \notin D$, then it follows that (v_{10}, w) is in D, since (v_4, w) is dominated by D. But this is a contradiction, since the only way to dominate both (v_7, w) and (v_{12}, w) is then from $H_B \cup H_C$. If $(v_4, w) \in D$ and $(v_3, w) \notin D$, then (v_{11}, w) and (v_{10}, w) must both belong to D, since D dominates (v_3, w) and (v_8, w), respectively. This also contradicts the assumption that $SC \times \{w\}$ is not dominated by $D \cap ((B \cup C) \times \{w\})$. A similar argument disposes of the situation where $|D \cap (T_2 \times \{w\})| = 1$, even if $|D \cap (T_1 \times \{w\})| = 1$. Thus, Case 1 is not possible.

Thus assume $w \in \mathcal{L}_1$ satisfies at least one of the subcases of Case 2. If (v_4, w) belongs to a component M of F_1 such that M has order at least two, then M can be dominated in $\gamma(M) \leq \frac{1}{2}|M|$ elements. In this case $\gamma(M)$ of the vertices of M could be counted towards dominating H_{v_1} and $\gamma(M)$ towards dominating those missing levels of SC. The same situation exists if (v_6, w) belongs to such a component of F_2.

If neither F_1 nor F_2 contains a vertex from G_w in a component of order at least two, then either subcase (a) or subcase (c) holds. If (a) holds and, say $(v_3, w), (v_4, w) \in D$, then one of these can be counted towards dominating H_{v_1} and the other towards dominating $SC \times \{w\}$. If (c) holds, then (v_j, w) is dominated by $D \cap H_{v_j}$, and so the vertex in $D \cap (T_j \times \{w\})$ can be counted towards dominating $SC \times \{w\}$.

We have now shown that $|D| \geq 5\gamma(H) = \gamma(G)\gamma(H)$, and so G satisfies Vizing's conjecture. This graph G is a special case of a class of graphs which are defined as follows in [10].

Let k, t and r be nonnegative integers, not all of which are zero. Consider a graph G with $\gamma(G) = d = k + t + r + 1$ and such that $V(G)$ can be partitioned into $S \cup SC \cup BC \cup C$, where $S = S_1 \cup S_2 \cup \ldots \cup S_k$, $BC = B_1 \cup B_2 \cup \ldots \cup B_t$, and $C = C_1 \cup C_2 \cup \ldots \cup C_r$. Each of $SC, B_1, \ldots, B_t, C_1, \ldots, C_r$ induces a clique. Every vertex of SC (special clique) has at least one neighbor outside SC, whereas each of $B_1, \ldots, B_t$ (the buffer cliques), say B_i, has at least one vertex, say b_i, which has no neighbors outside B_i. Each $S_i \in \{S_1, S_2, \ldots, S_k\}$ is star-like in that it contains a star centered at a vertex v_i which is adjacent to each vertex in $T_i = S_i - \{v_i\}$. The vertex v_i has no neighbors besides those in T_i. Although other pairs of vertices in T_i may be adjacent (and hence S_i does not necessarily induce a star), S_i does **not** induce a clique nor can more edges be added in $\langle S_i \rangle$ without lowering the domination number of G. Furthermore, there are no edges between vertices in S and vertices in C. For ease of reference we will say such a graph G is of *Type $\mathcal{X}$*.

It should be noted that a graph of Type $\mathcal{X}$ need not have a clique having the properties of SC, and any of t, r or k is allowed to be 0. However, if such an SC is not in G, then $\gamma(G) = d = k + t + r$. Also, if SC is not present and $\mathcal{BC}$ is empty, but $\mathcal{S}$ as well as $\mathcal{C}$ are not empty, then the graph is disconnected. The special clique SC can not be the only one of these which is nonempty since by definition its vertices must have neighbors outside SC. The graph G in Figure 7.9 is an example of a Type $\mathcal{X}$ graph, although it has special properties which a general Type $\mathcal{X}$ graph need not have (such as all vertices of C having neighbors in the special clique).

Theorem 7.31 [10] *If G is a spanning subgraph of a Type $\mathcal{X}$ graph G' such that $\gamma(G) = \gamma(G')$, then Vizing's conjecture is true for G.*

Every decomposable graph can be realized as a Type $\mathcal{X}$ graph with $\mathcal{S} = \emptyset$. Thus Theorem 7.11 is a corollary of Theorem 7.31. This along with the graph of Figure 7.9 show that the collection of graphs called the A-class by Barcalkin and German is a proper subclass of the graphs satisfying Theorem 7.31. In addition, the following corollary of Theorem 7.31 generalizes Corollary 7.12.

Corollary 7.32 [10] *If G is a graph such that $\gamma(G) \leq \rho_2(G) + 1$, then Vizing's conjecture is true for G.*

7.7   Open Questions and Related Conjectures

Although we have described a number of results concerning the conjecture of V. G. Vizing on the domination of Cartesian products of graphs, it seems to us that a proof of the conjecture is not on the immediate horizon. We end this chapter with several open questions. Most of these are suggested by one or more of the approaches to the conjecture which we summarized in this chapter.

1. Does there exist a constant $c > 0$ such that for every pair of graphs G and H, $\gamma(G \,\square\, H) \geq c\gamma(G)\gamma(H)$?

2. If $\chi(\overline{G}) = 4$, does G satisfy Vizing's conjecture?

3. If $\rho_2(G) = \gamma(G) - 2$, does G satisfy Vizing's conjecture?

4. Can we prove Vizing's conjecture for every G such that $\gamma(G) \leq 3$?

5. Are there easily stated sufficient conditions that can be placed on G so that $\gamma(G \,\square\, H)$ is strictly larger than $\gamma(G)\gamma(H)$ for every connected H of order at least two?

Related to Vizing's conjecture are many similar problems involving other graph products and graphical parameters. In [14] Nowakowski and Rall prove some Vizing-type inequalities and provide counterexamples to many others. Besides γ, they investigate Γ (upper domination), ir and IR (lower and upper irredundance), i and β_0 (independent domination number and independence number), γ_t and Γ_t (lower and upper total domination), χ and ψ (chromatic and achromatic number), and d and ad (domatic and adomatic number). The following conjectures from [14] remain open.

6. $ir(G \,\square\, H) \geq ir(G)ir(H)$.

7. $i(G \times H) \geq i(G)i(H)$.

8. $\Gamma(G \times H) \geq \Gamma(G)\Gamma(H)$.

9. $\Gamma(G \,\square\, H) \geq \Gamma(G)\Gamma(H)$.

Bibliography

[1] A. M. Barcalkin and L. F. German, The external stability number of the Cartesian product of graphs. *Bul. Akad. Štiince RSS Moldoven.* (1979) no.1 5–8. [MR 80j:05096]

[2] G. Chen, W. Piotrowski and W. Shreve, A partition approach to Vizing's conjecture. *J. Graph Theory* 21 (1996) 103–111.

[3] E. J. Cockayne, Domination in undirected graphs – a survey. In *Theory and Applications of Graphs in America's Bicentennial Year.* Edited by Y. Alavi and D. R. Lick Springer-Verlag, Berlin (1978) 141–147.

[4] M. El-Zahar and C. M. Pareek, Domination number of products of graphs. *Ars Combin.* 31 (1991) 223–227.

[5] R. J. Faudree, R. H. Schelp and W. E. Shreve, The domination number for the product of graphs. *Congr. Numer.* 79 (1990) 29–33. [MR 92j:05100]

[6] J. F. Fink, M. S. Jacobson, L. F. Kinch and J. Roberts, On graphs having domination number half their order. *Period. Math. Hungar.* 16 (1985) 287–293. [MR 87e:05085]

[7] B. L. Hartnell, On determining the 2-packing and domination numbers of the Cartesian product of certain graphs. manuscript (1994).

[8] B. L. Hartnell, private communication (1995).

[9] B. L. Hartnell and D. F. Rall, On Vizing's conjecture. *Congr. Numer.* 82 (1991) 87–96. [MR 92k:05071]

[10] B. L. Hartnell and D. F. Rall, Vizing's conjecture and the one-half argument. *Discuss. Math. - Graph Theory* 15 (1995) 205–216.

[11] W. Imrich and H. Izbicki, Associative products of graphs. *Monatsh. Math.* 80 (1975) 277–281. [MR 53 #7864]

[12] M. S. Jacobson and L. F. Kinch, On the domination number of products of graphs: I. *Ars Combin.* 18 (1983) 33–44. [MR 87a:05087]

[13] M. S. Jacobson and L. F. Kinch, On the domination of the products of graphs II: trees. *J. Graph Theory* 10 (1986) 97–106. [MR 87e:05056]

[14] R. J. Nowakowski and D. F. Rall, Associative graph products and their independence, domination and coloring numbers. *Discuss. Math. - Graph Theory* 16 (1996) 53–79.

[15] C. Payan and N. H. Xuong, Domination-balanced graphs. *J. Graph Theory* 6 (1982) 23–32.

[16] D. F. Rall, private communication (1993).

[17] V. G. Vizing, The Cartesian product of graphs. *Vyčisl. Sistemy* 9 (1963) 30–43. [MR 35 # 81]

[18] V. G. Vizing, Some unsolved problems in graph theory. *Uspehi Mat. Nauk* 23 (1968) no.6 (144) 117–134. [MR 39 # 1354]

Chapter 8

Algorithms

Dieter Kratsch

Friedrich-Schiller-Universität

Fakultät für Mathematik und Informatik

07740 Jena, Germany

In the last quarter of the century there has been a growing interest in the design and analysis of algorithms in different branches of mathematics, science and engineering. In particular, within the past 15 years there has appeared a wealth of literature on algorithms in graph theory, a significant part of which is related to domination in graphs. Since there are well over one hundred papers on domination or domination-related algorithms, this chapter will necessarily have to focus on a limited collection of these algorithms.

The intention of this chapter on domination algorithms for special graph classes is twofold. The major goal is to demonstrate some of the main techniques for designing efficient algorithms for domination problems. For this purpose we consider some particular examples and focus on five graph classes and five well researched variants of the domination problem. Another goal is to present a fairly comprehensive bibliography of algorithms for domination problems. We refer the reader who is unfamiliar with the design and analysis of algorithms to [57].

8.1 Introduction

We consider efficient algorithms for the domination problems DOMINATING SET, CONNECTED DOMINATING SET, TOTAL DOMINATING SET, INDEPENDENT DOM-INATING SET and DOMINATING CLIQUE. For all five problems the input is a graph $G = (V, E)$ and possibly real vertex weights. The output is a dominating set/connected dominating set/total dominating set/independent dominating set/dominating clique of minimum cardinality (respectively minimum weight), if such a set exists.

191

The corresponding decision versions, 'Given a graph $G = (V, E)$ and a positive integer k, does G have a dominating set/connected dominating set/total dominating set/independent dominating set/dominating clique of size at most k?', are all NP-complete as can be seen in Chapter 9.

One of the avenues for research on NP-complete graph problems is to consider their algorithmic complexity when they are restricted to special graph classes. The motivation for this research direction can be addressed to two main sources. The original motivation was to find graph classes $\mathcal{G}$ with nice structural properties, that enable the design of polynomial time algorithms for NP-complete graph problems when the input graphs are restricted to the class $\mathcal{G}$. Originally 'small classes' such as interval graphs and permutation graphs were considered. This led researchers to look for larger and larger graph classes, for which polynomial time domination algorithms can still be designed. Recent examples are the classes of AT-free graphs, dually chordal graphs and homogeneously orderable graphs ([59], [36], [37]).

Another stream of research is to design algorithms for domination problems that are as fast as possible. Indeed the major goal is to design optimal algorithms. This type of research has created a lot of papers and also interesting new techniques for the design of algorithms on graph classes such as interval graphs, permutation graphs and trapezoid graphs. Very often these algorithms exploit some particular representation of the input graph (that differs from the standard representation by adjacency lists). Throughout this chapter we consider various examples, among them the well-known representation of an interval graph by an interval model (see Section 8.6).

In particular the design of faster and faster algorithms requires a careful study of what we mean when we say that 'algorithm $\mathcal{A}$ solves problem $\mathcal{P}$ on graph class $\mathcal{G}$'. (For example, consider an algorithm solving the DOMINATING SET problem on permutation graphs.)

An algorithm $\mathcal{A}$ solving problem $\mathcal{P}$ on $\mathcal{G}$ typically consists of the following steps.

Recognition Check whether the input graph $G = (V, E)$ belongs to the class $\mathcal{G}$. Reject if $G \notin \mathcal{G}$. Possibly compute a (suitable) representation of G.

Problem Solve the problem $\mathcal{P}$ for the graph G, possibly using a representation of G.

Of course, for measuring the time of algorithm $\mathcal{A}$ we should sum up the amount of time for both steps. In fact, as long as the second step is the most time consuming part of the algorithm or our goal is 'only' to show the existence of a polynomial time algorithm, we do not have to worry about these details. However if we start 'hunting for fast algorithms', where the second step is faster than any known recognition algorithm for $\mathcal{G}$, the situation changes.

In the early papers the authors either considered the overall time for the two steps or did not care about the first step. Nowadays there is no common agreement. Sometimes this makes it difficult to compare algorithms in different papers.

Fortunately there is a reasonable approach used in the majority of papers. The authors present an algorithm for the second step and they consider the running time for this essential part. Additionally, they point out the running time for the first step by mentioning the corresponding recognition algorithm. This assists the reader in comparing algorithms for the same problem on the same graph class, that differ only in the assumptions concerning the input, as is the case for such graph classes as interval graphs, permutation graphs and trapezoid graphs. (See also Section 8.2 and 8.6.)

The input of the weighted case of a domination problem usually consists of a graph $G = (V, E)$ and a weight function w that assigns a real number $w(v)$, called the weight of v, to each vertex v of the graph G. For any $S \subseteq V$ the weight of S is defined to be $w(S) = \sum_{v \in S} w(v)$. The minimum weight dominating set problem is to locate a dominating set of minimum weight. The weighted cases of other domination problems are defined similarly.

Weighted cases are usually considered or mentioned for one of the following two reasons. Either the presented algorithm for the cardinality case works for the weighted case as well or there is already an optimal (or near optimal) algorithm for the cardinality case, and thus researchers turn to the weighted case.

G. Manacher and T. Mankus have shown the following theorem in [117]. The theorem justifies that the weighted case of the dominating set, total dominating set and connected dominating set problem can be restricted to nonnegative real vertex weights, since the negative weights can be incorporated in an easy fashion.

Theorem 8.1 *Let $\mathcal{P}$ be an extendible search problem (ESP), i.e., every vertex superset of a solution is again a solution. Let $\mathcal{A}$ be an algorithm solving $\mathcal{P}$ when the vertex weights are nonnegative. Then the following algorithm solves $\mathcal{P}$ for arbitrary real vertex weights.*

(1) *Let $G = (V, E)$ be the input graph and N, Z and P the sets of negative-weight, zero-weight and positive-weight vertices.*

(2) *Use the algorithm $\mathcal{A}$ to solve the problem $\mathcal{P}$ on the graph G, for which all vertices in N obtain weight zero. Let D^* be a minimum weight solution.*

(3) *Then $(D^* \cap P) \cup N \cup Z$ is a minimum weight solution for the original weight function.*

Our collection of efficient algorithms for domination problems on special graph classes shows that structural properties of special graph classes are important tools for the design of efficient algorithms. We shall demonstrate how domination algorithms on special graph classes exploit particular representations of the input graphs such as intersection models and vertex orderings.

We refer the reader to [29, 84] for definitions and properties of special classes of graphs not given in this chapter. Some of the graph classes that we consider have interesting real world applications. Various examples are given in [84, 138].

8.2   Permutation Graphs

Permutation graphs have been a favorite graph class for researchers designing polynomial time domination algorithms since the early eighties [38, 39, 63, 80]. Most of these algorithms were designed using dynamic programming on a permutation diagram or a (defining) permutation of the given permutation graph. These algorithms (usually for the cardinality case) were given to show that the corresponding NP-complete domination problem is solvable by a polynomial time algorithm when restricted to permutation graphs. Typically the running time of these classical algorithms is not faster than $O(n^2)$, which was the running time of the best known recognition algorithm for permutation graphs for a long time. This $O(n^2)$ recognition algorithm for permutation graphs is given by J. Spinrad in [141].

Nowadays the aim of the research is to design faster and faster algorithms and the weighted case seems to be the more attractive one. Recently R. McConnell and J. Spinrad have given an $O(n + m)$ recognition algorithm for permutation graphs that also computes a permutation diagram or a permutation, if the input is a permutation graph [119].

It is worth noting that domination algorithms for permutation graphs with time bounds like $O(n \log^2 n)$ [12], $O(n \log n)$ [11] or $O(n \log \log n)$ [150] appeared long before this significant improvement. Of course such time bounds require the assumption that the graph is given by a permutation diagram or a permutation.

This is an excellent example for continuing our discussion of Section 8.1 concerning different input models for domination algorithms and the way of measuring the running time of such algorithms. Assuming that the graph is the input, any domination algorithm takes time $\Omega(n + m)$. Nevertheless there are at least two good reasons for looking more carefully at the whole algorithm. First, the input might indeed be a permutation in practical applications. Probably more important, even if the input is a graph, it is reasonable to split the whole algorithm into parts and try to speed up some essential part. In the way *Recognition* and *Problem* are considered separately, it makes a lot of sense to consider that part of the algorithm, that starts with a model as the input, separately. Notice that a speed up of the recognition algorithm may lead to a speed up in the (worst case) running time of the whole algorithm. Another point is that a faster algorithm for the *Problem* part can be expected to give practically faster algorithms. For example, a permutation graph algorithm with running time $O(n \log \log n)$ seems to be better than an $O(n + m)$ algorithm, since a permutation graph has $O(n^2)$ edges on average.

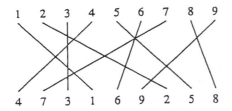 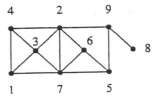

Figure 8.1: Permutation diagram and graph $G[\pi]$ for $\pi = [4, 7, 3, 1, 6, 9, 2, 5, 8]$.

Although domination algorithms for permutation graphs became faster and faster and some of them are already optimal by now, the major technique for designing them is still dynamic programming. An interesting new idea for designing fast dynamic programming algorithms for permutation graphs, that often leads to $O(n + m)$ algorithms, is the *ordered cross pair* method introduced by Liang et al. [110, 113, 137].

Let π be a permutation on $V_n = \{1, \ldots, n\}$. We think of it as the sequence $\pi = [\pi(1), \ldots, \pi(n)]$.

Definition 8.2 *For each permutation π on V_n, we can construct its* inversion graph $G[\pi] = (V, E)$ *with vertex set $V = \{1, \ldots, n\}$ and edge set E:*

$$\{i, j\} \in E \Leftrightarrow (i - j)(\pi^{-1}(i) - \pi^{-1}(j)) < 0.$$

(Notice that $\pi^{-1}(i)$ denotes the position of i in $\pi = [\pi(1), \ldots, \pi(n)]$.)
An undirected graph G is a permutation graph *if there is a permutation π such that $G \cong G[\pi]$.*

Permutation graphs are intersection graphs, which is illustrated by the permutation diagram [84].

Definition 8.3 *Let π be a permutation of $1, \ldots, n$. The* permutation diagram *can be obtained as follows. Write the numbers $1, \ldots, n$ horizontally from left to right. Underneath, write the numbers $\pi(1), \ldots, \pi(n)$, also horizontally from left to right. Draw a straight line, henceforth called* line*, joining the two 1's, the two 2's, etc. (An example is given in Figure 8.1.)*

For each $i \in V_n$, the line joining the two i's gets the label i. In this way we identify a vertex with a corresponding line in the diagram (which is very convenient). Notice that two vertices i and j of $G[\pi]$ are adjacent if and only if the lines i and j intersect. Furthermore, we say that line i is *left of* line j if $i < j$ and $\pi^{-1}(i) < \pi^{-1}(j)$. Hence if line i is left of line j then the lines i and j do not intersect.

Dynamic programming on the permutation diagram is an old idea for designing polynomial time domination algorithms on permutation graphs [39, 63, 80]. The important new idea of the domination algorithms by Liang et al. are the ordered cross pairs that enable $O(n + m)$ dynamic programming algorithms on the permutation diagram. Naturally $O(n + m)$ algorithms require efficient subroutines even for parts that can easily be implemented to meet the overall time bound when time bounds like $O(n^2)$ or $O(n^3)$ are to be established.

Definition 8.4 *Any two lines i and j in V_n, $i \leq j$, are called an* ordered cross pair, *denoted X_{ij}, if $i = j$ or i intersects (or crosses) j.*

Notice that there are exactly $n+m$ ordered cross pairs. A dynamic programming algorithm over all ordered cross pairs may run in $O(n+m)$ time, if the additional amount of work is not too large.

As an example, we shall consider the $O(n + m)$ algorithm for the minimum weight independent dominating set problem of D. Liang and C. Rhee, given in [113]. Former algorithms for this problem have running times $O(n^3)$ [80], $O(n^2)$ [39], $O(n \log^2 n)$ [12] and $O(n \log n)$ [11].

Since there is an $O(n+m)$ recognition algorithm for permutation graphs that also computes a permutation diagram, if the input is a permutation graph [119], we may assume that the input is a permutation graph G together with a permutation diagram of G.

For an ordered cross pair X_{ij} (thus $i \leq j$), let $V_{ij} = \{k : k \leq j$ and $\pi^{-1}(k) \leq \pi^{-1}(i)\}$. Hence V_{ij} contains the vertices i and j and all the vertices corresponding to lines for which none of the endpoints on the top and the bottom line is on the right of the endpoints of line i and line j.

Suppose that $w(1), w(2), \ldots, w(n)$ are the real vertex weights of G. Then let D_{ij} be any minimum weight independent dominating set of the graph $\langle V_{ij} \rangle$. (Whenever $\langle V_{ij} \rangle$ has more than one minimum weight independent dominating set, we may choose an arbitrary one among them.) Notice that $D_{\pi(n)n}$ is a minimum weight independent dominating set of the graph G, since $V_{\pi(n)n} = V_n$.

For each X_{ij}, let $V'_{ij} = \{i' : i' < i$ and $\pi^{-1}(i') < \pi^{-1}(j)\}$. Hence V'_{ij} is the set of those vertices that are neither adjacent to i nor to j and for which the corresponding line is left of i and left of j. Define $bound(X_{ij}) = \emptyset$ if $V'_{ij} = \emptyset$. Otherwise, define $bound(X_{ij})$ to be the ordered cross pair $X_{i_b j_b}$ satisfying (i) $i_b, j_b \in V'_{ij}$, (ii) j_b is the largest number in V'_{ij} and (iii) $\pi^{-1}(i_b) \geq \pi^{-1}(k)$ for all $k \in V'_{ij}$. In other words, j_b has the rightmost top endpoint and i_b has the rightmost bottom endpoint amongst all lines in V'_{ij}. Thus $bound(X_{ij})$ is a uniquely determined ordered cross pair, if $V'_{ij} \neq \emptyset$.

Lemma 8.5 *For each $j \in V_n$, let $X_{i_b j_b} = bound(X_{jj})$. Then $D_{jj} = D_{i_b j_b} \cup \{j\}$.*

Proof. Note that j is an isolated vertex in $\langle V_{jj} \rangle$ and that $V_{jj} \setminus \{j\} = V_{i_b j_b}$. Thus D is an independent dominating set of $\langle V_{jj} \rangle$ if and only if $j \in D$ and $D \setminus \{j\}$ is an independent dominating set of $\langle V_{i_b j_b} \rangle$. □

For each ordered cross pair X_{ij} with $i < j$, define the clique of X_{ij}, denoted by $CLIQ(X_{ij})$, to be the set of those lines k satisfying (i) $i \leq k \leq j$, (ii) $\pi^{-1}(j) \leq \pi^{-1}(k) \leq \pi^{-1}(i)$ and (iii) there is no line k' with $k < k' < j$ and $\pi^{-1}(k) < \pi^{-1}(k') < \pi^{-1}(i)$. Thus $CLIQ(X_{ij})$ is a clique in the graph $\langle V_{ij} \rangle$ containing those lines k of V_{ij} that are maximal in the sense that there is no line in V_{ij} that is right of k. In particular, $i, j \in CLIQ(X_{ij})$.

Lemma 8.6 *Let X_{ij} be an ordered cross pair with $i < j$. Then there is a $k \in CLIQ(X_{ij})$ such that D_{kk} is a minimum weight independent dominating set of $\langle V_{ij} \rangle$. Hence $D_{ij} = D_{kk}$ where D_{kk} is a minimum weight set amongst all sets D_{ss} with $s \in CLIQ(X_{ij})$.*

Proof. Consider any minimum weight independent dominating set D_{ij} of $\langle V_{ij} \rangle$. Then $|D_{ij} \cap CLIQ(X_{ij})| \leq 1$, since $CLIQ(X_{ij})$ is a clique and D_{ij} is an independent set in $\langle V_{ij} \rangle$.

Let k be the rightmost line in D_{ij}. Since D_{ij} is a dominating set, there is no line $k' \in D_{ij}$ with $k < k'$ and $\pi^{-1}(k) < \pi^{-1}(k')$. Thus $k \in CLIQ(X_{ij})$. Hence the set D_{kk} is a minimum weight independent dominating set of $\langle V_{ij} \rangle$. □

Lemmas 8.5 and 8.6 enable us to compute a minimum weight independent dominating set using dynamic programming. Thereby D_{ij} is computed for each ordered cross pair X_{ij}. The ordered cross pairs are scanned in the following order: X_{ij} is scanned before $X_{i'j'}$ if $j < j'$, or $j = j'$ and $i > i'$.

As is typical for the design of $O(n + m)$ algorithms, there are some details to be fixed for guaranteeing that the algorithm indeed meets the time bound. For an efficient handling of the sets $CLIQ(X_{ij})$ we use the line $div(X_{ij})$ defined as follows: $div(X_{ii})$ is the line i and if $i < j$ then $div(X_{ij})$ is max $CLIQ(X_{ij} \setminus \{j\})$, i.e., the line with the second rightmost top endpoint in $CLIQ(X_{ij})$.

Lemma 8.7 *Let $i < j$ and $t = div(X_{ij})$. Then D_{ij} is a set of minimum weight among D_{jj} and D_{it}.*

Proof. By Lemma 8.6, $D_{ij} = D_{kk}$ for some D_{kk} that has minimum weight amongst all sets D_{ss} with $s \in CLIQ(X_{ij})$. Thus either D_{jj} or some set D_{ss} with $s \in CLIQ(X_{ij}) \setminus \{j\}$ has minimum weight amongst all sets D_{ss} with $s \in CLIQ(X_{ij})$. Suppose $D_{ij} = D_{ss}$ for some $s \in CLIQ(X_{ij}) \setminus \{j\}$. Therefore $CLIQ(X_{ij}) \setminus \{j\} = CLIQ(X_{it})$ and Lemma 8.6 imply that $D_{ij} = D_{it}$. □

The algorithm uses so-called crossing lists $TLCL_j$, $BLCL_j$ and $BRCL_j$ for all $j \in V_n$.

$TLCL_j$ (Top Left Crossing List) denotes the ordered list $(k_0, k_1, \ldots, k_r)$ such that $k_0 < k_1 < \ldots < k_{r-1} < k_r = j$ and $\{k_0, k_1, \ldots, k_{r-1}\}$ is the set of all lines crossing j, that have their top endpoint left of the top endpoint of j.

$BLCL_j$ and $BRCL_j$ are similar ordered lists, that are used for computing $bound(X_{ij})$ and $div(X_{ij})$. $BLCL_j$ (Bottom Left Crossing List) denotes the ordered list $(k_0, k_1, \ldots, k_r)$ such that $\pi^{-1}(k_0) < \pi^{-1}(k_1) < \ldots < \pi^{-1}(k_{r-1}) <$

$\pi^{-1}(k_r) = \pi^{-1}(j)$ and $\{k_0, k_1, \ldots, k_{r-1}\}$ is the set of all lines crossing j, that have their bottom endpoint left of the bottom endpoint of j. $BRCL_j$ (Bottom Right Crossing List) denotes the ordered list $(k_0, k_1, \ldots, k_r)$ such that $\pi^{-1}(j) = \pi^{-1}(k_0) < \pi^{-1}(k_1) < \ldots < \pi^{-1}(k_r)$ and $\{k_0, k_1, \ldots, k_{r-1}\}$ is the set of all lines crossing j, that have their bottom endpoint right of the bottom endpoint of j.

We refer the reader to the original paper for $O(n+m)$ procedures computing all crossing lists, $bound(X_{ij})$ for all ordered cross pairs and $div(X_{ij})$ for all ordered cross pairs.

Algorithm 1

Input: A permutation graph $G = (V, E)$, a permutation diagram of G and real vertex weights $w(1), w(2), \ldots, w(n)$.

Output: A minimum weight independent dominating set of G.

Compute the crossing lists $TLCL_j$, $BLCL_j$ and $BRCL_j$
for all $j \in V_n$;
Compute $bound(X_{ij})$ and $div(X_{ij})$ for all ordered cross pairs X_{ij};
FOR $j = 1$ **TO** n **DO**
    BEGIN
    IF $bound(X_{jj}) = \emptyset$ **THEN** $D_{jj} := \{j\}$
    ELSE
        BEGIN
        $X_{i_b j_b} := bound(X_{ij})$;
        $D_{jj} := D_{i_b j_b} \cup \{j\}$
        END
    Let $TLCL_j = (k_0, k_1, \ldots, k_r)$;
    FOR $s = r - 1$ **DOWNTO** 0 **DO**
        BEGIN
        $t := div(X_{k_s j})$;
        $D_{k_s j}$ is a set of minimum weight among D_{jj} and $D_{k_s t}$;
        END
    END
Output $D_{\pi(n)n}$.

Theorem 8.8 *Given a permutation graph $G = (V, E)$ with real vertex weights, Algorithm 1 computes a minimum weight independent dominating set of G in time $O(n + m)$.*

The $O(n + m)$ algorithm for the minimum weight connected dominating set problem given in [110] and the $O(n + m)$ algorithm for the minimum weight dominating set problem in [137] also apply the ordered cross pair method.

Among the domination algorithms for permutation graphs that take the permutation and the permutation diagram, respectively, as input model, we mention the $O(n \log \log n)$ algorithm computing a minimum cardinality dominating set [150] and the $O(n)$ algorithm computing a minimum cardinality connected dominating set [95].

It is an interesting open question whether there is an $O(n)$ algorithm computing a minimum cardinality dominating set for permutation graphs, if the graph is given by a permutation diagram.

The many efficient algorithms for domination problems on the class of permutation graphs have encouraged researchers to study superclasses of permutation graphs. One of these superclasses is the class of *trapezoid graphs* which is also a subclass of the cocomparability graphs. Trapezoid graphs have an intersection model like the permutation diagram, a so-called *trapezoid diagram*. It seems likely that algorithms as efficient as for permutation graphs can be obtained for trapezoid graphs. In that respect it is worth mentioning that E. Köhler has designed an $O(n)$ algorithm to compute a minimum cardinality connected dominating set of a trapezoid graph, if the graph is given by a trapezoid diagram [101].

Furthermore, E. Elmallah and L. Stewart have presented domination algorithms for *polygon graphs* [74, 75] and J. Keil has considered the algorithmic complexity of domination problems on *circle graphs* [98], two other superclasses of the permutation graphs.

8.3 Cocomparability Graphs

The class of cocomparability graphs is a well researched superclass of the class of permutation graphs. Domination problems on cocomparability graphs were considered for the first time by Stewart et al. [105]. The authors obtained polynomial time algorithms computing a dominating/total dominating/connected dominating set of minimum cardinality and an algorithm computing a minimum weight independent dominating set on cocomparability graphs. All these algorithms work on a certain vertex ordering called cocomparability labeling, and most of them are designed by dynamic programming.

H. Breu and D. Kirkpatrick improved this as follows. They give $O(n m^2)$ algorithms computing a minimum cardinality dominating set and a minimum cardinality total dominating set and an $O(n^{2.376})$ algorithm computing a minimum weight independent dominating set on cocomparability graphs [40] (see also [7]).

It is worth mentioning that the problem 'Given a cocomparability graph G, does G have a dominating clique?' is NP-complete [105]. Furthermore the weighted cases of the problems DOMINATING SET, TOTAL DOMINATING SET and CONNECTED DOMINATING SET are NP-complete on cocomparability graphs

[47].

An $O(n^3)$ algorithm computing a minimum cardinality connected dominating set of a connected cocomparability graph has been given in [105]. We consider an algorithm for this problem given by H. Breu and D. Kirkpatrick [40] (see also [7]).

Definition 8.9 *A graph* $G = (V, E)$ *is a* comparability graph *if there is a transitive orientation F of the edges of G, hence for every $\{u, v\} \in E$ either*

$$(u, v) \in F \text{ or } (v, u) \in F, \text{ and}$$

$$(u, v) \in F \land (v, w) \in F \to (u, w) \in F.$$

A graph $G = (V, E)$ is a *cocomparability graph* if its complement $\overline{G}$ is a comparability graph. The class of cocomparability graphs contains the classes of interval graphs, permutation graphs and trapezoid graphs.

The following definition reveals the relation between cocomparability graphs and partially ordered sets that has been used in various algorithms for cocomparability graphs. For more information on partially ordered sets we refer to [149].

Definition 8.10 *The graph $G = (V, E)$ is a* cocomparability graph *if there is a partially ordered set $P_G = (V, \prec_{P_G})$ such that $u, v \in V$ are adjacent in G if and only if u and v are incomparable in P_G, i.e., neither $u \prec_{P_G} v$ nor $v \prec_{P_G} u$.*
A cocomparability labeling *of a graph $G = (V, E)$ is an ordering $v_1, v_2, \ldots, v_n$ of V satisfying for each i, j and k: If $i < j < k$ and $\{v_i, v_k\} \in E$, then $\{v_i, v_j\} \in E$ or $\{v_j, v_k\} \in E$.*

Notice that $v_1, v_2, \ldots, v_n$ is a cocomparability labeling of a graph $G = (V, E)$ if and only if $v_1, v_2, \ldots, v_n$ is a topological sort of $\overline{G}$ with respect to the transitive orientation F if and only if $L = v_1, v_2, \ldots, v_n$ is a linear extension of a partially ordered set P_G. Thus a graph $G = (V, E)$ is a cocomparability graph if and only if it has a cocomparability labeling.

There is an $O(n^{2.376})$ recognition algorithm for comparability graphs and thus for cocomparability graphs [141]. This has been improved by R. McConnell and J. Spinrad in the following sense. There is an $O(n + m)$ algorithm constructing an orientation of any given graph G such that the orientation is a transitive orientation of G if and only if G has a transitive orientation (i.e., G is a comparability graph) [119]. Unfortunately, the best algorithm for testing whether the orientation is indeed transitive has running time $O(n^{2.376})$.

Furthermore there is an $O(n + m)$ algorithm that computes for a given cocomparability graph G a linear extension of a partially ordered set P_G that has comparability graph $\overline{G}$ [119]. Clearly this is exactly a cocomparability labeling of G. Consequently, recognizing cocomparability graphs is 'relatively expensive' but the model for all the domination algorithms can be computed very quickly.

The following theorem is the basis of the $O(n^3)$ algorithm in [105].

Theorem 8.11 *Any connected cocomparability graph G has a minimum car-dinality connected dominating set S such that the induced subgraph $\langle S \rangle$ is a chordless path $p_1, p_2, \ldots, p_k$.*

The definition of a cocomparability labeling implies various properties that are very helpful for locating a minimum cardinality connected dominating set.

Let $v_1, v_2, \ldots, v_n$ be a cocomparability labeling of a cocomparability graph $G = (V, E)$. For vertices $u, w \in V$ we write $u < w$, if u appears before w in the labeling, i.e., $u = v_i$ and $w = v_j$ implies $i < j$. For $i \leq j$ the set $\{v_k : i \leq k \leq j\}$ is denoted by $[v_i, v_j]$. Then $\{i, j\} \in E$ implies that every vertex v_k with $i < k < j$ is adjacent to v_i or to v_j, thus $\{v_i, v_j\}$ dominates $[v_i, v_j]$. This can be generalized as follows: Let $S \subseteq V$ where $\langle S \rangle$ is connected. Then S dominates $[\min(S), \max(S)]$, where $\min(S)$ (respectively $\max(S)$) is the vertex of S with the smallest (respectively largest) index in the labeling.

The following lemma has been given in [105].

Lemma 8.12 *Let $S \subseteq V$ be a minimum cardinality connected dominating set for a cocomparability graph $G = (V, E)$ and $v_1, v_2, \ldots, v_n$ a cocomparability labeling of G. Assume $\langle S \rangle$ is a chordless path $p_1, p_2, \ldots, p_k$. Then every vertex $x < \min(S)$ is dominated by $\{p_1, p_2\}$ and every vertex $y > \max(S)$ is dominated by $\{p_{k-1}, p_k\}$.*

The following approach enables an elegant way of locating a chordless path of minimum cardinality that dominates the cocomparability graph. We say that v_i is a *source vertex* if v_i is a minimal element of P_G, and we say that v_j is a *sink vertex* if v_j is a maximal element of P_G. Then $v_1, v_2, \ldots, v_n$ is a *canonical cocomparability labeling* if $v_1, v_2, \ldots, v_r$, $1 \leq r < n$, are the source vertices and $v_s, v_{s+1}, \ldots, v_n$, $1 < s \leq n$, are the sink vertices. This corresponds to a linear extension of P_G for which all minimal elements appear first and all maximal elements appear last. Thus every cocomparability graph G has a canonical cocomparability labeling. Furthermore, given any cocomparability labeling a canonical one can be computed in time $O(n + m)$.

From now on we assume that $v_1, v_2, \ldots, v_n$ is a canonical cocomparability labeling. Since the source vertices of G form a clique, any source vertex v_i dominates $[v_1, v_i]$. Analogously, since the sink vertices of G form a clique, any sink vertex v_j dominates $[v_j, v_n]$. Therefore the vertex set of every path between a source vertex and a sink vertex is dominating.

The following theorem given in [40] highlights the key property.

Theorem 8.13 *Every connected cocomparability graph $G = (V, E)$ satisfying $\gamma_c(G) \geq 3$ has a minimum cardinality connected dominating set which is the vertex set of a shortest path between a source and a sink vertex of G.*

Proof. Let $v_1, v_2, \ldots, v_n$ be a canonical cocomparability labeling of G. By Theorem 8.11 there is a minimum cardinality connected dominating set S of G

such that $\langle S \rangle$ is a chordless path P : $p_1, p_2, \ldots, p_k$, $k \geq 3$. We construct a chordless path P'' between a source vertex and a sink vertex of G, that has the same number of vertices as the path P.

Let $p_1 = v_i$ and $p_2 = v_j$. First observe that $p_2 = v_j$ cannot be a source vertex, otherwise $N[p_2] \supseteq [1, v_j]$ implying that $\{p_2, p_3, \ldots, p_k\}$ is also a connected dominating set of G, a contradiction. If p_1 is a source vertex then P starts at a source vertex. In this case we proceed with the path $P' = P$ (possibly) rearranging p_{k-1}, p_k.

Suppose $p_1 = v_i$ is not a source vertex. Then p_1 is not a minimal element of P_G, and there is a minimal element u of P_G with $u \prec_{P_G} p_1$. Hence u is a source vertex of G with $\{u, p_1\} \notin E$. Since $v_1, v_2, \ldots, v_n$ is a canonical cocomparability labeling, and since p_1 and p_2 are not source vertices, we get $u < p_1$ and $u < p_2$. Since $\{v_i, v_j\} \in E$ and by Lemma 8.12, $\{v_i, v_j\}$ dominates $[1, \mathtt{max}(\{v_i, v_j\})]$. Consequently, $\{u, p_2\} \in E$.

Consider the set $S' = \{u, p_2, \ldots, p_k\}$. Since $\{u, p_2\}$ dominates $[1, v_j]$, S' is a dominating set. Since P : $p_1, p_2, \ldots, p_k$ is a chordless path, $t \geq 3$ implies $\{p_t, u\} \notin E$. Thus S' induces the chordless path P' : $u, p_2, \ldots, p_k$.

Similarly, starting from P' the vertex p_k can be replaced, if necessary. Vertex p_{k-1} is not a sink vertex. If p_k is a sink vertex then $S'' = S'$ and $P'' = P'$. Otherwise we replace p_k by a sink vertex v satisfying $\{v, p_k\} \notin E$ to obtain S'' and P''.

The subgraph $\langle S'' \rangle$ induces a chordless path between a sink and a source vertex. The vertex set of any path between a sink and a source vertex is a dominating set. By construction S'' is a minimum cardinality connected dominating set. Consequently, S'' is the vertex set of a shortest path between a source vertex and a sink vertex of G. $\square$

Thus computing a minimum cardinality connected dominating set of a connected cocomparability graph G reduces to computing a shortest path between a source and a sink vertex of G, if $\gamma_c(G) \geq 3$.

Algorithm 2

Input: A connected cocomparability graph $G = (V, E)$ and
            a canonical cocomparability labeling $v_1, v_2, \ldots, v_n$ of G.
Output: A minimum cardinality connected dominating set of G.

1. Check whether G has a minimum connected dominating set D of
    cardinality at most 2. If so, output D and stop.
2. Construct a new graph G' by adding two new vertices s and t
    to G such that s is adjacent exactly to the source vertices
    of G and t is adjacent exactly to the sink vertices of G.
3. Compute a shortest path P : $s, p_1, p_2, \ldots, p_k, t$
    between s and t in G' by BREADTH-FIRST-SEARCH.
4. Output $\{p_1, p_2, \ldots, p_k\}$.

The correctness of Algorithm 2 follows immediately from Theorem 8.13. The 'almost linear' running time of the algorithm follows from the well-known fact that BREADTH-FIRST-SEARCH is a linear time procedure.

Theorem 8.14 *Given as input a connected cocomparability graph $G = (V, E)$ with a canonical cocomparability labeling $v_1, v_2, \ldots v_n$, Algorithm 2 outputs a minimum cardinality connected dominating set of G. The running time of the algorithm is $O(nm)$. However all parts of the algorithm except checking for a connected dominating set of cardinality two can be done in time $O(n + m)$.*

It is clearly unsatisfactory that the straightforward test for a connected dominating set of cardinality two dominates the overall running time. The crux is that there are even permutation graphs for which each minimum cardinality connected dominating set of cardinality two contains neither a source nor a sink vertex (see [95, 101]). It seems that minimum cardinality dominating sets of this type cannot be found by a shortest path approach. It is an open question whether Step 1 of Algorithm 2 can be implemented in a more efficient way.

We mention that the $O(n)$ algorithms computing a minimum cardinality connected dominating set for permutation graphs [95] and trapezoid graphs [101] both rely on Theorem 8.13.

D.G. Corneil, S. Olariu and L. Stewart have done a lot of research on *asteroidal triple-free graphs*, usually called AT-free graphs [59, 61]. They are defined as those graphs not containing an *asteroidal triple*, i.e., a set of three vertices such that between any two of the vertices there is a path avoiding the neighbourhood of the third.

The AT-free graphs form a superclass of the cocomparability graphs. They are a 'large class of graphs' with nice structural properties and some of them are related to domination. One of the major theorems on the structure of AT-free graphs states that every connected AT-free graph has a *dominating pair*, i.e., a pair of vertices u, v such that the vertex set of each path between u and v is a dominating set. An $O(n + m)$ algorithm computing a dominating pair for a given connected AT-free graph has been presented in [61]. This can be used to obtain an $O(n + m)$ algorithm computing a dominating path for connected AT-free graphs (see also [60]).

An $O(n^3)$ algorithm computing a minimum cardinality connected dominating set for connected AT-free graphs is given in [14]. An $O(n + m)$ algorithm computing a minimum cardinality connected dominating set in connected AT-free graphs with diameter greater than three is given in [61].

8.4 Dually Chordal Graphs

The research on domination problems for dually chordal graphs is to some extent motivated by searching for larger and larger graph classes with polynomial time domination algorithms. F. Dragan and A. Brandstädt contributed significantly to the research on dually chordal graphs [31, 34, 36, 70].

Dually chordal graphs form a large class of graphs that contains the strongly chordal graphs which we consider in Section 8.5. We mention that the weighted versions of the five domination problems that we consider are all NP-complete when restricted to dually chordal graphs. Hence it is unlikely that the efficient algorithms for weighted domination problems on strongly chordal graphs (see Section 8.5) can be extended to dually chordal graphs.

Fortunately the maximum neighbourhood ordering, a vertex ordering that each dually chordal graph possesses by definition, is very useful for designing efficient algorithms in the cardinality case and for so-called r-domination problems, another variant of domination (see e.g. [34, 70]).

We consider an $O(n + m)$ algorithm solving the minimum cardinality dominating clique problem on dually chordal graphs. This is a special case of the $O(n+m)$ algorithm for the r-dominating clique problem presented in [70]. This greedy algorithm also extends the $O(n+m)$ algorithm of [104] for the minimum cardinality dominating clique problem on strongly chordal graphs.

Definition 8.15 *A vertex $u \in N[v]$ is a* maximum neighbour *of v in the graph $G = (V, E)$ if $N[w] \subseteq N[u]$ for all $w \in N[v]$.*

An ordering $v_1, v_2, \ldots, v_n$ of the vertex set V of a graph $G = (V, E)$ is a maximum neighbourhood ordering *of G if for all $i \in \{1, 2, \ldots, n\}$ the vertex v_i has a maximum neighbour in the graph $G_i = \langle \{v_i, v_{i+1}, \ldots, v_n\} \rangle$, i.e., for all $i \in \{1, 2, \ldots, n\}$ there is a vertex $u_i \in N_i[v_i]$ such that $N_i[w] \subseteq N_i[u_i]$ for all $w \in N_i[v_i]$. Here $N_i[v]$ denotes the closed neighbourhood of the vertex v in the graph G_i.*

Then a graph is called *dually chordal* if it admits a maximum neighbourhood ordering. It is worth noting that dually chordal graphs have various characterizations in terms of hypergraphs (see e.g. [36]).

There is an $O(n + m)$ time recognition algorithm for dually chordal graphs, that also computes a maximum neighbourhood ordering of the input graph, if it is dually chordal [69] (see also [34, 70]). This algorithm is based on *MCS*, the maximum cardinality search algorithm of [145].

Thus we may assume that a dually chordal graph $G = (V, E)$ is given with a maximum neighbourhood ordering $v_1, v_2, \ldots, v_n$ and that for each vertex v_i, $i \in \{1, 2, \ldots, n\}$, there is a pointer to a maximum neighbour $mn(v_i)$ of the vertex v_i in the induced subgraph G_i. (These pointers are provided by the $O(n + m)$ algorithm computing a maximum neighbourhood ordering given in [69].)

Algorithm 3

Input: A dually chordal graph $G = (V, E)$ and
a maximum neighbourhood ordering $v_1, v_2, \ldots, v_n$ of G.

Output: A minimum cardinality dominating clique of G, if there
is one.
"NO DOMINATING CLIQUE" otherwise.

Initially, $D := \emptyset$.
FOR $i = 1$ **TO** n **DO**
    BEGIN
       IF v_i has no neighbour in D **THEN** $D := D \cup \{mn(v_i)\}$
    END
IF D is a clique **THEN** output D
ELSE output *"NO DOMINATING CLIQUE"*

This simple greedy algorithm scans the vertices in the order $v_1, v_2, \ldots, v_n$. If there is a dominating clique, it finds one of minimum cardinality in the following way. A vertex is inserted in the potential dominating clique D only if this is necessary to dominate the currently scanned vertex v_i. Moreover, in that case the algorithm inserts the maximum neighbour of v_i in D.

Theorem 8.16 *Given as input a dually chordal graph $G = (V, E)$ with a maximum neighbourhood ordering $v_1, v_2, \ldots v_n$, Algorithm 3 computes a minimum cardinality dominating clique, if there is one. Otherwise it outputs "NO DOMINATING CLIQUE". The running time of Algorithm 3 is $O(n + m)$.*

The final set D is a minimum cardinality dominating clique if and only if G has a dominating clique.

Proof. The running time of the algorithm is $O(n + m)$ since for each vertex v_i the amount of time for scanning v_i is $O(d(v_i))$.

By construction, the final set D is a dominating set, since whenever a current vertex v_i is not yet dominated by D, then the algorithm adds the maximum neighbour $mn(v_i)$ to D. Hence the final D cannot be a clique, if G has no dominating clique. Thus the algorithm is correct, if G has no dominating clique.

Assume that G has a dominating clique. The following claim establishes the correctness of the algorithm. Then $N[v] \subseteq N[u]$ implies that G has a dominating clique D with $v \notin D$. To see this, observe that $\tilde{D}$ is a dominating clique and $v \in \tilde{D}$ implies that $D = (\tilde{D} \setminus \{v\}) \cup \{u\}$ is a dominating clique.

Whenever a vertex v_i is scanned and $mn(v_i)$ is inserted in D, then the above claim applied to any neighbour of v_i and $mn(v_i)$ shows that there is a minimum dominating clique of G that contains $mn(v_i)$ instead of any other neighbour of v_i. Hence the final set D is a minimum cardinality dominating clique. $\square$

The algorithmic use of maximum neighbourhood orderings has also been considered in [31, 34, 69]. Among others, efficient algorithms for the r-domination and the connected r-domination problem on dually chordal graphs are presented.

Recently A. Brandstädt, F. Dragan and F. Nicolai introduced the *homogeneously orderable graphs*, a graph class containing the dually chordal graphs, the distance-hereditary graphs and the homogeneous graphs [37]. Efficient algorithms for the r-dominating clique and the connected r-domination problem on homogeneously orderable graphs are presented in [71].

8.5 Strongly Chordal Graphs

Pioneering work on domination algorithms has been done by M. Farber [77, 79, 78, 80, 152]. In contrast to many other papers in the area he used the framework of linear programming for designing efficient algorithms. We refer the reader unfamiliar with linear programming, duality of linear programs and the conditions of complementary slackness to [127].

We consider Farber's $O(n + m)$ algorithm to compute a minimum weight dominating set on strongly chordal graphs [79]. The important tool for designing efficient algorithms on strongly chordal graphs is the strong elimination ordering.

Definition 8.17 *A strong elimination ordering of a graph $G = (V, E)$ is an ordering $v_1, v_2, \ldots, v_n$ of V satisfying the following conditions for each i, j, k and l:*

(a) *If $i < j < k$ and $\{v_i, v_j\} \in E$, $\{v_i, v_k\} \in E$, then $\{v_j, v_k\} \in E$.*

(b) *If $i < j < k < l$ and $\{v_i, v_k\} \in E$, $\{v_i, v_l\} \in E$, $\{v_j, v_k\} \in E$, then $\{v_j, v_l\} \in E$.*

A graph is called *strongly chordal*, if it admits a strong elimination ordering. Given a graph $G = (V, E)$, it can be checked in time $O(\min(m \log n, n^2))$ whether G is strongly chordal, and a strong elimination ordering can be computed within the same timebound if G is strongly chordal [126, 142].

We assume that a strongly chordal graph $G = (V, E)$ is given together with a strong elimination ordering $v_1, v_2, \ldots, v_n$ of G. By Theorem 8.1, we may assume that the weights $w_1, w_2, \ldots, w_n$ of the vertices are nonnegative reals. We will locate a minimum weight dominating set of G by solving an associated linear program. Following the original paper, for a strong elimination ordering $v_1, v_2, \ldots, v_n$ of a graph $G = (V, E)$, we write $i \sim j$, if $\{v_i, v_j\} \in E$ or $i = j$. Then the linear program can be formulated as follows.

$$P(G, w): \quad \text{Minimize} \quad \sum_{i=1}^{n} w_i x_i,$$

$$\text{subject to} \quad \sum_{i \sim j} x_i \geq 1 \quad \text{for each } j \in \{1, 2, \ldots, n\},$$

$$x_i \geq 0 \qquad \text{for each } i \in \{1, 2, \ldots, n\}.$$

There is a one-to-one correspondence between feasible $0-1$ solutions to $P(G, w)$ and dominating sets of G. Moreover, an optimal $0-1$ solution to $P(G, w)$ corresponds to a dominating set of G, that has minimum weight. The algorithm to be presented solves $P(G, w)$ as well as the following dual problem of $P(G, w)$.

$$D(G, w): \quad \text{Maximize} \quad \sum_{j=1}^{n} y_j,$$

$$\text{subject to} \quad \sum_{j \sim i} y_j \leq w_i \quad \text{for each } i \in \{1, 2, \ldots, n\},$$

$$y_j \geq 0 \quad \text{for each } j \in \{1, 2, \ldots, n\}.$$

To simplify the presentation of the algorithm we define a function h and a family of sets associated with the linear programs $P(G, w)$ and $D(G, w)$. For each i,

$$h(i) = w_i - \sum_{j \sim i} y_j, \quad \text{and} \quad T_i = \{k \ : \ k \sim i \text{ and } y_k > 0\}.$$

The algorithm is of the primal-dual type and has two stages. Stage one finds a feasible solution $y_1, y_2, \ldots, y_n$ of the dual problem $D(G, w)$ by scanning the vertices in the order $v_1, v_2, \ldots, v_n$. Stage two finds a feasible $0-1$ solution $x_1, x_2, \ldots, x_n$ of $P(G, w)$ by scanning the vertices in the order $v_n, v_{n-1}, \ldots, v_1$.

Algorithm 4

Input: A strongly chordal graph $G = (V, E)$, a strong elimination ordering $v_1, v_2, \ldots v_n$ of G and nonnegative real vertex weights $w_1, w_2, \ldots, w_n$.

Output: A minimum weight dominating set of G.

Initially, $T = \{1, 2, \ldots, n\}$, each $y_j = 0$ and each $x_i = 0$.

Stage one: **FOR** $j = 1$ **TO** n **DO** $y_j := \min\{h(k) \ : \ k \sim j\}$;

Stage two: **FOR** $i = n$ **DOWNTO** 1 **DO**

           IF $h(i) = 0$ **AND** $T_i \subseteq T$ **THEN**

                BEGIN

                    $x_i := 1$;

                    $T := T \setminus T_i$;

                END

Output $\{x_i \ : \ x_i = 1\}$.

Theorem 8.18 *Given as input a strongly chordal graph $G = (V, E)$ with a strong elimination ordering $v_1, v_2, \ldots v_n$ and nonnegative vertex weights $w_1, w_2, \ldots, w_n$. Algorithm 4 has running time $O(n+m)$ and the final values of $x_1, x_2, \ldots, x_n$ and $y_1, y_2, \ldots, y_n$ are optimal solutions to $P(G, w)$ and $D(G, w)$, respectively. Hence the output of Algorithm 4 is a minimum weight dominating set of G.*

Proof. The running time of the algorithm is $O(n + m)$ since in each stage every vertex is scanned once and the amount of time for scanning a vertex v is $O(d(v))$.

By the duality theorem of linear programming, in order to show that the algorithm finds optimal solutions to $P(G, w)$ and $D(G, w)$, it suffices to show that these solutions are feasible and that they satisfy the conditions of complementary slackness. For a feasible solution $x_1, x_2, \ldots, x_n$ of $P(G, w)$ and a feasible solution $y_1, y_2, \ldots, y_n$ of $D(G, w)$ the conditions of complementary slackness are

- for each i, $x_i > 0$ implies $\sum_{j \sim i} y_j = w_i$ (i.e., $h(i) = 0$), and

- for each j, $y_j > 0$ implies $\sum_{i \sim j} x_i = 1$.

Notice that $h(i)$ is the slack in the dual constraint associated with vertex v_i, and that T_i is the set of those constraints in $P(G, w)$ which contain x_i and must be at equality to satisfy the conditions of complementary slackness.

(i) Feasibility of dual solution. The instructions in Stage one guarantee that $y_j \geq 0$ and $h(j) \geq 0$ for each j. To see this, notice that each y_j is chosen such that all slacks $h(i)$, $i \in \{1, 2, \ldots, n\}$, remain nonnegative.

(ii) Feasibility of primal solution. Since $x_i \in \{0, 1\}$ for each i, it suffices to show that for each j, there is an $i \sim j$ with $x_i = 1$.

Consider an arbitrary j. By the choice of y_j in Stage one, there is some $k \sim j$ such that $h(k) = 0$ and $\max T_k \leq j$. If $x_k = 1$, we are done. Otherwise, by the algorithm, T_k was not a subset of T when v_k was scanned in Stage two. Since in Stage two vertices are scanned in the order $v_n, v_{n-1}, \ldots, v_1$, there is some $l > k$ such that $x_l = 1$ and $T_l \cap T_k \neq \emptyset$. Suppose $i \in T_l \cap T_k$. Then $i \leq j$ since $\max T_k \leq j$. Altogether we get $i \sim l$, $i \sim k$, $k \sim j$, $k < l$ and $i \leq j$.

We claim that this implies $l \sim j$ since $v_1, v_2, \ldots, v_n$ is a strong elimination ordering. This can be seen as follows. Clearly $l \sim j$ if $i = j$. Thus we may assume $i < j$. If $i \leq l$ then property (a) implies $l \sim j$, since $i \leq j$, $i \sim j$ and $i \sim l$. If $l < i$ then $k < l < i < j$, thus $k \sim i$, $k \sim j$ and $l \sim i$ implies $l \sim j$ by property (b).

The claim implies $x_l = 1$ and $l \sim j$. Consequently the primal solution is feasible.

(iii) Complementary slackness. If $x_i > 0$, then $x_i = 1$ and so $h(i) = 0$, i.e., $\sum_{j \sim i} y_j = w_i$.

Suppose $y_j > 0$. It is clear from the instructions of Stage two that $x_i = x_k = 1$ for $i \sim j$ and $k \sim j$ would imply $T_i \cap T_k = \emptyset$, contradicting $j \in T_i \cap T_k$. Thus, $\sum_{i \sim j} x_i \leq 1$. Equality follows from the feasibility of the primal solution. $\square$

M. Farber also gives an $O(n + m)$ algorithm computing a minimum weight independent dominating set, if the graph is given with a strong elimination ordering [79]. Furthermore G.J. Chang has presented in [42] an $O(n + m)$ algorithm computing a minimum cardinality total dominating set for strongly chordal graphs (if the strongly chordal graph is given together with a simple elimination ordering).

8.6 Interval Graphs

Interval graphs are a well-known class of graphs. They form a subclass of chordal graphs and of strongly chordal graphs. Interval graph are very useful for modelling real world problems. There are applications in various areas, among them molecular biology and archaeology (see e.g. [84, 138]).

Concerning domination algorithms, M. Farber constructed a simple $O(n + m)$ algorithm for computing a strong elimination ordering of a given interval graph [77]. Hence there are $O(n + m)$ algorithms computing a minimum weight dominating set and a minimum weight independent dominating set on interval graphs, as a consequence of M. Farber's algorithms presented in the previous section (see also [77, 79]).

Nevertheless there remains a natural question that seems worth studying. What happens if not the interval graph, but a representation of the graph is the input? A compact representation of the graph, that has only length $O(n)$, might allow $O(n)$ algorithms.

Definition 8.19 *A graph $G = (V, E)$ is an* interval graph *if there is a one-to-one correspondence between a finite collection of closed intervals of the real line and the vertex set V such that two vertices u and v are adjacent if and only if their corresponding intervals have a nonempty intersection.*

Then the collection of intervals that represents the graph $G = (V, E)$ is an *interval model* of G. It is worth noting that every interval graph has an interval model in which no two endpoints coincide [84].

We assume that an interval graph $G = (V, E)$ is presented by a particular interval model, that we call interval model $\mathcal{I}$. The coordinates of $\mathcal{I}$ are obtained as follows. The endpoints of the intervals are labeled from left to right by $1, 2, \ldots, 2n$. Furthermore the intervals are labeled by $1, 2, \ldots n$ in increasing order of their right endpoints. For algorithmic reason we assume that an interval model $\mathcal{I}$ is given by a list containing for each interval i its left endpoint a_i and its right endpoint b_i. Throughout this section we always assume that an interval model of an interval graph is of this type. (See Figure 8.2).

Let us see how an interval model $\mathcal{I}$ relates to other representations of an interval graph. Consider any interval model of an interval graph G in which all labels of endpoints are of size $O(n)$. Then sorting the labels of the endpoints of the intervals using bucket sort, we can obtain an interval model $\mathcal{I}$ of G in time $O(n)$. Thus the endpoints are then labeled $1, 2, \ldots, 2n$ from left to right.

Given a graph G, there are various $O(n + m)$ interval graph recognition algorithms that also compute an interval model, if the input is indeed an interval

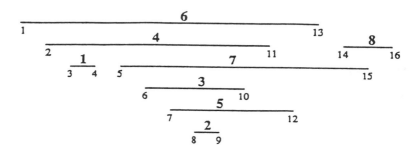

Figure 8.2: Interval model $\mathcal{I}$.

graph. The first one has been given by Booth and Lueker in [26]. All these recognition algorithms compute an interval model of the given interval graph G that can be transformed into an interval model $\mathcal{I}$ of G within $O(n)$ time.

Nevertheless it is necessary to mention that another type of interval model is used frequently in the literature. This interval model allows arbitrarily large coordinates of the endpoints of the intervals. Thus sorting the endpoints requires time $\Theta(n \log n)$. Using an $O(n \log n)$ sorting algorithm to sort the endpoints of all intervals of such a model, we obtain an interval model $\mathcal{I}$ in time $O(n \log n)$.

Now we consider a domination algorithm for interval graphs. M.S. Chang presents $O(n)$ algorithms for computing a minimum weight dominating set, a minimum weight independent dominating set and a minimum weight connected dominating set for interval graphs in [48]. His algorithms require as input an interval model $\mathcal{I}$ of the interval graph G. Thus, with the notation of Section 8.1, *Recognition* takes $O(n + m)$ time, but the *Problem* can be solved faster, namely in time $O(n)$.

We consider in detail the $O(n \log \log n)$ algorithm for computing a minimum weight total dominating set presented by M.S. Chang in [48]. Again the interval graph is given by an interval model $\mathcal{I}$. Thus the intervals are labeled $1, 2, \ldots, n$ and the endpoints are labeled $1, 2, \ldots, 2n$ from left to right. (See Figure 8.2).

The design and analysis of the algorithm are done using the interval model $\mathcal{I}$ and each vertex obtains the label of its corresponding interval. In this way we identify the graph with the given interval model. Furthermore, we identify each vertex with its corresponding interval and every interval i has a given real vertex weight $w(i)$. By Theorem 8.1 we may assume that all vertex weights are nonnegative.

We start with some definitions and notation. Let $\mathcal{I}$ be the interval model given as input for the graph $G = (V, E)$. By definition, $1 \le a_i < b_i \le 2n$ for $1 \le i \le n$. We say that interval i is *left* of interval j if $b_i < a_j$. Let S be a nonempty set of intervals in $\mathcal{I}$, thus $S \subseteq \{1, 2, \ldots, n\}$. Motivated by the fact that intervals are labeled in increasing order of their right endpoints, we denote the interval with right endpoint $\min\{b_i \ : \ i \in S\}$ by $first(S)$ and the interval with right endpoint $\max\{b_i \ : \ i \in S\}$ by $last(S)$. Notice that $first(\{1, 2, \ldots, n\}) = 1$.

Definition 8.20 *Let* $e \in \{1, 2, \ldots, 2n\}$ *be any endpoint of an interval in* $\mathcal{I}$. *Then* $in(e)$ *is the interval having endpoint* e. *We denote by* $p(e)$ *the largest left endpoint of an interval whose right endpoint is less than* e, *if there is one. Otherwise* $p(e) = 0$.
For $i \in \{1, 2, \ldots, n\}$, *we define* $V(i) = \{j \ : \ j \in \{1, 2, \ldots, n\}$ *and* $a_j < b_i\}$.

Hence $V \setminus V(i)$ is the set of those intervals l for which i is left of l, thus interval i has nonempty intersection with any interval $l \in V \setminus V(i)$. Consequently no vertex of a set $S \subseteq \{1, 2, \ldots, i\}$ has a neighbour in $V \setminus V(i)$.

The algorithm is designed by dynamic programming. It scans the endpoints of the intervals in increasing order and computes some partial solutions.

Definition 8.21 *A set* S *of intervals in* $\mathcal{I}$ *is a* total partial dominating set *if the vertex set* S *dominates* $V(last(S))$ *and* $\langle S \rangle$ *has no isolated vertex.*

Definition 8.22 *A set* S *of intervals in* $\mathcal{I}$ *is an* intermediate total partial dominating set *if the vertex set* S *dominates* $V(last(S))$ *and* $\langle S \rangle$ *has exactly one isolated vertex which is* $last(S)$.

Let $i \in \{1, 2, \ldots, n\}$. Then let $TPD(i)$ be the collection of all total partial dominating sets S that contain the interval $first(\{1, 2, \ldots, n\}) = 1$ and for which $last(S) = i$. Let $ITPD(i)$ be the collection of all intermediate total partial dominating sets S that contain the interval $first(\{1, 2, \ldots, n\}) = 1$ and for which $last(S) = i$. Furthermore $XTPD(i) = TPD(i) \cup ITPD(i)$. Notice that any set $S \in XTPD(i)$ contains the intervals 1 and i.

Definition 8.23 *Let* $i \in \{1, 2, \ldots, n\}$. *We define*

$$X(i) = \bigcup_{\{j \ : \ a_i < b_j < b_i\}} XTPD(j)$$

$$K(i) = \{\{j\} \ : \ j \in \{1, 2, \ldots, n\} \text{ and } a_i < a_j < b_j < b_i\}$$

$MTPD(i)$, $MITPD(i)$, $MXTPD(i)$, $MX(i)$ and $MK(i)$ denotes any minimum weight set in the collection $TPD(i)$, $ITPD(i)$, $XTPD(i)$, $X(i)$ and $K(i)$, respectively, if the collection is nonempty. Otherwise it denotes a set of infinite weight. The next lemma guarantees the correctness of the algorithm. In fact the algorithm of Chang is essentially based on the insights of this lemma and its proof.

Lemma 8.24 *Let* G *be an interval graph with interval model* $\mathcal{I}$. *Then the following four statements are true:*

(1) $TPD(1) = \emptyset$, $ITPD(1) = \{\{1\}\}$ *and* $XTPD(1) = \{\{1\}\}$.

(2) *Let* $i \in \{2, 3, \ldots, n\}$ *and* $a_i < b_1$. *Then* $ITPD(i) = \emptyset$.

(3) *Let* $i \in \{2, 3, \ldots, n\}$ *and* $a_i > b_1$. *Then*
$ITPD(i) = \{A \cup \{i\} \; : \; A \in TPD(j) \text{ with } p(a_i) < b_j < a_i\}$.

(4) *Let* $i \in \{2, 3, \ldots, n\}$. *Then*
$MTPD(i) = Min(\{MX(i) \cup \{i\}, MITPD(i) \cup MK(i)\})$.

Proof. By Definitions 8.21 and 8.22 and since each set of $TPD(i)$ as well as each set of $ITPD(i)$ must contain the intervals 1 and i, we obtain (1) and (2).

Consider (3). Let $S \in ITPD(i)$ and $k = last(S \setminus \{i\})$. By definition, $\{1, i\} \subseteq S$ and i is an isolated vertex of $\langle S \rangle$, implying $b_k < a_i$. By definition, S dominates $V(i) = \{j \; : \; a_j < b_i\}$. Thus $b_k < a_i$ implies that $S \setminus \{i\}$ dominates $V(k) \subseteq V(i)$ and $\{i\}$ dominates $V(i) \setminus V(k)$. Consequently there cannot be an interval l with $b_k < a_l < b_l < a_i$. Thus $S \in ITPD(i)$ implies $S \setminus \{i\} \in TPD(k)$ with $p(a_i) < b_k < a_i$. On the other hand, if $A \in TPD(j)$ and $p(a_i) < b_j < a_i$ then it is easy to see that $A \cup \{i\} \in ITPD(i)$. Thus we obtain (3).

Consider (4). Suppose $TPD(i) \neq \emptyset$. We claim that there exists a minimum weight set S in $TPD(i)$ such that, either $S \setminus \{i\} \in X(i)$ or there exists an interval $l \in S$ with $\{l\} \in K(i)$ and $S \setminus \{l\} \in ITPD(i)$.

Let S be a set of minimum cardinality among the minimum weight sets in $TPD(i)$. (This choice is made since G might have zero weight vertices.) By definition, S dominates $V(i)$. We denote by $C(i)$ the set of all intervals $l > 1$ for which $\{l\} \in K(i)$. Thus $C(i)$ is the set of all intervals, that are properly contained in interval i, except interval 1.

Case 1: $C(i) \neq \emptyset$.

Obviously $S \setminus C(i)$ is a dominating set of $V(i)$. Since all weights are non-negative, we have $w(S \setminus C(i)) \leq w(S)$. Since $|S \setminus C(i)| < |S|$, the choice of S implies $S \setminus C(i) \notin TPD(i)$. Consequently, $\langle S \setminus C(i) \rangle$ has isolated vertices. Now $S \in TPD(i)$ implies that the only isolated vertex of $\langle S \setminus C(i) \rangle$ is i. Thus $\{1\} \notin K(i)$. Consequently $S \setminus C(i) \in ITPD(i)$.

Furthermore $|S \cap C(i)| = 1$. Otherwise $S' = S \setminus \{t\}$ for some $t \in C(i)$ would be a set in $TPD(i)$ with $w(S') \leq w(S)$ and $|S'| < |S|$, contradicting the choice of S. Consequently there is an $l \in S$ with $\{l\} \in K(i)$ and $S \setminus \{l\} \in ITPD(i)$.

Case 2: $C(i) = \emptyset$.

If $\{1\} \in K(i)$, then $S = \{1, i\}$ by the choice of S. Hence $S \setminus \{i\} \in MITPD(1)$. Therefore $a_i < b_1 < b_i$ implies $S \setminus \{i\} \in X(i)$.

Now suppose $\{1\} \notin K(i)$, i.e., $K(i) = \emptyset$. Let $k = last(S \setminus \{i\})$. Since $\langle S \rangle$ has no isolated vertices and since no interval in $\mathcal{I}$ is contained in the interval i, we get $a_k < a_i < b_k < b_i$. This implies that $S \setminus \{i\}$ dominates $V(k)$.

Thus, if $\langle S \setminus \{i\} \rangle$ has no isolated vertex, then $S \setminus \{i\} \in TPD(k)$. Otherwise, if $\langle S \setminus \{i\} \rangle$ has isolated vertices then each of them must be adjacent to i, since

$\langle S \rangle$ has no isolated vertex. Furthermore, $K(i) = \emptyset$ implies that k is the only isolated vertex of $\langle S \setminus \{i\} \rangle$. Consequently, $S \setminus \{i\} \in ITPD(k)$.

Therefore we always have $S \setminus \{i\} \in XTPD(k)$. Thus $a_i < b_k < b_i$ implies $S \setminus \{i\} \in X(i)$. This completes the proof of the claim.

It is a matter of routine to verify the following two assertions. If $A \in X(i)$, then $A \cup \{i\} \in TPD(i)$. If $A \in ITPD(i)$ and $K(i) \neq \emptyset$, then $A \cup \{l\} \in TPD(i)$ for any $\{l\} \in K(i)$. Thus we obtain (4). $\square$

The remainder of this section is an attempt to provide the reader with most of the necessary details to understand the involved algorithm, which basically works as follows. Partial solutions belonging to $TPD(i)$ and $MTPD(i)$ are computed by a dynamic programming that scans the endpoints of the intervals in increasing order and uses the rules of Lemma 8.24. Nevertheless various involved data structures and several tricky ideas are needed to design an algorithm running in time $O(n \log \log n)$.

For technical reason the algorithm works on an extended interval model $\mathcal{I}'$, obtained by adding intervals $-1, 0, 2n+1, 2n+2$ to $\mathcal{I}$. The algorithm is applied to a graph G' obtained from the original graph G by adding two new connected components that are both isomorphic to K_2. Hence any total dominating set of G' contains $\{-1, 0, 2n+1, 2n+2\}$. Furthermore D is a minimum weight total dominating set of G if and only if $D \cup \{-1, 0, 2n+1, 2n+2\}$ is a minimum weight total dominating set of G'. We need a procedure `Contained`$(\mathcal{I})$ to compute $MK(i)$ for $i \in \{2, 3, \ldots n\}$. This procedure is the only part of the algorithm that cannot be performed in time $O(n)$ up to now.

The following technical comments are necessary to understand Algorithm 5. The algorithm maintains two lists L_1 and L_2. Thereby $pred_{L_2}(b_i)$ denotes the largest number in L_2 which is less than b_i. The algorithm also maintains a collection of sets B_k, that form a partition of the set of all left endpoints of intervals, during each step of the algorithm. This partition of the set of all left endpoints has to be maintained such that the operations `FIND` and `UNION` can be executed quickly.

To guarantee the overall running time $O(n \log \log n)$ all the sets $MTPD(i)$, $MITPD(i)$ and $MXTPD(i)$ for $i \in \{-1, 0, \ldots, n+2\}$ should be stored using pointers. This is a standard technique and can be done as follows.

We use arrays $MTPD[-1..n+2]$, $MITPD[-1..n+2]$ and $MXTPD[-1..n+2]$. Each entry of an array is either the sign ∞, indicating an infinite set, or a record. The record consists of an integer and a pointer. The integer is from the set $\{-1, 0, \ldots, n+2\}$ or an integer out of this range, say -2, as empty sign. The pointer points to one entry of one of the three arrays $MTPD[-1..n+2]$, $MITPD[-1..n+2]$ and $MXTPD[-1..n+2]$ or it is NIL.

If the corresponding set $MTPD(i)$ and $MITPD(i)$, respectively, is a singleton, then the value of the pointer is NIL and the set is $\{i\}$.

Now suppose that the corresponding collection of sets $TPD(i)$, $ITPD(i)$

and $XTPD(i)$ is nonempty. Then there are three instructions in Algorithm 5 that are implemented via pointers.

Consider $MITPD(i) := \{i\} \cup MTPD(head_{L_1})$. Then i is the integer in $MITPD(i)$ and the pointer is set to $MTPD(head_{L_1})$.

Consider $MTPD(i) := \text{Min}(\{MXTPD(k) \cup \{i\}, MITPD(i) \cup MK(i)\})$. If $MTPD(i) = MXTPD(k) \cup \{i\}$, then i is the integer in $MTPD(i)$ and the pointer is set to $MXTPD(k)$. Otherwise, if $MTPD(i) = MITPD(i) \cup MK(i)$ then the integer is l, if $\{l\} = MK(i)$, and the pointer is set to $MITPD(i)$.

Consider $MXTPD(i) := \text{Min}(\{MTPD(i), MITPD(i)\})$. $MXTPD(i)$ has a pointer to $MTPD(i)$ and $MITPD(i)$, respectively, and the integer is -2 (i.e., empty).

Notice that this implementation ensures that for any of these sets all its elements can be listed in time $O(n)$. For example, the elements of $MTPD(n+2)$ can be listed as follows. Start at the array entry $MTPD(n+2)$ and then follow the pointers and list all integers in the record of array entries passed during this procedure.

Furthermore $WT(i)$, $WIT(i)$ and $WX(i)$ is the minimum weight of a set in $TPD(i)$, $ITPD(i)$ and $XTPD(i)$, respectively, if the collection is nonempty. Otherwise this weight is ∞. Now we are ready to describe the algorithm.

Algorithm 5

Input: An interval model $\mathcal{I}$ of an interval graph $G = (V, E)$, which has no isolated vertex. For each interval $i \in \{1, 2, \ldots n\}$, a left endpoint a_i, a right endpoint b_i and a nonnegative real weight $w(i)$.

Output: A minimum weight total dominating set of G.

Add four intervals $-1, 0, n+1, n+2$ with $w(-1) = w(0) = w(n+1) = w(n+2) = 0$ and $a_{-1} = -3$, $a_0 = -2$, $b_{-1} = -1$, $b_0 = 0$, $a_{n+1} = 2n+1$, $a_{n+2} = 2n+2$, $b_{n+1} = 2n+3$ and $b_{n+2} = 2n+4$, creating a new interval model $\mathcal{I}'$.

Initialize $WIT(0) = \infty$, $WT(-1) = \infty$, $WIT(-1) = 0$, $WX(-1) = 0$.

Initialize two lists $L_1 := -1$ and $L_2 := b_{-1}, b_0, b_1, \ldots, b_{n+2}$.

Compute $MK(i)$ for all $i \in \{2, 3, \ldots, n\}$ using Contained($\mathcal{I}$);

Scan the endpoints of $\mathcal{I}'$ in increasing order to partition the set of all left endpoints of intervals into subsets $B_k = \{a_j : b_{k-1} < a_j < b_k\}$ for $k = -1, 0, \ldots, n+2$, where $b_{-2} = -4$.

FOR $e := 0$ **TO** $2n$ **DO**
    BEGIN
    IF $e = a_i$ **THEN**
        BEGIN
        $MITPD(i) := \{i\} \cup MTPD(head_{L_1})$;

$$WIT(i) := w(i) + WT(head_{L_1})$$
END
   ELSE $[e = b_i]$
      BEGIN
      FIND the set B_k that contains a_i.
      IF $k = i$ **THEN** $WT(i) := \infty$
      ELSE
         BEGIN
         $MTPD(i) := \text{Min}(\{MXTPD(k) \cup \{i\}, MITPD(i) \cup MK(i)\});$
         $WT(i) := \min\{w(i) + WX(k), w(MK(i)) + WIT(i)\}$
         END
      $MXTPD(i) := \text{Min}(\{MTPD(i), MITPD(i)\});$
      $WX(i) := \min\{WT(i), WIT(i)\}$
      WHILE $b_{head_{L_1}} < a_i$
         remove the current element from the head of L_1;
      WHILE $(L_1 \neq \emptyset$ **AND** $WT(tail_{L_1}) > WT(i))$
         remove the current element from the tail of L_1;
      Append i to the tail of L_1;
      WHILE $WX(in(pred_{L_2}(b_i))) > WX(i)$ **DO**
         BEGIN
         UNION $B_i := B_i \cup B_{pred_{L_2}(b_i)}$;
         remove $pred_{L_2}(b_i)$ from L_2.
         END
      END
   END
Output $MTPD(n+2) \setminus \{-1, 0, 2n+1, 2n+2\}$.

Certainly this is technically the most complicated algorithm of this chapter. A good way of learning how it works is to take some input data, such as that in Figure 8.2, and run the algorithm (on a sheet of paper).

What is important to learn from this section is that the design of fast domination algorithms quickly leads to problems that have little to do with graph theory or structural properties of graph classes. Data structures and basic algorithmic problems become important.

To see the correctness of the algorithm observe that the lists L_1 and L_2 are maintained such that the following two properties are fulfilled. If j is an element of L_1 and a_i is the currently scanned left endpoint, then $p(a_i) < b_j < a_i$. If the right endpoint e_1 is an element of L_2, $e_1 < e_2$ and e_2 is the currently scanned endpoint, then there does not exist a right endpoint e_3 such that e_3 is an element of L_2, $e_1 < e_3 < e_2$ and $WX(in(e_1)) > WX(in(e_3))$.

There are two interesting features of Algorithm 5 that we would like to mention. The first is the UNION-FIND data structure for maintaining the B_k

sets that form a partition of the set of all left endpoints of intervals. In our special case it can be handled in time $O(n)$ [48]. We refer the interested reader to [57] for more information on the UNION-FIND problem.

Using the result on the UNION-FIND problem, it is not hard to show that the running time of the algorithm is $O(n)$, except the amount of time for the procedure Contained($\mathcal{I}$). This procedure computes for each interval $i \in \{2, 3, \ldots n\}$ of $\mathcal{I}$ the label $MK(i)$ of a minimum weight interval contained in i, if there is one. M.S. Chang presents an $O(n \log \log n)$ implementation of this procedure that uses an interesting data structure of van Emde Boas given in [76].

It is not impossible that there is an $O(n)$ implementation for the procedure Contained($\mathcal{I}$). Maybe it will use an involved data structure from Computational Geometry.

Theorem 8.25 *Given as input an interval model $\mathcal{I}$ of an interval graph $G = (V, E)$, which has no isolated vertex, and nonnegative vertex weights $w(1), w(2), \ldots, w(n)$, Algorithm 5 computes in time $O(n \log \log n)$ a minimum weight total dominating set of G.*

In [48] M.S. Chang has also shown how to design efficient algorithms for weighted domination problems on circular-arc graphs.

8.7   Other Domination Algorithms

Much more research on algorithms for domination problems or problems closely related to domination has been done. In this section we mention some of this research to give the interested reader an impression of the various different algorithms for domination-type problems.

8.7.1   Trees and partial k-trees

Many of the first polynomial time algorithms for domination problems dealt with trees or series-parallel graphs. By now it is well-known that a large number of well-studied NP-complete graph problems are not only solvable in polynomial time on trees. There are different general approaches showing that a variety of intractable graph problems can be solved in polynomial time when restricted to *partial k-trees*, for some fixed k, also known as *graphs of bounded treewidth* [4, 5, 6, 16, 22]. For an overview we recommend [23].

Certainly these impressive results have to be known before working on some particular domination problem for classes of small treewidth such as trees, series-parallel graphs and Halin graphs.

Typically the abovementioned algorithms are not practical since there is a huge constant hidden in the 'big oh'. In this respect the work in [146] and [147] is worth mentioning. Its aim is the design of 'practical' algorithms for problems on graphs of bounded treewidth, with emphasis to domination type problems.

8.7.2 Domatic number, perfect domination and irredundance

Efficient algorithms computing the domatic number have been given on graph classes, that are domatically full. Indeed most results are based on the fact that strongly chordal graphs are domatically full [79]. There is a linear time algorithm to compute an optimal domatic partition for strongly chordal graphs in [128]. Fast algorithms to compute an optimal domatic partition for interval graphs are given in [118, 116, 136].

Perfect domination has been studied by various authors in the last years. Often the techniques are similar to those used for the corresponding domination problem without the perfectness condition. Weighted perfect domination problems for chordal graphs and split graphs are considered in [49]. An $O(n^2)$ algorithm for weighted independent perfect domination on cocomparability graphs is given in [47]. Linear time algorithms for perfect domination on interval graphs and $O(nm + n^2)$ algorithms for perfect domination on circular-arc graphs are presented in [50]. Further algorithms are given in [46, 88, 155].

An algorithm solving the irredundance problem on circular-arc graphs is given in [85].

8.7.3 More variants of domination

There are linear time algorithms for computing the private domination number on trees [91, 120]. A linear time algorithm computing the upper total domination number for trees is given in [82] and upper fractional domination is considered in [51]. In [140] a linear time algorithm is given to solve the general R-domination problem.

In [53] a linear time algorithm is given that finds a universal minimal dominating function for a given tree, if one exists. In [86] the authors present a linear time algorithm for finding two minimum dominating sets with minimum possible intersection for trees.

Linear time algorithms computing the k-domination number on trees and series-parallel graphs are given in [96]. Furthermore edge domination is considered in [154]. A linear time algorithm computing the efficiency of trees is given in [17] and this has been extended to partial k-trees in [148]. In [87] the authors present a set of recurrence relations for series-parallel graphs and show that this particular set of recurrence relations can be used to evaluate many subset parameters such as the minimum domination, maximum independence, minimum redundance, maximum efficient domination, and maximum packing numbers.

8.7.4   Parallel algorithms

There is an increasing number of papers on efficient parallel algorithms for domination problems on special graph classes.

Efficient parallel algorithms have been presented for the dominating set and the dominating clique problem on strongly chordal graphs [65], for connected, independent and total domination on interval graphs [2] and for the minimum weight dominating set problem on interval graphs [19]. Furthermore there are efficient parallel algorithms for the minimum dominating set and the minimum total dominating set problem on circular-arc graphs [21, 135]. An NC algorithm computing a minimum weight dominating set on cographs is given in [1].

More parallel domination algorithms can be found in [3, 9, 81, 90].

8.8   Complexity Summary

We summarize the current state of our knowledge about the algorithmic complexity of a collection of domination problems in a table (Figure 8.3). For definitions and properties of graph classes we refer again to [29, 84]. We also refer to the ancestor of our table presented in [63].

The following easy observations are worth mentioning. Assume a graph class $\mathcal{G}$ is a subclass of the graph class $\mathcal{G}'$. Then any polynomial time algorithm $\mathcal{A}$ solving problem $\mathcal{P}$ on $\mathcal{G}'$ clearly also solves $\mathcal{P}$ on $\mathcal{G}$. Furthermore if problem $\mathcal{P}$ remains NP-complete when restricted to $\mathcal{G}$ then it also remains NP-complete on $\mathcal{G}'$. Consequently it is important for understanding the relation between all the results summarized in the table, that one knows about the containment relations between the various graph classes. To assist the reader we list the fundamental containment relations:

bipartite $\subset$ comparability,

interval $\subset$ strongly chordal $\subset$ chordal,

split $\subset$ chordal,

strongly chordal $\subset$ dually chordal $\subset$ homogeneously orderable,

permutation $\subset$ cocomparability $\subset$ AT-free,

distance-hereditary $\subset$ homogeneously orderable,

permutation $\subset$ k-polygon $\subset$ circle.

graph class	domination	independent domination	connected domination	total domination	minimum dominating clique
bipartite	**NP-c**[67]	**NP-c**[62]	**NP-c**[129]	**NP-c**[129]	**IP**[39]
comparability	**NP-c**[67]	**NP-c**[62]	**NP-c**[129]	**NP-c**[129]	**IP**[39]
interval	**IP**[79][48]	**IP**[79] [48]	**IP**[134][48]	**IP**[134] [48]	**IP**[39]
strongly chordal	**IP**[79]	**IP**[79]	**IP**[152]	**IP**[42]	**IP**[104]
split	**NP-c**[18][62]	**IP**[78]	**NP-c**[107]	**NP-c**[107]	**NP-c**[39]
chordal	**NP-c**[25]	**IP**[78]	**NP-c**[107]	**NP-c**[108]	**NP-c**[39]
dually chordal	**IP**[72][34]	**NP-c**[32]	**IP**[69][32]	**IP**[106]	**IP**[34][70]
permutation	**IP**[80][150][137]	**IP**[80] [11] [113]	**IP**[56][95]	**IP**[38] [63][106]	**IP**[39]
cocomparability	**IP**[105][40]	**IP**[105] [40]	**IP**[105] [40]	**IP**[105][40]	**NP-c**[105]
AT-free	**IP**[103]	**IP**[41]	**IP**[14][61]	**IP**[103]	**NP-c**[105]
distance hereditary	**IP**[125]		**IP**[35]	**IP**[106]	**IP**[68]
homogeneously orderable		**NP-c**[32]	**IP**[71]		**IP**[71]
k-polygon (fixed $k \geq 3$)	**IP**[75]	**IP**[74]	**IP**[74]	**IP**[106]	**IP**[98]
circle	**NP-c**[98]		**NP-c**[98]	**NP-c**[98]	**IP**[98]
partial k-tree (fixed $k \geq 1$)	**IP**[6][146]	**IP**[5] [146]	**IP**[5]	**IP**[5][146]	**IP**[5]

Figure 8.3: Abbreviations: **NP-c**=NP-complete, **IP**=polynomial time solvable.

Bibliography

[1] G. S. Adhar and S. Peng, Parallel algorithms for cographs and parity graphs with applications. *J. Algorithms* 11 (1990) 252–284.

[2] G. S. Adhar and S. Peng, Parallel algorithms for finding connected, independent and total domination in interval graphs. Manuscript (1992).

[3] G. S. Adhar and S. Peng, Mixed domination in trees: a parallel algorithm. *Congr. Numer.* 100 (1994) 73–80.

[4] S. Arnborg, Efficient algorithms for combinatorial problems on graphs with bounded decomposability - a survey. *BIT* 25 (1985) 2–23.

[5] S. Arnborg, J. Lagergren and D. Seese, Easy problems for tree-decomposable graphs. *J. Algorithms* 12 (1991) 308–340.

[6] S. Arnborg and A. Proskurowski, Linear time algorithms for NP-hard problems restricted to partial k-trees. *Discrete Appl. Math.* 23 (1989) 11–24.

[7] K. Arvind, H. Breu, M. S. Chang, D. G. Kirkpatrick, F. Y. Lee, Y. D. Liang, K. Madhukar, C. Pandu Rangan and A. Srinivasan, Efficient algorithms in cocomparability and trapezoid graphs. Manuscript (1996).

[8] K. Arvind and C. Pandu Rangan, Connected domination and Steiner set on weighted permutation graphs. *Inform. Process. Lett.* 41 (1992) 215–220.

[9] K. Arvind and C. Pandu Rangan, Transitive reduction and efficient poly-log algorithms on permutation graphs. Manuscript.

[10] T. Asano, Dynamic programming on intervals. *Internat. J. Comput. Geom. Appl.* 3 (1993) 323–330.

[11] M. J. Atallah and S. R. Kosaraju, An efficient algorithm for maxdominance, with applications. *Algorithmica* 4 (1989) 221–236.

[12] M. J. Atallah, G. K. Manacher and J. Urrutia, Finding a minimum independent dominating set in a permutation graph. *Discrete Appl. Math.* 21 (1988) 177–183.

[13] G. Bacsó and Zs. Tuza, Dominating cliques in P_5-free graphs. *Period. Math. Hungar.* 21 (1990) 303–308.

[14] H. Balakrishnan, A. Rajaraman and C. Pandu Rangan, Connected domination and Steiner set on asteroidal triple-free graphs. In *Proc. Workshop on Algorithms and Data Structures (WADS'93)*, Eds. F. Dehne, J. R. Sack, N. Santoro, and S. Whitesides, Springer-Verlag, Berlin, *Lecture Notes in Comput. Sci.* 709 (1993) 131–141.

[15] R. Bar-Yehuda and S. Moran, On approximation problems related to the independent set and vertex cover problems. *Discrete Appl. Math.* 9 (1984) 1–10.

[16] M. W. Bern, E. L. Lawler and A. L. Wong, Linear-time computation of optimal subgraphs of decomposable graphs. *J. Algorithms* 8 (1987) 216–235.

[17] P. J. Bernhard, S. T. Hedetniemi and D. P. Jacobs, Efficient sets in graphs. *Discrete Appl. Math.* 44 (1993) 99–108.

[18] A. A. Bertossi, Dominating sets for split and bipartite graphs. *Inform. Process. Lett.* 19 (1984) 37–40.

[19] A. A. Bertossi and M. A. Bonuccelli, Some parallel algorithms on interval graphs. *Discrete Appl. Math.* 16 (1987) 101–111.

[20] A. A. Bertossi and A. Gori, Total domination and irredundance in weighted interval graphs. *SIAM J. Discrete Math.* 1 (1988) 317–327.

[21] A. A. Bertossi and S. Moretti, Parallel algorithms on circular-arc graphs. *Inform. Process. Lett.* 33 (1990) 275–281.

[22] H. L. Bodlaender, Dynamic programming on graphs with bounded treewidth. In *International Colloquium on Automata, Languages and Programming (ICALP'88)*, Eds. T. Lepisto and A. Salomaa, Springer-Verlag, Berlin, *Lecture Notes in Comput. Sci.* 317 (1988) 105–118.

[23] H. L. Bodlaender, A tourist guide through treewidth. *Acta Cybernet.* 11 (1993) 1–23.

[24] M. A. Bonuccelli, Dominating sets and domatic number of circular arc graphs. *Discrete Appl. Math.* 12 (1985) 203–213.

[25] K. S. Booth and J. H. Johnson, Dominating sets in chordal graphs. *SIAM J. Comput.* 11 (1982) 191–199.

[26] K. S. Booth and G. S. Lueker, Testing for the consecutive ones property, interval graphs, and graph planarity using PQ-tree algorithms. *J. Comput. Systems Sci.* 13 (1976) 335–379.

[27] A. Brandstädt, The computational complexity of feedback vertex set, Hamiltonian circuit, dominating set, Steiner tree, and bandwidth on special perfect graphs. *J. Inform. Process. Cybernet.* 23 (1987) 471–477.

[28] A. Brandstädt, On the domination problem for bipartite graphs. *Topics in Combinatorics and Graph Theory*, Eds. R. Bodendieck and R. Henn, Physica, Heidelburg (1990) 145–152.

[29] A. Brandstädt, Special graph classes - a survey. Technical Report SM-DU-199, Universität Duisburg (1991).

[30] A. Brandstädt, On improved bounds for permutation graph problems. In *Proceedings of WG'92*, Ed. E. W. Mayr, Springer-Verlag, Berlin, *Lecture Notes in Comput. Sci.* 657 (1993) 1–10.

[31] A. Brandstädt and H. Behrendt, Domination and the use of maximum neighbourhoods. Technical Report SM–DU–204, Universität Duisburg (1992).

[32] A. Brandstädt, V. D. Chepoi and F. F. Dragan, The algorithmic use of hypertree structure and maximum neighbourhood orderings. Technical Report SM–DU–244, Universität Duisburg (1993).

[33] A. Brandstädt, V. D. Chepoi and F. F. Dragan, Clique r-domination and clique r-packing problems on dually chordal graphs. Technical Report SM–DU–251, Universität Duisburg (1994).

[34] A. Brandstädt, V. D. Chepoi and F. F. Dragan, The algorithmic use of hypertree structure and maximum neighbourhood orderings. In *20th International Workshop "Graph-Theoretic Concepts in Computer Science"(WG'94)*, Springer-Verlag, Berlin, *Lecture Notes in Comput. Sci.* 903 (1995) 65–80.

[35] A. Brandstädt and F. F. Dragan, A linear-time algorithm for connected r-domination and Steiner tree on distance-hereditary graphs. Technical Report SM–DU–261, University of Duisburg (1994).

[36] A. Brandstädt, F. F. Dragan, V. D. Chepoi and V. I. Voloshin, Dually chordal graphs. In *19th International Workshop "Graph-Theoretic Concepts in Computer Science"(WG'93)*, Springer-Verlag, Berlin, *Lecture Notes in Comput. Sci.* 790 (1993) 237–251.

[37] A. Brandstädt, F. F. Dragan and F. Nicolai, Homogeneously orderable graphs and the Steiner Tree problem. In *21st International Workshop "Graph-Theoretic Concepts in Computer Science"(WG'95)*, Springer-Verlag, Berlin, *Lecture Notes in Comput. Sci.* 1017 (1995) 381–395.

[38] A. Brandstädt and D. Kratsch, On the restriction of some NP-complete graph problems to permutation graphs. In *Proceedings FCT'85*, Ed. L. Budach, Springer-Verlag, Berlin, *Lecture Notes in Comput. Sci.* 199 (1985) 53–62.

[39] A. Brandstädt and D. Kratsch, On domination problems on permutation and other graphs. *Theoret. Comput. Sci.* 54 (1987) 181–198.

[40] H. Breu and D. G. Kirkpatrick, Algorithms for dominating and Steiner set problems in cocomparability graphs. Manuscript (1993).

[41] H. J. Broersma, T. Kloks, D. Kratsch and H. Müller, Independent sets in asteroidal triple-free graphs. Manuscript (1996).

[42] G. J. Chang, Labelling algorithms for domination problems in sunfree chordal graphs. *Discrete Appl. Math.* 22 (1988) 21–34.

[43] G. J. Chang, Total domination in block graphs. *Oper. Res. Lett.* 8 (1989) 53–57.

[44] G. J. Chang and G. L. Nemhauser, r-domination of block graphs. *Oper. Res. Lett.* 1(6) (1982) 214–218.

[45] G. J. Chang and G. L. Nemhauser, The k-domination and k-stability on sun-free chordal graphs. *SIAM J. Algebraic Discrete Methods* 5 (1984) 332–345.

[46] G. J. Chang, C. Pandu-Rangan and S. R. Coorg, Weighted independent perfect domination on cocomparability graphs. In *Proceedings ISAAC'93*, Springer-Verlag, Berlin, *Lecture Notes in Comput. Sci.* 766 (1993) 506–514.

[47] M. S. Chang, Weighted domination on cocomparability graphs. In *Proceedings ISAAC'95*, Springer-Verlag, Berlin, *Lecture Notes in Comput. Sci.* 1004 (1995) 122–131.

[48] M. S. Chang, Efficient algorithms for the domination problems on interval and circular-arc graphs. To appear in *SIAM J. Comput.*

[49] M. S. Chang and Y. C. Liu, Polynomial algorithms for the weighted perfect domination problems on chordal graphs and split graphs. *Inform. Process. Lett.* 48 (1993) 205–210.

[50] M. S. Chang and Y. C. Liu, Polynomial algorithms for weighted perfect domination problems on interval and circular-arc graphs. *J. Inform. Sci. Engineering* 10 (1994) 549-568.

[51] G. Cheston and G. H. Fricke, Classes of graphs for which upper fractional domination equals independence, upper domination, and upper irredundance. *Discrete Appl. Math.* 55 (1994) 241-258.

[52] E. J. Cockayne, S. E. Goodman and S. T. Hedetniemi, A linear algorithm for the domination number of a tree. *Inform. Process. Lett.* 4 (1975) 41-44.

[53] E. J. Cockayne, G. MacGillivray and C. M. Mynhardt, A linear algorithm for universal minimal dominating functions in trees. *J. Combin. Math. Combin. Comput.* 10 (1991) 23-31.

[54] C. J. Colbourn, J. M. Keil and L. K. Stewart, Finding minimum dominating cycles in permutation graphs. *Oper. Res. Lett.* 4 (1985) 13-17.

[55] C. J. Colbourn and L. K. Stewart, Dominating cycles in series-parallel graphs. *Ars Combin.* 19A (1985) 107-112.

[56] C. J. Colbourn and L. K. Stewart, Permutation graphs: connected domination and Steiner trees. *Discrete Math.* 86 (1990) 179-189.

[57] T. H. Cormen, C. E. Leiserson and R. L. Rivest, *Introduction to Algorithms.* MIT Press, Cambridge, Massachusetts, 1990.

[58] D. G. Corneil and J. M. Keil, A dynamic programming approach to the dominating set problem on k-trees. *SIAM J. Algebraic Discrete Methods* 8 (1987) 535-543.

[59] D. G. Corneil, S. Olariu and L. Stewart, Asteroidal triple-free graphs. In *Proceedings of the 19th International Workshop on Graph-Theoretic Concepts in Computer Science (WG'93)*, Springer-Verlag, Berlin, *Lecture Notes in Comput. Sci.* 790 (1994) 211-224.

[60] D. G. Corneil, S. Olariu and L. Stewart, A linear time algorithm to compute a dominating path in an AT-free graph. *Inform. Process. Lett.* 54 (1995) 253-258.

[61] D. G. Corneil, S. Olariu and L. Stewart, A linear time algorithm to compute dominating pairs in asteroidal triple-free graphs. In *Proceedings of the 22nd International Colloquium on Automata, Languages and Programming (ICALP'95)*, Springer-Verlag, Berlin, *Lecture Notes in Comput. Sci.* 944 (1995) 292-302.

[62] D. G. Corneil and Y. Perl, Clustering and domination in perfect graphs. *Discrete Appl. Math.* 9 (1984) 27–39.

[63] D. G. Corneil and L. K. Stewart, Dominating sets in perfect graphs. *Discrete Math.* 86 (1990) 145–164.

[64] M. B. Cozzens and L. L. Kelleher, Dominating cliques in graphs. *Discrete Math.* 86 (1990) 101–116.

[65] E. Dahlhaus and P. Damaschke, The parallel solution of domination problems on chordal and strongly chordal graphs. *Discrete Appl. Math.* 52 (1994) 261–273.

[66] P. Damaschke, H. Müller and D. Kratsch, Domination in convex and chordal bipartite graphs. *Inform. Process. Lett.* 36 (1990) 231–236.

[67] A. K. Dewdney, Fast turing reductions between problems in NP; chapter 4; reductions between NP-complete problems. Technical Report 71, Dept. Computer Science, University of Western Ontario (1981).

[68] F. F. Dragan, Dominating cliques in distance-hereditary graphs. In *"Algorithm Theory – SWAT'94 " 4th Scandinavian Workshop on Algorithm Theory*, Springer-Verlag, Berlin, *Lecture Notes in Comput. Sci.* 824 (1994) 370–381.

[69] F. F. Dragan, HT-graphs: centers, connected r-domination and Steiner trees. *Comput. Sci. J. Moldova* (Kishinev) 1 (1993) 64–83.

[70] F. F. Dragan and A. Brandstädt, Dominating cliques in graphs with hypertree structure. In *International Symposium on Theoretical Aspects of Computer Science (STACS'94)*, Springer-Verlag, Berlin, *Lecture Notes in Comput. Sci.* 775 (1994) 735–746.

[71] F. F. Dragan and F. Nicolai, r–domination problems on homogeneously orderable graphs. In *Proceedings of FCT'95*, Springer-Verlag, Berlin, *Lecture Notes in Comput. Sci.* 965 (1995) 201–210.

[72] F. F. Dragan, C. F. Prisacaru and V. D. Chepoi, Location problems in graphs and the Helly property (in Russian). *Discrete Math.* (Moscow) 4 (1992) 67–73.

[73] J. Dunbar, S. T. Hedetniemi, M. A. Henning and A. A. McRae, On the algorithmic complexity of minus domination in graphs. *Discrete Appl. Math.* 68 (1996) 73–84.

[74] E. S. Elmallah and L. K. Stewart, Domination in polygon graphs. *Congr. Numer.* 77 (1990) 63–76.

[75] E. S. Elmallah and L. K. Stewart, Independence and domination in polygon graphs. *Discrete Appl. Math.* 44 (1993) 65–77.

[76] P. van Emde Boas, Preserving order in a forest in less than logarithmic time and linear space. *Inform. Process. Lett.* 6 (1977) 80–82.

[77] M. Farber, *Applications of linear programming duality to problems involving independence and domination.* Ph.D. Dissertation, Simon Fraser University (1981) TR-81-13.

[78] M. Farber, Independent domination in chordal graphs. *Oper. Res. Lett.* 1 (1982) 134–138.

[79] M. Farber, Domination, independent domination, and duality in strongly chordal graphs. *Discrete Appl. Math.* 7 (1984) 115–130.

[80] M. Farber and J. M. Keil, Domination in permutation graphs. *J. Algorithms* 6 (1985) 309–321.

[81] M. R. Fellows and M. N. Hoover, Perfect domination. *Australas. J. Combin.* 3 (1991) 141–150.

[82] G. H. Fricke, E. O. Hare, D. P. Jacobs and A. Majumdar, On integral and fractional total domination. *Congr. Numer.* 77 (1990) 87–95.

[83] G. H. Fricke, M. A. Henning, O. R. Oellermann and H. C. Swart, An efficient algorithm to compute the sum of two distance domination parameters. *Discrete Appl. Math.* 68 (1996) 85–91.

[84] M. C. Golumbic, *Algorithmic Graph Theory and Perfect Graphs.* Academic Press, New York, 1980.

[85] M. C. Golumbic and R. C. Laskar, Irredundancy in circular arc graphs. *Discrete Appl. Math.* 44 (1993) 79–89.

[86] D. L. Grinstead and P. J. Slater, On minimum dominating sets with minimum intersection. *Discrete Math.* 86 (1990) 239–254.

[87] D. L. Grinstead and P. J. Slater, A recurrence template for several parameters in series-parallel graphs. *Discrete Appl. Math.* 54 (1994) 151–168.

[88] D. L. Grinstead, P. J. Slater, N. A. Sherwani and N. D. Holmes, Efficient edge domination problems in graphs. *Inform. Process. Lett.* 48 (1993) 221–228.

[89] J. H. Hattingh, M. A. Henning and P. J. Slater, On the algorithmic complexity of signed domination in graphs. *Australas. J. Combin.* 12 (1995) 101–112.

[90] X. He and Y. Yesha, Efficient parallel algorithms for r-dominating set and p-center problems on trees. *Algorithmica* 5 (1990) 129–145.

[91] S. M. Hedetniemi, S. T. Hedetniemi and D. P. Jacobs, Private domination: theory and algorithms. *Congr. Numer.* 79 (1990) 147–157.

[92] S. M. Hedetniemi, S. T. Hedetniemi and R. C. Laskar, Domination in trees: models and algorithms. *Graph Theory with Applications to Algorithms and Computer Science*, Eds. Y. Alavi et al., Wiley, (1984) 423–442.

[93] S. T. Hedetniemi, R. C. Laskar and J. Pfaff, A linear algorithm for finding a minimum dominating set in a cactus. *Discrete Appl. Math.* 13 (1986) 287–292.

[94] W. Hsu and K. Tsai, Linear time algorithms on circular-arc graphs. *Inform. Process. Lett.* 40 (1991) 123–129.

[95] O. H. Ibarra and Q. Zheng, Some efficient algorithms for permutation graphs. *J. Algorithms* 16 (1994) 453–469.

[96] M. S. Jacobson and K. Peters, Complexity questions for n-domination and related parameters. *Congr. Numer.* 68 (1989) 7–22.

[97] D. S. Johnson, The NP-completeness column: an ongoing guide. *J. Algorithms* 6 (1985) 291–305,434–451.

[98] J. M. Keil, The complexity of domination problems in circle graphs. *Discrete Appl. Math.* 42 (1993) 51–63.

[99] J. M. Keil and D. Schaefer, An optimal algorithm for finding dominating cycles circular-arc graphs. *Discrete Appl. Math.* 36 (1992) 25–34.

[100] T. Kikuno, N. Yoshida and Y. Kakuda, A linear algorithm for the domination number of a series-parallel graph. *Discrete Appl. Math.* 5 (1983) 299–311.

[101] E. Köhler, Connected domination on trapezoid graphs in $O(n)$ time. Manuscript (1996).

[102] D. Kratsch, Finding dominating cliques efficienctly, in strongly chordal graphs and undirected path graphs. *Discrete Math.* 86 (1990) 225–238.

[103] D. Kratsch, Domination and total domination in asteroidal triple-free graphs. Technical Report, Math/Inf/96/25, F.-Schiller-Universität, Jena (1996).

[104] D. Kratsch, P. Damaschke and A. Lubiw, Dominating cliques in chordal graphs. *Discrete Math.* 128 (1994) 269–275.

[105] D. Kratsch and L. Stewart, Domination on cocomparability graphs. *SIAM J. Discrete Math.* 6(3) (1993) 400–417.

[106] D. Kratsch and L. Stewart, Total domination and transformation. Manuscript (1996).

[107] R. C. Laskar and J. Pfaff, Domination and irredundance in split graphs. Technical Report 430, Dept. Mathematical Sciences, Clemson University (1983).

[108] R. C. Laskar, J. Pfaff, S. M. Hedetniemi and S. T. Hedetniemi, On the algorithmic complexity of total domination. *SIAM J. Algebraic Discrete Methods* 5 (1984) 420–425.

[109] E. L. Lawler and P. J. Slater, A linear time algorithm for finding an optimal dominating subforest of a tree. *Graph Theory with Applications to Algorithms and Computer Science*, Wiley, New York (1985) 501–506.

[110] Y. D. Liang, Permutation graphs: connected domination and steiner trees revisited. Manuscript (1992).

[111] Y. D. Liang, Dominations in trapezoid graphs. *Inform. Process. Lett.* 52 (1994) 309–315.

[112] Y. D. Liang, Steiner set and connected domination in trapezoid graphs. *Inform. Process. Lett.* 56 (1995) 165–171.

[113] Y. D. Liang and C. Rhee, Linear algorithms for two independent set problems in permutation graphs. In *Proc. of 22nd Computer Science Conference* (1994) 90–93.

[114] Y. Liang, C. Rhee, S. K. Dall and S. Lakshmivarahan, A new approach for the domination problem on permutation graphs. *Inform. Process. Lett.* 37 (1991) 219–224.

[115] E. Loukakis, Two algorithms for determining a minimum independent dominating set. *Internat. J. Comput. Math.* 15 (1984) 213–229.

[116] T. L. Lu, P. H. Ho and G. J. Chang, The domatic number problem in interval graphs. *SIAM J. Discrete Math.* 3 (1990) 531–536.

[117] G. K. Manacher and T. A. Mankus, Incorporating negative-weight vertices in certain vertex-search graph algorithms. *Inform. Process. Lett.* 42 (1992) 293–294.

[118] G. K. Manacher and T. A. Mankus, Finding a domatic partition of an interval graph in time $O(n)$. *SIAM J. Discrete Math.* 9 (1996) 167–172.

[119] R. M. McConnell and J. P. Spinrad, Modular decomposition and transitive orientation. Manuscript (1995).

[120] A. A. McRae and S. T. Hedetniemi, Finding n-independent dominating sets. *Congr. Numer.* 85 (1991) 235–244.

[121] S. Mitchell, S. T. Hedetniemi and S. Goodman, Some linear algorithms on trees. *Congr. Numer.* 14 (1975) 467–483.

[122] M. Moscarini, Doubly chordal graphs, Steiner trees, and connected domination. *Networks* 23 (1993) 59–69.

[123] H. Müller and A. Brandstädt, The NP-completeness of STEINER TREE and DOMINATING SET for chordal bipartite graphs. *Theoret. Comput. Sci.* 53 (1987) 257–265.

[124] K. S. Natarajan and L. J. White, Optimum domination in weighted trees. *Inform. Process. Lett.* 7 (1978) 261–265.

[125] F. Nicolai and T. Szymzcak, Domination and homogeneous sets – a linear time algorithm for distance-hereditary graphs. Technical Report SM–DU–336, Universität Duisburg (1996).

[126] R. Paige and R. E. Tarjan, Three partition refinement algorithms. *SIAM J. Comput.* 6 (1987) 973–989.

[127] C. H. Papadimitriou and K. Steiglitz, *Combinatorial Optimization: Networks and Complexity.* Prentice Hall, Englewood Cliffs, New Jersey, 1982.

[128] S. L. Peng and M. S. Chang, A simple linear time algorithm for the domatic partition problem on strongly chordal graphs. *Inform. Process. Lett.* 43 (1992) 297-300.

[129] J. Pfaff, R. Laskar and S. T. Hedetniemi, NP-completeness of total and connected domination, and irredundance for bipartite graphs. Technical Report 428, Dept. Mathematical Sciences, Clemson University (1983).

[130] J. Pfaff, R. Laskar and S. T. Hedetniemi, Linear algorithms for independent domination and total domination in series-parallel graphs. *Congr. Numer.* 45 (1984) 71–82.

[131] A. Proskurowski, Minimum dominating cycles in 2-trees. *Internat. J. Comput. Inform. Sci.* 8 (1979) 405–417.

[132] A. Proskurowski and M. M. Syslo, Minimum dominating cycles in outerplanar graphs. *Internat. J. Comput. Inform. Sci.* 10 (1981) 127–139.

[133] G. Ramalingam and C. Pandu Rangan, Total domination in interval graphs revisited. *Inform. Process. Lett.* 27 (1988) 17–21.

[134] G. Ramalingam and C. Pandu Rangan, A unified approach to domination problems in interval graphs. *Inform. Process. Lett.* 27 (1988) 271–274.

[135] A. S. Rao and C. Pandu Rangan, Optimal parallel algorithms on circular-arc graphs. *Inform. Process. Lett.* 33 (1989) 147–156.

[136] A. S. Rao and C. Pandu Rangan, Linear algorithm for domatic number problem on interval graphs. *Inform. Process. Lett.* 33 (1989-90) 29–33.

[137] C. Rhee, Y. D. Liang, S. K. Dhall and S. Lakshmivarahan, An $O(n + m)$ algorithm for finding a minimum-weight dominating set in a permutation graph. *SIAM J. Comput.* 25 (1996) 404–419.

[138] F. S. Roberts, *Graph Theory and Its Applications to Problems of Society.* CBMS-NSF Monograph, Vol. 29, SIAM Publications, 1978.

[139] M. Skowrońska and M. M. Syslo, Dominating cycles in Halin graphs. *Discrete Math.* 86 (1990) 215–224.

[140] P. J. Slater, R-domination in graphs. *J. Assoc. Comput. Mach.* 23 (1976) 446–450.

[141] J. Spinrad, On comparability and permutation graphs. *SIAM J. Comput.* 14 (1985) 658–670.

[142] J. Spinrad, Doubly lexical ordering of dense $0 - 1$ matrices. *Inform. Process. Lett.* 45 (1993) 229–235.

[143] R. Sridhar and S. S. Iyengar, Efficient parallel algorithms for domination problems on strongly chordal graphs. Manuscript.

[144] A. Srinivasan and C. Pandu Rangan, Efficient algorithms for the minimum weighted dominating clique problem on permutation graphs. *Theoret. Comput. Sci.* 91 (1991) 1–21.

[145] R. E. Tarjan and M. Yannakakis, Simple linear time algorithms to test chordality of graphs, test acyclicity of hypergraphs, and selectively reduce acyclic hypergraphs. *SIAM J. Comput.* 3 (1984) 566–579.

[146] J. A. Telle, Complexity of domination-type problems in graphs. *Nordic J. Comput.* 1 (1994) 157–171.

[147] J. A. Telle and A. Proskurowski, Practical algorithms on partial k-trees with an application to domination-type problems. In *Proc. Workshop on Algorithms and Data Structures (WADS'93)*, Eds. F. Dehne, J. R. Sack, N. Santoro and S. Whitesides, Springer-Verlag, Berlin, *Lecture Notes in Comput. Sci.* 709 (1993) 610–621.

[148] J. A. Telle and A. Proskurowski, Efficient sets in partial k-trees. *Discrete Appl. Math.* 44 (1993) 109–117.

[149] W. T. Trotter, *Combinatorics and Partially Ordered Sets: Dimension Theory.* The John Hopkins University Press, Baltimore, Maryland, 1992.

[150] K. Tsai and W. L. Hsu, Fast algorithms for the dominating set problem on permutation graphs. *Algorithmica* 9 (1993) 109–117.

[151] C. Tsouros and M. Satratzemi, Tree search algorithms for the dominating vertex set problem. *Internat. J. Computer Math.* 47 (1993) 127–133.

[152] K. White, M. Farber and W. Pulleyblank, Steiner trees, connected domination and strongly chordal graphs. *Networks* 15 (1985) 109–124.

[153] T. V. Wimer, *Linear algorithms on K-terminal graphs.* Ph.D. Dissertation, Clemson University (1987).

[154] M. Yannakakis and F. Gavril, Edge dominating sets in graphs. *SIAM J. Appl. Math.* 38 (1980) 264–272.

[155] C. Yen and R. C. T. Lee, The weighted perfect domination problem. *Inform. Process. Lett.* 35 (1990) 295–299.

Chapter 9

Complexity Results

Stephen T. Hedetniemi
Computer Science
Clemson University
Clemson, SC   29634 USA

Alice A. McRae
Dolores A. Parks
Mathematical Sciences
Appalachian State University
Boone, NC   28607 USA

Abstract. We present original NP-completeness results and new and simpler proofs of existing NP-completeness results for domination-related parameters (see Tables 9.2 and 9.3). All of the proofs of these results are no longer than two pages; most are less than one page long. The great majority of these results use reductions from the Exact Cover by 3 Sets Problem and use a common graph construction. The full generality of this technique has not been fully developed, and we suggest a number of "open" problems that may be shown NP-complete by applying similar techniques. We conclude by providing a compendium of other original NP-completeness proofs not contained in this chapter but which use similar proof constructions.

9.1   Introduction

Perhaps the fascination with domination-related parameters is in their application to so many real world problems, problems that demand "solutions" in a reasonable amount of time. Knowing whether or not a problem is NP-complete is essential in determining an approach for finding a solution. Even if a problem is known to be NP-complete, it may be possible to find a polynomial algorithm for a restricted set of instances from a particular application. For example, domination problems that are NP-complete for general graphs are usually polynomial when restricted to trees. When restricted to other graph classes, such as bipartite graphs or chordal graphs, many problems remain NP-complete. A collection of polynomial-time domination algorithms are given in Chapter 8, and in this chapter we survey known NP-completeness results for several domination-related parameters.

We focus on a general NP-complete construction which can be applied to a number of domination parameters. In Section 9.2, an NP-completeness construction designed by Yen and Lee [44] for perfect domination is studied. We generalize this NP-completeness construction in Section 9.3 for other vertex graph parameters by defining several conditions that if met will assure that the parameter is NP-complete. In the following sections, we give many NP-completeness proofs that use this construction technique. Most of the results hold for bipartite graphs and chordal graphs.

In *Computers and Intractability: A Guide to the Theory of NP-Completeness* [22], Garey and Johnson outline four steps involved in proving that a decision problem is NP-complete. The first step is to prove that the problem resides in the class NP. This involves showing that the problem can be solved in non-deterministic, polynomial time. For a problem to be in NP, there should exist a witness for any *yes* instance of the problem. This witness could be given as input to a deterministic polynomial time algorithm to verify that a given instance has a *yes* answer. For example, this chapter looks at decision problems for finding minimum vertex sets that have some property P, such as perfect dominating. A witness for this type of problem is often a property-P set. To show that a problem is in NP, an argument is given that there is a deterministic polynomial time algorithm that can verify that a given set has property P. In most NP-completeness proofs, this argument is considered trivial, and usually the statement is given that the problem is "clearly in NP." However, it is important to remember that this is not always the case.

The second step of an NP-completeness proof is to select a known NP-complete problem, and define a transformation from this problem to the problem you wish to prove NP-complete. The third step is to show that an instance of the first problem is a *yes* instance if and only if the constructed instance of the second problem is a *yes* instance. The remaining step of an NP-completeness proof is to show that this transformation can be carried out in polynomial time. Garey and Johnson [22] provide a list of the most common NP-complete problems used in making such transformations. One of these is the Exact Cover by 3-Sets Problem(X3C), first shown to be NP-complete by Karp [33]. All of the proofs presented in this chapter apply transformations from X3C. This problem can be stated in the following form.

EXACT COVER BY 3-SETS (X3C)
INSTANCE: A finite set X with $|X| = 3q$ and a collection C of 3-element subsets of X.
QUESTION: Does C contain an exact cover for X, that is, a subcollection $C' \subseteq C$

such that every element of X occurs in exactly one member of C'? Note that if C' exists, then its cardinality is precisely q.

Although m usually is used to denote the number of edges in a graph, in this chapter m is more often used to denote the number of clauses $C_1, C_2, \ldots, C_m$ in an instance of X3C. To simplify the proofs in this chapter, we have added the requirement to the instance of X3C that each element appears in at least two subsets. Clearly, this slight modification does not change the complexity of X3C.

9.2   Perfect Domination

A set $S \subseteq V$ is a *perfect dominating set* if for every vertex $v \in V - S$, $|N(v) \cap S| = 1$. The decision problem for perfect dominating sets can be stated as follows.

PERFECT DOMINATING SET
INSTANCE: A graph G and positive integer k.
QUESTION: Does G have a perfect dominating set of cardinality k or less?

Theorem 9.1 [44] *PERFECT DOMINATING SET is NP-complete.*

Proof. The following is a slight variation of the proof in [44].

The Perfect Dominating Set Problem is in NP. A set S, $|S| \leq k$, could be given as a witness to a *yes* instance. In $O(V + E)$ time, the neighborhoods of all vertices $v \in V - S$ could be checked to ensure that $|N(v) \cap S| = 1$.

A graph $G = (V, E)$ and positive integer k will be constructed from an instance of X3C, such that the X3C instance will have an exact cover if and only if G has a perfect dominating set of cardinality at most k. Let $X = \{x_1, x_2, \ldots x_{3q}\}$ and $C = \{C_1, C_2, \ldots, C_m\}$ be an arbitrary instance of X3C, where $|X| = 3q$ and $|C| = m$. The construction of G is accomplished by creating a vertex x_i for each element $x_i \in X$, and a component consisting of a P_2 with vertices labeled c_j and s_j for every subset $C_j \in C$. Additional communication edges $E' = \bigcup_{j=1}^{m} \{(x_i, c_j) | x_i \in C_j\}$ join the element vertices with the subset components. As the last step, let $k = m$. Figure 9.1 illustrates this construction.

It remains to show that C has an exact cover if and only if G has a perfect dominating set of cardinality at most k. Suppose C' is an exact cover for C. Then it is easy to verify that $S = \{c_j | C_j \in C'\} \cup \{s_j | C_j \notin C'\}$ is a perfect dominating set for G, with $|S| = m$.

Now suppose that S, $|S| \leq m$, is a perfect dominating set for G. In each P_2, either s_j is in S, or s_j must have exactly one neighbor in S, namely c_j. Therefore, at least one vertex in each P_2 must be in S. Because each element vertex x_i is adjacent to some c_j, it follows that none of the element vertices x_i

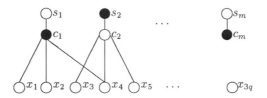

Figure 9.1: The Yen-Lee construction.

can be in S. Therefore, $C' = \{C_j | c_j \in S\}$ will be an exact cover for C. For if some element x_i is not covered exactly once in C', then the corresponding vertex x_i would not be dominated exactly once in G, and S would not be a perfect dominating set. Since the graph G will have $3q + 2m$ vertices and $4m$ edges, this is clearly a polynomial transformation. □

Corollary 9.2 *PERFECT DOMINATING SET is NP-complete even when restricted to bipartite graphs.*

Proof. The graph constructed in Theorem 9.1 is bipartite. □

Corollary 9.3 *PERFECT DOMINATING SET is NP-complete even when restricted to chordal graphs.*

Proof. If a complete subgraph is formed among the element vertices in the graph in Figure 9.1, then the resulting graph will be chordal. The same argument used in proving Theorem 9.1 holds for this graph as well. □

9.3   A Generalized Proof

The construction for perfect domination in Theorem 9.1 replaced each element x_i of the X3C instance with a tree, rooted at x_i and each subset C_j with a tree, rooted at c_j. The "element" tree, which we will refer to as T_1, consisted of a single vertex, and the "subset" tree T_2 consisted of a P_2. Communication edges E' were added to G connecting the root of each subset tree to the roots of the three element trees corresponding to the elements that appeared in the subset. This type of construction is applicable to several other vertex minimization problems, in which one seeks a minimum cardinality vertex set S such that S has a given property P.

Theorem 9.4 [37] *An NP-completeness proof can be given, using the construction technique described above, for any vertex minimization problem, corresponding to a property P, that meets the following three conditions. This result will hold even when the problem is restricted to bipartite graphs.*

1. The problem is in NP.

2. Trees T_1 and T_2 can be found such that in the graph G, as constructed above, the following criteria are met:

   (a) For any property P-set S in G, $|S \cap V(T_1)| \geq x$; for some fixed nonnegative integer x.

   (b) If $|S \cap V(T_1)| = x$, then r_1, the root of T_1, must be joined to exactly one vertex $w \in S - V(T_1)$.

   (c) For any property P-set S in G, $|S \cap V(T_2)| \geq y$; for some fixed nonnegative integer y.

3. We can find vertex sets $S_1 \subseteq V(T_1)$, $|S_1| = x$, $S_2 \subseteq V(T_2)$, $|S_2| = y$, $S_{r,2} \subseteq V(T_2)$, $|S_{r,2}| = y$, such that given an exact cover for the X3C instance, then a property P-set S can be found for G by making S the union of

   (a) all S_1 vertices in the T_1 element components of G,

   (b) all $S_{r,2}$ vertices in the T_2 subset components where the corresponding subset is part of the exact cover, and

   (c) all S_2 vertices in the T_2 subset components where the corresponding subset is not part of the exact cover.

Proof. We follow Garey and Johnson's four steps for proving a problem NP-complete.

Step 1. The problem is in NP. This is part of the hypothesis.

Step 2. X3C is chosen as the known NP-complete problem. For the transformation, each element x_i of the arbitrary X3C instance I is replaced by an element tree T_1. Label the root of each tree x_i. Each subset C_j is replaced by a subset tree T_2. Label the root of this tree c_j. Add communication edges E' between the roots of the trees, connecting the roots of the subset trees to the roots of the element trees corresponding to the three elements that appear in the subset. Let G be the bipartite graph so constructed; cf. Figure 9.2. Finally, set $k = my + 3qx$.

Step 3. We must show that C has an exact cover C' if and only if G has a property P-set S, with $|S| \leq my + 3qx$. Suppose C has an exact cover C'. Then a property P-set S of cardinality $my + 3qx$ can be found by taking the S_1 vertex set in each of the element trees, the $S_{r,2}$ vertex set for all subset trees

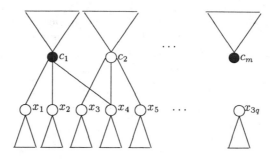

Figure 9.2: A general NP-completeness construction.

rooted at c_j where $C_j \in C'$, and the S_2 vertex set for the subset trees rooted at c_j where $C_j \notin C'$. Since $|S_1| = x$, $|S_2| = |S_{r,2}| = y$ and there are 3q of the S_1 sets taken, q of the $S_{r,2}$ sets taken, and $m - q$ of the S_2 vertex sets taken, $|S| = 3qx + qy + (m-q)y = 3qx + my$. Condition 3 above confirms that S is a property P-set.

Conversely, suppose that G has a property P-set S, with $|S| \leq my + 3qx$. By Condition 2(c) for the T_2 trees, it follows that at least my vertices must be chosen in the subset trees altogether. By Condition 2(a) for the T_1 trees, it follows that at least $3qx$ vertices must be chosen in the element trees altogether. This accounts for all the vertices. So by Condition 2(b) for the T_1 trees, each root must be adjacent to exactly one c_j vertex that is in S. The set $C' = \{C_j | c_j \in S\}$ will be an exact cover for C.

Step 4. Graph G will require $3q * |V(T_1)| + m * |V(T_2)|$ vertices and $3q * |E(T_1)| + m * (|E(T_2)| + 3)$ edges. Thus, this transformation is polynomial in q and m. □

9.4   Applying the Proof

In this section, five NP-completeness proofs are given that are based on the general NP-completeness proof developed in the previous section. The problems are Efficient Dominating Set, Total Efficient Dominating Set, Minimum Maximal Strong Stable Set, Minimum Maximal Open Strong Stable Set, Minimum 1-Maximal Nearly Perfect Set. Although the NP-completeness of Efficient Domination Set was established in [2], and 1-Maximal Nearly Perfect Set in [14], the proofs given in this chapter are original.

A set $S \subseteq V$ is an *efficient dominating set* of a graph $G = (V, E)$ if for every vertex $v \in V$, $|N[v] \cap S| = 1$ [2]. As noted in [2], if a graph has an efficient dominating set, then every such set has cardinality $\gamma(G)$.

EFFICIENT DOMINATING SET
INSTANCE: A graph G and a positive integer k.
QUESTION: Does G have an efficient dominating set of cardinality k or less?

Corollary 9.5 [2] *EFFICIENT DOMINATING SET is NP-complete, even when restricted to bipartite graphs or chordal graphs.*

Proof. EFFICIENT DOMINATING SET is in NP. A set S, $|S| \leq k$, could be given as a witness to a *yes* instance. In $O(V + E)$ time, the neighborhoods of all vertices v in V could be checked to ensure that $|N[v] \cap S| = 1$.

Figure 9.3 shows a T_1 tree, rooted at r_1, and two T_2 trees, rooted at r_2, that have the properties specified in Theorem 9.4. It is possible that $V(T_1) \cap S = \emptyset$ for an efficient dominating set S of G. If no vertices in T_1 are in an efficient dominating set, then r_1 must be adjacent to exactly one vertex in S, otherwise r_1 would not be dominated exactly once. Thus, the condition in 2(a) of Theorem 9.4 is satisfied for $x = 0$.

Figure 9.3: Efficient domination trees.

Each tree T_2 has to have at least one of its vertices in S, otherwise the non-root vertex is not dominated. Thus, the value $y = 1$ satisfies condition 2(c) in the statement of Theorem 9.4. The tree T_1 and two trees T_2 in Figure 9.3 show the S_1 (the empty set), S_2 and $S_{r,2}$ sets, respectively. The shaded vertices represent members of the set S. All vertices in the T_2 trees are dominated exactly once, and each r_1 vertex will be dominated by the root of the subset tree that corresponds to the subset covering the element in the solution $C' \subseteq C$ to the X3C instance. In this case, $k = my + 3qx = m(1) + 3q(0) = m$.

Forming a clique among the roots of the T_1 trees will produce a chordal graph. The arguments given above hold for this chordal graph as well. $\square$

A set $S \subseteq V$ is a *total efficient dominating set* if for every vertex $v \in V$, $|N(v) \cap S| = 1$ [23].

TOTAL EFFICIENT DOMINATING SET
INSTANCE: A graph G and a positive integer k.
QUESTION: Does G have a total efficient dominating set of cardinality k or less?

Corollary 9.6 [37] *TOTAL EFFICIENT DOMINATING SET is NP-complete, even when restricted to bipartite graphs or chordal graphs.*

Proof. TOTAL EFFICIENT DOMINATING SET is in NP. A set S, $|S| \leq k$, could be given as a witness to a *yes* instance. In $O(V + E)$ time, the neighborhoods of all vertices v in V could be checked to ensure that $|N(v) \cap S| = 1$.

Figure 9.4 shows a T_1 tree, rooted at r_1, and two T_2 trees, rooted at r_2, that have the properties specified in Theorem 9.4. There may be some total efficient dominating sets for G that would not have any vertices in $V(T_1)$. If no vertices in T_1 are in the total efficient dominating set, then r_1 must be adjacent to exactly one vertex in S, otherwise r_1 would not be dominated. Thus, $x = 0$.

Figure 9.4: Total efficient domination trees.

Each tree T_2 has to have at least two of its vertices in S. One of these must be vertex b, otherwise vertex a will not be dominated. Also either vertex a or r_2 must be in S so that vertex b is dominated. Thus, $y = 2$.

The T_1 tree and two T_2 trees in Figure 9.4 show the S_1, S_2 and $S_{r,2}$ sets, respectively. All vertices in the T_2 trees are dominated exactly once, and the r_1 vertex will be dominated by the root of the subset tree that corresponds to the subset covering the element in the solution $C' \subseteq C$ to the X3C instance. In this case, $k = m(2) + 3q(0) = 2m$.

Forming a clique among the roots of the T_1 trees will produce a chordal graph. The arguments given above will hold for this graph as well. $\square$

A set $S \subseteq V$ is a *strong stable set* (or a 2-packing) if for every vertex $v \in V$, $|N[v] \cap S| \leq 1$. A P-set S is maximal if no proper superset S' of S has property

P. A strong stable set is maximal if and only if for every vertex $v \in V - S$ there exists a $w \in S$, such that $d(v, w) \leq 2$, where $d(v, w)$ denotes the distance between vertices v and w.

MINIMUM MAXIMAL STRONG STABLE SET
INSTANCE: A graph G and a positive integer k.
QUESTION: Does G have a maximal strong stable set of cardinality k or less?

Corollary 9.7 [37] *MINIMUM MAXIMAL STRONG STABLE SET is NP-complete, even when restricted to bipartite graphs or chordal graphs.*

Proof. MINIMUM MAXIMAL STRONG STABLE SET is in NP. A set S, $|S| \leq k$, could be given as a witness to a *yes* instance. In $O(V + E)$ time, the neighborhoods of all vertices in V could be checked to verify that S is a strong stable set, and that every vertex not in S is within distance two of some vertex in S.

Figure 9.5 shows tree T_1, rooted at r_1, and two T_2 trees, rooted at r_2, that have the properties specified in Theorem 9.4. Let S be a maximal strong stable set for G. Then T_1 does not have to have any of its vertices in S. If no vertices in T_1 are in S, then r_1 must be adjacent to exactly one vertex in S. If r_1 is not adjacent to any vertices in S, then the nonroot vertex of T_1 could be added to S, contradicting the maximality of S. Furthermore, r_1 cannot be adjacent to more than one vertex in S, since S is a strong stable set. Thus, for the T_1 trees, $x = 0$.

Figure 9.5: Strong stability trees.

Tree T_2 has to have at least one of its vertices in S, otherwise there would be no S-vertices within distance two of vertex a. Thus, $y = 1$.

The T_1 tree and the two T_2 trees in Figure 9.5 show the S_1, S_2 and $S_{r,2}$ sets, respectively. The distance between any two vertices in S is at least two and no other vertices can be added to S without destroying the strong stability property. In this case, $k = m(1) + 3q(0) = m$.

Forming a clique among the roots of the T_1 trees will produce a chordal graph. The same arguments apply to this graph. □

A set $S \subseteq V$ is an *open strong stable set* if for every vertex $v \in V$, $|N(v) \cap S| \leq 1$. An open strong stable set is maximal if and only if for every vertex $v \in V - S$, there exists a $w \in S$ such that there is a path of length 2 from v to w.

MINIMUM MAXIMAL OPEN STRONG STABLE SET
INSTANCE: A graph G and positive integer k.
QUESTION: Does G have a maximal open strong stable set of cardinality k or less?

Corollary 9.8 [37] *MINIMUM MAXIMAL OPEN STRONG STABLE SET is NP-complete, even when restricted to bipartite graphs or chordal graphs.*

Proof. MINIMUM MAXIMAL OPEN STRONG STABLE SET is in NP. A set S, $|S| \leq k$, could be given as a witness to a *yes* instance. In $O(V + E)$ time, the neighborhoods of all vertices in V could be checked to ensure S is an open strong stable set, and that every vertex $v \in V - S$ is distance two from some vertex in S.

Figure 9.6 shows tree T_1, rooted at r_1, and two T_2 trees, rooted at r_2, that have the properties specified in Theorem 9.4 for maximal open strong stability. Tree T_1 does not have to have any of its vertices in S. If no vertices in T_1 are in the maximal open strong stable set, then r_1 must be adjacent to exactly one vertex in S. Thus, $x = 0$.

Figure 9.6: Open strong stability trees.

Tree T_2 has to have at least two of its vertices in S. Either vertex a or vertex c must be in a maximal open strong stable set, otherwise vertex a is in $V - S$, and no member of S is distance two from a. The same argument says that either vertex b or r_2 must be in S. Thus, $y = 2$.

The T_1 tree and two T_2 trees in Figure 9.6 show the S_1, S_2 and $S_{r,2}$ sets, respectively. Every vertex not in S will be distance two from some vertex in

S (the r_1 vertices will be distance two from some c vertex). No vertex will have more than one S-neighbor in its open neighborhood. In this case, $k = m(2) + 3q(0) = 2m$.

Forming a clique among the roots of the T_1 trees produces a chordal graph. The same arguments apply to this graph. $\square$

A set $S \subseteq V$ is a *nearly perfect set* if for every vertex $v \in V - S$, $|N(v) \cap S| \leq 1$. A P-set S is 1-maximal if for every vertex v in $V - S$, $S \cup \{v\}$ is not a P-set. In [14] it is shown that a nearly perfect set is 1-maximal if and only if every vertex u in $V - S$ is adjacent to a vertex $v \neq u$ in $V - S$ which is adjacent to exactly one vertex in S.

1-MAXIMAL NEARLY PERFECT SET
INSTANCE: A graph G and a positive integer k.
QUESTION: Does G have a 1-maximal nearly perfect set of cardinality k or less?

Corollary 9.9 [14] *1-MAXIMAL NEARLY PERFECT SET is NP-complete, even when restricted to bipartite graphs or chordal graphs.*

Proof. 1-MAXIMAL NEARLY PERFECT SET is in NP. A set S, $|S| \leq k$, could be given as a witness to a *yes* instance. In $O(V + E)$ time, it could be determined that S is nearly perfect, and in $O(V + E)$ time it also can be determined that all vertices $v \in V - S$ are adjacent to some other vertex $w \in V - S$ that is adjacent to exactly one vertex in S.

Figure 9.7 shows tree T_1, rooted at r_1, and two T_2 trees, rooted at r_2, that have the properties specified in Theorem 9.4 for 1-maximal nearly perfect sets. Tree T_1 does not have to have any of its vertices in S. If no vertices in T_1 are in the 1-maximal nearly perfect set S, then r_1 must be adjacent to exactly one vertex in S, otherwise the non-root vertex could be added to S. Thus, $x = 0$.

Tree T_2 has to have at least three of its vertices in S. Vertex a must be in every 1-maximal nearly set, otherwise $b \notin S$ and $c \in S$. But then vertex b does not satisfy the maximality condition for S. For vertex e to satisfy the maximality condition for S, either c, e, or r_2 must be in S. For vertex b to satisfy the maximality condition, either b or d must be in S. Thus, $y = 3$.

The T_1 tree and two T_2 trees in Figure 9.7 show the S_1, S_2 and $S_{r,2}$ sets, respectively. Every vertex not in S will be adjacent to at most one vertex in S. Also every vertex not in S is adjacent to at least one other vertex not in S that is adjacent to exactly one vertex in S, since by assumption every element $x \in X$ appears in at least two subsets. In this case, $k = m(3) + 3q(0) = 3m$.

Forming a clique among the roots of the T_1 trees produces a chordal graph. The same arguments from Theorem 9.4 apply to this graph. $\square$

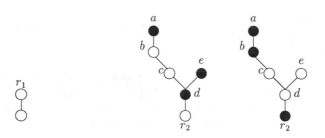

Figure 9.7: Nearly perfect trees.

9.5   A Variation of the Proof

For many vertex minimization problems, it is not easy to find T_1 trees, where T_1 can have x vertices chosen only if the root is joined to exactly one vertex in S. If we relax that criteria and allow the root to be joined to one or more vertices in S, then finding a T_1 tree may be much easier. Unfortunately, in this case all c_j vertices could be chosen for S, and given a property P set for G, we could not distinguish between the c_j vertices to find an exact cover for C. To allow for a more flexible T_1 and to prevent all c_j vertices from being added to S, we now look for a slightly different T_2, a tree that will have one more vertex chosen for S, if the root is in S.

Theorem 9.10 [37] *An NP-completeness proof can be given for any vertex minimization problem involving a property P that meets the following three conditions. The result will hold even when restricted to bipartite graphs.*

1. The decision problem is in NP.

2. Trees T_1 and T_2 can be found that meet the following criterion.

    (a) For any property P-set S in G, $|S \cap V(T_1)| \geq x$, for some fixed nonnegative integer x.

    (b) If $|S \cap V(T_1)| = x$, then r_1 must be joined to at least one vertex $w \in S - V(T_1)$.

    (c) For any property P-set S in G, if $r_2 \notin S$, then $|S \cap V(T_2)| \geq y$ for some fixed nonnegative integer y.

    (d) For any property P-set S in G, if $r_2 \in S$, then $|S \cap V(T_2)| \geq y + 1$.

3. We can find vertex sets $S_1 \subseteq V(T_1)$, with $|S_1| = x$, and $S_2 \subseteq V(T_2)$, with $|S_2| = y$ if $r_2 \notin S_2$, and $S_{r,2} \subseteq V(T_2)$, with $|S_{r,2}| = y + 1$, if $r_2 \in S_{r,2}$ such that given an exact cover for the X3C instance, then a property P-set S can be found for G by making S the union of all S_1 vertices in the T_1 element trees of G, all $S_{r,2}$ vertices in the T_2 subset trees where the corresponding subset is part of the exact cover, and all S_2 vertices in the T_2 subset trees where the corresponding subset is not part of the exact cover.

Proof. As before, Garey and Johnson's four steps for proving a problem NP-complete are followed.

Step 1. The problem is in NP. This is in the hypothesis.

Step 2. X3C is chosen as the known NP-complete problem. For the transformation each element x_i of the arbitrary X3C instance is replaced with an element tree T_1. Label the root of this tree x_i. Each subset C_j is replaced with a subset tree T_2. Label the root of this tree with c_j. Add communication edges E' between the roots of the trees, connecting the roots of the subset trees to the roots of the element trees corresponding to the elements that appear in the subset. Set $k = my + q + 3qx$.

Step 3. We must show that C has an exact cover C' if and only if G has a property P-set of cardinality $\leq my + q + 3qx$.

Suppose C has an exact cover C'. Then a property P-set S of cardinality $my + q + 3qx$ can be found by taking the S_1 vertex set in each of the element trees, the $S_{r,2}$ vertex set for all subset trees rooted at c_j where $C_j \in C'$, and the S_2 vertex set for the subset trees rooted at c_j where $C_j \notin C'$. Since $|S_1| = x$, $|S_2| = y$ and $|S_{r,2}| = y + 1$ and there are 3q of the S_1 sets taken, q of the $S_{r,2}$ sets taken, and $m - q$ of the S_2 vertex sets taken, $|S| = 3qx + q(y + 1) + (m - q)y = 3qx + q + my$. Condition 3 assures that this is a property P-set.

Now suppose that G has a property P-set S, with $k = |S| \leq my + q + 3qx$. By Conditions 2(c) and 2(d) for the T_2 trees, it follows that at least my of the non-root vertices must be chosen in the subset trees altogether. Let t be the number of T_1 trees with exactly x vertices chosen. Since there are 3q subset trees, $t \leq 3q$. By Condition 2 for the T_1 trees, at least $\lceil t/3 \rceil$ r_2 vertices must be in S. Therefore, $my + q + 3qx \geq |S| \geq tx + (3q - t)(x + 1) + my + \lceil t/3 \rceil = 3qx + 3q - t + \lceil t/3 \rceil + my \geq 3qx + q + my$. Therefore, $t = 3q$, and the number of r_2 vertices in S is precisely q. Since each T_1 has only x vertices in S, each root must be adjacent to at least one of the q r_2 vertices that are in S. Since each of the q r_2 vertices is adjacent to exactly three of the r_1 vertices, at most 3q of the r_1 vertices can be "covered" by r_2 vertices that are in S. It follows that each r_1 vertex will be adjacent to exactly one of the r_2 vertices in S. Therefore, $C' = \{C_j | c_j \in S\}$ will be an exact cover for C.

Step 4. The constructed graph has $3q * |V(T_1)| + m * |V(T_2)|$ vertices and $3q * |E(T_1)| + m * (|E(T_2)| + 3)$ edges. This is polynomial in q and m. Chordal results can be obtained if forming a clique among the r_2 vertices does not destroy property P over the union of the S_1, S_2, and $S_{r,2}$ sets to form S. □.

9.6   Applying the Variation

In this section, several more vertex minimization constructions are described that use the variation of the general NP-completeness proof presented in the previous section. Several of these results are not new. Dominating set has been proved NP-complete for many different graph classes, including bipartite graphs [10] and chordal graphs [5]. Total dominating set also has been shown NP-complete for bipartite graphs [40] and chordal graphs [34]. Independent dominating set is NP-complete for bipartite graphs [9] [31] but polynomial for chordal graphs [19]. Minimum maximal irredundant set has been shown NP-complete for both bipartite graphs [40] and chordal graphs [34]. Odd dominating set [42] and weak vertex-edge dominating set [39] have been shown NP-complete in general.

DOMINATING SET
INSTANCE: A graph G and a positive integer k.
QUESTION: Does G have a dominating set of cardinality k or less?

Corollary 9.11 [8] *DOMINATING SET is NP-complete, even when restricted to bipartite graphs or chordal graphs.*

Proof. DOMINATING SET is in NP. A set S, $|S| \leq k$, could be given as a witness to a *yes* instance and the condition $|N[v] \cap S| \geq 1$ for every vertex $v \in V$ can be verified in $O(V + E)$ time.

Figure 9.8 shows tree T_1, rooted at r_1, and two T_2 trees, rooted at r_2, that have the properties specified in Theorem 9.10 for dominating sets. Tree T_1 does not have to have any of its vertices in S; i.e., $x = 0$. If no vertices in T_1 are in the dominating set, then r_1 must be adjacent to at least one vertex in S, otherwise r_1 would not be dominated.

Tree T_2 has to have at least one of its vertices in S; i.e., $y = 1$. If $r_2 \in S$, then T_2 will have at least two of its vertices in S. The T_1 tree and two T_2 trees in Figure 9.8 show the S_1, S_2 and $S_{r,2}$ sets, respectively. It is easy to see that S will dominate all of the vertices in G. In this case, $k = 3q(0) + q + (1)m = q + m$.

Forming a clique among the roots of the T_2 trees will produce a chordal graph. The same argument holds for this class of graphs as well. □

TOTAL DOMINATING SET
INSTANCE: A graph G and a positive integer k.
QUESTION: Does G have a total dominating set of cardinality k or less?

Figure 9.8: Domination trees.

Corollary 9.12 [40] [34] *TOTAL DOMINATING SET is NP-complete, even when restricted to bipartite graphs and chordal graphs.*

Proof. TOTAL DOMINATING SET is in NP. A set S, $|S| \leq k$, could be given as a witness to a *yes* instance and in $O(V+E)$ time, the condition $|N(v) \cap S| \geq 1$ can be verified.

Figure 9.9 shows tree T_1, rooted at r_1, and two T_2 trees, rooted at r_2, that have the properties specified in Theorem 9.10 for total dominating sets. Tree T_1 does not have to have any of its vertices in S; i.e., $x = 0$. If no vertices in T_1 are in a total dominating set, then r_1 must be adjacent to at least one vertex in S, otherwise r_1 would not be dominated.

Figure 9.9: Total domination trees.

Tree T_2 has to have at least two of its vertices in S; i.e., $y = 2$, since one vertex must be chosen from $N(a)$ and another from $N(b)$. Note $N(a) \cap N(b) = \emptyset$. If r_2 is in S, then three vertices must be chosen from T_2, since $r_2 \notin N(a)$ and $r_2 \notin N(b)$. The T_1 tree and two T_2 trees in Figure 9.9 show the S_1, S_2 and $S_{r,2}$ sets, respectively. Since all r_1 vertices will be adjacent to an r_2 vertex in an $S_{r,2}$ set, then S and all vertices in the T_2 trees are totally dominated. Thus, S will be a total dominating set. In this case, $k = 3q(0) + q + 2m = q + 2m$.

Forming a clique among the roots of the T_2 trees will produce a chordal graph. The same argument holds for this class of graphs. □

A set $S \subseteq V$ is an *independent dominating set* if for every vertex $v \in V - S$, $|N(v) \cap S| \geq 1$ and for every vertex $v \in S, |N(v) \cap S| = 0$.

INDEPENDENT DOMINATING SET
INSTANCE: A graph G and a positive integer k.
QUESTION: Does G have an independent dominating set of cardinality k or less?

Corollary 9.13 [9] [31] *INDEPENDENT DOMINATING SET is NP-complete, even when restricted to bipartite graphs.*

Proof. INDEPENDENT DOMINATING SET is in NP. A set S, $|S| \leq k$, could be given as a witness to a *yes* instance and can be verified to be both independent and dominating in $O(V + E)$ time.

Figure 9.10 shows tree T_1, rooted at r_1, and two T_2 trees, rooted at r_2, that have the properties specified in Theorem 9.10 for independent dominating sets. Tree T_1 does not have to have any of its vertices in S; i.e., $x = 0$. If no vertices in T_1 are in an independent dominating set, then r_1 must be adjacent to at least one vertex in S, otherwise r_1 would not be dominated.

Figure 9.10: Independent domination trees.

Each T_2 tree has to have at least one of its vertices in S; i.e., $y = 1$, since one vertex must be chosen to dominate a or b. If r_2 is in S, then two vertices must be chosen from T_2.

The T_1 tree and two T_2 trees in Figure 9.10 show the S_1, S_2 and $S_{r,2}$ sets, respectively. It is easy to see that S will be dominating and that no two vertices in S will be adjacent. In this case, $k = 3q(0) + q + (1)m = q + m$. □

A set $S \subseteq V$ is an *odd dominating set* if for every vertex $v \in V - S, |N(v) \cap S|$ is odd.

ODD DOMINATING SET
INSTANCE: A graph G and positive integer k.
QUESTION: Does G have an odd dominating set of cardinality k or less?

Corollary 9.14 [42] *ODD DOMINATING SET is NP-complete, even when restricted to bipartite graphs or chordal graphs.*

Proof. ODD DOMINATING SET is in NP. A set S, $|S| \leq k$, could be given as a witness to a *yes* instance and can be verified as an odd dominating set in $O(V + E)$ time.

Figure 9.11 shows tree T_1, rooted at r_1, and two T_2 trees, rooted at r_2, that have the properties specified in Theorem 9.10 for odd domination. Tree T_1 does not have to have any of its vertices in S; i.e., $x = 0$. If no vertices in T_1 are in the odd dominating set, then r_1 must be adjacent to at least one vertex in S, otherwise r_1 would be dominated zero times.

Tree T_2 has to have at least one of its vertices in S; i.e., $y = 1$, because either a must be in S or vertex b must be chosen to dominate a. If r_2 is in S, then two vertices must be chosen from the T_2.

The T_1 tree and two T_2 trees in Figure 9.11 show the S_1, S_2 and $S_{r,2}$ sets, respectively. Each vertex not in S is dominated exactly once, and so S is an odd dominating set. In this case, $k = 3q(0) + q + m = q + m$.

Figure 9.11: Odd dominating trees.

Forming a clique among the roots of the T_2 trees will produce a chordal graph, but the r_2 vertices that are not in S will be dominated an odd number of times only if q is even. To handle the case where q is odd, add one P_2 to the graph, form a clique among the roots of the T_2 trees and one of the vertices in the P_2 and let $k = q + m + 1$. The result is straightforward. □

A set $S \subseteq V$ is a *weak vertex-edge dominating set* if for every edge $uv \in E$, $|(N[u] \cup N[v]) \cap S| \geq 1$. In this case, we say that the edge uv is weakly dominated.

WEAK VERTEX-EDGE DOMINATING SET
INSTANCE: A graph G and a positive integer k.

QUESTION: Does G have a weak vertex-edge dominating set of cardinality k or less?

Corollary 9.15 [39] *WEAK VERTEX-EDGE DOMINATING SET is NP-complete, even when restricted to bipartite graphs or chordal graphs.*

Proof. WEAK VERTEX-EDGE DOMINATING SET is in NP. A set S, $|S| \leq k$, could be given as a witness to a *yes* instance and verified in $O(V + E)$ time to be a weak vertex-edge dominating set.

Figure 9.12 shows tree T_1, rooted at r_1, and two T_2 trees, rooted at r_2, that have the properties specified in Theorem 9.10 for weak vertex edge domination. It is possible that T_1 does not have any of its vertices in S; i.e., $x = 0$. This happens only if r_1 is adjacent to at least one vertex in S, otherwise the edge in T_1 would not be weakly dominated. If r_1 is adjacent to at least one vertex in S, then S_1 is the empty set.

Figure 9.12: Weak vertex-edge domination trees.

Tree T_2 has to have at least one of the vertices a, b, or c in S, because the edge (b,c) must be weakly dominated; i.e., $y = 1$. If r_2 is in S, then two vertices must be chosen from T_2. The two T_2 trees in Figure 9.12 indicate the S_2 and $S_{r,2}$ sets, respectively. In this case, $k = 3q(0) + q + m = q + m$.

Forming a clique among the roots of the T_2 trees will produce a chordal graph. The same argument holds for this class of graphs. □

Let S be a set. The private S-neighbors of a vertex $v \in S$ are the vertices in the set $N[v] - N[S - v]$. A set $S \subseteq V$ is an *independent set* if for every vertex $v \in S$, $|N[v] \cap S| = 1$. A set $S \subseteq V$ is a *2-maximal independent set* if it is a maximal independent set and there do not exist vertices $w \in S$, $x, y \in V - S$, $x \neq y$, such that $S - \{w\} \cup \{x, y\}$ is an independent set.

Lemma 9.16 *A maximal independent set S is 2-maximal if and only if there do not exist vertices w, x, and y, such that $w \in S$, $xy \notin E$, and x and y are private S-neighbors of w.*

Proof. Suppose S is maximal independent and there exist vertices x, y, and z, such that $x \in S$, $yz \notin E$, and y and z are private S-neighbors of x. Then $S' = S - \{x\} \cup \{y, z\}$ would be an independent set and S would not have been 2-maximal. On the other hand, suppose S is a maximal independent set, but not a 2-maximal independent set. It is well known that a set is maximal independent if and only if it is an independent dominating set. Since S is not 2-maximal independent, it must be possible to remove some w from S, and then add vertices x and y. When w is removed from S, the vertices in N[w] may no longer be dominated but the rest of the vertices in the graph will be dominated by the same vertex as in S. For x and y to be added to S requires x and y to be undominated. So $x, y \in N(w)$, and were private S-neighbors of w. Also $\{x, y\} \notin E$ since x and y are both in an independent set.

2-MAXIMAL INDEPENDENT SET
INSTANCE: A graph G and positive integer k.
QUESTION: Does G have a 2-maximal independent set of cardinality k or less?

Corollary 9.17 [37] *2-MAXIMAL INDEPENDENT SET problem is NP-complete, even when restricted to bipartite graphs.*

Proof. 2-MAXIMAL INDEPENDENT SET is in NP. A set S, $|S| \le k$, could be given as a witness to a *yes* instance. A brute force $O(E * V^3)$ algorithm could try all combinations of removing a vertex w and adding non-adjacent vertices x and y to S, in order to determine that S is 2-maximal independent.

Figure 9.13 shows tree T_1, rooted at r_1, and two T_2 trees, rooted at r_2, that have the properties specified in Theorem 9.10 for 2-maximal independence. T_1 must have at least one vertex in S; i.e., $x = 1$. Otherwise vertex c could not be dominated. If T_1 has only one vertex in S, then r_1 must be adjacent to at least one vertex in $S - V(T_1)$ (otherwise we could remove the S-vertex from T_1 and add vertices r_1 and c to S). If r_1 is adjacent to at least one vertex in S, then vertex b would dominate T_1 and would not have two private S-neighbors. Therefore, S_1 would be a set satisfying Condition (3) for element trees.

Tree T_2 has to have at least two of its vertices in S; i.e., $y = 2$. Vertex h must be dominated by either g or h, and vertex e must be dominated by either r_2, e, or f. Suppose r_2 is in S, and only two vertices are chosen from the T_2. In that case, vertex g would also have to be in S. But g could be removed and f and h added, so S would not be 2-maximal. Therefore, if r_2 is in S, then two other vertices from T_2 would also have to be in S. The two T_2 trees in Figure 9.13 indicate the S_2 and $S_{r,2}$ sets, respectively. Note that the S-members in these trees will dominate the trees, and that no S-member has two or more non-adjacent private S-neighbors. In this case, $k = 3q(1) + q + 2m = 4q + 2m$. $\square$

A set $S \subseteq V$ is an *irredundant set* if for every vertex $v \in S$, $N[v] - N[S - v] \neq \emptyset$. In other words, every $v \in S$ has a private S-neighbor.

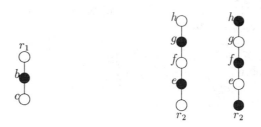

Figure 9.13: 2-maximal independence trees.

MINIMUM MAXIMAL IRREDUNDANT SET
INSTANCE: A graph G and a positive integer k.
QUESTION: Does G have a maximal irredundant set of cardinality k or less?

Corollary 9.18 [34] [40] *MINIMUM MAXIMAL IRREDUNDANT SET is NP-complete, even when restricted to bipartite graphs or chordal graphs.*

Proof. MINIMUM MAXIMAL IRREDUNDANT SET is in NP. A set S, $|S| \leq k$, could be given as a witness to a *yes* instance and verified in at most $O(V * E)$ time to be a maximal irredundant set.

Figure 9.14 shows tree T_1, rooted at r_1, and two T_2 trees, rooted at r_2, that have the properties specified in Theorem 9.10 for maximal irredundance. At least one vertex from T_1 must be chosen for S, otherwise z could be added to S without destroying any private neighbor; i.e., $x = 1$. If only one vertex in T_1 is chosen, this vertex cannot be x or r_1. Otherwise, if x or r_1 is the chosen vertex, then z could be added to S, with x and z serving as private S-neighbors. Again this contradicts the maximality of S. If y or z is the only vertex in S chosen from a T_1 tree, then r_1 must be joined to some S-vertex in $S - V(T)$. Otherwise x could be added to S, having r_1 as a private neighbor. So $x = 1$ and r_1 must be adjacent to at least one outside S-vertex, if only one vertex is chosen from the T_1.

Tree T_2 has to have at least two of its vertices in S; i.e., $y = 2$. At least one vertex from a, b, or c must be in S, otherwise a could be added to S, using b as its private neighbor. If only one vertex from a, b, or c is chosen, then the chosen vertex can use b as its private neighbor, and d can be added to S, using e as its private neighbor. So T_2 has to have at least two of its vertices in S. If $r_2 \in S$, then three vertices must be chosen from the T_2. The two T_2 trees in Figure 9.14 show the S_2 and $S_{r,2}$ sets, respectively. For the S_2 set, no other vertex can be added that would have a private neighbor. In the $S_{r,2}$ case, it is easy to see that no other vertices could be added to S without destroying irredundance.

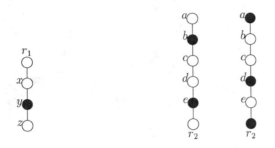

Figure 9.14: Irredundance trees.

If S_1 in T_1 is the set shown in Figure 9.14 and r_1 is joined to an outside S-vertex, then no other vertex in T_1 can be added to S. Note that x cannot be added because it would not have a private neighbor given the S_2 and $S_{r,2}$ sets. In this case, $k = 3q(1) + q + m(2) = 4q + 2m$.

Forming a clique among the roots of the T_2 trees will produce a chordal graph. The r_2 vertices in S will have private neighbors in the T_1 trees. The same argument can be applied to this class of graphs. $\square$

A set $S \subseteq V$ is a *closed open irredundant set* if for every vertex $v \in S$, $N[v] - N(S - v) \neq \emptyset$. For a vertex $v \in S$, the vertices in $N[v] - N(S - v)$ will be referred to as co-private neighbors (closed open private S-neighbors) of v.

MINIMUM MAXIMAL CLOSED OPEN IRREDUNDANT SET
INSTANCE: A graph G and a positive integer k.
QUESTION: Does G have a maximal closed open irredundant set of cardinality k or less?

Corollary 9.19 [37] *MINIMUM MAXIMAL CLOSED OPEN IRREDUNDANT SET is NP-complete, even when restricted to bipartite graphs or chordal graphs.*

Proof. MINIMUM MAXIMAL CLOSED OPEN IRREDUNDANT SET is in NP. A set S, $|S| \leq k$, could be given as a witness to a *yes* instance and verified in at most $O(V * E)$ time to be a maximal closed open irredundant set.

Figure 9.15 shows tree T_1, rooted at r_1, and two T_2 trees, rooted at r_2, that have the properties specified in Theorem 9.10 for closed open irredundance.

Consider tree T_1. At least one vertex from v, u, z must be chosen for S, otherwise v could be added to S without destroying a unique co-private neighbor and using u or itself as its co-private neighbor. If only one vertex is chosen from T_1, then w could be added to S. So at least two vertices from a T_1 component

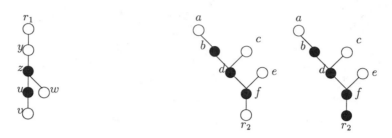

Figure 9.15: Closed open irredundance trees.

must be in S; i.e., $x = 2$. Also we will show that if only two vertices from a T_1 component are in S, then r_1 must be adjacent to a vertex in $S - V(T_1)$. We examine two cases.

Case 1. Suppose there is only one S-vertex among v, u, z, and w. Since we have shown that one of v, u, or z must be in S, then w is not in S. The only reason w could not be added to S would be that w would not have a co-private neighbor or that it would destroy all co-private neighbors of some vertex in S. If w does not have a co-private neighbor, then $z \in S$ (otherwise w uses itself as a co-private neighbor) and $y \in S$ (otherwise w uses z as a co-private neighbor). But in this case u could also be added to S, using z as a co-private neighbor, and in this case three vertices in T_2 are in S. So if Case 1 holds, then vertex w destroys all the co-private neighbors for some vertex in S, and this vertex must be y, implying that y cannot use r_1 as a co-private neighbor. This must be because r_1 is adjacent to a vertex in S.

Case 2. Two S-vertices are chosen from v, u, z, and w. No other vertices in T_1 are in S. Then consider why y cannot be added to S. One reason could be because y would have no co-private neighbor. In this case r_1 would have to be joined to a vertex in $S - V(T_1)$ (otherwise y uses r_1 as a co-private neighbor). The only other reason is that y destroys a co-private neighbor. If y destroys r_1 as a co-private neighbor, then r_1 would have to be adjacent to an S-member. Vertex y cannot destroy itself as a co-private neighbor for z, because since $r_1 \notin S$, $y \in N[z] - N(S - z)$. If y destroys z as a co-private neighbor then $w \in S$ (u could use v as a co-private neighbor). But in this case, v could be added to S, using itself as a co-private neighbor. So it follows that if only two vertices in T_1 are in S, then r_1 must be adjacent to an outside S-vertex.

Consider T_2. Notice if b, d, or f is in S, then it can get its co-private neighbor from a, c, or e, respectively. One of the two vertices a or b must be in S, otherwise a could be added using itself as a co-private neighbor. Using the same reasoning, at least one of c or d must be in S. If neither e nor f is in S, then r_2 must be using f as a co-private neighbor. In that case $d \notin S$, $c \in S$, and $a, b \in S$. Therefore, at least three of the non-root vertices are in S; i.e., $y = 3$. If $r_2 \in S$, then four vertices in T_2 must be taken.

Figure 9.15 shows the S_1, S_2 and $S_{r,2}$ sets, respectively. All that remains is to show that S is closed open irredundant and will be maximal if r_1 is adjacent to one r_2 vertex in S. Vertices u and z can use each other as co-private neighbors. Neither v ,w, nor y can be added because none would have a co-private neighbor. For the S_2 and $S_{r,2}$ sets, no other vertices can be added that would have a co-private neighbor. In this case, $k = 3q(2) + q + 3m = 7q + 3m$.

Forming a clique among the roots of the T_2 trees will produce a chordal graph. The same argument holds for this class of graphs as well. □

A set $S \subseteq V$ is an *open open irredundant set* if for every vertex $v \in S$, $N(v) - N(S - v) \neq \emptyset$. For a vertex $v \in S$, the vertices in $N(v) - N(S - v)$ will be referred to as oo-private neighbors (open open private S-neighbors) of v.

MINIMUM MAXIMAL OPEN OPEN IRREDUNDANT SET
INSTANCE: A graph G and a positive integer k.
QUESTION: Does G have a maximal open open irredundant set of cardinality k or less?

Corollary 9.20 [37] *MINIMUM MAXIMAL OPEN OPEN IRREDUNDANT SET is NP-complete, even when restricted to bipartite graphs or chordal graphs.*

Proof. MINIMUM MAXIMAL OPEN OPEN IRREDUNDANT SET is in NP. A set S, $|S| \leq k$, could be given as a witness to a *yes* instance and verified to be a maximal open open irredundant set in at most $O(V * E)$ time.

Figure 9.16 shows tree T_1, rooted at r_1, and two T_2 trees, rooted at r_2, that have the properties specified in Theorem 9.10 for open open irredundance. Consider tree T_1. At least one vertex from v, u, z must be chosen for S, otherwise v could be added to S, using u as its oo-private neighbor. If only one vertex is chosen from T_1, then any neighbor of the S-vertex at v, u, or z could be added, with these two S-vertices using each other as oo-private neighbors. So at least two vertices from a T_1 component must be in S; i.e., $x = 2$. We must also show that if only two vertices from a T_1 component are in S, then r_1 must be adjacent to a vertex in $S - V(T_1)$. We examine two cases.
Case 1. Suppose only one vertex is taken from v, u, z. If this vertex is u then S is not maximal because v could be added. If this vertex is v or z, then y must be in S and have only z as an oo-private neighbor (otherwise u could be added to S). Vertex y can have z as a unique oo-private neighbor only if r_1 is adjacent to some vertex in $S - V(T_1)$.
Case 2. Two S-vertices are chosen from v, u, z. Then consider why y cannot be added to S. One reason would be because y would have no oo-private neighbor.

In this case r_1 would have to be joined to a vertex in $S - V(T_1)$ (otherwise y uses r_1 as a oo-private neighbor). The only other reason is that y destroys an oo-private neighbor. If y destroys r_1 as a oo-private neighbor, then r_1 would have to be adjacent to an outside S-member. If $y \in S$, then y cannot destroy z as a unique oo-private neighbor because u can use v as a oo-private neighbor. So it follows that if only two vertices in T_1 are in S, then r_1 must be adjacent to an outside S-vertex.

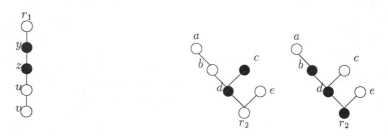

Figure 9.16: Open open irredundance trees.

Consider T_2. Notice if b, d, or r_2 is in S, then it can get its oo-private neighbor from a, e, or f, respectively. If neither a nor b is in S then d must be in S (otherwise a could be added) and c must be in S, otherwise b could be added. If neither d nor c is in S, then both a and b would be in S. So at least two non-root vertices must be chosen for S in T_2; i.e., $y = 2$. If $r_2 \in S$, then three vertices in T_2 must be taken.

Figure 9.16 shows the S_1, S_2 and $S_{r,2}$ sets. It is easy to see that S_1 is open open irredundant and will be maximal if r_1 is adjacent to one r_2 vertex in S. Vertices y and z can use each other as oo-private neighbors. Neither v, u nor r_1 can be added because none would have an oo-private neighbor. For the S_2 and $S_{r,2}$ sets, it is easy to see that no other vertices can be added to S and still maintain open open irredundance. In this case, $k = 3q(2) + q + 2m = 7q + 2m$.

Forming a clique among the roots of the T_2 trees will produce a chordal graph. The same argument holds for this class of graphs as well. $\square$

A set $S \subseteq V$ is an *open irredundant set* if for every vertex $v \in S$, $N(v) - N[S-v] \neq \emptyset$. For a vertex $v \in S$, the vertices in $N(v) - N[S-v]$ will be referred to as open private neighbors (open private S-neighbors) of v.

MINIMUM MAXIMAL OPEN IRREDUNDANT SET
INSTANCE: A graph G and a positive integer k.

QUESTION: Does G have a maximal open irredundant set of cardinality k or less?

Corollary 9.21 [37] *MINIMUM MAXIMAL OPEN IRREDUNDANT SET is NP-complete, even when restricted to bipartite graphs or chordal graphs.*

Proof. MINIMUM MAXIMAL OPEN IRREDUNDANT SET is in NP. A set S, $|S| \leq k$, could be given as a witness to a *yes* instance and verified to be a maximal open irredundant set in at most $O(V * E)$ time.

Figure 9.17 shows tree T_1, rooted at r_1, and two T_2 trees, rooted at r_2, that have the properties specified in Theorem 9.10 for open irredundance. Consider tree T_1. At least one vertex from w, y, z must be chosen for S, otherwise z could be added to S, using y as its open private neighbor. We must show that if only one vertex from a T_1 component is in S, then r_1 must be adjacent to a vertex in $S - V(T_1)$. Let us assume that only one vertex from T_1 is in S. The reason that x cannot be added to S must be either that x would have no open private neighbor or that it would destroy a unique private neighbor. Vertex r_1 could be a unique private neighbor only to a vertex in $S - V(T_1)$. If x cannot use r_1 as a private open neighbor, then either w is in S or r_1 is adjacent to a vertex in $S - V(T_1)$. If no vertex from $S - V(T_1)$ is adjacent to vertex r_1, then y could be added to S, using z as its open private neighbor. So if only one S-vertex is chosen from a T_1 component, then r_1 will be adjacent to an outside S-vertex; i.e., $x = 1$.

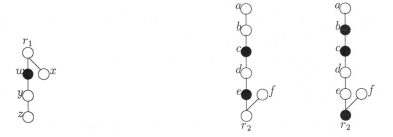

Figure 9.17: Open irredundance trees.

Consider T_2. Notice if $r_2 \in S$, then it can get its open private neighbor from f, since f could not also be in S. One of a or b or c is in S; if not, a could be added. If only one of the vertices in T_2 is in S, then vertex d could be added. So at least two non-root vertices must be chosen for S in T_2; i.e., $y = 2$. If $r_2 \in S$, then three vertices in T_2 must be taken. Figure 9.17 shows the S_1, S_2 and $S_{r,2}$

sets. Set S_1 will be maximal if each r_1 is adjacent to at least one r_2 vertex in S and at least one r_2 vertex in $V - S$. Note that the version of Exact Cover by 3-Sets defined in this chapter required each element to be in at least two sets. For the S_2 and $S_{r,2}$ sets, it is easy to see that no other vertices can be added and still maintain open irredundance. In this case, $k = 3q(1) + q + 2m = 4q + 2m$. □

9.7 Extending the Result

In this section we look at two NP-completeness proofs that suggest further extensions of the general NP-completeness proof. The first proof will lead to a discussion of functions on vertices, the second motivates extending the proof to edge parameters.

9.7.1 Double dominating multiset

A *double dominating multiset* is a multiset S such that for every vertex $v \in V - S, |N[v] \cap S| \geq 2$. Since S is a multiset, a vertex may appear in a set S more than once. Double Domination was first suggested by Domke [11].

DOUBLE DOMINATING MULTISET
INSTANCE: A graph G and a positive integer k.
QUESTION: Does G have a double dominating multiset of cardinality k or less?

Corollary 9.22 *DOUBLE DOMINATING MULTISET is NP-complete, even when restricted to bipartite graphs or chordal graphs.*

Proof. DOUBLE DOMINATING MULTISET is in NP. A multiset S, $|S| \leq k$, could be given as a witness to a *yes* instance and verified to be a double dominating multiset in $O(V + E)$ time.

Figure 9.18 shows tree T_1, rooted at r_1, and two T_2 trees, rooted at r_2. At least one of the vertices in the T_1 tree must be in a multiset S; i.e., $x = 1$, because if the non-root vertex is not in S, then r_1 would have to be in S twice. If only one vertex from T_1 is in S, then this must be the non-root vertex, and r_1 must be adjacent to an r_2 vertex in S, in order to be dominated twice.

Tree T_2 has to have at least two of its vertices in S; i.e., $y = 2$, since if vertex a and vertex b are not both in S, then vertex d must be in S twice. If r_2 is in S, then at least three vertices must be chosen from T_2.

The T_1 tree and two T_2 trees in Figure 9.19 show the S_1, S_2 and $S_{r,2}$ sets, respectively. The vertices that show a darkened circle within a circle represent vertices that are in the multiset S twice. Since all r_1 vertices will be adjacent

Figure 9.18: Double domination trees.

to an r_2 vertex in an $S_{r,2}$ set, S will be a double dominating multiset. In this case, $k = 3q(1) + 2m + q = 4q + 2m$.

Forming a clique among the roots of the T_2 trees will produce a chordal graph. The same argument holds for this class of graphs as well. □

9.7.2   Strong matching

A *strong matching* is a set S of vertices in a graph $G = (V, E)$ such that the subgraph $< S >$ induced by S consists of disjoint K_2's.

MINIMUM MAXIMAL STRONG MATCHING
INSTANCE: A graph G and a positive integer k.
QUESTION: Does G have a maximal strong matching of cardinality k or less?

Corollary 9.23 [37] *MINIMUM MAXIMAL STRONG MATCHING is NP-complete, even when restricted to bipartite graphs.*

Proof. MINIMUM MAXIMAL STRONG MATCHING is in NP. A set S, $|S| \leq k$, could be given as a witness to a *yes* instance and verified in polynomial time to satisfy $< S >$ consists of disjoint K_2's.

Figure 9.19 shows tree T_1, rooted at r_1, and two T_2 trees, rooted at r_2. We first note that at least one of the vertices in $N[r_1]$ must be in S, otherwise S is not maximal, and this S-vertex, as part of a K_2 must be adjacent to another vertex in S. So each r_1 is dominated by a K_2 in S.

Tree T_2 has to have at least two of its vertices in S; i.e., $y = 2$, since one vertex must be chosen from $N[a]$ and a neighbor of it to complete the K_2. Note that these K_2's cannot dominate the r_1 vertices. If r_2 is in S, then at least two K_2's must be chosen that have vertices in T_2.

The T_1 tree and two T_2 trees in Figure 9.19 show the S_1, S_2 and $S_{r,2}$ sets, respectively. Since all r_1 vertices will be adjacent to an r_2 vertex in an $S_{r,2}$

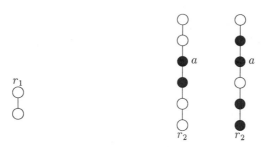

Figure 9.19: Strong matching trees.

set, then no other vertices in the T_1 or T_2 trees can be in S. In this case, $k = 3q(0) + 2q + 2m = 2q + 2m$. □

9.7.3 Minus parameters

The parameters discussed in previous sections can be redefined in terms of $\{0,1\}$-valued functions on the vertices of a graph: a vertex has a value 1 if it is in a set S or a value 0 if it is not in S. By defining set parameters in terms of functions, it is possible to identify new parameters by generalizing the two-valued function definition. For example, a dominating function is a two-valued function f defined on the vertices of a graph $G = (V, E)$, $f : V \rightarrow \{0,1\}$, such that for every $v \in V, f(N[v]) \geq 1$, where $f(N[v])$ is the sum of the function values over the closed neighborhood of v. Above we showed that double domination is NP-complete for bipartite and chordal graphs. Double domination could be considered a $\{0,1,2\}$-valued function on the vertices of a graph. Fractional domination and irredundance [11] [12], signed domination [15], and minus domination [16] are several other examples of generalizations of the domination and irredundance parameters. In this section we consider defining other domination parameters in terms of $\{-1,0,1\}$-valued functions on the vertices of a graph.

A function $f : V \rightarrow \{-1,0,1\}$ is a *minus dominating function* if for every vertex $v \in V$, $f(N[v]) \geq 1$. In [16], it is shown that a minus dominating function is minimal if and only if for every vertex v with $f(v) > -1$ there exists a $w \in N[v]$, such that $f(N[w]) = 1$. It is well known that a dominating set is minimal if and only if it is an irredundant dominating set. Using the minimality condition for a minus dominating function, a *minus irredundant function* can be defined. A function $f : V \rightarrow \{-1,0,1\}$ is a *minus irredundant function* if for every vertex v with $f(v) > -1$, there exists a $w \in N[v]$, such that $f(N[w]) = 1$.

In defining some of the domination parameters in terms of {-1,0,1}-valued functions, there may be more than one natural way of re-defining the parameter. The definition of minus irredundant function can serve as an example. A function is a minimal dominating function if (1) it is a dominating function and (2) for every vertex v with $f(v) > -1$, there exists a vertex $w \in N[v]$ such that $f(N[w]) \leq 1$. Since no vertices in a dominating function can have neighborhood sums less than 1, there is no difference between "≤ 1" and "$= 1$" in the minimality statement. However, in the definition of irredundant function, there is a difference. In defining fractional irredundance [21], the distinction was made by introducing fractional irreducibility. Their generalization can be used to define another minus parameter. A function $f : V \to \{-1, 0, 1\}$ is a *minus irreducible function* if for every vertex v with $f(v) > -1$, there exists a $w \in N[v]$ such that $f(N[w]) \leq 1$.

Other domination parameters may yield natural (-1,0,1) function definitions that change the properties of the parameter. Consider independent set. A function $f : V \to \{-1, 0, 1\}$ is a *minus independent function* if for every vertex v, if $f(v) = 1$ then $f(N(v)) = 0$. An independent set is maximal if and only if it is an independent dominating set. However, using the previous definitions, it is possible to have a maximal minus independent function that is not a minus dominating function, see Figure 9.20. Of course, the definition could require that the independent function is also dominating. A function $f : V \to \{-1, 0, 1\}$ is a *minus independent dominating function* if for every vertex v, $f(N[v]) \geq 1$ and if $f(v) = 1$ then $f(N(v)) = 0$. Note that all $\{0,1\}$-valued maximal independent functions will also be minimal minus dominating functions and maximal minus independent dominating functions.

$$-1 \quad 1 \quad 1 \quad -1$$

Figure 9.20: A maximal minus independent function that is not dominating.

There are numerous minus parameters that could be defined from the dominating set parameters discussed in earlier sections. For example, Table 9.1 defines six parameters that are generalizations of the definition for minus domination. In [37], a generalized NP-completeness proof is given for minus parameters in graphs. Several results are given in the tables at the end of the chapter.

Table 9.1: Minus functions.

f(N[v])	≤ 1	= 1	≥ 1
f(N[v])	Minus 2-packing	Minus efficient domination	Minus domination
f(N(v))	Minus Open 2-packing	Minus total efficient domination	Minus total domination

9.7.4 Edge parameters

Let $G = (V, E)$ be a graph. The *open neighborhood* of an edge $e \in E$, denoted $N(e)$, is the set of all edges in E that share a vertex with e. The open neighborhood of an edge set S, denoted $N(S)$, is the union of the open neighborhoods of all edges in S. The *closed neighborhood* of an edge $e \in E$, denoted $N[e]$, is the open neighborhood of e unioned with $\{e\}$. The closed neighborhood of an edge set S, denoted $N[S]$, is the union of the closed neighborhoods of all edges in S. A vertex u is *saturated*, or covered, by any edge (u, v). With these definitions, many of the parameters discussed in the introduction can be redefined as edge parameters. For example, a set $S \subseteq E$ is an *edge dominating set* of a graph $G = (V, E)$ if for every edge $e \in E$, $|N[e] \cap S| \geq 1$. Given a graph, $G = (V, E)$, the *line graph* $L(G) = (V', E')$, has $V' = E$ and $E' = \{e_1 e_2 | e_1$ and e_2 are adjacent in $G\}$. Alternately, the edge versions of the parameters can be thought of as the vertex versions of the problem applied to the line graph of G. Tables 9.2 and 9.3 also show references for several parameters when restricted to line graphs. In [37], a generalized NP-completeness construction is given for edge parameters.

9.7.5 Approximation results

Some NP-complete problems, such as 2-dimensional bin packing have polynomial approximation algorithms that will guarantee that solutions found by the algorithms will be within a fixed factor f of an optimal solution [22]. Other NP-complete problems fall into a different category. Problems such as the Unconstrained Traveling Salesman have no such polynomial approximation algorithms, unless P=NP [22]. Irving [31] has shown that independent domination falls into this second category. For most NP-complete problems, the complexity of a corresponding approximation problem has not been determined. The method Irving uses to get the result for independent domination takes an existing NP-completeness construction for an original decision problem and adds multiple copies of some of the components. This strategy seems like it should be useful in getting similar results for many NP-complete problems. In [37], Irving's result is generalized for several other domination-related parameters. Some interesting recent developments on approximation for domination-related parameters can be found in [38] [1] [35] [3].

9.8   An Overview of the Results

Tables 9.2 and 9.3 summarize the NP-completeness results for the domination parameters listed in this chapter. Where a result is known by the authors, a citation is given. The notation "NPh" refers to NP-hard. Table 9.4 gives a description of the problems listed in the two tables.

Some problems in the chart have polynomial solutions while other closely related problems are NP-complete. Identifying general underlying attributes that appear in NP-complete problems but not in related polynomial problems will make it easier to determine the complexity of new problems.

Another area of study is to find other NP-completeness constructions that could be generalized for a wide variety of problems. For example, Hedetniemi and Laskar define two transformations, VV and $VV+$, from a graph G to a bipartite graph G' in [28, 29]. These transformations can be used to show that many problems involving open neighborhoods are NP-complete for bipartite graphs, given that the corresponding closed neighborhood problem is NP-complete for arbitrary graphs. An example of a general local replacement method is shown by Stockmeyer in [41] where an NP-completeness result for 3-Colorability is restricted to planar graphs by replacing edge crossings in an arbitrary graph with a planar component. This same approach has proven useful in generating many NP-completeness results for planar graphs. The new approximation results may be useful models, since there seem to be relatively few approximation results known for domination-related parameters.

These tables certainly are not complete. First, many entries in the table remain empty, especially in the minus and approximation columns. Second, and perhaps even more interesting, are the columns and rows that do not appear in the table. For example, there are many columns that could be added to this table for different classes of graphs, such as planar graphs, undirected path graphs, or strongly chordal graphs. There are other possible columns for different functions on vertices or edges, e.g. signed functions or real-valued functions. An expanded table should help us learn more about the boundary between P and NP.

Table 9.2: NP-completeness results I (starred entries in this chapter).

	ANY GRAPH	BIPAR. GRAPH	CHORD. GRAPH	LINE GRAPH	LINE G. OF BIPAR. GRAPH	MINUS
i	NP [22]	NP [31] [9]	P [19]	NP [43]	NP [43]	
γ	NP [8]	NP [10]	NP [5]	NP [43]	NP [43]	NP [13]
ir	NP [40]	NP [40]	NP [34]	NP [37]	NP [37]	
oir	NP *	NP *	NP *	NP [37]	NP [37]	
$ooir$	NP *	NP *	NP *	NP [37]	NP [37]	
$coir$	NP *	NP *	NP *	NP [37]	NP [37]	
p_2	NP *	NP *	NP *	NP [37]	NP [37]	NP [37]
$p_{o,2}$	NP *	NP *	NP *	NP [37]	NP [37]	NP [37]
γ_p	NP [44]	NP [44]	NP [44]	NP [37]	NP [37]	
np	NP [13]	NP [13]	NP [13]	NP [37]	NP [37]	
γ_t	NP [40]	NP [40]	NP [34]	NP [37]	NP [37]	NP [37]
β_2	NP *	NP *		NP [27]	NP [27]	
F	NP [2]	NP [2]	NP *	NP [37]	NP [37]	NP [37]
F_t	NP *	NP *	NP *	NP [37]	NP [37]	NP [37]
dd	NP *	NP *	NP *	NP [37]	NP [37]	
γ_o	NP [42]	NP *	NP *	NP [37]	[37]	

Table 9.3: NP-complete results II.

	ANY GRAPH	BIPAR. GRAPH	CHORD. GRAPH	LINE GRAPH	LINE G. OF BIPAR. GRAPH	MINUS
β	NP [22]	P [25]	P [24]	P [17]	P [17]	
Γ	NP [26]	P [8]	P [32]	NP [37]	NP [37]	NP [13]
IR	NP [20]	P [8]	P [32]	NP [37]	NP [37]	
OIR	NP [18]			NP [37]	NP [37]	
$OOIR$	NP [6]	NP [6]		NP [37]	NP [37]	
$COIR$	NP [20]	NP [20]		NP [37]	NP [37]	
P_2	NP [7] [30]	NP [7]	NP [7]	NP [6]	NP [6]	NP [37]
$P_{o,2}$	NP [36]	NP [36]	NP [37]			NP [37]
Γ_p	NPh [37]	NPh [37]				
Γ_t	NP [32]	NP [32]		NP [37]	NP [37]	NP [37]

Table 9.4: Description of parameters.

Parameter	Description
i	minimum independent dominating set
γ	minimum dominating set
ir	minimum maximal irredundant set
oir	minimum maximal open irredundant set
$ooir$	minimum maximal open open irredundant set
$coir$	minimum maximal closed open irredundant set
p_2	minimum maximal strong stable set
$p_{o.2}$	minimum maximal open strong stable set
γ_p	minimum perfect dominating set
np	minimum 1-maximal nearly perfect set
γ_t	minimum total dominating set
β_2	minimum 2-maximal independent set
F	efficient domination number
F_t	total efficient domination number
dd	minimum double dominating multiset
γ_o	minimum odd dominating set
β	maximum independent set
Γ	maximum minimal dominating set
IR	maximum irredundant set
OIR	maximum open irredundant set
$OOIR$	maximum open open irredundant set
$COIR$	maximum closed open irredundant set
P_2	maximum strong stable set
$P_{o,2}$	maximum open strong stable set
Γ_p	maximum minimal perfect dominating set
Γ_t	maximum minimal total dominating set

Bibliography

[1] O. Arora, C. Lund, R. Motwani and M. Sudan, Proof verification and intractability of approximation problems. In *Proc. of the 33rd IEEE Symp. on Foundations of Computer Science* (1992).

[2] D. W. Bange, A. E. Barkauskas and P. J. Slater, Efficient dominating sets in graphs. In *Applications of Discrete Mathematics*, Eds. R. D. Ringeisen and F. S. Roberts, SIAM Philadelphia (1988) 189–199.

[3] M. Bellare, S. Goldwasser, C. Lund and A. Russell, Efficient probabilistically checkable proofs and applications to approximation. In *Proc. of the 25th ACM Symp. on the Theory of Computing* (1993) 294–304.

[4] B. Bollobás and E. J. Cockayne, Graph theoretic parameters concerning domination, independence, and irredundance. *J. Graph Theory* 3(3) (1979) 241–249.

[5] K. S. Booth and J. H. Johnson, Dominating sets in chordal graphs. *SIAM J. Comput.* 11 (1982) 191–199.

[6] K. Cameron, Induced matchings. *Discrete Appl. Math.* 24 (1989) 97–102.

[7] G. J. Chang and G. L. Nemhauser, Covering, packing, and generalized perfection. *SIAM J. Alg. Discrete Meth.* 6(1) (1985) 109–132.

[8] E. J. Cockayne, O. Favaron, C. Payan and A. Thomason, Contributions to the theory of domination, independence, and irredundance in graphs. *Discrete Math.* 33(3) (1981) 249–258.

[9] D. G. Corneil and Y. Perl, Clustering and domination in perfect graphs. *Discrete Appl. Math.* 9 (1984) 27–39.

[10] A. K. Dewdney, Fast turing reductions between problems in NP; chapter 4; reductions between NP-complete problems. Technical Report 71, Dept. Computer Science, University of Western Ontario (1981).

[11] G. S. Domke, *Variations of Colorings, Coverings and Packings of Graphs.* Ph.D. Dissertation, Clemson University (1988).

[12] G.S. Domke, S.T. Hedetniemi and R. Laskar, Fractional packings, covering and irredundance in graphs. *Congr. Numer.* 66 (1988) 227–238.

[13] J. Dunbar, W. Goddard, S. T. Hedetniemi, M. A. Henning and A. McRae, The algorithmic complexity of minus domination in graphs. *Discrete Math.* 68 (1996) 73–84.

[14] J. E. Dunbar, F. Harris, S. M. Hedetniemi, S.T. Hedetniemi, R. Laskar and A. McRae, Nearly perfect sets in graphs. *Discrete Math.* 138 (1995) 229–246.

[15] J. E. Dunbar, S. T. Hedetniemi, M. A. Henning and A. McRae, Minus domination in graphs. Manuscript.

[16] J. E. Dunbar, S. T. Hedetniemi, M. A. Henning and P. J. Slater, Signed domination in graphs. In *Proc. Seventh Internat. Conf. in Graph Theory, Combinatorics, Algorithms, and Applications* Kalamazoo, MI (1994) 311–322.

[17] J. Edmunds, Paths, trees, and flowers. *Canad. J. Math.* 17 (1965) 449–467.

[18] S. Even, O. Goldreich and P. Tong, On the NP-completeness of certain network testing problems. Technical Report 230, Technion, Dept. of Computer Science, Haifa, Israel (1981).

[19] M. Farber, Independent domination in chordal graphs. *Oper. Res. Lett.* 1(4) (1982) 134–138.

[20] M. Fellows, G. Fricke, S. T. Hedetniemi and D. Jacobs, The private neighbor cube. *SIAM J. Discrete Math.* 7(1) (1994) 41–47.

[21] G. Fricke, S. M. Hedetniemi and D. P. Jacobs, When is an irredundant function on a graph maximal? Manuscript (1988).

[22] M.R. Garey and D.S. Johnson, *Computers and Intractability: A Guide to the Theory of NP-Completeness.* W.H. Freeman and Company, New York, NY, 1979.

[23] H. Gavlas, K. Schultz and P. J. Slater, Efficient open domination in graphs. To appear in *Scientia.*

[24] F. Gavril, Algorithms for minimum colorings, maximum clique, minimum coverings by cliques, and maximum independent set of a chordal graph. *SIAM J. Comp.* 1 (1972) 180–187.

[25] F. Harary, *Graph Theory.* Addison-Wesley, Reading, MA, 1969.

[26] S. M. Hedetniemi, S. T. Hedetniemi and D. P. Jacobs, Private domination:theory and algorithms. *Congr. Numer.* 79 (1990) 147–157.

[27] S. M. Hedetniemi, S. T. Hedetniemi and A. A. McRae, Matchings, matchability, and 2-maximal matching. Manuscript (1993).

[28] S. T. Hedetniemi and R. Laskar, A bipartite theory of graphs: I. *Congr. Numer.* 55 (1986) 5–14.

[29] S. T. Hedetniemi and R. Laskar, A bipartite theory of graphs: II. *Congr. Numer.* 64 (1988) 137–146.

[30] D. S. Hochbaum and D. B. Schmoys, A best possible heuristic for the k-center problem. *Math. Oper. Res.* 10(2) (1985) 180–184.

[31] R. Irving, On approximating the minimum independent dominating set. *Inform. Proc. Lett.* 37(4) (1991) 197–200.

[32] M.S. Jacobson and K. Peters, Chordal graphs and upper irredundance, upper domination and independence. *Discrete Math.* 86 (1990) 59–69.

[33] R. M. Karp, Reducibility among combinatorial problems. In *Complexity of Computer Combinatorial Problems*, Eds. R. E. Miller and J. W. Thatcher, Plemum Press New York, NY (1972).

[34] R. Laskar, J. Pfaff, S.M. Hedetniemi and S. T. Hedetniemi, On the algorithmic complexity of total domination. *SIAM J. Alg. Discrete Meth.* 5 (1984) 420–425.

[35] C. Lund and M. Yannakakis, On the hardness of approximation minimization problems. In *Proc. of the 25th ACM Symp. on the Theory of Computing* (1993) 286–293.

[36] A. Majumdar, *Neighborhood Hypergraphs: A Framework for Covering and Packing Parameters in Graphs.* Ph.D. Dissertation, Clemson University (1992).

[37] A. A. McRae, *Generalizing NP-completeness Proofs for Bipartite Graphs and Chordal Graphs.* Ph.D. Dissertation, Clemson University (1994).

[38] C. Papadimitriou and M. Yannakakis, Optimization, approximation and complexity classes. In *Proc. of the 20th ACM Symp. on the Theory of Computing* (1988) 229–234.

[39] K. Peters, *Theoretical and Algorithmic Results on Domination and Connectivity.* Ph.D. Disseration, Clemson University (1986).

[40] J. Pfaff, R. Laskar and S. T. Hedetniemi, NP-completeness of total and connected domination and irredundance for bipartite graphs. Technical Report 428, Clemson University, Dept. of Math. Sci. (1983).

[41] L. J. Stockmeyer, Planar 3-colorability is NP-complete. *SIGACT News* 5(3) (1973) 19–25.

[42] K. Sutner, Linear cellular automata and the Garden-of-Eden. *Math. Intelligencer* 11(2) (1989) 49–53.

[43] M. Yannakakis and F. Gavril, Edge dominating sets in graphs. *SIAM J. Appl. Math.* 38(3) (1980) 364–372.

[44] C. C. Yen and R. C. T. Lee, The weighted perfect domination problem. *Inform. Process. Lett.* 35(6) (1990) 295–299.

Chapter 10

Domination Parameters of a Graph

E. Sampathkumar

Department of Mathematics

University of Mysore

Mysore, 570006, India

10.1   Introduction

It is clear from the content of this and the accompanying book [15] that domination in graphs, along with its many variations, provides an extremely rich area of study. In India, interest in modern domination in graphs was triggered by the monograph-cum-thesis of Walikar [47]. This chapter surveys several topics in the field introduced since then. In particular, the following are discussed: (1) connected domination; (2) strong and weak domination, and domination balance; (3) least domination number; (4) dominating strength and weakness; (5) set and global set domination; (6) point-set and global point-set domination; (7) neighborhood numbers; (8) independent, perfect, and connected neighborhood numbers; (9) domination and neighborhood critical, fixed, free and totally free vertices and edges; and (10) mixed domination. Most proofs are omitted, but a few of the simpler ones are incorporated in order to illustrate the use of some of the concepts.

The results contained herein include many relationships among graphical invariants of graph G. When the graph G is clear from the context, the argument G is omitted from the designation of the invariant.

10.2   Connected Domination

The connected domination number γ_c of a connected graph G has received a
great deal of attention, and many of the results are surveyed in this section. A
straightforward relationship exists among it, the domination number γ, and the
total domination number γ_t when the graph is connected, of order n, and with
maximum degree $\Delta < n-1$. In this case $\gamma \le \gamma_t \le \gamma_c$. As we show in Proposition
10.5 this triple has been characterized. Sampathkumar and Walikar [45] were
the first to study connected domination, and this seminal paper includes the
results of our first proposition.

Proposition 10.1

1. *For a tree of order $n \ge 3$ with e endvertices, $\gamma_c = n - e$.*

2. *Let H be a connected spanning subgraph of a graph G. Then $\gamma_c(G) \le \gamma_c(H)$.*

3. *For any connected graph of order $n \ge 3, \gamma_c \le n - 2$.*

4. *Let G be a connected graph of order n and size m. Then*

$$\frac{n}{\Delta + 1} \le \gamma_c \le 2m - n$$

with $\gamma_c = 2m - n$ if and only if G is path.

Hedetniemi and Laskar [17] have found several other major results involving
γ_c. Not surprisingly, one of these is that the decision problem associated with
determining γ_c for an arbitrary graph is NP-complete. Another was motivated
by work of Nieminen [30] who showed that, for any connected graph of order n,
$\gamma + \epsilon_F = n$. Here ϵ_F is the maximum number of pendant edges in any spanning
forest of the graph. The following corresponding Hedetniemi and Laskar result
for γ_c involves ϵ_T, the maximum number of pendant edges in any spanning tree
of a connected graph.

Proposition 10.2 [17] *For any connected graph, $\gamma_c + \epsilon_T = n$.*

Other results obtained in [17] appear in the next proposition. Recall that a
spider is a tree, having at most one vertex of degree three or more and β_1 is the
edge independence or matching number of the graph.

Proposition 10.3 *For any connected graph,*

1. *$diam - 1 \le \gamma_c \le n - \Delta$,*

2. *$\gamma_c \le 2\beta_1$, and*

3. if G is a tree, then $\gamma_c = n - \Delta$ if and only if G is a spider.

Additional upper bounds have been obtained by Duchet and Meyniel [12].

Proposition 10.4 *For any connected graph,*

1. $\gamma_c \leq 2\beta_0 - 1$ and

2. $\gamma_c \leq 3\gamma - 2$.

As noted, for a connected graph G if $\Delta < n - 1$ (equivalently, if $\gamma \geq 2$) then $\gamma \leq \gamma_t \leq \gamma_c$. Dankelmann, Slater, Smithdorf, and Swart [11] characterized the collection of triples (a, b, c) for which there exists a connected graph H having domination number $\gamma(H) = a$, total domination number $\gamma_t(H) = b$, and connected domination number $\gamma_c(H) = c$. They noted that, for the triples that are realizable, H can be required to be a tree.

Proposition 10.5 *Given integers a, b, c with $2 \leq a \leq b \leq c$ there exists a connected graph H with $\gamma(H) = a$, $\gamma_t(H) = b$, and $\gamma_c(H) = c$ if and only if*

(1) $b \leq 2a$,

(2) $c \leq 3a - 2$, and

(3) $c \leq \begin{cases} 2b - 2 & \text{if } b \text{ is even} \\ 2b - 3 & \text{if } b \text{ is odd}. \end{cases}$

If so, such an H exists where H is a tree.

Another result by Duchet and Meyniel [12] involves the *contraction* or *Hadwiger number* $h(G)$ of a connected graph G. This number is the maximum order of any complete graph obtained from G by iterated contractions of edges.

Proposition 10.6 [12] *The deletion of a connected dominating set from any connected graph G decreases the value of $h(G)$.*

It is known [18] that $\gamma + \overline{\gamma} \leq n + 1$ for any graph of order n. A similar Nordhaus-Gaddum type of result holds for γ_c.

Proposition 10.7 [17] *If both G and $\overline{G}$ are connected, then $\gamma_c + \overline{\gamma_c} \leq n + 1$.*

Proof. From Proposition 10.3 we have $\gamma_c \leq n - \Delta$ and $\overline{\gamma_c} \leq n - \overline{\Delta}$. Therefore,

$$\begin{aligned} \gamma_c + \overline{\gamma_c} &\leq (n - \Delta) + (n - \overline{\Delta}) \\ &= 2n - (\Delta + \overline{\Delta}) \\ &= 2n - (\Delta + n - 1 - \delta) \\ &= n + 1 + \delta - \Delta \\ &\leq n + 1 \end{aligned}$$

since $\delta - \Delta \leq 0$. $\square$

Laskar and Peters [26] carry Proposition 10.7 a step further by showing that $\gamma_c + \overline{\gamma_c} = n + 1$ if and only if $G = C_5$. Thus, if $G \neq C_5$, then $\gamma_c + \overline{\gamma_c} \leq n$. In this same paper, they also characterize the graphs for which $\gamma_c + \overline{\gamma_c} = n$.

Analogous to the domatic number $d(G)$ of an arbitrary graph G is the *connected domatic number* $d_c(G)$ of a connected graph G. It is the largest order of a partition of $V(G)$ into connected dominating sets. Paulraj and Arumugam [22] and Hedetniemi and Laskar [17] provide the results given in the following proposition.

Proposition 10.8

1. [22] *If both G and $\overline{G}$ are connected, then $d_c + \overline{d_c} \leq n - 2$.*

2. [17] *For a connected graph G, $\gamma_c + d_c \leq n + 1$, with equality if and only if $G = K_p$.*

In addition, the graphs for which equality holds in Proposition 10.8(1) are characterized in [22]. Motivated by the concept of connected domination, Paulraj Joseph and Arumugam [21, 23] introduced three more domination parameters, i.e., block domination, connected cutfree domination, and 2-edge connected domination.

10.3   Strong and Weak Domination and Domination Balance

Certain practical situations are concerned with the relative degrees of adjacent vertices of a graph. For example, consider a model of a network of roads with the vertices corresponding to intersections. Suppose a street joins intersections u and v with $deg(u) > deg(v)$, that is, more streets meet at u than do at v. This is an indication that traffic is heavier at u than at v, and it makes sense for a traffic control scheme to give preference to vehicles traveling from u to v rather than vice versa. The subject of this section, all of whose results, unless otherwise stated, are taken from Sampathkumar and Pushpa Latha [43], is a theory which can be applied to such situations.

Let $G = (V, E)$ be a graph and $u, v \in V$. Then u *strongly dominates* v and v *weakly dominates* u if (i) $uv \in E$ and (ii) $deg(u) \geq deg(v)$. A set $S \subseteq V$ is a *strong dominating set* of G if every vertex in $V - S$ is strongly dominated by at least one vertex in S. The *strong domination number* γ_S of G is the minimum cardinality of such a set. A *weak dominating set* and the *weak domination number* γ_W are defined similarly. The following proposition presents properties of vertices in a minimal strong (weak) dominating set.

Proposition 10.9 *Let* S *be a minimal strong (weak) dominating set of graph* $G = (V, E)$. *Then for each* $v \in S$ *one of the following holds:*

1. *no vertex in* S *strongly (weakly) dominates* v, *or*

2. *there exists a vertex* $u \in V - S$ *such that* v *is the only vertex in* S *which strongly (weakly) dominates* u.

For any graph G, let $\overline{\gamma}_S = \gamma_S(\overline{G})$ and $\overline{\gamma}_W = \gamma_W(\overline{G})$. The next result relates these and other parameters for a tree. We define a *support* in a tree to be a vertex adjacent to a pendant vertex.

Proposition 10.10 *Let* T *be tree with* $n \geq 3$ *vertices,* e *pendant vertices, and* b *supports. Then*

1. $b \leq \gamma \leq \gamma_S \leq \gamma_W \leq n - e,$

2. $e \leq \gamma_W \leq n - b,$ *and*

3. *if* $T \neq K_{1,n}$ *then*

    (a) $\overline{\gamma}_S = 2,$

    (b) $\overline{\gamma}_W \leq n - e,$

    (c) $b + 2 \leq \gamma_S + \overline{\gamma}_S \leq n,$ *and*

    (d) $\gamma_W + \overline{\gamma}_W \leq 2n - 3.$

In [43] it was conjectured that for any tree T, $i(T) \leq \gamma_W(T)$. This conjecture was proven by Hattingh and Laskar [14]. Furthermore, they showed that for any tree T, $\gamma_W(T) \leq \beta_0(T)$, constructed an infinite class of trees for which the differences $\gamma_W - i$ and $\beta_0 - \gamma_W$ become arbitrarily large, and proved that the decision problem corresponding to the computation of γ_W is NP-complete.

A related concept is the *independent strong (weak) domination number* γ_{iS} (γ_{iW}) of G. This number is the minimum cardinality of a strong (weak, respectively) dominating set that is independent. Clearly, for any graph G, $i \leq \gamma_{iS}$ and $i \leq \gamma_{iW}$. Allan and Laskar [7] have shown that $\gamma = i$ when G does not have $K_{1,3}$ as an induced subgraph. As we see next, similar results, shown in [43], hold for γ_S and γ_W.

Proposition 10.11 *If a graph does not contain* $K_{1,3}$ *as an induced subgraph, then*

1. $\gamma_S = \gamma_{iS}$ *and*

2. $\gamma_W = \gamma_{iW}$.

The notions of strong and weak dominating sets come together in the following concepts. A set $S \subseteq V$ is $s-full$ ($w-full$) if $V - S$ is a weak (strong) dominating set. For a detailed study of *full-sets* (or "*enclaveless sets*" [46]) in hypergraphs and their interrelationships with domination, independence and colorings in hypergraphs, see [1, 2, 4]. Further, G is *domination balanced* if there exists a strong dominating set S_1 and a weak dominating set S_2 such that $S_1 \cap S_2 = \emptyset$. The final proposition of this section characterizes domination balanced graphs.

Proposition 10.12 [43] *For any graph G, the following statements are equivalent:*

1. *G is domination balanced,*

2. *there exists a strong dominating set S which is s-full, and*

3. *there exists a weak dominating set S which is w-full.*

Proof. Suppose $G = (V, E)$ is domination balanced so there is a strong dominating set S_1 and a weak dominating set S_2 such that $S_1 \cap S_2 = \emptyset$. Since $S_2 \subseteq V - S_1$ and S_2 is a weak dominating set, $V - S_1$ also is a weak dominating set so S_1 is s-full. This shows (1) implies (2). To show (2) implies (3), assume S is a strong dominating set which is s-full, that is, $V - S$ is a weak dominating set. Then it is immediate that $V - S$ is w-full. Finally, for (3) implying (1), let S be a weak dominating set which is w-full, meaning $V - S$ is a strong dominating set and G is domination balanced. $\square$

10.4   The Least Domination Number

This section deals with modified versions of both vertex covers and dominating sets. Here all vertex covers S must include any isolated vertices, and among these we are interested in ones for which the subgraph induced by S has the smallest vertex cover. Similarly, among all the dominating sets D, we want the ones for which the subgraph induced by D has the smallest domination number. Unless otherwise stated, all results from this section are taken from Sampathkumar [34]. A series of definitions makes the above ideas precise.

A set $S \subseteq V$ of a graph $G = (V, E)$ is a *total vertex cover* if S is a vertex cover containing all isolates of G. The number $\alpha_t(G)$ is the minimum cardinality of a total vertex cover. A total vertex cover S is a *least vertex cover* if for any total vertex cover S_1, $\alpha_t(<S>) \leq \alpha_t(<S_1>)$. The *least vertex covering number* $\alpha_L(G)$ of G is the minimum cardinality of a least vertex cover.

In an analogous fashion, a dominating set D of G is a *least dominating set* if $\gamma(<D>) \leq \gamma(<D_1>)$ for any dominating set D_1 of G. The *least domination number* $\gamma_L(G)$ of G is the minimum cardinality of a least dominating set.

Exact values of α_L and γ_L are known for paths P_n and cycles C_n.

Proposition 10.13 *Let* $n \geq 3$. *Then*

1. $\alpha_L(P_n) = (n-1) - \lceil (n-1)/4 \rceil$,

2. $\alpha_L(C_n) = n - \lceil n/4 \rceil$, *and*

3. $\gamma_L(P_n) = \gamma_L(C_n) = n - 2\lceil n/5 \rceil$.

There is a general relationship between the least and the total domination numbers.

Proposition 10.14 *If a graph has no isolates, then* $\gamma_L \leq \gamma_t$.

Proof. Let $G = (V, E)$ be a graph without isolates and D be a minimum total dominating set such that $\gamma(< D >)$ is least. We claim that D is a least dominating set. To see this, let D_1 be any dominating set. If $< D_1 >$ has no isolates, then it is a total dominating set. Hence, by our choice of D, we have $\gamma(< D >) \leq \gamma(< D_1 >)$. Suppose then that $< D_1 >$ has k isolates u_1, $u_2, \ldots, u_k$ and let v_i be a vertex of $V - D$ adjacent to u_i. The v_i are not necessarily distinct. Then $D_2 = D_1 \cup \{v_1, v_2, \ldots, v_k\}$ is a total dominating set and $\gamma(< D >) \leq \gamma(< D_2 >) \leq \gamma(< D_1 >)$ showing D to be a least dominating set. Since $|D| = \gamma_t$, the result holds. $\square$

In [34] it was conjectured that for any graph, $\gamma_L \leq \alpha_L$. Both Favaron [13] and Zverovich [48] independently proved this conjecture. Another conjecture from [34] is that for any connected graph, $\gamma_L \leq 3n/5$. Favaron [13] disproved this conjecture and presented some applications of least domination to broadcasting.

10.5 Dominating Strength and Weakness

It is sometimes useful in a communication network linking cities to have a dominating set D for which it is known to how many other cities every member of D can transmit a message. Two situations have application in this and other frameworks. One involves knowledge of the minimum number of vertices each member of D can reach, while the other deals with the maximum such number. The strength and weakness, respectively, of a graph are defined here to handle these situations. The source for this section is Sampathkumar and Pushpa Latha [44].

Let D be a minimal dominating set of a graph G. The *strength* $s(D)$ of D is defined by $s(D) = min\{deg(v) : v \in D\}$ and the *dominating strength* of G is $d_s(G) = max\{s(D) : D$ is a minimal dominating set$\}$. The *weakness* $w(D)$ and *dominating weakness* $d_w(G)$ are similarly defined by $w(D) = max\{deg(v) : v \in D\}$ and $d_w(G) = min\{w(D) : D$ is a minimal dominating set$\}$, respectively. For the graph of Figure 10.1 we have $\gamma = 5$, $d_s = 3$ where an applicable dominating set is $\{e, f, g, j, k, l\}$, and $d_w = 2$ illustrated by dominating set $\{a, b, c, d, h, i, m, n, o, p\}$.

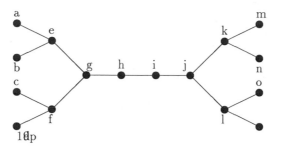

Figure 10.1: Illustration of dominating strength and weakness.

Even though, for any vertex v in a graph, there always exists a minimal dominating set containing v, it is not necessarily true that there is a minimal dominating set with strength or weakness $deg(v)$. However, the minimum and maximum degrees of the graph do form bounds since, for any dominating set $D \subseteq V$, it is clear that $\delta \leq s(D) \leq w(D) \leq \Delta$.

Domination numbers for dominating sets of strength or weakness k are defined in a natural manner. Let k be the degree of a vertex in G. The *domination number of strength k*, $\gamma_{sk}(G)$, is the minimum cardinality of a dominating set of strength k, if such a set exists. Similarly we define $\gamma_{wk}(G)$, the *domination number of weakness k*. Whenever we speak of γ_{sk} or γ_{wk} below, it is understood that they exist. For any graph, $\gamma_{s\delta}$ and $\gamma_{w\Delta}$ always exist. For the graph of Figure 10.2(a), $\gamma = \gamma_{s3} = \gamma_{w3} = 2$ with $\{a,c\}$ being an appropriate dominating set, and $\gamma_{s1} = 3$ as can be seen from the set $\{b,c,d\}$. In Figure 10.2(b) we have $\gamma_{w4} = 4$ with set $\{a,b,c,d\}$ and $\gamma_{w5} = 5$ with set $\{b,c,d,e,f\}$.

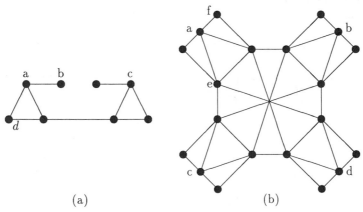

(a)   (b)

Figure 10.2: Illustration of domination numbers of strength and weakness k.

Some further definitions are necessary before developing bounds for γ_{sk} and γ_{wk}. A γ-*set* is a minimum dominating set and a γ_{sk}-*set* (γ_{wk}-*set*) is a minimum dominating set of strength (weakness) k. We denote the number of vertices in G of degree at most k (at least k) by M_k (N_k).

Proposition 10.15 *For any graph,*

1. $\frac{n}{1+w(D)} \leq \gamma \leq n - \Delta$ *where D is any* $\gamma - set$,

2. $\frac{n}{1+w(D)} \leq \gamma_{wk} \leq M_k$ *where D is any* $\gamma_{wk} - set$, *and*

3. $\frac{n}{1+w(D)} \leq \gamma_{sk} \leq N_k$ *where D is any* $\gamma_{sk} - set$.

The *domination number of maximum strength*, $\gamma_{ms}(G)$, is the minimum cardinality of a minimal dominating set of strength $d_s(G)$. The *domination number of minimum weakness*, $\gamma_{mw}(G)$, is similarly defined. For the tree in Figure 10.1, $\gamma_{ms} = 6$ and $\gamma_{mw} = 10$ which can be verified using the dominating sets mentioned earlier when discussing this figure. We shall establish some bounds for γ_{ms} and γ_{mw} after the following definitions. Let $V(G) = \{v_1, v_2, \ldots, v_n\}$ and $deg(v_i) = d_i$ with $d_1 \geq d_2 \geq \cdots \geq d_n$. Define integers k and t by $k = min\{r : \{v_1, v_2 \ldots, v_r\}$ is a dominating set$\}$, and $t = max\{r : \{v_r, v_{r+1} \ldots, v_n\}$ is a dominating set$\}$.

Proposition 10.16 *For any graph G,*

1. $\gamma_{ms} \leq k$ *and* $\gamma_{mw} \leq n - t + 1$,

2. $\gamma_{ms} \leq n - |S|$ *and* $\gamma_{mw} \leq n - |T|$ *where* $S = \{v \in V : deg(v) < d_s(G)\}$ *and* $T = \{v \in V : deg(v) > d_w(G)\}$,

3. $\gamma_{ms} \leq n - \Delta$ *and* $\gamma_{mw} \leq n - \delta$, *and*

4. *if G is connected, then* $\gamma_{ms} \leq n/2$ *and* $\gamma_{ms} \leq \alpha_0$.

It is natural to define the *independent domination number of maximum strength* i_{ms} (*minimum weakness* i_{mw}) as the minimum cardinality of an independent dominating set of maximum strength (minimum weakness). As we see next, $\gamma = \gamma_{ms}$ and $i = i_{ms}$ for trees with small diameters.

Proposition 10.17 *Consider any tree.*

1. *If diam* ≤ 6, *then* $\gamma = \gamma_{ms}$.

2. *If diam* ≤ 7, *then* $i = i_{ms}$.

A graph $G = (V, E)$ is said to be *ds-perfect* if, for each k in $L = \{deg(v) : v \in V\}$, γ_{sk} exists. A *dw-perfect* graph is defined similarly. The graph in Figure 10.2(b) is both *ds*-perfect and *dw*-perfect. The following proposition characterizes *dw*-perfect trees.

Proposition 10.18 *A tree is dw-perfect if and only if every nonpendant vertex is a support.*

Proof. Assume tree T is dw-perfect. Then T must have a minimal dominating set of weakness one, implying the pendant vertices dominate all vertices of degree greater than one. For the converse assume each vertex with degree greater than one is a support. Then γ_{w1} exists. Now let v be any nonpendant vertex and define the set D to be composed of v and all pendant vertices not adjacent to v. Then D is a minimal dominating set of weakness $deg(v)$ implying T is dw-perfect. $\square$

The characterization of ds-perfect trees appears to be a much more difficult problem. We conclude this section with the following general open problem.

Problem 10.19 *Characterize ds-perfect and dw-perfect graphs.*

10.6 Set and Global Set Domination

In this section we discuss a special kind of dominating set for connected graphs. The concept is considered first in its own right, and then it is applied simultaneously to both a graph and its complement.

10.6.1 Set domination

Let $G = (V, E)$ be a connected graph. A set $D \subseteq V$ is a *set dominating set (sd-set)* if, for any set $T \subseteq V - D$, there exists a non-empty set $S \subseteq D$ such that the subgraph $<S \cup T>$ is connected. Note that a set dominating set is a dominating set since T can be composed of any single vertex. The *set-domination number* γ_s is, as one would expect, the minimum cardinality of an sd-set. This concept was introduced by Sampathkumar and Pushpa Latha [41], and unreferenced results in the current subsection are taken from this source.

Clearly, any connected dominating set is an sd-set, so we have $\gamma \leq \gamma_s \leq \gamma_c$. In general, $\gamma_s \neq \gamma_c$, as can be seen from the fact that, for $n \geq 5$, $\gamma_s(C_n) = n - 3$ while $\gamma_c(C_n) = n - 2$. However, the two invariants are equal if the graph has any cutvertices.

Proposition 10.20 *For any graph having cutvertices, $\gamma_s = \gamma_c$.*

It follows that one should now consider blocks. The next result characterizes blocks for which $\gamma_s < \gamma_c$, and also gives the structure of the subgraphs induced by any γ_s-set in this case.

Proposition 10.21 *For a block B, $\gamma_s < \gamma_c$ if and only if the subgraph $<D>$ is disconnected for each γ_s-set D of B. Further, if $\gamma_s < \gamma_c$, then, for every γ_s-set D, each component in $<D>$ is either trivial or a path.*

Newman-Wolfe, Dutton, and Brigham [29] have studied two similar but different concepts which are defined as follows. A path P with terminal vertices u and v in a graph $G = (V, E)$ is a *connecting path through* $V' \subseteq V$ if $V(P) - V' = \{u, v\}$ and $V(P) \cap V' \neq \emptyset$. Furthermore, V' is a *strong connecting set (SCS)* if every pair of vertices of $V - V'$ has a connecting path through V', and a *weak connecting set (WCS)* if every pair of non-adjacent vertices of $V - V'$ has a connecting path through V'. The size of a smallest SCS is denoted by α_s, and a smallest WCS by α_w. Clearly any SCS is an sd-set. The converse is not true, however, as can be seen by the fact that any two nonadjacent vertices of the cycle C_5 form an sd-set that is not a SCS. In addition, every sd-set is a WCS, but not conversely. In fact, every sd-set is a dominating set, whereas, a WCS need not be so. The following relation is shown in [29]: $\alpha_w \leq \alpha_s \leq \gamma_c \leq \alpha_w + 1$. Corresponding results involving the set domination number are given next.

Proposition 10.22

1. $\alpha_w \leq \gamma_s \leq \alpha_s$

2. $\gamma \leq \gamma_s \leq \alpha_s \leq \gamma_c \leq \gamma_s + 1$

It is known that every γ_c-set of a graph contains all of its cutvertices. The fact that this is not the case for γ_s-sets is shown in [41], so it becomes interesting to find sufficient conditions for a cutvertex to be in every γ_s-set. The next proposition presents two of them.

Proposition 10.23 *Let v be a cutvertex of a graph G. Then v is in every γ_s-set of G if either*

1. $G - v$ has at least three components, or

2. $G - v$ has exactly two components G_1 and G_2, and neither $<G_1 \cup \{v\}>$ nor $<G_2 \cup \{v\}>$ is a path.

In view of Proposition 10.23, it is of interest to ask if any γ_s-set contains all cutvertices, a question settled by the next result.

Proposition 10.24 *For any graph, there exists a γ_s-set containing all cutvertices.*

The following result is a direct consequence. The notation *cut(G)* refers to the number of cutvertices in graph G.

Proposition 10.25 *Let T be a tree with $n \geq 3$ vertices and e pendant vertices. Then $cut(T) = \gamma_s = \alpha_s = \gamma_c = n - e$.*

Next we present some bounds involving γ_s.

Proposition 10.26 *For any graph,*

  1. $diam - 1 \le \gamma_s$ *and*

  2. $\frac{n}{1+\Delta} \le \gamma_s$ *with equality if and only if* $\Delta = n - 1$.

A graph may not have an independent sd-set, as is illustrated by the cycle C_{10}. The final result of this subsection presents a necessary condition for such a set to exist.

Proposition 10.27 *If a graph has an independent sd-set, then diam ≤ 4 and this bound is sharp.*

Proof. Suppose D is an independent sd-set of graph $G = (V, E)$ and let u and v be two vertices. If $u, v \in V - D$, they must have a common neighbor in D so $d(u, v) \le 2$. If $u, v \in D$, there are vertices u_1 and v_1 in $V - D$ such that $u_1 u$ and $v_1 v$ are edges. Again u_1 and v_1 have a common neighbor in D so $d(u, v) \le 4$. Finally let $u \in D$ and $v \in V - D$. As before, there is a vertex $u_1 \in V - D$ such that $u_1 u$ is an edge and we have $d(u, v) \le 1 + d(u_1, v) \le 3$. Thus, in all cases, $d(u, v) \le 4$ and hence $diam(G) \le 4$. Sharpness is shown by the graph constructed from two copies of C_4 joined at a vertex. Then $diam = 4$. The set composed of the cutvertex and the two vertices of distance four forms an independent sd-set. $\square$

Problem 10.28 *Characterize the graphs G (of diameter ≤ 4) having an independent γ_s-set.*

10.6.2   Global set domination

The notion of domination has been applied simultaneously to both a graph and its complement by defining the *global domination number* γ_g of a graph $G = (V, E)$ as the minimum cardinality of a set $D \subseteq V$ such that D is a dominating set of both G and $\overline{G}$ (see [33]). The invariant γ_g is discussed in Chapter 11 of this volume. Sampathkumar and Pushpa Latha [42] have extended the global concept to set domination of connected graphs and all results of this subsection, unless otherwise noted, are taken from this reference.

An sd-set of a graph G is a *global sd-set* if it also is an sd-set of $\overline{G}$. The *global set domination number* γ_{sg} of G is the minimum cardinality of a global sd-set. Some elementary bounds for γ_{sg} follow.

Proposition 10.29 *For any graph of order $n \ge 4$, $2 \le \gamma_{sg} \le n - 2$.*

These bounds are sharp. The upper bound is attained for paths P_n with $n \ge 4$ and the cycle C_5, while the lower bound is reached by the tree with degree sequence 3, 3, 1, 1, 1, 1. Graphs for which $\gamma_{sg} = 2$ have been characterized.

Proposition 10.30 *For a graph G of order $n \geq 4$, $\gamma_{sg} = 2$ if and only if one of G or $\overline{G}$ has a bridge both of whose endvertices are cutvertices and $diam(G)$ $= diam(\overline{G}) = 3$.*

Rall [32] has shown that $\gamma = \gamma_g$ for graphs with diameter at least five. A similar result holds for γ_s and γ_{sg}. For many graphs, γ and γ_g, as well as γ_s and γ_{sg}, differ by at most two.

Proposition 10.31 *For a graph with cutvertices, $\gamma_{sg} \leq \gamma_s + 1$.*

Proposition 10.32 *Let $diam = k$. Then*

1. *$\gamma_s = \gamma_{sg}$ if $k \geq 5$,*

2. *$\gamma_g \leq \gamma + 1$ and $\gamma_{sg} \leq \gamma_s + 1$*
   when $k = 4$, and

3. *$\gamma_g \leq \gamma + 2$ and $\gamma_{sg} \leq \gamma_s + 2$*
   when $k = 3$.

Note that when $k = 2$, the difference between γ and γ_g, as well as between γ_s and γ_{sg}, may be as large as we please. We close this subsection with the following open problem.

Problem 10.33 *Characterize minimal sd-sets for a graph G.*

(Editorial note: See also Chapter 11 of this book.)

10.7   Point-Set and Global Point-Set Domination

Point-set domination for connected graphs resembles set-domination, but is more restrictive since the set in $S \subseteq D$ is limited to a single vertex. Again we consider both the basic concept for a graph and a global version.

10.7.1   Point-set domination

The definitions for point-set domination of connected graph $G = (V, E)$ parallel those for set-domination. A set $D \subseteq V$ is a *point-set dominating set (psd-set)* if, for every set $T \subseteq V - D$, there exists a vertex $v \in D$ such that the subgraph $<T \cup \{v\}>$ is connected. The *point-set domination number* $\gamma_p(G)$ of G is the minimum cardinality of a *psd*-set. This concept first appeared in Sampathkumar and Pushpa Latha [40] and all results of this subsection, unless otherwise stated, are from this source. The first series of propositions outlines basic properties of *psd*-sets.

Proposition 10.34 *Let D be a psd-set of graph $G = (V, E)$. Then any two vertices in $V - D$ are at a distance of at most two.*

An interesting characterization of *psd*-sets due to Acharya and Gupta [6] is given next.

Proposition 10.35 *[6] A set $D \subseteq V$ of graph $G = (V, E)$ is a psd-set if and only if every independent set $W \subseteq V - D$ is contained in $N(u) \cap (V - D)$ for some $u \in D$.*

Since any *psd*-set is a *sd*-set which is in turn a dominating set, we have, for any graph, that $\gamma \leq \gamma_s \leq \gamma_p$. Other bounds follow.

Proposition 10.36 *For any graph, $\gamma_p \leq n - \Delta$ with equality attained for trees, block graphs and cacti other than C_5.*

Proposition 10.37 *If a graph G has cutvertices, then $\gamma_p = min\{n - \Delta, n - k\}$, where $k = max\{|V(B)| - \gamma_p(B)\}$ and the maximum is taken over all blocks B of G.*

Proof. From Proposition 10.36, $\gamma_p \leq n - \Delta$. Furthermore, for any graph $G = (V, E)$, a subset of V consisting of all vertices except for those in one block B along with a minimum *psd*-set of B constitutes a *psd*-set of G, implying $\gamma_p \leq n - k$. Thus $\gamma_p \leq min\{n - \Delta, n - k\}$.

Now let D be a *psd*-set of G of cardinality γ_p. First assume that all vertices of $V - D$ belong to a single block B. If $B \cap D$ is a *psd*-set for block B, then $\gamma_p = n - (|V(B)| - |B \cap D|) \geq n - (|V(B)| - \gamma_p(B)) \geq n - k$ so $\gamma_p = n - k$ and the theorem holds.

Thus suppose $B \cap D$ is not a *psd*-set for B. Then there must be a set $T \subseteq B - (B \cap D) = V - D$ for which there is no vertex $z \in B \cap D$ such that $< T \cup \{z\} >$ is connected. If $< T >$ is not connected, let u and v be vertices of T that are in different components of $< T \cup \{z\} >$. Since D is a *psd*-set for G, there is a w in $D - B$ such that $< u, w, v >$ is connected. Now B is a block, so there exists a uv-path in B which together with (u, w, v) forms a cycle. Since any two vertices lying on a common cycle must belong to the same block, it follows that $w \in B$, a contradiction.

Then $< T >$ must be connected, so no vertex of T has a neighbor in $B \cap D$. If for some $u \in T$, there is a vertex $v \in V - D$ and $v \notin N[u]$, then there must be a vertex $w \in D - B$ such that (u, w, v) is a path. Using the above argument, we again reach a contradiction. Hence each $u \in T$ must be adjacent to every other vertex in $V - D$. Furthermore, since D is a *psd*- set for G, it is a dominating set for G implying that each $u \in T$ has at least one neighbor in D. Thus $\Delta \geq deg(u) \geq |V - D| - 1 + 1 = n - \gamma_p$ so $\gamma_p = n - \Delta$ and the theorem holds.

The only remaining possibility is that $V - D$ includes vertices in different blocks. Since the distance between two vertices is at most two by Proposition 10.34, it follows that all vertices of $V - D$ are adjacent to a single cutvertex w. Thus $\gamma_p \geq n - deg(w) \geq n - \Delta$, implying $\gamma_p = n - \Delta$ and completing the proof. $\square$

The following result is similar to Proposition 10.27 for sd-sets.

Proposition 10.38 *For any graph with an independent psd-set, diam ≤ 4 and this bound is sharp.*

This diameter restriction also is a necessary condition for γ to equal γ_p.

Proposition 10.39 *If $\gamma = \gamma_p$ for a graph, then diam ≤ 4.*

One can generalize the concept of point-set domination. Let $r \geq 1$ be an integer. A set $D \subseteq V$ is an *r-point-set dominating set (r-psd-set)* of graph $G = (V, E)$ if, for every set $T \subseteq V - D$, there exists a non-empty set $S \subseteq D$ containing at most r vertices such that the subgraph $<S \cup T>$ is connected. The *r-point-set-domination number* γ_{rp} of G is the cardinality of a smallest r-point-set dominating set. Since every $r-psd$-set is an sd-set, we have $\gamma_s \leq \gamma_{rp}$. The next proposition generalizes a result on γ_p. Here k is defined by $k = max\{|N(S) \cap (V - S)|\}$, where the maximum is taken over all sets $S \subseteq V$ such that $|S| \leq r$ and $<S>$ is connected.

Proposition 10.40 *For any graph, $\gamma_{rp} \leq n - k$ with equality if the graph is a tree.*

10.7.2   Global point-set domination

Recently, Pushpa Latha [31] has extended the global concept to point-set domination, and this reference is the source for the material in the current subsection. Again, all graphs are assumed to be connected and have connected complements.

Let $G = (V, E)$ be a graph. A set $D \subseteq V$ is a *global psd-set* if it is a *psd*-set of both G and $\overline{G}$. The *global point-set domination number* γ_{pg} of G is defined in the usual way as the smallest cardinality of a global *psd*-set. Using Proposition 10.35, it is straightforward to characterize global *psd*-sets.

Proposition 10.41 *For a graph $G = (V, E)$, a set $D \subseteq V$ is a global psd-set if and only if both of the following are true:*

1. *for every independent set W in $V - D$, there exists a $u \in D$ such that $W \subseteq N(u) \cap (V - D)$ in G; and*

2. *for every set $S \subseteq V - D$ such that $< S >$ is complete, there exists a $v \in D$ such that $S \cap N(v) = \emptyset$ in G.*

The next proposition shows basic bounds for γ_{pg}.

Proposition 10.42 *For a graph of order $n \geq 5$, $3 \leq \gamma_{pg} \leq n-2$ and the bounds are sharp.*

The three following results present additional relationships for γ_{pg} and show that this invariant has properties similar to those of γ_{sg}. The first of the three resembles a fact proven by Sampathkumar [33] which says that γ and γ_g differ by at most one for any graph with cutvertices.

Proposition 10.43 *For any graph with cutvertices, $\gamma_p \leq \gamma_{pg} \leq \gamma_p + 1$.*

Corollary 10.44 *For a tree, a block graph or a cactus, $n - \Delta \leq \gamma_{pg} \leq n - \Delta + 1$.*

Proposition 10.45 *For any graph with diameter at least four, $\gamma_{pg} = \gamma_p$.*

We conclude this subsection with a characterization of which trees of diameter at least three have each of the values for γ_{pg} allowed by Corollary 10.44.

Proposition 10.46 *For any tree T,*

1. *$\gamma_{pg} = n - \Delta + 1$ if $diam(T) = 3$, and*

2. *$\gamma_{pg} = n - \Delta$ if $diam(T) \geq 4$.*

10.8 Neighborhood Numbers

This section introduces the notions of neighborhood sets and neighborhood numbers. There are both vertex and edge versions of these concepts, and they are related, respectively, to the domination and edge domination numbers.

10.8.1 Vertex neighborhood numbers

A set S of vertices in a graph G is a *(vertex) neighborhood set (n-set)* if $G = \bigcup_{v \in S} < N[v] >$. The *(vertex) neighborhood number* $n_0(G)$ of G is the minimum cardinality of an *n*-set. Sampathkumar and Neeralagi [36] first introduced $n_0(G)$, and obtained the following results involving n_0 and the covering and independence numbers.

Proposition 10.47 *[36] For a graph without isolates, $\gamma \leq n_0 \leq \alpha_0$. Furthermore, if the graph has no triangles, $n_0 = \alpha_0$.*

Proposition 10.48 *[36] For any bipartite graph without isolates, $n_0 = \alpha_0 = \beta_1$. Further, if the graph has a 1-factor, then $n_0 = \alpha_0 = \beta_0 = \alpha_1 = \beta_1 = n/2$.*

Corollary 10.49 [36] *For a regular bipartite graph, $n_0 = n/2$.*

As an important application of Corollary 10.49, we have that $n_0 = 2^{n-1}$ for the n-cube Q_n. Jayaram, Kwong, and Straight [20] have made several contributions to the area of neighborhood numbers. The first two of their results deal with existence questions.

Proposition 10.50 [20] *For any positive integers r, s, and t with $2 \leq r \leq s \leq t$, there exists a graph having $\gamma = r$, $n_0 = s$, and $\alpha_0 = t$.*

In the proof to the next proposition we need the fact, also shown in [20], that $n_0(G \cdot uv) \leq n_0(G)$ where $G \cdot uv$ is the contraction of G along edge uv. Furthermore, $n_0(G \cdot uv) = n_0(G)$ if $N[u] = N[v]$.

Proposition 10.51 [20] *For any positive integers p, k, and n with $p \leq k \leq n - p$, there exists a graph G of order n such that $n_0 = p$ and G has a minimal n-set of cardinality k.*

Proof. If $k \leq n-2p$, let $G = (V, E)$ be the complete tripartite graph $K_{n,k,n-p-k}$, and if $k = n - p$, let $G = K_{p,n-p}$. In each case it is straightforward to check that $n_0 = p$ and the partite set with cardinality k is a minimal n-set. Now suppose $n - 2p < k < n - p$ and define G as follows. Let V be partitioned as $\{X, Y\}$ with $X = \{x_1, x_2, \ldots, x_{n-k}\}$ and $|Y| = k$, and

$$E = \{xy | x \in X, y \in Y\} \cup \{x_i x_{p+i} | 1 \leq i \leq n - p - k\}.$$

It can be seen that Y is a minimal n-set with cardinality k and $\{x_1, x_2, \ldots, x_p\}$ is a minimal n-set of cardinality p. Note that $N[x_i] = N[x_{p+i}]$ for $1 \leq i \leq n-p-k$. Contracting each edge $x_i x_{p+i}$ produces a graph isomorphic to $K_{p,k}$ so $n_0(G) = p$ by the remark preceding the proposition and the proof is complete. $\square$

Jayaram, et. al. [20] have defined a graph $P(G)$ derived from a partition $P = \{V_1, V_2, \ldots, V_l\}$ of the vertex set of graph G. This new graph has vertex set P, and V_i is adjacent to V_j if and only if $v_i v_j$ is an edge for some $v_i \in V_i$ and $v_j \in V_j$. The next result settles a conjecture by Acharya.

Proposition 10.52 [20] *For graph $G = (V, E)$, let P be the partition of V induced by the equivalence relation defined by $u * v$ if and only if $N[u] = N[v]$. Then $n_0(P(G)) = n_0(G)$.*

For yet another area of application of n-sets, let $\Omega(G)$ be the intersection graph of the collection of minimal n-sets of graph G, that is, $V(\Omega(G)) = \{S \subseteq V$: S is a minimal n-set$\}$ and $E(\Omega(G)) = \{S_1 S_2 : S_1 \cap S_2 \neq \emptyset\}$. The graphs $\Omega(P_n)$ and $\Omega(C_n)$ are given by the next proposition.

Proposition 10.53 [20]

1. Let $n \geq 2$ and function f be defined recursively by $f(2) = f(3) = 2$, $f(4) = 3$, and $f(n) = f(n-2) + f(n-3)$ for $n \geq 5$. Then $\Omega(P_n) = K_{f(n)} - x$ where x is an edge.

2. Let $n \geq 4$ and function g be defined recursively by $g(4) = 2$, $g(5) = g(6) = 5$, and $g(n) = g(n-2) + g(n-3)$ for $n \geq 4$. Then $\Omega(C_n) = \begin{cases} K_{g(n)} & \text{if } n \text{ is odd} \\ K_{g(n)} - x & \text{if } n \text{ is even} \end{cases}$ where x is an edge.

Further results concerning $n_0(G)$ can be found in [3, 19, 24, 28]. In an interesting application, Kwong and Straight [25] consider $N(n)$, the maximum neighborhood number among all connected graphs of order n. Its value is determined for $n \leq 15$, and the authors conjecture that $N(n) \leq \lfloor 9n/13 \rfloor$ in general.

10.8.2   Edge neighborhood numbers

An edge analogue of $n_0(G)$ is introduced by Sampathkumar and Neeralagi in [37] which serves as the source for all results in this subsection. It is defined as follows. For an edge $x = uv$ in graph $G = (V, E)$, let $N[x] = N[u] \cup N[v]$. A set $T \subseteq E$ is an *edge-neighborhood set (en-set)* if $G = \bigcup_{x \in T} < N[x] >$. The *edge-neighborhood number* $n_0'(G)$ of G is the minimum cardinality of an en-set. Some known results on $n_0'(G)$ are presented in the remaining propositions of this section. Here $\gamma'(G)$ is the edge domination number of G.

Proposition 10.54 *For any graph G,*

1. $\gamma/2 \leq n_0' \leq n_0 \leq min(\alpha_0, \alpha_1)$,

2. $n_0' \leq \gamma' \leq min(\alpha_0, \alpha_1, \beta_1)$,

3. $n_0' \leq \gamma' \leq \lfloor n/2 \rfloor$, and

4. $\frac{1}{2} min(3, n+1) \leq n_0'(G) + n_0'(\overline{G}) \leq 2\lfloor n/2 \rfloor$.

Proposition 10.55 *Let G be a graph such that its line graph $L(G)$ has no isolates. Then $n_0'(G) \leq \gamma'(G) \leq n_0(L(G))$.*

Proposition 10.56 *For any bipartite graph without four cycles, $n_0' = \gamma'$.*

10.9   Neighborhood Number Variations

In this section we consider three special cases of the vertex neighborhood numbers discussed in Section 10.8. The cases are defined as follows for an n-set S of a graph G:

1. S is an *independent n-set (IN-set)* if S is independent,

2. S is a *perfect n-set (PN-set)* if, for all $u, v \in S$ such that $u \neq v$, the subgraphs $<N[u]>$ and $<N[v]>$ are edge disjoint, and

3. S is a *connected n-set* if $<S>$ is connected.

Sampathkumar and Neeralagi [39] define the *independent neighborhood number* $n_i(G)$, the *perfect neighborhood number* $n_p(G)$, and the *connected neighborhood number* $n_c(G)$ of graph G as the minimum cardinality of an IN-set, a PN-set, and a connected n-set, respectively. All results in this section appear in [39]. The first presents an existence condition involving two of these neighborhood numbers, along with n_0 and the vertex independence number β_0.

Proposition 10.57 *Let a, b, c, d, and n be integers with $1 \leq a \leq b \leq c \leq d \leq n$; $d = b + c - a$; and $n = b + c - 1$. Then there exists a graph of order n such that $n_0 = a$, $n_i = b$, $n_p = c$, and $\beta_0 = d$.*

We note that a graph may not have an IN-set or a PN-set. This can be seen by any odd cycle of length $n \geq 5$ which has neither. Furthermore, any PN-set is an IN-set, but the converse is not necessarily true. We shall call a graph an IN-graph (PN-graph) if it has an IN-set (PN-set). Every bipartite graph, for example, is a PN-graph and hence an IN-graph. Necessary conditions for PN-graphs follow.

Proposition 10.58 *For any PN-graph,*

1. *$n_o \leq n_i \leq n_p \leq \beta_0$ and*

2. *for every odd cycle C_n, $n \geq 5$, there exists a vertex v such that the subgraph $<N[v]>$ contains an odd number of edges of the C_n.*

The next proposition considers characterization questions associated with IN-graphs.

Proposition 10.59

1. *A graph $G = (V, E)$ is an IN-graph if and only if there exists an n-set S such that $V - S$ is a vertex cover.*

2. An IN-graph $G = (V, E)$ is a PN-graph if and only if there exists an IN-set S such that every edge in the subgraph $<V - S>$ belongs to exactly one triangle of G.

Proof. Assume G is an IN-graph meaning it has an independent n-set S. Then S is an independent dominating set so $V - S$ is a vertex cover. Conversely, if S is an n-set such that $V - S$ is a vertex cover, then S is independent and G is an IN-graph, completing the proof of (1).

Now assume G is an IN-graph which also is a PN-graph and hence has a PN-set S. It is easy to see that S is independent and hence is an IN-set. Suppose $<V - S>$ contains an edge xy which is in two triangles of G. Then the third vertices of each of these triangles must be in S. But then the closed neighborhoods of these third vertices will contain xy, contradicting the fact that S is a PN-set. Notice that xy must be in at least one triangle since S is an n-set, so xy is in exactly one triangle. For the converse we assume G has an IN-set (and hence is an IN-graph) S such that every edge of $<V - S>$ is in exactly one triangle. This means that, for any vertices $u, v \in S$, $<N[u]>$ and $<N[v]>$ have no common edge. Thus S is a PN-set so G is a PN-graph and the proof to (2) is complete. $\square$

We conclude this section with several results on the connected neighborhood number.

Proposition 10.60 Let G be a connected graph with size m. Then

1. $\gamma_c \leq n_c \leq 2\beta_1$;

2. any cutvertex of G belongs to each connected n-set of G and $k \leq n_c \leq n - e + \lceil e/2 \rceil$, where k is the number of cutvertices in G and e is the number of pendant vertices in any spanning tree of G;

3. $n_c = n - 1$ if and only if $G = K_2$ or $G = C_n$ with $n \geq 4$; and

4. if G is not K_2 or C_n with $n \geq 4$, then

    (a) $n_c \leq n - 2$ and

    (b) $n_c \leq 2m - n$ with equality if and only if G is path.

10.10 Domination and Neighborhood Critical, Fixed, Free, and Totally Free Vertices and Edges

A vertex or edge is critical with respect to an invariant if removal of the vertex or edge changes the value of the invariant. Domination related invariants are natural ones to consider in this context. Here we shall be concerned with the domination number γ and the neighborhood number n_0.

Some general definitions which can be applied to any of a number of graphical invariants are given first. We shall employ e to represent an element, either a vertex or an edge, of a graph G and t to be the cardinality of a set of elements with some prescribed property. Then a *t-set* of G is a set with that property. For example, for $t = \gamma$, a t-set is a γ-set, that is, a minimum dominating set. For any such t, we say an element e of G is

1. *t-critical* if $t(G - e) \neq t(G)$,

2. t^+-*critical* if $t(G - e) > t(G)$,

3. t^--*critical* if $t(G - e) < t(G)$,

4. *t-fixed* if e belongs to every t-set,

5. *t-free* if e belongs to some t-set but not to all t-sets, and

6. *t-totally free* if e belongs to no t-set.

Sampathkumar and Neeralagi [38] have studied the relationship among such types of elements when $t = \gamma$ or $t = n_0$; all results of this section are taken from this reference unless otherwise noted. The symbol t will be employed here to represent both γ and n_0. Subsection 10.10.1 considers the case when e is a vertex and Subsection 10.10.2 when e is an edge.

10.10.1 The vertex case

Since t can decrease by at most one when a vertex v is deleted, it follows that v is t^--critical if and only if $t(G - v) = t(G) - 1$. The first result gives information about critical vertices.

Proposition 10.61 *Every t^--critical vertex of graph G belongs to a t-set of G, and every t^+-critical vertex of G is t-fixed.*

A critical vertex of any type may be fixed or free. It is easy to determine this for a γ^--critical vertex.

Proposition 10.62 *A γ^--critical vertex is γ-fixed if it is isolated and γ-free otherwise.*

Bauer, Harary, Nieminen, and Suffel [9] obtained the following results which include characterizations of γ^+-critical vertices.

Proposition 10.63 [9]

1. *If a cutvertex v is γ-fixed, then v is γ^+-critical.*

2. *A vertex v in a tree with $n \geq 3$ vertices is γ^+-critical if and only if v is γ-fixed.*

3. *A vertex v in a graph G is γ^+-critical if and only if*

   (a) $deg(v) > 0$,

   (b) v is γ-fixed, and

   (c) there is no dominating set for $<V - N[v]>$ having $\gamma(G)$ vertices which also dominates $N(v)$.

Other properties of γ^--critical and γ^+-critical vertices may be found in [10, 15]. The final result of this subsection relates n_0-critical and α_0-critical vertices.

Proposition 10.64 *For any graph without isolates and triangles,*

1. *any n_0-critical vertex is α_0-critical and*

2. *any α_0-critical vertex which is not adjacent to a pendant vertex is n_0-critical.*

10.10.2 The edge case

This subsection deals with the characterization of t-critical edges and investigates whether the endvertices of such an edge are t-fixed, t-free or t-totally free. Notice that a γ-critical edge x of graph G is always γ^+-critical and $\gamma(G - x) = \gamma(G) + 1$, while an n_0-critical edge may be either n_0^--critical or n_0^+-critical. The first characterization results appear next.

Proposition 10.65 *Let G be a graph. Then*

1. *an edge x is γ-critical (n_0^+-critical) if and only if there is no dominating set of $G - x$ with $\gamma(G)$ ($n_0(G)$) vertices and*

2. *an edge x is n_0^--critical if and only if there is no n-set with $n_0(G - x)$ vertices.*

Proof. Suppose edge x is γ-critical, so $\gamma(G - x) = \gamma(G) + 1$, and there exists a dominating set of $G - x$ with $\gamma(G)$ vertices. Then $\gamma(G - x) \leq \gamma(G)$, a contradiction. Conversely, if there is no dominating set of $G - x$ with $\gamma(G)$ vertices, then $\gamma(G - x) \neq \gamma(G)$ and x is γ-critical. Similar arguments can be applied to prove the statements involving n_0^+-critical and n_0^--critical edges. □

Corollary 10.66 *Let x be a γ-critical (n_0^+-critical) edge of a graph. Then any γ-set (n_0-set) of the graph contains exactly one of the endvertices of x.*

It follows from Corollary 10.66 that, if an edge $x = uv$ is t-critical, then either (i) u is t-fixed and v is t-totally free or (ii) both u and v are t-free. The final result of this subsection relates n_0-critical and α_0-critical edges in graphs without triangles.

Proposition 10.67 *Let x be an edge in a triangle free graph. Then*

1. *x is n_0^--critical if and only if x is α_0-critical and x is not a pendant edge, and*

2. *x is n_0^+-critical if and only if x is a pendant edge and, if x does not form a component by itself, then x is not α_0-critical.*

10.11 Mixed Domination

One can think of vertices dominating edges and vice versa. These situations are related to a concept called mixed domination. For a formal definition, let $G = (V, E)$ be a graph without isolates. Let $v \in V$, $x = ab \in E$, and $N[x] = N[a] \cup N[b]$. Then vertex v *m-dominates* edge x if $x \in <N[v]>$, and edge x *m-dominates* vertex v if $v \in N[x]$. Clearly, if v m-dominates x, then x m-dominates v, but not conversely.

Based on these concepts, one can define three parameters of G. A set $S \subseteq V$ is a *ve-dominating set (ved-set)* if every edge of G is m-dominated by a vertex in S. Similarly, a set $T \subseteq E$ is an *ev-dominating set (evd-set)* if every vertex of G is m-dominated by an edge in T. As usual, the *ve-domination number* $\gamma_{ve}(G)$ and the *ev-domination number* $\gamma_{ev}(G)$ are defined as the minimum cardinality of a *ved*-set and an *evd*-set, respectively.

Sampathkumar and Kamath [35] define a set $D \subseteq V \cup E$ as a *mixed dominating set (md-set)* if every element not in D is m-dominated by an element of D. The *mixed domination number* $\gamma_m(G)$ is, of course, the minimum cardinality of an *md*-set. In keeping with previous terminology; a *ved*-set, *evd*-set, or *md*-set with minimum cardinality is called a γ_{ve}-set, a γ_{ev}-set, or a γ_m-set; respectively.

As an example of these concepts, let P_6 be the path $v_1 v_2 v_3 v_4 v_5 v_6$ and $e_i = v_i v_{i+1}$ for $1 \leq i \leq 5$. Then $\{v_2, v_4, v_6\}$ is a γ_{ve}-set, $\{e_2, e_5\}$ is a γ_{ev}-set, and $\{v_2, v_4, e_2, e_5\}$ is a γ_m-set.

Laskar and Peters [27] define related invariants in the following way. Let $v \in V$ and $x \in E$. Then

1. v and x *weakly dominate* each other if $v \in N[x]$, and

2. v and x *strongly dominate* each other if $x \in <N[v]>$.

The *vertex-edge weak domination number* $\gamma_{01}(G)$ of graph G is the minimum cardinality of a set of vertices weakly dominating all the edges of G and the *edge-vertex weak domination number* $\gamma_{10}(G)$ is the minimum cardinality of a set of edges weakly dominating all the vertices of G. The *vertex-edge strong domination number* $s\gamma_{01}(G)$ and the *edge-vertex strong domination number* $s\gamma_{10}(G)$ are defined analogously.

It is straightforward to show that, for any graph, $\gamma_{01} \leq s\gamma_{01} = \gamma_{ve} = n_0$, $\gamma_{10} = \gamma_{ev} \leq s\gamma_{10}$, and $\gamma_{ev} \leq n_0'$. Other results concerning $s\gamma_{01}$ and $s\gamma_{10}$ are given in the next proposition.

Proposition 10.68 [27]

1. *For any graph without isolates, $\gamma \leq s\gamma_{10} \leq \alpha_1$ with $s\gamma_{10} = \alpha_1$ if the graph is triangle free.*

2. *For a bipartite graph without isolates, $s\gamma_{10} = \alpha_1 = \beta_0$ and $s\gamma_{01} + s\gamma_{10} = n$.*

3. *For a bipartite graph having a 1-factor, $s\gamma_{01} = \alpha_0 = \beta_0 = s\gamma_{10} = \alpha_1 = \beta_1 = n/2$ which implies, for the k-cube Q_k, that $s\gamma_{10} = 2^{k-1}$.*

4. *For any graph, $s\gamma_{10} \leq n - \delta$. Further, if the graph is connected and not complete or an odd cycle, then $s\gamma_{10} \leq n - \delta - 1$.*

One can verify easily that $\gamma_{ev} = 1$ for K_n and $K_{r,s}$, where $r, s \geq 1$, and that $\gamma_{ev} = \lfloor n/4 \rfloor$ for path P_n and cycle C_n. Furthermore, $\gamma_m = 2$ if $\Delta = |V| - 1 \geq 2$, and $\gamma_m(K_{r,s}) = min(r, s) + 1$. The next proposition presents exact values of γ_m for paths and cycles.

Proposition 10.69 [35]

1. $\gamma_m(P_n) = \begin{cases} \lfloor 2n/3 \rfloor & \text{if } n \equiv 0, 2 \ (mod \ 3) \\ \lceil 2n/3 \rceil & \text{if } n \equiv 1 \ (mod \ 3) \end{cases}$

2. $\gamma_m(C_n) = n - \lceil (n - 1)/4 \rceil$.

We conclude this section with a collection of relationships involving the parameters discussed.

Proposition 10.70 [35] *For any graph G of order n and size m,*

1. $\gamma_{ev} \leq \gamma_{ve} \leq \gamma_m \leq \gamma_{ve} + \gamma_{ev}$;

2. $\gamma \leq \gamma_{ve}$ *and* $\gamma \leq \gamma_{ev}$;

3. *if T is a tree with $n \geq 2$ vertices, r pendant vertices, and $s > 1$ supports, then $\gamma_m(T) \leq n + s - r - 1$; and*

4. $\gamma_m \leq m + n - 2\Delta + 1.$

Proof of (1). The inequalities are justified in sequence by the following facts:

1. replacing each vertex of a γ_{ve}-set by an edge incident to the vertex produces an *evd*-set,

2. replacing each edge in a γ_m-set by one of its endvertices produces a *ved*-set, and

3. the union of a *ved*-set and an *evd*-set is an *md*-set.

□

Arumugam and Thuraiswamy [8] have continued the study of the mixed domination number by obtaining its bounds and Nordhaus-Gaddum type results. They have also investigated the notions of connected, total and independent mixed domination.

(Editorial note: Several papers have been published on "mixed domination" which is defined differently than the mixed domination given in this section. In this concept, a vertex dominates adjacent vertices and incident edges, and an edge dominates adjacent edges and its two endvertices. "Mixed parameters" defined by a minimum cardinality of vertices and edges which cover V, E, or $V \cup E$ are surveyed in [16].)

10.12   Concluding Remarks

We do not claim that this survey is exhaustive. Because of space limitations we have been able to provide only a rather small sample of the ongoing work on the above topics. To the authors whose research on these topics we have not mentioned, we offer our sincere apologies.

Acknowledgement.
The author is extremely grateful to Professor Robert Brigham for going through the manuscript very carefully and suggesting many changes in its presentation. In fact, he has put the manuscript into Latex after effecting the changes - which is a rather extraordinary help. My sincere apologies to you Professor Brigham for all the trouble that I gave you indirectly, and thank you again for the trouble you took for my sake.

Bibliography

[1] B. D. Acharya, Full sets in hypergraphs. *Sankhyā* 44 (1992) 1–6.

[2] B. D. Acharya, Interrelations among the notions of independence, domination and full sets in a hypergraph. *Nat. Acad. Sci. Lett.* 13 (1990) 421–422.

[3] B. D. Acharya, On a relation between neighborhood number and Dilworth number. *Proc. Nat. Acad. Sci. India Sect. A* 57 (1987) 600-603.

[4] B. D. Acharya, Stable set covers, chromaticity and kernels of hypergraphs. Private communication (1996).

[5] B. D. Acharya and P. Gupta, On point-set domination in graphs. Preprint (1996).

[6] B. D. Acharya and P. Gupta, private communication.

[7] R. B. Allan and R. Laskar, On domination and independent domination numbers of a graph. *Discrete Math.* 23 (1978) 73-76.

[8] S. Arumugam and A. Thuraiswamy, Mixed domination in graphs. Preprint (1996).

[9] D. Bauer, F. Harary, J. Nieminen, and C. L. Suffel, Domination alteration sets in graphs. *Discrete Math.* 47 (1983) 153-161.

[10] J. Carrington, F. Harary, and T. W. Haynes, Changing and unchanging the domination number of a graph. *J. Combin. Math. Combin. Comput.* 9 (1991) 57-63.

[11] P. Dankelmann, P. J. Slater, V. Smithdorf, and H. C. Swart, A note on domination, total domination, and connected domination triples of graphs. Preprint (1996).

[12] P. Duchet and H. Meyniel, On Hadwiger's number and stability number. *Ann. Discrete Math.* 13 (1982) 71-74.

[13] O. Favaron, Least domination in a graph. Preprint (1996).

[14] J. H. Hattingh and R. Laskar, On weak domination in graphs. To appear in *Ars Combin.*

[15] T. W. Haynes, S. T. Hedetniemi, and P. J. Slater, *Fundamentals of Domination in Graphs.* Marcel Dekker, Inc., 1997.

[16] S. M. Hedetniemi, S. T. Hedetniemi, R. Laskar, A. McRae, and A. Majumdar, Domination, independence and irredundance in total graphs: a brief survey. *Graph Theory, Combinatorics, and Applications*, Eds. Y. Alavi and A. Schwenk, John Wiley and Sons, Inc. (1995) 671–683.

[17] S. T. Hedetniemi and R. Laskar, Connected domination in graphs. *Graph Theory and Combinatorics*, Eds. B. Bollobàs, Academic Press (London, 1984) 209-218.

[18] F. Jaeger and C. Payan, Relations du type Nordhaus-Gaddum pour le nombre d'absorption d'un graphe simple. *C. R. Acad. Sci. Paris* A, 274 (1972) 728-730.

[19] S. R. Jayaram, The nomatic number of a graph. *Nat. Acad. Sci. Lett.* 10 (1987) 23-25.

[20] S. R. Jayaram, Y. H. H. Kwong, and H. J. Straight, Neighborhood sets in graphs. *Indian J. Pure Appl. Math.* 22 (1991) 259-268.

[21] Paulraj Joseph and S. Arumugam, On connected cutfree domination in graphs. *Indian J. Pure Appl. Math.* 23 (1992) 643–647.

[22] Paulraj Joseph and S. Arumugam, On the connected domatic number of a graph. *J. Ramanujan Math. Soc.* 9 (1994) 69-77.

[23] Paulraj Joseph and S. Arumugam, On 2-edge connected domination. To appear in *Internat. J. Management Systems.*

[24] V. R. Kulli and S. C. Singarakanti, Further results on the neighborhood number of a graph. *Indian J. Pure Appl. Math.* 23 (1992) 575-577.

[25] Y. H. H. Kwong and H. J. Straight, An extremal problem involving neighborhood number. Preprint (1996).

[26] R. Laskar and K. Peters, Connected domination of complementary graphs. Technical Report 429, Department of Mathematical Sciences, Clemson University (1983).

[27] R. Laskar and K. Peters, Vertex and edge domination parameters in graphs. *Congr. Numer.* 48 (1985) 291-305.

[28] P. S. Neeralagi, Complete domination and neighborhood numbers in a graph. *J. Math. Phys. Sci.* 27 (1993) 295-303.

[29] R. E. Newman-Wolfe, R. D. Dutton, and R. C. Brigham, Connecting sets in graphs-a domination related concept. *Congr. Numer.* 67 (1988) 67-76.

[30] J. Nieminen, Two bounds for the domination number of a graph. *J. Inst. Math. Appl.* 14 (1974) 183-187.

[31] L. Pushpa Latha, The global point-set domination number of a graph. To appear in *Indian J. Pure Appl. Math.*

[32] D. F. Rall, Dominating a graph and its complement. *Congr. Numer.* 80 (1991) 89-95.

[33] E. Sampathkumar, The global domination number of a graph. *J. Math. Phys. Sci.* 23 (1989) 377-385.

[34] E. Sampathkumar, The least point covering and domination numbers of a graph. *Discrete Math.* 86 (1990) 137-142.

[35] E. Sampathkumar and S. S. Kamath, Mixed domination in graphs. *Sankhyā* 54 (1992) 399-402.

[36] E. Sampathkumar and P. S. Neeralagi, The neighborhood number of a graph. *Indian J. Pure Appl. Math.* 16 (1985) 126-136.

[37] E. Sampathkumar and P. S. Neeralagi, The line neighborhood number of a graph. *Indian J. Pure Appl. Math.* 17 (1986) 142-149.

[38] E. Sampathkumar and P. S. Neeralagi, Domination and neighborhood critical fixed, free and totally free points. *Sankhyā* 54 (1992) 403-407.

[39] E. Sampathkumar and P. S. Neeralagi, Independent, perfect and connected neighborhood numbers of a graph. *J. Combin. Inform. System Sci.* 19 (1994) 149-156.

[40] E. Sampathkumar and L. Pushpa Latha, Point-set domination number of a graph. *Indian J. Pure Appl. Math.* 24 (1993) 225-229.

[41] E. Sampathkumar and L. Pushpa Latha, Set domination in graphs. *J. Graph Theory* 18 (1994) 489-495.

[42] E. Sampathkumar and L. Pushpa Latha, The global set domination number of a graph. *Indian J. Pure Appl. Math.* 25 (1994) 1053-1057.

[43] E. Sampathkumar and L. Pushpa Latha, Strong, weak domination and domination balance in a graph. To appear in *Discrete Math.*

[44] E. Sampathkumar and L. Pushpa Latha, Dominating strength and weakness of a graph. Submitted for publication.

[45] E. Sampathkumar and H. B. Waliker, The connected domination number of a graph. *J. Math. Phys. Sci.* 13 (1979) 607-613.

[46] P. J. Slater, Enclaveless sets and MK-systems. *J. Res. Nat. Bur. Standards* 82 (1977) 197-202.

[47] H. B. Walikar, B. D. Acharya and E. Sampathkumar, Recent developments in the theory of domination in graphs. *MRI Lecture Notes in Math.* 1 (1979).

[48] I. E. Zverovich, Proof of a conjecture in domination theory. Preprint (1996).

Chapter 11

Global Domination

Robert C. Brigham
Department of Mathematics
University of Central Florida
Orlando, FL 32816 USA

Julie R. Carrington
Department of Mathematical Sciences
Rollins College
Winter Park, FL 32789 USA

11.1 Introduction

The notion of a dominating set of a graph has been extended in a natural way to a collection of vertices which simultaneously dominates two or more edge disjoint graphs having the same vertex set. This concept was introduced independently by Sampathkumar [13] (who coined the term global domination used here) and Brigham and Dutton [3] (under the name factor domination). The following defines global dominating sets and related concepts.

Definition 11.1 *Let $G = (V, E)$ be a graph having spanning subgraphs $F_i = (V, E_i)$, $1 \leq i \leq k$, where E_1, E_2, ..., E_k partition E. The F_i are called factors of G. Then $D_g \subseteq V$ is a global dominating set (GDS) if D_g is a dominating set for each F_i, $1 \leq i \leq k$. The cardinality of a smallest such set, designated by $\gamma_g(F_1, F_2, \ldots, F_k)$, is the global domination number of the factoring. A dominating set D_i for a factor F_i is a local dominating set (LDS) and a minimum such set has cardinality designated by γ_i, called the local domination number of F_i.*

If the context makes the factors clear, the notation $\gamma_g(F_1, F_2, \ldots, F_k)$ is reduced to γ_g. Two observations are immediate from the definition.

Observation 11.2

(i) *Determining whether γ_g is at most M is an NP-complete problem* (restrict the problem to normal domination, that is, $k = 1$).

(ii) *For any factoring of graph G, $max\{\gamma_1, \gamma_2, \ldots, \gamma_k\} \leq \gamma_g \leq \gamma_1 + \gamma_2 + \cdots + \gamma_k$.*

Both Sampathkumar and Brigham and Dutton concentrated on the important special case of when there are two factors: a graph G and its complement $\overline{G}$. Rall [12] also has worked in the two factor area. Carrington [5] studied the general case extensively, and much of her work is included in this chapter. In addition, the global concept has been applied to other domination related invariants by Dunbar, Laskar, Monroe, Rall, and Wallis [8, 9, 10, 12].

Sections 11.2 and 11.3 review some of the known results about global domination, Section 11.4 briefly indicates possible applications, Section 11.5 deals with problems arising when attempting to characterize which sets of domination values are achievable by some graph with appropriate factorings, Section 11.6 discusses special cases, and Section 11.7 lists some open problems.

11.2   Some Early Results

Unless otherwise stated, all results reported in this section come from [3] or [13] or both. Here we use $\kappa(G)$ for vertex connectivity, $\omega(G)$ for clique number, and Iso for the set of vertices isolated in at least one factor. Throughout this chapter, the parameter G will often be omitted.

It is noteworthy that the global domination concept can be transformed into a standard domination problem by constructing a graph H from disjoint copies of G, F_1, F_2, $\ldots$, F_k and, then, by joining vertex u of the copy of G to vertex u of the copy of F_i and to all vertices of this F_i which are adjacent to u in F_i, for all $u \in V(G)$ and $i = 1, 2, \ldots, k$. Then $\gamma(H) = \gamma_g(F_1, F_2, \ldots, F_k)$. Another general result is that $\gamma_g = n$ if $k > \Delta$ since every vertex will be isolated in at least one factor. Otherwise, $\gamma_g \geq max\{k, \gamma + k - 2\}$. Of course, $\gamma + k - 2$ is the larger value unless $\gamma = 1$. It is also true that $\gamma_g \geq nk/(\Delta + k)$ which generalizes the well known fact that $\gamma \geq n/(\Delta + 1)$. Another familiar result is $\gamma + \varepsilon = n$ where ε is the maximum number of end edges in a spanning forest of G [11]. The parameter ε can be generalized to ε_g, the cardinality of the largest set of vertices X such that, in each F_i, there is a spanning forest in which X is independent and each vertex of X has degree one. Then $\gamma_g + \varepsilon_g = n$. Other general upper bounds based on the set Iso are $\gamma_g \leq \alpha_0 + |Iso|$ and $\gamma_g \leq n - \gamma + |Iso|$.

Observation 11.2(i) indicates that the global domination decision problem is NP-complete. Carrington [5] and Carrington and Brigham [6] have shown that the problem remains NP-complete even when the factors are greatly simplified, namely, when each of the factors is a path; when there are four factors and each of them is a caterpillar; or when there are four factors and each of them is a forest all of the components of which are K_1 or K_2. The restricted problem of a small fixed k when all factors are paths remains open. It is interesting in view of the trivial nature of the $k = 1$ case that we have been unable to settle the status of the problem even when there are only two paths.

All other preliminary work has been restricted to the case $k = 2$ and factors

G and $\overline{G}$. In this special situation only, we abuse notation and write $\gamma_g(G, \overline{G})$ as $\gamma_g(G)$, so then $\gamma_g(G) = \gamma_g(\overline{G})$. The value of $\gamma_g(G)$ has been computed for several common families of graphs. Here P_n, C_n, and W_n are the path, the cycle, and the wheel, respectively, on n vertices and $K_{n_1, n_2, \dots, n_r}$ is the complete r-partite graph.

$$\gamma_g(K_n) = n$$

$$\gamma_g(P_n) = \begin{cases} 2 & \text{if } n = 2, 3 \\ \lceil n/3 \rceil & \text{if } n \geq 4 \end{cases}$$

$$\gamma_g(C_n) = \begin{cases} 3 & \text{if } n = 3, 5 \\ \lceil n/3 \rceil & \text{otherwise} \end{cases}$$

$$\gamma_g(W_n) = \begin{cases} 4 & \text{if } n = 4 \\ 3 & \text{otherwise} \end{cases}$$

$$\gamma_g(K_{n_1, n_2, \dots, n_r}) = r.$$

A wealth of bounds on the global domination number of complementary factors exists. Several from [3] and [13] are listed here, and others can be found in the references. Observation $11.2(ii)$ shows bounds in terms of the domination numbers of the factors. In addition to these we have $\gamma_g < n$ if G is not complete or empty, $\gamma_g \leq \gamma + 1$ for any graph with a pendant vertex, $\gamma \leq \gamma_g \leq \gamma + 1$ if G is triangle free, and either $\gamma_c \leq \gamma_g$ or $\overline{\gamma_c} \leq \gamma_g$. Many other graphical invariants can be employed to establish bounds for γ_g. Some that depend on the minimum and maximum degrees are: $\gamma_g \leq max\{\Delta + 1, \overline{\Delta} + 1\} = max\{n - \overline{\delta}, n - \delta\}$ or this bound minus one if G is not complete, empty, or an odd cycle; if $\gamma_g > max\{\gamma, \overline{\gamma}\}$ then

$$\gamma_g \leq min\{\Delta + 1, \overline{\Delta} + 1\};$$

and

$$\gamma_g \leq \begin{cases} \delta + 2 & \text{if } \delta = \overline{\delta} \leq 2 \\ max\{\delta + 1, \overline{\delta} + 1\} & \text{otherwise} \end{cases}.$$

Bounds which involve various covering and independence numbers include

$$\frac{2m - n(n - 3)}{2} \leq \gamma_g \leq n - \beta_0 + 1$$

if G has no isolated vertex, and

$$\gamma_g \leq max\{\beta_1 + 1, \beta_1 + |Iso|\}$$

if $\beta_1 \leq \alpha_1 - 2$. It follows from this last bound that, if neither G nor $\overline{G}$ has an isolated vertex, then $\gamma_g \leq min\{\beta_1 + 1, \overline{\beta_1} + 1\}$. Other results in the same vein are $\gamma_g \leq max\{\chi, \overline{\chi}\}$, $\gamma_g \leq min\{\omega + \gamma - 1, \overline{\omega} + \overline{\gamma} - 1\}$, and $\gamma_g \leq max\{n - \kappa - 1, n - \overline{\kappa} - 1\}$. Finally, for graphs without isolated vertices: $i(G) + \gamma_g \leq n + 1$ and hence $\gamma + \gamma_g \leq n + 1$.

Some work has concentrated on when $\gamma_g = max\{\gamma, \overline{\gamma}\}$, that is, when γ_g assumes its lower bound. Sufficient conditions which assure this are $diam + \overline{diam} \geq 7$, $max\{rad, \overline{rad}\} \geq 3$, and either G or $\overline{G}$ is disconnected. Related results when both G and $\overline{G}$ are connected are $\gamma_g \leq max\{4, \gamma + 1, \overline{\gamma} + 1\}$ if $diam + \overline{diam} = 6$, $\gamma_g \leq max\{\gamma + 2, \overline{\gamma} + 2\}$ if $diam + \overline{diam} = 5$, and $\gamma_g \leq min\{\delta + 1, \overline{\delta} + 1\}$ if $diam = \overline{diam} = 2$.

The inequality $\gamma \leq \gamma_g \leq \gamma + 1$ mentioned above for triangle free graphs applies, of course, to trees. Both Rall [12] and Brigham and Dutton [3] have characterized those trees which achieve the upper value.

Theorem 11.3 *Let T be a tree. Then $\gamma_g(T) = \gamma + 1$ if and only if (1) T is a star with at least two vertices, or (2) T has radius two and contains a vertex of degree at least two all of whose neighbors have degree at least three.*

We conclude this section with the following theorem [3] which shows that every combination of values for γ, $\overline{\gamma}$, and γ_g permitted by the basic restriction (ii) of Observation 11.2 can occur. This is interesting in view of work discussed below which proves that not all such permissible values for the γ_i and γ_g can occur when the number of factors is large enough.

Theorem 11.4 *For any integers a, b, and c such that $2 \leq a \leq b \leq c \leq a + b$, there exists a graph G for which $\gamma = a$, $\overline{\gamma} = b$, and $\gamma_g = c$.*

11.3 Global Interpretations of Other Domination Invariants

The field of domination in graphs has grown into a study of many related but distinct concepts. Global generalizations of some of them have been defined by Dunbar, Laskar, and Monroe [9], Dunbar and Laskar [8], Dunbar, Laskar, and Wallis [10], Rall [12], and Sampathkumar [13] for the case of complementary factors. Results in this section, unless otherwise stated, are taken from these references.

Recall that, $\Gamma(G)$ is the order of a largest minimal dominating set; $ir(G)$ and $IR(G)$ are the smallest and largest orders, respectively, of a maximal irredundant set; and $\gamma_t(G)$ and $\Gamma_t(G)$ are the smallest and largest orders, respectively, of a minimal total dominating set. In order to define related global parameters we need the following concepts: a set $S \subseteq V(G)$ is a *global total dominating set* if S is a total dominating set of both G and $\overline{G}$, it is a *global irredundant set* if each $x \in S$ is irredundant in either G or $\overline{G}$, and it is a *universal irredundant set* if each $x \in S$ is irredundant in both G and $\overline{G}$. We now define $\Gamma_g(G)$ as the largest order of a minimal global dominating set; $\gamma_{gt}(G)$ and $\Gamma_{gt}(G)$ as the smallest and largest orders, respectively, of a minimal global total dominating set; $ir_g(G)$ and $IR_g(G)$ as the smallest and largest orders respectively, of a maximal global irredundant set; and $IR_u(G)$ as the largest order of a maximal universal irredundant set. The *global domatic number*, $d_g(G)$, is the maximum order of a partition of $V(G)$ such that each member of the partition is a global dominating set.

As is the case with γ_g, $\Gamma_g(G) = \Gamma_g(\overline{G})$. Furthermore, if a graph has diameter at least five, then $\gamma_g = \gamma$ and $\Gamma_g = \Gamma$. The latter equality is extended further to trees with diameter three or four. In the same vein, if neither G nor $\overline{G}$ has isolated vertices and the diameter of G is at least five, then $\gamma_{gt} = \gamma_t$ and $\Gamma_{gt} = \Gamma_t$. For any graph,

$$d_g \leq min\{\delta + 1, \overline{\delta} + 1\};$$

$$d_g \leq (n+1)/2;$$

and, if G is a tree, $d_g \leq 2$. In addition, $\gamma_g + d_g \leq n+1$ with equality if and only if G is complete or empty.

The inequality $ir_g \leq \gamma_g \leq \Gamma_g \leq IR_g$ is a fundamental one relating four of the global invariants and parallels a corresponding result for the nonglobal versions [7]. We also have $IR_u \leq IR \leq IR_g$, which raises the question of which graphs have $IR = IR_g$. This is known to be true for bipartite graphs other than K_2, cycles with at least six vertices, and graphs for which $\Delta < IR$. On the other hand, if $IR < IR_g$, then $IR_g \leq \Delta + 1$. Graphs with $IR_g \geq 3$ are equivalent to those containing an induced subgraph isomorphic to K_3, $\overline{K_3}$, C_6, $\overline{C_6}$, or C_5.

Determining whether $IR_u \geq M$, where M is a positive integer, is an NP-complete problem. Nevertheless, the value of IR_u is known for many special situations. For example, $IR_u(P_n) = 2$ for $n > 3$, $IR_u(K_{1,s}) = 1$, $IR_u(K_{r,s}) = 2$ if $2 \leq r \leq s$, $IR_u(K_n) = 1$, and $IR_u(K_{r,r} \text{ less a one factor}) = r$. Furthermore, $IR_u = 2$ for all trees except $K_{1,s}$. On the other hand, $IR_u(G) \geq 2$ for any graph with diameter at least three. Finally, $IR_u \leq \Delta + 1$ and equality is achieved for connected graphs with at least three vertices if and only if the graph is $K_{r,r}$ minus a one factor.

Little is known about other global domination invariants in the general case of an arbitrary graph and $k \geq 2$. From Carrington [5] we have that $\gamma_{gt} \geq k+1$ if k is odd and $\gamma_{gt} \geq k+2$ if k is even. A GDS which is connected in every factor is a global connected dominating set and a minimum such set contains γ_{gc} nodes. It is known that $\gamma_{gc} \geq 2k$.

A set is a k-dominating set if for each vertex $v \in V - S$, v has at least k neighbors in S. The cardinality of a minimum k-dominating set is the k-domination number $\gamma^k(G)$. The following relationship between the k-domination number and global domination was shown in [4]. For a graph G with size $m \geq k$, γ^k is equal to the minimum value of γ_g taken over all k-factorings of G.

11.4   Applications

Several applications have been postulated for global domination. One possibility is a communication network, modeled by G, with k edge disjoint subnetworks, represented by the factors F_i. The subnetworks might be required for reasons of security, redundancy, or limitation of recipients for different classes of messages. The number γ_g then represents the minimum number of "master" stations required so that a message issued simultaneously from all masters reaches all desired recipients after traveling over only one communication link, no matter which subnetworks are active.

Graph partitioning is important in the implementation of parallel algorithms. In this light, communication networks and their underlying graphs are often partitioned, either to impose a particular structure [2, 14] or to reflect routes taken by messages as they proceed through a network [1, 2]. Carrington [5] illustrates two applications where the partition of a graph is important. Only the most elementary of introductions to these applications can be given here; [5] must be examined for details and other references.

The first of the applications involves the solution to binary constraint satisfaction problems in which vertices represent variables and edges refer to bidirectional binary constraints between the two joined variables. As a simple example, consider a proper coloring of the vertices of a graph with colors taken from the set $\{red, blue, yellow\}$. Then each edge can be interpreted as the constraint $\{\{red, blue\}, \{red, yellow\}, \{blue, yellow\}\}$. In general, the major work in such problems is in the checking of the constraints, that is, in using the information associated with the edges, rather than in labeling the vertices. It is here that a graph with a factoring can be employed in connection with a parallel processing system. The vertices reside in shared memory and each factor corresponds to the edges assigned to a single processor. In such a system a minimum LDS of a factor is a minimum set of vertices such that, once all vertices in the set are labeled, all other unlabeled vertices are constrained on *one* processor by an already labeled one. Similarly, a minimum GDS is a minimum set of vertices such that any vertex not in the set is constrained on *every* processor by some vertex in the set.

The second application relates to the problem of multicast messages in a communication network, where a single source sends messages to many receivers. There are several approaches to creating factors in such an environment, and the significance of $LDS's$ and $GDS's$ is dependent upon the approach. The factorings might be based on edges which connect specific vertices that are designated to receive messages from each other or that use the same subset of a data base, or on edges which are carrying messages in any single time unit, or on edges which are involved in the traversal of a single message. Some of these factorings make sense only if factors do not include all of the vertices of the underlying graph or if edges can appear in more than a single factor. In all cases, the advantages of $LDS's$ and $GDS's$ lie in the fact that vertices not in such sets are just one communication hop away from vertices in the sets.

11.5 Concerning a Characterization

In this section we consider the problem of determining, given a collection of integers γ_1, γ_2, ..., γ_k, and γ_g, if there is a graph which can be factored into k factors in such a way that a minimum LDS of factor F_i has size $\gamma_i, 1 \leq i \leq k$, and a minimum GDS has size γ_g. Theorem 11.4 solves the problem completely when $k = 2$. In this case, given only the most basic conditions of $2 \leq \gamma_1 \leq \gamma_2 \leq \gamma_g \leq \gamma_1 + \gamma_2$, there is always a complete graph which can be factored into two factors in such a way that γ_1 vertices form a minimum dominating set of F_1, γ_2 vertices form a minimum dominating set of F_2, and γ_g vertices make up a minimum GDS.

For arbitrary k, however, the situation becomes more complicated. An ideal characterization would depend only on the given values of the local and global domination numbers, as is the case when $k = 2$. It would not require information about the involved graph, its factors, or the various dominating sets. Unfortunately, we have been unable to find such a simple characterization in the general case and strongly suspect one does not exist.

This section briefly sketches, without proofs, some of the reasons for our conclusion that simple restrictions on the local and global domination numbers will not provide enough information for a characterization, and that there is no one construction which serves for all possible sets of domination numbers which do allow a factoring. A complete characterization does indeed exist, but it involves conditions imposed on the structure of the factoring. These conditions are embodied in the notion of a *configuration*. The formal definition of a configuration and proof of the characterization theorem are lengthy and can be found in [5]. While this general theorem is extremely complicated, important special cases are not, and their simpler characterizations are presented in the next section.

It quickly became obvious that, in order to obtain a general result, we would have to consider the ways in which the D_i intersect D_g. This is necessary because every vertex in G must be dominated in every factor, F_i, both by a vertex in D_g and by one in D_i. These may be the same vertex if D_i intersects D_g, and must be different vertices otherwise. The following observation reflects the fact that, if $D_i \cap D_g = \emptyset$, some vertex $v \in D_i$ must dominate at least γ_g/γ_i vertices of D_g in F_i (we call this v's *share* in F_i) and there must be $k - 1$ vertices left in D_g to dominate v in the other $k - 1$ factors.

Observation 11.5 *If there is some LDS D_i such that $D_i \cap D_g = \emptyset$, then*

$$\gamma_g \geq \frac{(k-1)\gamma_i}{\gamma_i - 1}.$$

Notice, however, that if $D_i \cap D_g = I_i \neq \emptyset$, then $u \in I_i$ can dominate itself in all factors. Thus, in F_i, the construction can force such a u to dominate a sufficient number of vertices of D_g so that the share of $v \in D_i - I_i$ is reduced in F_i to a point where there are $k - 1$ vertices of D_g which do not have an edge to v in F_i. These $k - 1$ vertices then can be used to dominate v in the other factors.

There is a limit, however, to what can be accomplished with the vertices of I_i. Let $I = \bigcup_{i=1}^{k} I_i$. Each of the $\gamma_i - |I_i|$ vertices $v \in D_i - I_i$ can dominate no more than $\gamma_g - k + 1$ vertices of I in F_i if there is to be some vertex in D_g available to dominate v in each of the remaining $k - 1$ factors. Thus the vertices of I_i must dominate those vertices of $I - I_i$ which are undominated by any vertex of $D_i - I_i$. There are at least $|I| - |I_i| - (\gamma_i - |I_i|)(\gamma_g - k + 1)$ such vertices. The following lemma counts the edges needed for this purpose among vertices of I for all factors, and shows that this number can be no more than the total number of edges possible between vertices of I.

Lemma 11.6 *For any k-factoring of a graph G,*

$$\sum_{i=1}^{k} [|I| - |I_i| - (\gamma_i - |I_i|)(\gamma_g - k + 1)] \leq \frac{|I|(|I| - 1)}{2}.$$

A similar lemma, not given here, counts edges required among vertices of D_g. The two lemmas allow development of the following necessary condition for a factoring to occur. It implies a new restriction on the global and local domination numbers.

Theorem 11.7 *Let $\gamma_1, \gamma_2, \ldots, \gamma_k$, and γ_g be integers such that $k \geq 3$ and $2 \leq \gamma_1 \leq \gamma_2 \leq \cdots \leq \gamma_k \leq \gamma_g \leq \gamma_1 + \gamma_2 + \cdots + \gamma_k$. If G can be factored into k factors F_i such that, for all i, γ_i vertices form a minimum LDS for F_i and γ_g vertices form a minimum GDS, then*

$$\sum_{i=1}^{k} [k - 1 - (\gamma_i - 1)(\gamma_g - k + 1)] \leq \frac{k(k - 1)}{2}.$$

Unfortunately, Theorem 11.7 is not sufficient. For example, the integers $\gamma_g = k = 7$; $\gamma_1 = \gamma_2 = \gamma_3 = \gamma_4 = 2, \gamma_5 = 6$, and $\gamma_6 = \gamma_7 = 7$ fulfill the requirements of the theorem, but in the next section we show there is no graph which can be factored into seven factors having these domination numbers. Notice that, if the I_i are pairwise disjoint and each consists of a single vertex, then Lemma 11.6 reduces to Theorem 11.7. This suggests a factoring having this property, that is, each D_i intersects D_g in a single distinct vertex. The proof to the following theorem, which presents a sufficient condition, is based on this type of factoring.

Theorem 11.8 *Let* $\gamma_1, \gamma_2, \ldots, \gamma_k,$ *and* γ_g *be integers such that* $k \geq 3$ *and* $2 \leq \gamma_1 \leq \gamma_2 \leq \cdots \leq \gamma_k \leq \gamma \leq \gamma_1 + \gamma_2 + \cdots + \gamma_k.$ *If*

$$\sum_{i=1}^{r} [r - 1 - (\gamma_i - 1)(\gamma_g - k + 1)] \leq \frac{r(r-1)}{2}$$

for all $r,\ 1 \leq r \leq k,$ *then there is a graph* G *which can be factored into* k *factors such that, for all* $i,$ γ_i *vertices form a minimum LDS for* F_i *and* γ_g *vertices form a minimum GDS.*

Unfortunately, Theorem 11.8 does not give a necessary condition. Even when some set of global and local domination numbers can result in a graph with an appropriate factoring, that factoring does not always take the form stated above. For example, there is a graph which can be appropriately factored given the numbers $\gamma_g = k = 6,$ $\gamma_1 = \gamma_2 = \gamma_3 = 2, \gamma_4 = 3,$ and $\gamma_5 = \gamma_6 = 6,$ but not with the above configuration. Rather, each of the $D_i,$ $1 \leq i \leq 3,$ must be disjoint from the others and wholly contained in D_g as illustrated in Figure 11.1. It can be shown that this configuration is extendable to a legitimate factoring. Observe that these values do not satisfy the condition of Theorem 11.8, as can be seen when $r = 4.$

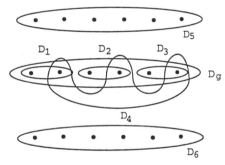

Figure 11.1: $\gamma_g = k = 6; \gamma_1 = \gamma_2 = \gamma_3 = 2; \gamma_4 = 3; \gamma_5 = \gamma_6 = 6.$

While, in this case, the $D_i, 1 \leq i \leq 3,$ must all be pairwise disjoint, there are other situations in which an appropriate factoring can be accomplished only if two or more of the D_i intersect. Just as allowing $D'_i s$ to intersect D_g generally creates more flexibility, so does permitting intersections between $D'_i s$ allow more edges to be used for domination with fewer vertices. Notice that in Figure 11.2(a), showing three $D'_i s$ or portions thereof, the three edges among u, v, and w contribute nothing to the domination properties of the factoring. In Figure 11.2(b), however, where the $D'_i s$ overlap, those edges become available and fewer are needed for the three sets to dominate each other. Thus, knowledge of the orders of the sets and the numbers of edges which could be constructed is not sufficient; we need information about how the sets are configured.

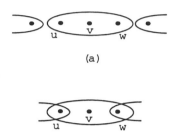

Figure 11.2: Intersecting $D_i's$.

The implication is that, in order to determine whether or not there is a factoring with the desired domination properties, we need some information about the actual global and local dominating sets and how they intersect. Specifically, so far we have seen a need for the configuration definition to include at least

(*i*)   a set of vertices, D_g, of order γ_g;

(*ii*) k possibly overlapping sets of vertices, D_i, of order $\gamma_i, 1 \leq i \leq k$;

(*iii*) k possibly empty sets $I_i = D_i \cap D_g, 1 \leq i \leq k$; and

(*iv*) k possibly empty sets of edges E_i' such that each member of E_i' is an edge between some vertex in I_i and a unique vertex in $D_g - I_i, 1 \leq i \leq k$.

We also need

(*v*)   for each D_i such that $D_i - I_i \neq \emptyset$, a labelling of the vertices of D_g with the labels $d_{ir}, 1 \leq r \leq \gamma_i - |I_i|, 1 \leq i \leq k$.

The purpose of item (*v*) is to allow determination of whether or not all vertices of D_g are dominated in F_i, either by using one of the edges of E_i' or by a vertex of $D_i - I_i$. The vertices in $D_i - I_i$ are labeled $v_{ir}, 1 \leq r \leq \gamma_i - |I_i|$, and are associated with the labels d_{ir} of D_g. The construction forces v_{ir} to dominate all vertices labeled d_{ir} in F_i.

As is detailed in [5], if we know that there is no LDS such that $|D_i - I_i| = 1$, then our configuration definition need not include any information about vertices not in D_g and so is now complete. In this case, the construction can assume $(D_i - I_i) \cap (D_j - I_j) = \emptyset$ for all i, j and it is a simple matter to ensure that in F_i any vertex v_{jr} is dominated either by a vertex in I_i or by both some vertex in the GDS and one in $D_i - I_i$.

In the general case, however, we still do not have enough information about the graph to determine whether or not the desired factoring exists. Consider Figure 11.3 in which $D_i - I_i = \{v_{i1}\}$.

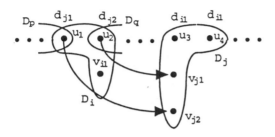

Figure 11.3: $D_i - I_i = \{v_{i1}\}$.

A situation which leads to difficulty is one in which u_1 dominates v_{j2} in F_p; u_2 dominates v_{j1} in F_q (these edges appear in the figure); in F_j, v_{j1} dominates u_1 and v_{j2} dominates u_2; and in F_i, v_{i1} dominates both u_3 and u_4. These edges in F_i and F_j are implied by the labeling. Notice that v_{i1} is the only vertex in D_i which does not already have edges to both v_{j1} and v_{j2} and so must dominate both in F_i. Now, however, there is no vertex to dominate v_{i1} in F_j. Although there is no factoring with the desired domination properties in this situation, there is one given the identical situation except that v_{i1} and v_{j1} are the same vertex, as in Figure 11.4. Now vertex $v_{i1} = v_{j1}$ dominates itself in both factors F_i and F_j and dominates v_{j2} in F_i. Therefore, we must be able to refer to this vertex as both v_{i1} and v_{j1}.

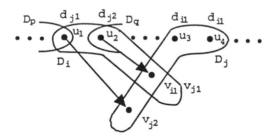

Figure 11.4: $D_i \cap D_j = \{v_{i1} = v_{j1}\}$.

Our configuration definition, then, must include information on vertices of $D_i - I_i$; specifically, we require two additional pieces of information:

(*vi*) a set of names associated with each vertex of $D_i - I_i$; and

(*vii*) an ordering of all vertices of $\bigcup_{i=1}^{k}(D_i - I_i)$.

If a vertex is in $LDS's$ $D_i, D_j, \ldots, D_m$, item (vi) specifies that the vertex will be named $\{v_{ih_i}, v_{jh_j}, \ldots, v_{mh_m}\}$, although the name is simplified to v_{ih_i} when considering D_i. The ordering in (vii) allows us to consider each vertex, v_{ih_i} in the specified order and construct for each a system of distinct representatives for D_j such that $j \neq i$ and $v_{ih_i} \notin D_j$. The representatives will be designated to dominate v_{ih_i} in their respective factors.

The above discussion hints at the complications involved in the development of a characterization theorem. The configuration is defined in such a way that the complications are controlled. The theorem itself then states that, given the domination parameters, there is an appropriate factoring if and only if there is a way of defining a configuration such that three relatively simple conditions are satisfied.

11.6   Subproblems

It is clear from the previous section that the general characterization theorem requires too much structural information to be of practical value in determining when a factoring exists with the desired domination properties. However, it is often the case that a complicated general problem has an easier solution when restricted to smaller instances or specific cases. For example, we have already mentioned that disallowing $|D_i - I_i| = 1$ permits us to simplify both the configuration definition and the characterization. We consider such subproblems in this section.

We make extensive use of Theorem 11.7 which is a necessary condition for assuring the existence of a factoring, and of Theorem 11.8 which is a sufficient condition. Both theorems, since they are based on nothing more than the integers involved, have the advantage of being simple. Of course, neither is a complete characterization. On the other hand, they have converses when k is sufficiently small. In fact, we know from Theorem 11.4 that, when k = 2, there is a complete graph which can be appropriately factored with no conditions on γ_g, γ_1, and γ_2 except the most basic ones that $2 \leq \gamma_1 \leq \gamma_2 \leq \gamma_g \leq \gamma_1 + \gamma_2$.

As opposed to Theorem 11.4, the graph which is constructed by the general characterization theorem and which is to be factored is not necessarily complete. However, there is a sufficient condition which guarantees a factoring of a complete graph. This condition is based on the domination numbers being sufficiently large to allow the $LDS's$ to be disjoint from D_g according to Observation 11.5. The proof can be found in [5].

Theorem 11.9 *Let* $k \geq 3$, $2 \leq \gamma_1 \leq \gamma_2 \leq \ldots \leq \gamma_k$, *and* $\gamma_k \geq \frac{(k-1)\gamma_1}{\gamma_1 - 1}$. *Then, for any value of* γ_g *such that* $\gamma_k \leq \gamma_g \leq \gamma_1 + \gamma_2 + \cdots + \gamma_k$, *the complete graph* K_n *can be factored into k factors such that* γ_g *vertices form a minimum GDS and* γ_i *vertices form a minimum LDS of* F_i, $1 \leq i \leq k$, *where* $n = \gamma_1 + \gamma_2 + \cdots + \gamma_k + max\{\gamma_g, \gamma_k + \gamma_1\}$.

The remainder of this section deals with situations resulting from restricting k, with characterizations developed for $k \leq 6$.

11.6.1 Special case of $k = 3$

As is the case with $k = 2$, when $k = 3$ we can find, for any set of integers which satisfy the basic conditions, some graph which can be factored with appropriate domination numbers.

Theorem 11.10 *For any set of four integers, $\gamma_g, \gamma_1, \gamma_2$, and γ_3, such that $\gamma_g \geq 3$ and $2 \leq \gamma_1 \leq \gamma_2 \leq \gamma_3 \leq \gamma_g \leq \gamma_1 + \gamma_2 + \gamma_3$, there is a graph G which can be factored in such a way that some γ_g vertices constitute a minimum GDS for the factoring, some γ_1 vertices form a minimum LDS for F_1, some γ_2 vertices form a minimum LDS for F_2, and some γ_3 vertices form a minimum LDS for F_3.*

Proof. We show that any set of integers which satisfies the hypothesis also satisfies the conditions of Theorem 11.8. To see this, consider a more general situation which will also prove useful later. Let $k \geq 3, \gamma_g \geq k$, and $2 \leq \gamma_1 \leq \gamma_2 \leq \ldots \leq \gamma_k \leq \gamma_g$. Examine cases corresponding to allowing the r of Theorem 11.8 to range from 1 to 3:

(a) $r = 1 : 0 - (\gamma_1 - 1)(\gamma_g - k + 1) < 0 = \frac{r(r-1)}{2}$.

(b) $r = 2 : 2 - [(\gamma_1 - 1)(\gamma_g - k + 1) + (\gamma_2 - 1)(\gamma_g - k + 1)] \leq 0 < \frac{r(r-1)}{2}$.

(c) $r = 3 : 6 - [(\gamma_1 - 1)(\gamma_g - k + 1) + (\gamma_2 - 1)(\gamma_g - k + 1) + (\gamma_3 - 1)(\gamma_g - k + 1)] \leq 3 = \frac{r(r-1)}{2}$.

Since these inequalities hold for any k, they hold for $k = 3$. Thus the hypothesis of Theorem 11.8 is satisfied and a factoring is possible. □

11.6.2 Special case of $k = 4$

We will show that Theorems 11.7 and 11.8 are equivalent for $k = 4$, so in this case the simple condition of Theorem 11.7 is sufficient as well as necessary.

Theorem 11.11 *For any set of five integers, $\gamma_g, \gamma_1, \gamma_2, \gamma_3$, and γ_4, such that $\gamma_g \geq 4$ and $2 \leq \gamma_1 \leq \gamma_2 \leq \gamma_3 \leq \gamma_4 \leq \gamma_g \leq \gamma_1 + \gamma_2 + \gamma_3 + \gamma_4$, there is a graph G which can be factored into four factors such that some γ_g vertices form a minimum GDS for the factoring and, for $1 \leq i \leq 4$, some γ_i vertices form a minimum LDS for F_i if and only if $\sum_{i=1}^{4}[3 - (\gamma_i - 1)(\gamma_g - 3)] \leq 6$.*

Proof. Consider the r of Theorem 11.8 as it varies from 1 to 4. For $r = 1, 2$, or 3 the condition of Theorem 11.8 holds by the proof of Theorem 11.10 and for $r = 4$ the condition of Theorem 11.8 becomes $\sum_{i=1}^{4}[3 - (\gamma_i - 1)(\gamma_g - 3)] \leq 6$. □

11.6.3   Special case of $k = 5$

As is the case when $k = 4$, Theorems 11.7 and 11.8 are equivalent when $k = 5$.

Theorem 11.12 *Given a set of six integers, $\gamma_g, \gamma_1, \gamma_2, \gamma_3, \gamma_4,$ and γ_5, such that $\gamma_g \geq 5$ and $2 \leq \gamma_1 \leq \gamma_2 \leq \gamma_3 \leq \gamma_4 \leq \gamma_5 \leq \gamma_g \leq \gamma_1 + \gamma_2 + \gamma_3 + \gamma_4 + \gamma_5$, there is a graph G which can be factored into five factors such that some γ_g vertices form a minimum GDS for the factoring and, for $1 \leq i \leq 5$, some γ_i vertices form a minimum LDS for F_i if and only if $\sum_{i=1}^{5}[4 - (\gamma_i - 1)(\gamma_g - 4)] \leq 10$.*

Proof. Again we use Theorem 11.8 and let r vary from 1 to 5. For $r = 1, 2,$ or 3 the condition of Theorem 11.8 holds by the proof of Theorem 11.10 and for $r = 5$ the formula reduces to our current condition. Thus we need only check the case where $r = 4$. Since $\sum_{i=1}^{5}[4 - (\gamma_i - 1)(\gamma_g - 4)] \leq 10$, we know that $\sum_{i=1}^{5}(\gamma_i - 1)(\gamma_g - 4) \geq 10$. If the condition of Theorem 11.8 is not satisfied when $r = 4$, then $\sum_{i=1}^{4}[3 - (\gamma_i - 1)(\gamma_g - 4)] > 6$ or $\sum_{i=1}^{4}(\gamma_i - 1)(\gamma_g - 4) < 6$ which is clearly not possible if $\gamma_g > 5$. If $\gamma_g = 5$, then we have $\sum_{i=1}^{4}(\gamma_i - 1) < 6$ and $\sum_{i=1}^{5}(\gamma_i - 1) \geq 10$. By subtraction, $\gamma_5 - 1 \geq 5$ or $\gamma_5 > 5$ which is a contradiction. □

11.6.4   Special cases of $k = 6$ and even k

Theorems 11.7 and 11.8 diverge for the first time when $k = 6$ although Theorem 11.7 is a sufficient as well as a necessary condition in this case. Define a *minimum set* of numbers to be an assignment of the smallest possible positive integer to each of $\gamma_g, \gamma_1, \gamma_2, \ldots, \gamma_k$ in turn such that $\gamma_g = k$, $2 \leq \gamma_1 \leq \gamma_2 \leq \ldots \leq \gamma_k \leq \gamma_g \leq \gamma_1 + \gamma_2 + \ldots + \gamma_k$, and Theorem 11.7 holds. When $k = 6$, the minimum set is $\gamma_g = 6$, $\gamma_1 = \gamma_2 = \gamma_3 = 2$, $\gamma_4 = 3$, and $\gamma_5 = \gamma_6 = 6$. Using these values $\sum_{i=1}^{4}[3 - (\gamma_i - 1)(\gamma_g - k + 1)] = 7 > 6$, and the condition of Theorem 11.8 is not satisfied. However, there is a factoring where the *LDS's* and *GDS* are positioned as in Figure 11.1. In fact, this is a special case of the following more general result, proven in [5].

Theorem 11.13 *For $k = 2t, t \geq 3$, there is a graph which can be factored such that the minimum set of integers, $\gamma_g, \gamma_1, \gamma_2, \ldots, \gamma_k$, is a set of global and local domination numbers for the factoring.*

We can now prove the following theorem.

Theorem 11.14 *Given a set of seven integers, $\gamma_g, \gamma_1, \gamma_2, \gamma_3, \gamma_4, \gamma_5,$ and γ_6, such that $\gamma_g \geq 6$ and $2 \leq \gamma_1 \leq \gamma_2 \leq \gamma_3 \leq \gamma_4 \leq \gamma_5 \leq \gamma_6 \leq \gamma_g \leq \gamma_1 + \gamma_2 + \gamma_3 + \gamma_4 + \gamma_5 + \gamma_6$, there is a graph G which can be factored into six factors such that some γ_g vertices form a minimum GDS for the factoring and, for $1 \leq i \leq 6$, some γ_i vertices form a minimum LDS for F_i if and only if $\sum_{i=1}^{6}[5 - (\gamma_i - 1)(\gamma_g - 5)] \leq 15$.*

Proof. Theorem 11.13 shows the result for a minimum set of values, so we assume we do not have such a set. Analogous to earlier arguments, the condition of Theorem 11.8 holds for $r = 1, 2, 3$, or 6 so we need be concerned only with $r = 4$ and $r = 5$.

Case 1: $\gamma_g \geq 7$. In this case, Theorem 11.8 holds for all values for the γ_i.

$$r = 4 : 12 - \sum_{i=1}^{4}(\gamma_i - 1)(\gamma_g - 5) \leq 4 < \frac{r(r-1)}{2} = 6.$$

$$r = 5 : 20 - \sum_{i=1}^{5}(\gamma_i - 1)(\gamma_g - 5) \leq 10 = \frac{r(r-1)}{2}.$$

Case 2: $\gamma_g = 6$. If Theorem 11.8 fails when $r = 4$, then $\sum_{i=1}^{4} \gamma_i < 10$ which can only happen in the case of the minimum set. If Theorem 11.8 fails when $r = 5$, then $\sum_{i=1}^{5} \gamma_i < 15$. Thus, $36 - \sum_{i=1}^{6} \gamma_i = 36 - \sum_{i=1}^{5} \gamma_i - \gamma_6 > 21 - \gamma_6 \geq 15$ since $\gamma_6 \leq 6$ and we have a contradiction to the condition of the theorem. □

11.6.5   Special cases of $k = 7$ and odd k

The case where $k = 7$ is the smallest for which Theorem 11.7 is not sufficient. This can be seen by examining the minimum set, i.e., $\gamma_g = 7, \gamma_1 = \gamma_2 = \gamma_3 = \gamma_4 = 2, \gamma_5 = 6$, and $\gamma_6 = \gamma_7 = 7$. In order to see that there is no graph which can be factored with these domination numbers, we first investigate possible arrangements when $\gamma_g = k$ and four or more $\gamma_i's$ equal 2.

Lemma 11.15 *If $k \geq 4, \gamma_g = k$, and $\gamma_i = 2$ for $1 \leq i \leq s$ where $s \geq 4$, then $D_i \subseteq D_g$ for $1 \leq i \leq s$, and $D_i \cap D_j = \emptyset$ for $1 \leq i \neq j \leq s$.*

Proof. We first show that $D_i \cap D_j = \emptyset$ for $1 \leq i \neq j \leq s$. Suppose there are fewer than four vertices among some four $D_i's$ of size 2. Without loss of generality, we take the four to be D_1, D_2, D_3, and D_4. Since $\gamma_g = k$, there is at least one other vertex, say x, in the graph. Then x must have four or more edges to vertices in this set in order to be dominated in factors F_1, F_2, F_3, and F_4, an impossibility. Therefore, there must be at least four vertices in $\bigcup_{i=1}^{4} D_i$. If there are exactly four, then in each of F_1, F_2, F_3, and F_4 there must be two edges among these four vertices for a total of eight edges, again an impossibility. A similar argument tells us that there must be 12 internal edges when there are five vertices and 16 when there are six vertices, a contradiction in either case. Suppose there are seven vertices in the four sets. Then, for some three distinct sets D_i, D_j, and $D_h, 1 \leq i, j, h \leq 4, D_i \cap D_j = \{v\}$ and D_h is disjoint from both D_i and D_j. If either vertex of D_i dominates both vertices of D_h in F_i, then that vertex will be undominated in F_h. A similar comment applies to the vertices of D_j. Thus, v must dominate one vertex of D_h in F_i and the other vertex of D_h in F_j, meaning that v is undominated in F_h. We conclude that we need eight vertices among these four $D_i's$ which means that $D_i \cap D_j = \emptyset$ for $i \neq j$.

Now we must show that, for each D_i of size two, where there are four or more of them, $D_i \subseteq D_g$. First, recall that Observation 11.5 says that, if $D_i \cap D_g = \emptyset$, then $\gamma_g \geq 2(k-1)$ when $\gamma_i = 2$. Since $\gamma_g = k < 2(k-1)$, it follows that $|D_i \cap D_g| \geq 1$. Suppose all of the first four such $D_i's$ intersect D_g in one vertex, giving the situation in Figure 11.5.

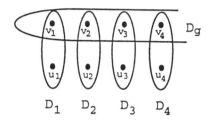

Figure 11.5: Four $D_i's$ of size 2, each intersecting D_g in one vertex.

Because $\gamma_g = k$, no vertex in any $D_i - I_i$ can dominate more than one vertex of D_g in F_i. In particular, in each $F_i, 1 \leq i \leq 4, u_i$, the vertex not in D_g, can dominate at most one of v_1, v_2, v_3, or v_4. Then, for each i, v_i must dominate at least the other two, resulting in the need for eight edges among these four v_i vertices, an impossibility.

Now suppose one LDS, say D_1, intersects D_g in one vertex and another, say D_2, intersects D_g in two vertices. Set D_3 intersects D_g in either one or two vertices and we consider both cases. First suppose $|D_3 \cap D_g| = 1$. Then we have the situation in Figure 11.6.

Once again, in F_1, u_1 can dominate at most one vertex of D_g and neither u_1 nor v_1 can dominate both vertices of either D_2 or D_3. The only possibility is that u_1 dominates u_3. Thus edge $u_1 u_3$ is forced to be in F_1. By a similar argument, in F_3, u_3 must dominate u_1 and edge $u_1 u_3$ must also be in F_3, a contradiction.

Next suppose $|D_3 \cap D_g| = 2$, giving the situation in Figure 11.7.

Because u_1 can dominate at most one vertex in D_2 and D_3 combined, v_1 must dominate both vertices of one of them, say D_2. But then, in F_2, there is no vertex to dominate v_1.

Since the same reasoning applies to any three $D_i's$ of order two under the conditions of the lemma, the lemma is proved. $\square$

An immediate consequence of the lemma is that, for $k = 7$, there is no graph which can be factored such that the minimum set constitutes a set of global and

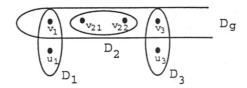

Figure 11.6: One D_i contained in D_g, two intersecting D_g in one vertex.

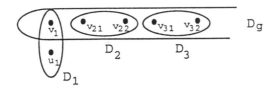

Figure 11.7: Two $D_i's$ contained in D_g, one intersecting D_g in one vertex.

local domination numbers. This justifies the earlier statement that Theorem 11.7 is not sufficient when $k \geq 7$.

We can generalize this result for any odd k.

Theorem 11.16 *For $k = 2t - 1 \geq 7$, there is no graph which can be factored in such a way that the minimum set of integers which satisfy the condition of Theorem 11.7 constitutes a set of global and local domination numbers for the factoring.*

Proof. It is easy to check that a minimum set of integers has $\gamma_g = k$ and $\gamma_1 = \gamma_2 = \cdots = \gamma_t = 2, \gamma_{t+1} = k - 1, \gamma_{t+2} = \gamma_{t+3} = \cdots = \gamma_k = k$. Since $t \geq 4$, Lemma 11.15 shows we must have the t $LDS's$ of order two as disjoint subsets of D_g, implying $\gamma_g \geq 2t = k + 1$, a contradiction. $\square$

11.7 Open Problems

We conclude this chapter with an indication of possible areas of further research. Papers [8, 9, 10] present problems, all still open as of this writing, related to the global domination concepts discussed in Section 11.3. A simple characterization of when a factoring exists, given the local and global domination numbers, remains as elusive as ever. It also would be helpful to extend the class of special cases for which reasonable characterizations can be found. Further complexity results would be interesting. For example, is determining whether the global domination number of two paths is at most M an NP-complete problem or a polynomial one? It is possible to extend the global idea to invariants other than domination related ones. Carrington [5] has produced some initial results and also has shown that many invariants do not extend in a natural way. Nevertheless, determining which ones do and finding relations between them would be of interest. As an example, with α_g^1 being the minimum number of edges such that all vertices are incident to at least one of the edges in all factors and β_g^1 being the maximum number of edges such that no two are adjacent in any factor, Carrington has developed the Gallai type relation $\alpha_g^1 + \beta_g^1 = kn$. Theorem 11.9 presents a result when the graph to be factored is constrained to be a complete graph. Studies might consider other graphs as the base graphs. One such approach by Caron, Brigham, and Carrington [4] employed the cycle as this base. A completely untapped area of study is when edges are allowed to appear in two or more factors. As alluded to in Section 11.4, some applications require this generalization.

Bibliography

[1] L. Berry, Graph theoretic models for multicast communications. *Comput. Networks and ISDN Systems* 20 (1990) 95-99.

[2] A. Bouloutas and P. M. Gopar, Some graph partitioning problems and algorithms related to routing in large computer networks. *Proceedings of Ninth International Conference on Distributed Computing Systems* (June, 1989) 110-117.

[3] R. C. Brigham and R. D. Dutton, Factor domination in graphs. *Discrete Math.* 86 (1990) 127-136.

[4] R. M. Caron, R. C. Brigham, and, J. R. Carrington, A minimum factoring of cycles. *Congr. Numer.* 97 (1993) 171-183.

[5] J. R. Carrington, *Global Domination of Factors of a Graph.* Ph.D. Dissertation, University of Central Florida (1992).

[6] J. R. Carrington and R. C. Brigham, Global domination of simple factors. *Congr. Numer.* 88 (1992) 161-167.

[7] E. J. Cockayne and S. T. Hedetniemi, Towards a theory of domination in graphs. *Networks* 7 (1977) 247-261.

[8] J. Dunbar and R. Laskar, Universal and global irredundancy in graphs. *J. Combin. Math. Combin. Comput.* 12 (1992) 179-185.

[9] J. Dunbar, R. Laskar, and T. Monroe, Global irredundant sets in graphs. *Congr. Numer.* 85 (1991) 65-72.

[10] J. Dunbar, R. Laskar, and C. Wallis, Some global parameters of graphs. *Congr. Numer.* 89 (1992) 187-191.

[11] J. Nieminen, Two bounds for the domination number of a graph. *J. Inst. Math. Applics.* 14 (1974) 183-187.

[12] D. F. Rall, Dominating a graph and its complement. *Congr. Numer.* 80 (1991) 89-95.

[13] E. Sampathkumar, The global domination number of a graph. *J. Math. Phys. Sci.* 23 (1989) 377-385.

[14] M. M. Theimer and K. A. Lantz, Finding idle machines in a workstation-based distributed system. *Proceedings of Eighth International Conference on Distributed Computing Systems* (June, 1988) 112-122.

Chapter 12

Distance Domination in Graphs

Michael A. Henning

Department of Mathematics

University of Natal

Private Bag X01

Pietermaritzburg, 3209 South Africa

Abstract. For $k \geq 1$ an integer, a set S of vertices of a graph $G = (V, E)$ is a k-dominating set of G if every vertex of $V - S$ is within distance k from some vertex of S. In this chapter, we survey some recent results concerning this distance version of domination in graphs.

12.1 Introduction

If S is a set of vertices of G and v is a vertex of G, then the *distance from v to S*, denoted by $d_G(v, S)$, is the shortest distance from v to a vertex of S. The kth power G^k of a connected graph G, where $k \geq 1$, is that graph with $V(G^k) = V(G)$ for which $uv \in E(G^k)$ if and only if $1 \leq d_G(u, v) \leq k$.

Let $k \geq 1$ be an integer, and let $G = (V, E)$ be a graph. The *(open) k-neighbourhood* $N_k(v)$ of a vertex v in G is defined in [24] as the set of all vertices of G different from v and at distance at most k from v in G. The *closed k-neighbourhood* $N_k[v]$ of v is defined as $N_k(v) \cup \{v\}$, i.e., $\{u \mid d(u, v) \leq k\}$. If $u \in N_k(v)$, then we say that u and v are *k-adjacent* vertices. The *k-degree*, $deg_k v$, of v in G is given by $|N_k(v)|$. Hence $N_1(v) = N(v)$ and $deg_1 v = deg v$. The minimum k-degree of G is given by $\delta_k(G) = \min\{deg_k v \mid v \in V\}$, and the maximum k-degree of G by $\Delta_k(G) = \max\{deg_k v \mid v \in V\}$. The *open (closed) k-neighbourhood of a set X of vertices*, denoted by $N_k(X)$ $(N_k[X])$ is the union of the open (closed) k-neighbourhoods $N_k(v)$ $(N_k[v])$ of vertices v in X.

For $k \geq 1$ an integer, a set $\mathcal{D}$ of vertices of a graph $G = (V, E)$ is defined in [35] to be a *k-dominating set* of G if every vertex in $V - \mathcal{D}$ is within distance k from some vertex of $\mathcal{D}$; equivalently, $\mathcal{D}$ is a *k-dominating set* of G if each $v \in V$ is either in $\mathcal{D}$ or k-adjacent to a vertex of $\mathcal{D}$. (That is, $N_k[\mathcal{D}] = V$.) The minimum cardinality among all k-dominating sets of G is called the k-*domination number* of G and is denoted by $\gamma_k(G)$. Observe that $\gamma(G) = \gamma_1(G)$. A k-dominating set of cardinality $\gamma_k(G)$ is called a γ_k-set. For the graph shown in Figure 12.1, $D = \{v_2, v_5, v_8\}$ is a 2-dominating set of G with $\gamma_2(G) = |D|$.

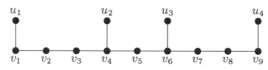

Figure 12.1: The graph G.

In this chapter, we survey some recent results concerning distance domination in graphs. This distance version of domination has been studied by, among others, Bascó and Tuza [4, 5], Bean, Henning, and Swart [6], Beineke and Henning [7], Bondy and Fan [9] , Chang [12], Chang and Nemhauser [13, 14, 15], Fraisse [24], Fricke, Hedetniemi, and Henning [25, 26], Fricke, Henning, Oellermann, and Swart [27], Gimbel and Henning [29], Hattingh and Henning [30, 31, 32], Henning, Oellermann, and Swart [35, 36, 37, 38, 39], Meir and Moon [42], Mo and Williams [43], Slater [46], Topp and Volkmann [49], Tuza [50], and Xin He and Yesha [51].

12.2   The Distance Domination Number

12.2.1   Introduction

In 1975 Meir and Moon [42] introduced the concept of a k-packing and a k-dominating set (called a "k-covering" in [42]) in a graph, and established various relations between the k-packing number and the k-domination number of a tree.

In 1976 Slater [46] considered the problem of finding a minimum k-dominating set (called a "k-basis" in [46]) in a graph. The application presented by Slater is in terms of communication networks. Consider a graph associated with a collection of cities where the vertices correspond to the cities and where two vertices are joined by an edge if there is a communication link between the corresponding cities. One may be interested in selecting a minimum number of cities as sites for transmitting stations so that every city either contains a transmitter or can receive messages from at least one of the transmitting stations through the links. If communication over paths of k links (but not of $k+1$ links) is adequate in quality and speed, the problem becomes that of determining a *minimum k-dominating set* in the graph. Actually [46] introduces the more general problem in which each vertex v_i has an associated value a_i, and we require that there be a transmitting station within distance a_i of v_i.

This concept of distance domination in graphs finds applications in many situations and structures which give rise to graphs. Consider, for instance, the following illustration. Let G be the graph associated with the road grid of a city where the vertices of G correspond to the street intersections and where two vertices are adjacent if and only if the corresponding street intersections are a block apart. A minimum k-dominating set in G may be used to locate a minimum number of facilities (such as utilities, police stations, waste disposal dumps, hospitals, blood banks, transmission towers) such that every intersection is within k city blocks of a facility.

In 1977 Lichtenstein [41] considered the following geometric problem: For a given (finite) subset P of points in $\mathbf{Z} \times \mathbf{Z}$ and a positive integer k, what is the smallest cardinality of a subset P' of P such that every point of $P - P'$ is within Euclidean distance k of some point in P' given that the graph with vertex set P' in which two points of P' are adjacent if and only if they are within Euclidean distance k of each other is connected? Of course, this problem is defined only if the graph with vertex set P and such that two points of P are joined by an edge if and only if they are within distance k of each other is itself connected. It was shown [41] that this optimization problem appears to be computationally difficult by showing that a corresponding decision problem is NP-complete. This geometric problem suggests several related graph problems, including the concept of distance domination, which have since then been introduced and studied.

The p-center problem studied in operations research is in some sense a dual problem to that of determining γ_k. The p-center problem prespecifies p as the cardinality of a subset D of vertices of a graph G, and seeks to minimize k such that $\gamma_k(G) \leq p$. There is much active operations research work on the subject of p-centers (and p-medians). Tansel, Francis, and Lowe [48] have written an excellent survey paper on the subject of p-centers (and p-medians).

12.2.2 Properties of k-dominating sets

We begin with properties of minimal k-dominating sets. The following proposition from [35] generalizes a classical result of Ore [45] concerning dominating sets.

Proposition 12.1 *For $k \geq 1$, let $\mathcal{D}$ be a k-dominating set of a graph G. Then $\mathcal{D}$ is a minimal k-dominating set of G if and only if each $d \in \mathcal{D}$ has at least one of the following two properties:*

P_1: *There exists a vertex $v \in V(G) - \mathcal{D}$ such that $N_k(v) \cap \mathcal{D} = \{d\}$;*

P_2: *The vertex d is at distance at least $k + 1$ from every other vertex of D in G.*

Proposition 12.1 is generalized even further by Slater (see Theorem 3 of [46]) in his paper on R-domination in graphs. We state next a useful observation, the proof of which is immediate.

Proposition 12.2 *If G is a connected graph, then $\gamma_k(G) = \gamma(G^k)$.*

Bollobás and Cockayne [8] established the following result.

Proposition 12.3 *If G is a connected nontrivial graph, then there exists a minimum dominating set $\mathcal{D}$ of G such that for each $d \in \mathcal{D}$, there exists a vertex $v \in V(G) - \mathcal{D}$ such that $N(v) \cap D = \{d\}$.*

An immediate consequence of Propositions 12.2 and 12.3 is that if G is a connected nontrivial graph, then there exists a minimum k-dominating set $\mathcal{D}$ of G such that for each $d \in \mathcal{D}$, there exists a vertex $v \in V(G) - \mathcal{D}$ such that $N_k(v) \cap D = \{d\}$. The following stronger result is proved in [39].

Proposition 12.4 *For $k \geq 1$, if G is a connected graph of order at least $k + 1$ with diam $G \geq k$, then there exists a minimum k-dominating set $\mathcal{D}$ of G such that for each $d \in \mathcal{D}$, there exists a vertex $v \in V(G) - \mathcal{D}$ at distance exactly k from d such that $N_k(v) \cap D = \{d\}$.*

In order to prove this result, we introduce some notation. Let S be a set of vertices of a connected graph G. We will call a nondecreasing sequence $\ell_1, \ell_2, \ldots, \ell_{|S|}$ of integers the *distance sequence* of S in G if the vertices of S can be labelled $v_1, v_2, \ldots, v_{|S|}$ so that $\ell_i = d_G(v_i, S - \{v_i\})$ for all i. For example, for the graph G given in Figure 12.1, the set $\{v_2, v_5, v_8\}$ has distance sequence $3, 3, 3$ in G, while the distance sequence of the set $\{u_1, v_5, u_4\}$ in G is $5, 5, 5$. (Observe that both $\{v_2, v_5, v_8\}$ and $\{u_1, v_5, u_4\}$ are 2-dominating sets of G.) As a further example, let G be obtained from a connected graph H by attaching a path of length k to each vertex of H. Then the distance sequence of $V(H)$ in G is the sequence $1, 1, \ldots, 1$ of length $|V(H)|$. (Observe that $V(H)$ is a minimum k-dominating set of G.) Suppose $s_1 : a_1, a_2, \ldots, a_p$ and $s_2 : b_1, b_2, \ldots, b_q$ are two nondecreasing sequences of positive integers. Then we say that s_1 *precedes* s_2 in *dictionary order* if either $p \leq q$ and $a_i = b_i$ for $1 \leq i \leq p$ or if there exists an i ($1 \leq i \leq min\{p, q\}$) such that $a_i < b_i$ and $a_j = b_j$ for $j < i$. We are now in a position to present a proof of Proposition 12.4.

Proof of Proposition 12.4. Among all minimum k-dominating sets of vertices of G, let $\mathcal{D}$ be one that has the smallest distance sequence in dictionary order. Let the distance sequence of $\mathcal{D}$ be given by $\ell_1, \ell_2, \ldots, \ell_{\gamma_k(G)}$, where $\mathcal{D} = \{v_1, v_2, \ldots, v_{\gamma_k(G)}\}$ and $\ell_i = d_G(v_i, \mathcal{D} - \{v_i\})$ for $1 \leq i \leq \gamma_k(G)$.

We show first that each vertex of $\mathcal{D}$ has property P_1. If this is not the case, then let i be the smallest integer such that the vertex v_i does not have property P_1. By Proposition 12.1, v_i has property P_2, and so $\ell_i \geq k + 1$. Now let $v'_i \in$

$N_k(v_i)$ and consider the set $\mathcal{D}' = (\mathcal{D} - \{v_i\}) \cup \{v_i'\}$. Necessarily $\mathcal{D}'$ is a minimum k-dominating set of G. Furthermore, the vertex v_i' is within distance k from some vertex of $\mathcal{D} - \{v_i\}$; consequently, $\ell_i' = d_G(v_i', \mathcal{D}' - \{v_i'\}) < \ell_i$. Now let j be the largest integer for which $\ell_j < \ell_i$, and consider the value $\ell_k' = d_G(v_k, \mathcal{D}' - \{v_k\})$ for each k with $1 \leq k \leq j$. Since $\ell_k < \ell_i$, a shortest path from the vertex v_k to a vertex of $\mathcal{D} - \{v_i\}$ does not contain v_i. It follows, therefore, that $\ell_k' \leq \ell_k$ for all k ($1 \leq k \leq j$). This, together with the observation that $\ell_i' < \ell_r$ for all $r > j$, implies that the distance sequence of $\mathcal{D}'$ precedes that of $\mathcal{D}$ in dictionary order. This produces a contradiction. Hence every vertex of $\mathcal{D}$ has property P_1.

For each vertex v_i of $\mathcal{D}$, let w_i be a vertex of $V(G) - \mathcal{D}$ at maximum distance from v_i in G satisfying $N_k(w_i) \cap \mathcal{D} = \{v_i\}$ ($1 \leq i \leq \gamma_k(G)$). We show that $d(v_i, w_i) = k$ for all i. If this is not the case, then let i be the smallest integer for which $d(v_i, w_i) < k$. Observe that every vertex of $V(G) - \mathcal{D}$ at distance greater than $k - 1$ from v_i is within distance k from some vertex of $\mathcal{D} - \{v_i\}$. Consider a shortest path from the vertex v_i to a vertex of $\mathcal{D} - \{v_i\}$ in G. Let v_i^* denote the vertex adjacent to v_i on such a path. Further, let $\mathcal{D}^* = (\mathcal{D} - \{v_i\}) \cup \{v_i^*\}$. Necessarily $\mathcal{D}^*$ is a minimum k-dominating set of G. Now let j be the largest integer for which $\ell_j < \ell_i$, and consider the value $\ell_k^* = d_G(v_k, \mathcal{D}^* - \{v_k\})$ for each k with $1 \leq k \leq j$. Necessarily, $\ell_k^* \leq \ell_k$ for all k ($1 \leq k \leq j$). Furthermore, $d_G(v_i^*, \mathcal{D}^* - \{v_i^*\}) = \ell_i - 1 < \ell_r$ for all $r > j$. It follows, therefore, that the distance sequence of $\mathcal{D}^*$ precedes that of $\mathcal{D}$ in dictionary order, a contradiction. Hence $d(v_i, w_i) = k$ for all i. $\square$

12.2.3 Algorithmic and complexity results

From a computational point of view the problem of finding $\gamma_k(G)$ appears to be very difficult. In fact there is no known efficient algorithm for solving this problem and a corresponding decision problem is NP-complete (see [15]). The computability of $\gamma_k(G)$ if G belongs to certain special classes of graphs (for example if G is bipartite or chordal of diameter $2k + 1$) has been considered, but even with these restrictions the problem remains NP-hard (see [15]). Further aspects of the computability of $\gamma_k(G)$ are discussed in [10], [15], [22] and [46]. Since the problem of computing $\gamma_k(G)$ appears to be a difficult one, it is desirable to find good bounds on this parameter. As an immediate corollary of Proposition 12.4, we have the following result which was established in [35].

Corollary 12.5 For $k \geq 1$, if G is a connected graph of order $n \geq k + 1$, then $\gamma_k(G) \leq \frac{n}{k+1}$.

The proof of Corollary 12.5 that is given in [35] suggests an algorithm that finds, for a connected graph G of order n, a k-dominating set of cardinality at most $n/(k+1)$. Clearly if G is a connected graph of order $n \leq k+1$, any vertex of G will k-dominate every vertex of G. For $n \geq k + 2$, we have the following algorithm from [38].

Algorithm 12.6 *Let $k \geq 1$ be an integer and G a connected graph of order $n \geq k + 1$.*

1. *Find a spanning tree T of G. Set $D_k \leftarrow \emptyset$.*

2. *If rad $T \leq k$, then let v be a central vertex of T, output $D_k \cup \{v\}$ and stop. Otherwise continue.*

3. *Let $d = diam\ T$ and find a path $u_0, u_1, ..., u_d$ of length d in T. Set $D_k \leftarrow D_k \cup \{u_k\}$ and let T be the component containing u_{k+1} in $T - u_k u_{k+1}$ and return to Step 2.*

The class of paths $P_{2(k+1)}$ and the class of cycles $C_{2(k+1)}$ illustrate that the bound in Corollary 12.5 can be attained. Furthermore, for any integers n, k and ℓ such that $k \geq 1$, $n \geq 2(k+1)$ and $2 \leq \ell \leq n/(k+1)$, the graph $G = H(n, k, \ell)$ obtained by attaching disjoint paths of length k to ℓ vertices of $K_{n-k\ell}$ is such that $\gamma_k(G) = \ell$ and $|V(G)| = n$.

12.2.4   Nordhaus-Gaddum-type results

In 1956 Nordhaus and Gaddum [44] established sharp bounds on the sum and product of the chromatic numbers of a graph and its complement. Since then such results have been given for several parameters. They include the following on the domination number due to Jaeger and Payan [40].

Proposition 12.7 *If G is a graph of order n, then $\gamma(G) + \gamma(\bar{G}) \leq n + 1$ and $\gamma(G)\gamma(\bar{G}) \leq n$, and these bounds are sharp.*

In [35] the results of Proposition 12.7 were extended to γ_k for $k \geq 2$.

Proposition 12.8 *Let $k \geq 2$ be an integer, and let G be a graph of order $n \geq k + 1$. Then $2 \leq \gamma_k(G) + \gamma_k(\bar{G}) \leq n + 1$ and $1 \leq \gamma_k(G)\gamma_k(\bar{G}) \leq n$.*

Proposition 12.9 *For an integer $k \geq 2$, if both G and $\bar{G}$ are connected graphs of order $n \geq k+1$, then $2 \leq \gamma_k(G) + \gamma_k(\bar{G}) \leq n/(k+1)+1$ and $1 \leq \gamma_k(G)\gamma_k(\bar{G}) \leq n/(k+1)$.*

The bounds in Proposition 12.9 are best possible in the following sense: If ℓ is an integer, $2 \leq \ell \leq n/(k+1)$, the graph $G = H(n, k, \ell)$, defined after Algorithm 12.6, has the property that $\gamma_k(G) = \ell$ and $\gamma_k(\bar{G}) = 1$. So, in particular, the upper bounds in Proposition 12.9 are sharp. The lower bounds in both Propositions 12.8 and 12.9 are attained by self-complementary graphs of diameter 2 and order $n \equiv 0, 1 \ (mod\ 4)$. Further, for any integer ℓ such that $2 \leq \ell \leq n$, the graph $G \cong K_{n-\ell+1} \cup (\ell - 1)K_1$ has order n, $\gamma_k(G) = \ell$ and $\gamma_k(\bar{G}) = 1$; whence $\gamma_k(G) + \gamma_k(\bar{G}) = \ell + 1$ and $\gamma_k(G)\gamma_k(\bar{G}) = \ell$. So the bounds of Proposition 12.8 are also best possible.

12.2.5   Well-k-dominated graphs

Finbow, Hartnell, and Nowakowski [23] defined a graph to be *well-dominated* if every minimal dominating set has the same cardinality. In this subsection we extend the definition of well-dominated graphs. Let $k \geq 1$ be an integer, and let $G = (V, E)$ be a graph. The *upper k-domination number* $\Gamma_k(G)$ of a graph G is the maximum cardinality taken over all minimal k-dominating sets of G. A graph is defined in [30] to be *well-k-dominated* if every minimal k-dominating set of the graph has the same cardinality. Hence G is well-k-dominated if and only if $\gamma_k(G) = \Gamma_k(G)$.

A parameter of interest here is the k-packing number defined by Meir and Moon [42]. A set S of vertices of a graph G is a *k-packing* of G if $d_G(x, y) > k$ for all pairs of distinct vertices x and y in S. The *k-packing number* $\rho_k(G)$ of G is the maximum cardinality of a k-packing set in G. An important result relating $\rho_{2k}(G)$ and $\gamma_k(G)$ where G is a connected block graph is the following proposition due to Domke, Hedetniemi, and Laskar [21], where a *block graph* is a graph in which each block is complete. A tree is a block graph where each block is K_2, the complete graph on two vertices.

Proposition 12.10 [21] *For $k \geq 1$ an integer, if G is a connected block graph, then $\rho_{2k}(G) = \gamma_k(G)$.*

In [30], block graphs that are well-k-dominated are characterized. Since a graph is well-k-dominated if and only if each of its components is well-k-dominated, we restrict ourselves to connected graphs. The following result of [30] extends a result due to Topp and Volkmann [49] from trees to connected block graphs.

Proposition 12.11 *Let G be a connected block graph. Then the following statements are equivalent:*

(1) $\gamma_k(G) = \rho_{2k}(G) = r$.

(2) One of the following statements holds:

> *(a) G has diameter at most k and $r = 1$.*

> *(b) There exists a decomposition of G into r subgraphs $G_1, G_2, \ldots, G_r$ in such a way that*

>> *(i) G_i is a connected block graph with diameter k $(i = 1, 2, \ldots, r)$,*

>> *(ii) for each $i \in \{1, 2, \ldots, r\}$, there exists $u_i \in V(G_i) - V(G_0)$ such that*
>> *$d_G(u_i, V(G_0)) = k$, where G_0 is the subgraph of G generated by the edges which do not belong to any of the subgraphs $G_1, G_2, \ldots, G_r$, and*

(iii) there is at most one edge with one end in $V(G_i)$ and the other end in $V(G_j)$ for $1 \leq i < j \leq r$;

(3) G is well-k-dominated.

If G is any connected graph, then the conditions given in (2) of Proposition 12.11 are easily seen to be sufficient for G to be well-k-dominated. That the conditions are not necessary for any connected graph G, may be seen by considering the graph H_k constructed as follows. Let T be a complete binary tree of height k in which every leaf is at level k (and so T has order $2^{k+1} - 1$). Let T_1 and T_2 be two (disjoint) copies of T. Finally, let H_k be obtained from T_1 and T_2 by inserting a 1-factor (a matching) between the corresponding leaves of T_1 and T_2. (Figure 12.2 shows the graphs H_1 and H_2.) Then H_k is well-2k-dominated with $\gamma_{2k}(H_k) = 2$, but H_k does not satisfy condition (2) of Proposition 12.11.

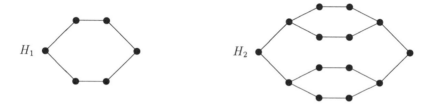

Figure 12.2: The graphs H_1 and H_2.

Corollary 12.12 *If G is a connected block graph, then $\gamma_{2k}(G) = \gamma_k(G)$ if and only if G is well-2k-dominated.*

12.2.6   Distance domination critical graphs

As Sumner [47] points out, "Graphs which are minimal or critical with respect to a given property frequently play an important role in the investigation of that property. Not only are such graphs of considerable interest in their own right, but also a knowledge of their structure often aids in the development of the general theory." In this subsection we consider graphs which are critical with respect to their k-domination number. We examine the effects on $\gamma_k(G)$ when G is modified by deleting a vertex. Unless otherwise stated, the results of this subsection are from [36].

Brigham, Chinn, and Dutton [11] define a vertex v of a graph G to be *critical* if $\gamma(G - v) < \gamma(G)$. The graph G is *vertex domination-critical* (or γ-*critical*) if each vertex is critical. For $k \geq 1$ an integer, a vertex v of a graph G is defined in

[36] to be *k-critical* if $\gamma_k(G - v) < \gamma_k(G)$. The graph G is *vertex k-domination-critical* (or γ_k-*critical*) if each vertex of G is k-critical. If G is γ_k-critical and $\gamma_k(G) = \ell$, we say G is *(γ_k, ℓ)-critical*. For example, the graphs G_2 and G_3 of Figure 12.3 are $(\gamma_2, 2)$-critical and $(\gamma_3, 2)$-critical, respectively. Further, for integers $k \geq 1$ and $\ell \geq 2$, the cycle $C_{(\ell-2)(2k+1)+1}$ is (γ_k, ℓ)-critical. Note that γ_1-critical graphs are vertex domination-critical graphs.

Figure 12.3: The graphs G_2 and G_3.

As pointed out in [11], vertex domination-critical graphs can be used to model multiprocessor networks. Similarly, γ_k-critical graphs can serve as models for multiprocessor networks in which a subset of processors (represented by an k-dominating set) can transmit messages to all remaining processors in at most k time units (where a time unit is the time it takes for a message to be sent between adjacent processors). These γ_k-critical networks have the desirable characteristics that any processor can be in a minimum set of 'k-dominating' processors and the failure of any processor leaves a network which requires one fewer 'dominating' processors.

We begin with some basic results.

Proposition 12.13 *If G is a γ_k-critical graph, then for every pair of distinct vertices u and v of G, $N_k[v] \not\subseteq N_k[u]$.*

The next result shows that $\delta_k(G)$ cannot be too small for a nontrivial connected γ_k-critical graph G.

Proposition 12.14 *If G is a nontrivial connected γ_k-critical graph, then $\delta_k(G) \geq 2k$.*

The following result establishes an upper bound for the order of a γ_k-critical graph in terms of its maximum k-degree and k-domination number.

Proposition 12.15 *If G is a graph of order n that contains an k-critical vertex, then $n \leq (\Delta_k(G) + 1)(\gamma_k(G) - 1) + 1$.*

That the bound of Proposition 12.15 is best possible may be seen by considering the infinite class of γ_k-critical graphs $G_{s,r}$, where $r \geq 2$, $s = 2\ell$ for some positive integer ℓ and $G_{s,r} \cong C_{(r-1)(sk+1)+1}^{k\ell}$ (that is, the $k\ell$th power of the cycle on $(r-1)(sk+1)+1$ vertices). Then $\gamma_k(G_{s,r}) = r$ while $\gamma_k(G_{s,r} - v) = r - 1$ for every vertex v of $G_{s,r}$. This examples serves to illustrate the existence of (γ_k, r)-critical graphs of connectivity s for every integer $s \geq 2$.

Proposition 12.16 *For an integer $k \geq 2$, if G is a $(\gamma_k, 2)$-critical graph of order n, then $\delta_k(G) = \Delta_k(G)$ and $n = \delta_k(G) + 2$.*

Propositions 12.14 and 12.16 yield the following lower bound on the order of a $(\gamma_k, 2)$-critical graph.

Corollary 12.17 *For an integer $k \geq 2$, if G is a $(\gamma_k, 2)$-critical graph of order n, then $n \geq 2k + 2$.*

The following question is posed in [36]: "For integers $k \geq 2$ and $\ell \geq 2$, is it true that if G is a (γ_k, ℓ)-critical graph of order n, then $n \geq (\delta_k(G)+1)(\gamma_k(G) - 1) + 1$?" If this question can be answered in the affirmative, then this result and Proposition 12.14, imply that $n \geq (2k + 1)(\ell - 1) + 1$ for any nonempty (γ_k, ℓ)-critical graph G of order n. Moreover, the cycle $C_{(\ell-1)(2k+1)+1}$ shows that this bound is attainable. Corollary 12.17 solves this problem in the special case of $(\gamma_k, 2)$-critical graphs.

Proposition 12.18 *Let $k \geq 2$ and $\ell \geq 2$ be integers. Then G is a (γ_k, ℓ)-critical graph if and only if G^k is a (γ, ℓ)-critical graph.*

We next describe a construction technique which can be employed to produce large classes of γ_k-critical graphs. We note that a graph is γ_k-critical if and only if each of its components is γ_k-critical. A similar statement holds for the blocks of G. Suppose H and G are nonempty graphs. Let u and w be two non-isolated vertices of H and G, respectively. Then $(H \cdot G)(u, w : v)$ denotes the graph obtained from H and G by identifying u and w in a vertex labeled v.

Proposition 12.19 *If u and w are two non-isolated vertices of two nonempty graphs H and G, respectively, then*

$$\gamma_k(H) + \gamma_k(G) - 1 \leq \gamma_k((H \cdot G)(u, w : v)) \leq \gamma_k(H) + \gamma_k(G).$$

Furthermore,

(1) If v is a critical vertex of $(H \cdot G)(u, w : v)$, then $\gamma_k((H \cdot G)(u, w : v)) = \gamma_k(H)+\gamma_k(G)-1$ and u and w are critical vertices of H and G, respectively.

(2) If u and w are critical vertices of H and G, respectively, then $\gamma_k((H \cdot G)(u, w : v)) = \gamma_k(H) + \gamma_k(G) - 1$ and v is a critical vertex of $(H \cdot G)(u, w : v)$.

Proposition 12.20 *Let H, G, u, w and v be as in the hypothesis of Proposition 12.19. Then $(H \cdot G)(u, w : v)$ is γ_k-critical if and only if H and G are both γ_k-critical.*

Next we investigate the relationship between the k-domination number of a graph and the k-domination number of its blocks.

Proposition 12.21 *A graph G is γ_k-critical if and only if each block of G is γ_k-critical. Further, if G is γ_k-critical with blocks G_1, G_2, ..., G_t, then*

$$\gamma_k(G) = \sum_{i=1}^{t} \gamma_k(G_i) - (t - 1).$$

As an illustration of Proposition 12.21, let B_1, B_2, ..., B_{2k+3} be $(2k + 2)$-cycles. For each i $(1 \leq i \leq 2k + 2)$ let u_i be a vertex of B_i and let $w_1, w_2, \ldots, w_{2k+2}$ be the vertices of B_{2k+3}. Let G be obtained by identifying u_i and w_i for $1 \leq i \leq 2k + 2$. Then B_1, B_2, ..., B_{2k+3} are the blocks of G. Since each B_i is $(\gamma_k, 2)$-critical, we know from Proposition 12.21 that G is γ_k-critical. Furthermore, $\gamma_k(G) = \sum_{i=1}^{2k+3} \gamma_k(B_i) - (2k + 2) = 2k + 4$. This examples serves to illustrate the existence of γ_k-critical graphs that contain cut-vertices.

Attempts to characterize γ_k-critical graphs have been unsuccessful. The following result shows that it is not possible to do so in terms of forbidden subgraphs.

Proposition 12.22 *For any graph G there is a (γ_k, ℓ)-critical graph H containing G as an induced subgraph.*

12.3 The Total Distance Domination Number

12.3.1 Introduction

Let $k \geq 1$ be an integer and let $G = (V, E)$ be a graph. A set $\mathcal{D}$ of vertices of G is defined in [35] to be a *total k-dominating set* of G if every vertex in V is within distance k from some vertex of $\mathcal{D}$ other than itself. The minimum cardinality among all total k-dominating sets of G is called the *total k-domination number* of G and is denoted by $\gamma_k^t(G)$. A total k-dominating set of cardinality $\gamma_k^t(G)$ is called a γ_k^t-set. We note that the parameter $\gamma_k^t(G)$ is defined only for graphs with no isolated vertex. Observe that $\gamma_t(G) = \gamma_1^t(G)$, where $\gamma_t(G)$ is the total domination number. For the graph shown in Figure 12.1, $T = \{v_2, v_3, v_7, v_8\}$ is a total 2-dominating set with $\gamma_2^t(G) = |T|$.

This concept of total distance domination in graphs finds applications in many situations and structures which give rise to graphs. Consider, for example, the facility location problem discussed in Subsection 12.2.1. For practical reasons it may be desirable that each facility be sited within k blocks of some other facility (for instance to cope with emergencies and breakdowns), in which case the use of a total k-dominating set of minimum cardinality is indicated. Corresponding applications to the design of computer networks and defense systems exist. For example, the problem of finding total k-dominating sets has potential applications to storage location problems in a computer network. Suppose G is a graph that models a multiprocessor computer network where the vertices of G represent processors and an edge of G indicates that the processors corresponding to its endvertices can communicate directly. The same data are to be stored at each member of a subset S of these processors so that any processor in the rest of the network can be sent this information in at most k time units (where a time unit is the time it takes for the data to be sent between adjacent processors). Furthermore, we wish to select S in such a way so that if a processor should lose its data due to failure, then it can obtain the data from another element of S in at most k time units. (We assume here that at most one of the elements of S will fail at any one time.) The problem of finding such a set S corresponds to the problem of finding a total k-dominating set of vertices of G, and an optimal solution to the problem has cardinality $\gamma_k^t(G)$.

12.3.2   Algorithmic and complexity results

As is the case for the k-domination number of a graph, it appears to be a computationally difficult problem to determine the total k-domination number of a graph. There is no known efficient algorithm for this purpose. However, the following result from [35] provides a sharp upper bound for the total k-domination number of a connected graph.

Proposition 12.23 *For an integer $k \geq 2$, if G is a connected graph of order n, then $\gamma_k^t(G) = 2$ if $2 \leq n \leq 2k$ and $\gamma_k^t(G) \leq 2n/(2k+1)$ if $n \geq 2k+1$.*

If $2 \leq n \leq 2k$, then any central vertex of G and some other vertex form a total k-dominating set of G. For $n \geq 2k+1$, an algorithm for finding a total k-dominating set D_k of vertices of a connected graph G of order n such that $|D_k| \leq 2n/(2k+1)$ may be found in [38]. This algorithm is based on the proof of Proposition 12.23 given in [35].

The bounds of Proposition 12.23 are best possible in the following sense: If n, k, and ℓ are integers satisfying $k \geq 2$ and $2 \leq \ell \leq 2n/(2k+1)$, then there is a graph G of order n with $\gamma_k^t(G) = \ell$. Let $G = J(n, k, \ell)$ be obtained by joining an endvertex of each of $\ell - 1$ disjoint paths of order $2k+1$ to a single vertex of $K_{n-2k(\ell-1)}$. Then $|V(G)| = n$ and $\gamma_k^t(G) = \ell$. In the case where $\ell = 2n/(2k+1)$, the graph $J(n, k, \ell)$ is necessarily isomorphic to P_{2k+1} and $\ell = 2$. However, there is a general class of graphs for which the upper bound in Proposition 12.23 is attained in the case $n \geq 2(k+1)$. Let $\ell \geq 2$ be any integer and H any connected graph of order ℓ. Join an endvertex of a path of order $2k + 1$ to every vertex of H to obtain a G of order n. Then $\gamma_k^t(G) = 2\ell = 2n/(2k + 1)$.

12.3.3 Nordhaus-Gaddum-type results

We now turn to the Nordhaus-Gaddum-type results with respect to the total k-domination number which were obtained in [35].

Proposition 12.24 *For integers $n \geq k + 1 \geq 2$, if G and $\bar{G}$ are connected graphs of order n, then $\gamma_k^t(G) + \gamma_k^t(\bar{G}) = 4$ and $\gamma_k^t(G)\gamma_k^t(\bar{G}) = 4$ if $n \leq 2k + 1$, while $4 \leq \gamma_k^t(G) + \gamma_k^t(\bar{G}) \leq 2n/(2k - 1) + 2$ and $4 \leq \gamma_k^t(G)\gamma_k^t(\bar{G}) \leq 4n/(2k + 1)$ if $n \geq 2k + 2$.*

Proposition 12.25 *For integers $n \geq k + 1 \geq 2$, if G and $\bar{G}$ are graphs of order n with no isolated vertices, then $\gamma_k^t(G) + \gamma_k^t(\bar{G}) \leq n + 2$ and $\gamma_k^t(G)\gamma_k^t(\bar{G}) \leq 2n$.*

The bounds given in Proposition 12.24 are best possible in the following sense. If $2 \leq \ell \leq 2n/(2k + 1)$, then the graph $J(n, k, \ell)$ defined in Section 12.3.2 is such that $\gamma_k^t(G) + \gamma_k^t(\bar{G}) = \ell + 2$ and $\gamma_k^t(G)\gamma_k^t(\bar{G}) = 2\ell$. The upper bounds given in Proposition 12.25 can be attained, for instance if n is even and G is the disjoint union of $n/2$ copies of K_2.

12.4 Independent Distance Domination

12.4.1 Introduction

The fact that every maximal independent set of vertices in a graph is also a minimal dominating set motivated Cockayne and Hedetniemi [17] in 1974 to initiate the study of '*independent domination*' in graphs. The independent domination number $i(G)$ has received considerable attention in the literature.

In this subsection, we consider two extensions of the definition of independent dominating sets in graphs. The first extension is the *independent k-domination number $id(k, G)$* of G, which is defined in [29] as the minimum cardinality among all independent k-dominating sets of G. Thus S is an independent 1-dominating set of G if and only if S is an independent dominating set of G. Hence $id(1, G) = i(G)$.

The second extension is the *k-independent domination number $i_k(G)$* of G. A set S of vertices of a graph G is defined in [35] to be *k-independent* in G if every vertex of S is at distance at least $k + 1$ from every other vertex of S in G. Furthermore, S is defined to be an *k-independent dominating set* of G if S is k-independent and k-dominating in G. The *k-independent domination number $i_k(G)$* of G is the minimum cardinality among all k-independent dominating sets of G. Hence 1-independent dominating sets of G are independent dominating sets of G and $i_1(G) = i(G)$. For the graph G of Figure 12.1, $S = \{v_3, v_7\}$ is a 3-independent dominating set of G with $|S| = i_3(G)$. This concept of independent distance domination in graphs finds applications in many situations and structures which give rise to graphs. Consider, for example, the facility location problem discussed in Subsection 12.2.1. To avoid interference and contamination, it may be required that no two facilities be within k blocks of each other, and facilities should then be sited at points corresponding to vertices in a minimum k-independent dominating set. We consider each extension in turn.

12.4.2 The independent k-domination number

In [29], it is shown that the decision problem corresponding to the problem of computing $id(k, G)$ is NP-complete, even when restricted to bipartite graphs, by demonstrating a polynomial time reduction from the decision problem **Independent Dominating Set** ($IDOM$). The decision problem $IDOM$ for the independent domination number of a graph is known to be NP-complete (see Garey and Johnson [28]), and remains NP-complete for the class of bipartite graphs, as shown by Corneil and Perl [19].

Since the problem of computing $id(k, G)$ appears to be a difficult one, it is desirable to find good upper bounds on this parameter. For $k \geq 2$, Beineke and Henning [7] established the following upper bound on $id(k, G)$ for a connected graph G.

Proposition 12.26 *For $k \geq 2$, if G is a connected graph of order $n \geq k$, then $id(k, G) \leq n/k$, and this bound is asymptotically best possible.*

That the bound given in Proposition 12.26 is in a sense best possible, may be seen by considering the connected graph G constructed as follows: For ℓ and b very large integers, let G be obtained from a complete graph on b vertices by attaching to each of its vertices ℓ disjoint paths of length k. (The graph G is shown in Figure 12.4.) Then $id(k, G) = (b-1)\ell + 1$ and $n = |V(G)| = b(k\ell + 1)$, so

$$\frac{id(k, G)}{n} = \frac{b\ell - \ell + 1}{bk\ell + b} = \frac{1 - \frac{1}{b} + \frac{1}{b\ell}}{k + \frac{1}{\ell}} \xrightarrow{b, \ell \to \infty} \frac{1}{k}.$$

If we restrict our attention to trees, then Meir and Moon [42] established the following upper bound on $id(k, T)$.

Proposition 12.27 *For $k \geq 1$, if T is a tree of order $n \geq k+1$, then $id(k, T) \leq n/(k+1)$, and this bound is sharp.*

That the bound given in Proposition 12.27 is sharp, may be seen by considering a tree T_k of order n obtained from a path on b vertices by attaching a path of length k to each vertex of the path. Then $id(k, T_k) = b = n/(k+1)$.

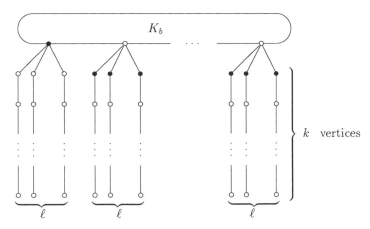

Figure 12.4: The graph G.

The following result in [29] improves on that of Proposition 12.26. The proof is constructive in that it constructs an independent k-dominating set of sufficiently small cardinality.

Theorem 12.28 *For $k \geq 1$, if $G = (V, E)$ is a connected graph of order $n \geq k + 1$, then*

$$id(k, G) \leq \frac{n + k + 1 - 2\sqrt{n}}{k}$$

and this bound is sharp.

Proof. The following paragraphs outline the proof. (The detailed proof may be found in [29].) Let $\mathcal{D} = \{v_1, \ldots, v_b\}$ be a minimum k-dominating set of G that satisfies the statement of Proposition 12.4. We introduce the following notation. For $i = 1, \ldots, b$, let $W_i = \{w \in V - \mathcal{D} \mid d(v_i, w) = k \text{ and } N_k(w) \cap \mathcal{D} = \{v_i\}\}$, $X_i = \{x \in V \mid x \text{ belongs to a } v_i\text{-}w \text{ path of length } k \text{ for some } w \in W_i\}$ and $U_i = \{u \in V \mid u \text{ is the vertex adjacent to } v_i \text{ on some } v_i\text{-}w \text{ path of length } k \text{ for some } w \in W_i\}$. By our choice of $\mathcal{D}$, we know that $W_i \neq \emptyset$ for all i. Hence $|X_i| \geq k + 1$ and $v_i \in X_i$ for all i. Furthermore, $X_i \cap X_j = \emptyset$ for $1 \leq i < j \leq b$. Hence, since $\mathcal{D}$ k-dominates V, we can partition V into sets $V_1, \ldots, V_b$, where each V_i induces a *connected* graph of radius at most k, and where $X_i \subseteq V_i$ and v_i k-dominates V_i. Let S be the set produced by the following algorithm.

Algorithm 12.29 :
Begin

    1. $S \leftarrow \emptyset$, $I \leftarrow \{1, \ldots, b\}$ and $i \leftarrow 1$.

    2. $S \leftarrow S \cup \{v_i\}$ and $I \leftarrow I - \{i\}$.

    3. **For** $j \in I$ **do**

        If $v_i v_j \in E$ **then**

        3.1. **For** $u \in U_j$ **do**

            If (*$d(u, w) = k - 1$ for some $w \in W_j$ satisfying $d(w, S) > k$*)
                then $S \leftarrow S \cup \{u\}$.

        End for
        3.2. $I \leftarrow I - \{j\}$.

    End for

    4. **If** $I = \emptyset$, **then** *continue; otherwise, let $i' \in I$, set $i \leftarrow i'$, and return to Step 2.*

    5. **If** S *k-dominates V,* **then** *stop; otherwise, continue.*

    6. $T \leftarrow \{t \in V \mid d(t, S) > k\}$.

    7. **For** $t \in T$ **do**

        7.1. **If** $t \in V_j$ **then**

            $u_t \leftarrow$ (*the vertex adjacent to v_j on some v_j-t path of length k in $\langle V_j \rangle$*)

        7.2. **If** $d(t, S) > k$ **then** $S \leftarrow S \cup \{u_t\}$.

    End for

End

    For $i = 1, \ldots, b$, let $|V_i| = n_i$. By the Pigeonhole Principle, at least one of the sets V_i contains at least n/b vertices. Relabeling the sets if necessary, we may assume that $n_1 \geq n/b$. For $i = 1, \ldots, b$, let $S_i = S \cap V_i$. Then (see [29]) for each $i = 1, \ldots, b$, $|S_i| \leq (n_i - 1)/k$. Furthermore (see [29]), the set S produced by Algorithm 12.29 is an independent k-dominating set of G. Thus

$$
\begin{aligned}
id(k,G) \le |S| &= |S_1| + \textstyle\sum_{i=2}^{b} |S_i| \\
&\le |\{v_1\}| + \textstyle\sum_{i=2}^{b} (n_i - 1)/k \\
&= 1 + ((n - n_1) - (b - 1))/k \\
&\le 1 + \left(n - \tfrac{n}{b} - b + 1\right)/k \qquad \text{(since } n_1 \ge \tfrac{n}{b}) \\
&= \tfrac{1}{k}(k + n + 1 - \tfrac{n}{b} - b).
\end{aligned}
$$

The last expression is maximized with $b = \sqrt{n}$. Thus

$$
id(k,G) \le |S| \le \frac{1}{k}(k + n + 1 - 2\sqrt{n}).
$$

That this upper bound on $id(k,G)$ is sharp may be seen by considering the graph G shown in Figure 12.4 with $b = k\ell + 1$. Then, as before, $id(k,G) = 1 + k\ell^2$ and $n = |V(G)| = (k\ell + 1)^2$. Thus

$$
\frac{1}{k}(k + n + 1 - 2\sqrt{n}) = 1 + k\ell^2 = id(k,G). \quad \square
$$

12.4.3   The k-independent domination number

We begin this section with the following useful observation, the proof of which is immediate.

Proposition 12.30 *If G is a connected graph, then $i_k(G) = i(G^k)$.*

The following results are from [25].

Proposition 12.31 *For each positive integer k, there exist a connected graph G so that every spanning tree T of G satisfies $i_k(T) < i_k(G)$.*

Proposition 12.32 *For $k \in \{1,2\}$, the tree T' obtained from a tree T by joining a new vertex to some vertex of T satisfies $i_k(T') \ge i_k(T)$.*

It is somewhat surprising that Proposition 12.32 is not true for $k \ge 3$ as shown in [25]. In [25] it is shown that the decision problem corresponding to the problem of computing $i_k(G)$ is NP-complete, even when restricted to bipartite graphs, by describing a polynomial transformation from the known NP-complete decision problem **One-In-Three 3SAT** (see [28]). Since the problem of computing $i_k(G)$ appears to be a difficult one, it is desirable to find good bounds on this parameter. The following lower bound on $i_k(G)$ in terms of the maximum k-degree $\Delta_k(G)$ is presented in [25].

Proposition 12.33 *For $k \geq 1$, if G is a graph of order n and maximum k-degree $\Delta_k(G) \geq 2k$, then*

$$i_k(G) \geq \frac{n}{(\frac{k+1}{k})\Delta_k - 1}.$$

Furthermore, we have equality if and only if all components of G are either paths or cycles on $\ell \equiv 0 \mod (2k+1)$ vertices, or have order exactly $2k+1$.

The following upper bound on $i_k(G)$ in terms of the maximum k-degree $\Delta_k(G)$ is presented in [25].

Proposition 12.34 *For $k \geq 1$, if G is a graph of order n, then $i_k(G) \leq n - \Delta_k(G)$, and this bound is sharp.*

To see that the bound in Proposition 12.34 is sharp, consider the graph G obtained from a star $K_{1,\ell}$, $\ell \geq 2$, by subdividing $\ell - 1$ of the edges k times and one edge $k - 1$ times. Then $i_k(G) = \ell$, $n = |V(G)| = \ell(k+1)$ and $\Delta_k(G) = k\ell$; consequently, $i_k(G) = n - \Delta_k(G)$. The following upper bound on $i_k(G)$ in terms of the order of the graph and the integer k may be found in [26].

Proposition 12.35 *For $k \geq 1$, if G is a connected graph of order $n > \lceil (k+1)/2 \rceil$, then $i_k(G) < n/\lceil (k+1)/2 \rceil$.*

The bound given in Proposition 12.35 is shown in [26] to be asymptotically best possible.

Proposition 12.36 *For $k \geq 1$, let $i_k(n) = \max\{ i_k(G) \,|\, G$ is a connected graph of order $n\}$. Then*

$$i_k(n)/n \to 1/\lceil (k+1)/2 \rceil \text{ as } n \to \infty.$$

The results of Proposition 12.35 are extended to ℓ-connected graphs as follows (see [26]).

Proposition 12.37 *For $k \geq 1$, if G is an ℓ-connected graph of order $n > \lfloor k/2 \rfloor \ell + 1$, then $i_k(G) < n/(\lfloor k/2 \rfloor \ell + 1)$, and this bound is asymptotically best possible.*

We close this subsection by investigating good upper bounds on $i_k(T)$ for a tree T. A simple observation suffices to show that for any connected bipartite graph G of order $n \geq 2$ (and hence for any tree), $i(G) \leq n/2$. This follows because every bipartite graph, being 2-colorable, is the union of two independent sets, each of which dominates the other.

Proposition 12.38 *If T is a tree of order n, then $i(T) \leq n/2$, and this bound is sharp.*

That the bound in Proposition 12.38 is sharp, may be seen by considering the tree T obtained from the union of two (disjoint) copies of $K_{1,t}$ by joining the two vertices of degree t $(t \geq 2)$ with an edge. As a corollary of Proposition 12.35 we have the following result.

Proposition 12.39 *For $k \geq 1$, if T is a tree of order $n > \lceil (k+2)/2 \rceil$, then $i_k(G) < n/\lceil (k+2)/2 \rceil$.*

That the bound given in Proposition 12.39 is asymptotically best possible may be seen as follows. For $k \geq 2$ an even integer and for ℓ a large (positive) integer, let $T'_{k,\ell}$ be a complete ℓ-ary tree of height $\lceil (k+1)/2 \rceil$ in which every leaf is at level $\lceil (k+2)/2 \rceil$. Further, let $T_{k,\ell}$ be the tree obtained from $T'_{k,\ell}$ by attaching a path of length $\lfloor (k+1)/2 \rfloor$ to every leaf of $T'_{k,\ell}$. Then (see [26]) the tree $T_{k,\ell}$ of order n is such that $i_k(T_{k,\ell}) \to n/\lceil (k+2)/2 \rceil$ as $\ell \to \infty$. For $k \geq 3$ an odd integer and for ℓ a large (positive) integer, let $T'_{k,\ell}$ be a complete ℓ-ary tree of height $(k+1)/2$ in which every leaf is at level $(k+1)/2$. Further, let $T_{k,\ell}$ be the tree of order n obtained from $T'_{k,\ell}$ by attaching a path of length $(k+1)/2$ to every leaf of $T'_{k,\ell}$. Then (see [26]) the tree $T_{k,\ell}$ is such that $i_k(T_{k,\ell}) \to n/\lceil (k+2)/2 \rceil$ as $\ell \to \infty$.

Proposition 12.40 *For every $\epsilon > 0$, there exists a tree T of order n such that $i_k(T)/n \geq 1/\lceil (k+2)/2 \rceil - \epsilon$ for n sufficiently large.*

12.5 The Distance Irredundance Number

Let $k \geq 1$ be an integer and let $G = (V, E)$ be a graph. For $x \in X \subseteq V$, if $N_k[x] - N_k[X - \{x\}] = \emptyset$, then x is said to be *k-redundant in X*. Equivalently, x is k-redundant in X if and only if $N_k[x] \subseteq N_k[X - \{x\}]$. A set X containing no k-redundant vertex is called *k-irredundant*. Equivalently, a set X of vertices in G is k-irredundant if for every vertex $x \in X$ there exists a vertex y which is within distance k from x but at distance greater than k from every vertex of $X - \{x\}$.

Proposition 12.41 *A set X of vertices of a graph $G = (V, E)$ is a k-irredundant set of G if every vertex $v \in X$ has at least one of the following two properties:*

P_1: *There exists a vertex $w \in V - X$ such that $N_k(w) \cap X = \{v\}$;*

P_2: *The vertex v is k-adjacent to no other vertex of X.*

If X is k-irredundant and $x \in X$, the set $N_k[x] - N_k[X - \{x\}]$ is called the set of *private k-neighbours of x* and is denoted by $PN_k(x)$. It is apparent that k-irredundance is an hereditary property and that any k-independent set is also k-irredundant. The *irredundance number of G*, denoted by $ir(G)$, is the minimum cardinality taken over all maximal irredundant sets of vertices of G. The *k-irredundance number of G*, denoted by $ir_k(G)$, is the minimum cardinality taken over all maximal k-irredundant sets of vertices of G. A k-irredundant set of cardinality $ir_k(G)$ is called an ir_k-set. The concept of k-irredundance was introduced in [31].

To illustrate the concept of k-irredundance in graphs, let G_k be the graph constructed as follows. let $P: u_1, u_2, \ldots, u_{2k+3}$ and $Q: v_1, v_2, \ldots, v_{2k+3}$ be two (disjoint) paths on $2k+3$ vertices. Let G_k be the graph obtained from $P \cup Q$ by joining u_{k+2} and v_{k+2} with an edge, and then subdividing this edge $2k-1$ times. Let the resulting u_{k+2}-v_{k+2} path be denoted by $u_{2k+2}, w_1, w_2, \ldots, w_{2k-1}, v_{k+2}$. Then $X = \{u_{k+2}, u_{k+3}, v_{k+2}, v_{k+3}\}$ is a k-irredundant set of G_k with $ir_k(G) = |X|$. (Note that $\gamma_k(G) = 5$.)

When $k = 1$, the definition of an k-irredundant set coincides with the notion of an irredundant set, introduced and first studied by Cockayne, Hedetniemi, and Miller [18]. Since then, results on irredundance have been presented by many authors (see [31]).

Bollobás and Cockayne [8] showed that $ir(G) \geq n/(2\Delta - 1)$ for a graph G of order n and maximum degree $\Delta \geq 2$. In [31], this result is generalized as follows.

Proposition 12.42 *Let $k \geq 1$ be an integer and let G be a graph of order n with maximum k-degree $\Delta_k \geq (3k+1)/2$. Then $ir_k(G) \geq (2n)/(4\Delta_k - 3k + 1)$.*

12.6   Relations Involving Distance Domination Parameters

In this section we investigate various relations involving distance domination parameters.

12.6.1   Bounds relating i_k and γ_k

Since every k-independent dominating set of a graph G is an k-dominating set of G, we have the following proposition.

Proposition 12.43 *For $k \geq 1$ and for every graph G, $\gamma_k(G) \leq i_k(G)$.*

We note that strict inequality may occur in Proposition 12.43. Consider for instance the graph G constructed as follows. For $k, t \geq 1$, recall that the double star $S(t, k)$ is obtained from the (disjoint) union of two stars $K_{1,k}$ and $K_{1,t}$ by joining a vertex of maximum degree in $K_{1,k}$ to a vertex of maximum degree in $K_{1,t}$. The graph G is obtained from the double star $S(2, 2)$ by subdividing each edge $k - 1$ times. Then G is a graph for which $\gamma_k(G) = 2$ and $i_k(G) = 3$.

Allan and Laskar [1] established the following sufficient condition for the independent domination number of a graph to equal its domination number.

Proposition 12.44 *If a graph G has no induced subgraph isomorphic to $K_{1,3}$, then $\gamma(G) = i(G)$.*

In order to present the next two results, we need to define a generalization of $K_{1,r}$ for $r \geq 3$. Let G be a graph that contains a k-independent set $\mathcal{I}_r$ of r vertices and a vertex v of G that is within distance k from every vertex of $\mathcal{I}_r$. Then we shall refer to a connected subgraph of G of minimum size that contains all the vertices in $\mathcal{I}_r \cup \{v\}$ as a k-generalized $K_{1,r}$ in G. The next result follows immediately from Propositions 12.30 and 12.44 and the fact that if a graph G contains no k-generalized $K_{1,3}$, then G^k contains no induced $K_{1,3}$.

Proposition 12.45 *For $k \geq 1$, if G is a graph containing no k-generalized $K_{1,3}$, then $\gamma_k(G) = i_k(G)$.*

Bollobás and Cockayne [8] established the next result.

Proposition 12.46 *If G is a graph containing no induced subgraph isomorphic to $K_{1,r+1}$ $(r \geq 2)$, then $i(G) \leq (r - 1)\gamma(G) - (r - 2)$.*

Propositions 12.46 may be generalized as in Proposition 12.47. The proof is immediate from Propositions 12.30 and Proposition 12.44 and the fact that if a graph G contains no n-generalized $K_{1,r+1}$ $(r \geq 2)$, then G^k contains no induced $K_{1,r+1}$. However in order to characterize the extremal graphs, a direct proof of this result is provided in [39].

Proposition 12.47 *For $k \geq 1$ and $r \geq 2$, if G is a graph containing no k-generalized $K_{1,r+1}$, then $i_k(G) \leq (r - 1)\gamma_k(G) - (r - 2)$.*

The following relation between i_k and γ_k is established in [37]. The proof presented below follows that of [39].

Proposition 12.48 *For $k \geq 1$, if G is a connected graph of order $n \geq k + 1$, then $i_k(G) + k\gamma_k(G) \leq n$.*

Proof. Among all the k-dominating sets of vertices of G with cardinality $\gamma_k(G)$, let $\mathcal{D}$ be one which comes first in dictionary order. Using the notation introduced in the proof of Proposition 12.4, let Q_i denote a v_i-w_i path of length k in G for each i with $1 \leq i \leq \gamma_k(G)$. We show that this collection $\{Q_1, Q_2, \ldots, Q_{\gamma_k(G)}\}$

of paths is disjoint. If this is not the case, then for some i and j with $1 \leq i < j \leq \gamma_k(G)$, we have $V(Q_i) \cap V(Q_j) \neq \emptyset$. This implies, however, that at least one of w_i and w_j is within distance k from both v_i and v_j, which produces a contradiction. Hence the collection $\{Q_1, Q_2, \ldots, Q_{\gamma_n(G)}\}$ of paths is disjoint. Let S be a minimum k-independent dominating set of vertices of G. Then S contains at most one vertex from each path Q_i ($1 \leq i \leq \gamma_n(G)$). Let W_i be a set of k vertices of Q_i that are not in S for all i. Then $(\cup_{i=1}^{\gamma_k(G)} W_i) \cap S = \emptyset$ and $|\cup_{i=1}^{\gamma_k(G)} W_i| = k\gamma_k(G)$. Hence we have $i_k(G) + k\gamma_k(G) = |S| + |\cup_{i=1}^{\gamma_k(G)} W_i| = |S \cup (\cup_{i=1}^{\gamma_k(G)} W_i)| \leq |V(G)| = n$. $\quad\square$

That the bound given in Proposition 12.48 is best possible, may be seen by considering a graph G of order n obtained from a connected graph H by attaching a path of length k to each vertex of H. Then $i_k(G) = \gamma_k(G) = |V(H)|$ and $i_k(G) + k\gamma_k(G) = (k+1)|V(H)| = n$.

12.6.2   Bounds relating i_k and γ_k^t

Allan, Laskar and Hedetniemi [3] established the following relationship between the independent domination number and total domination number of a graph.

Proposition 12.49 *If G is a connected graph of order $n \geq 3$, then $i(G) + \gamma_t(G) \leq n$.*

In [36] Proposition 12.49 is extended for all trees of sufficiently large order.

Proposition 12.50 *For an integer $k \geq 2$, if T is a tree of order $n \geq 2k + 1$, then $i_k(T) + k\gamma_k^t(T) \leq n$.*

In [27] an algorithm is presented for finding a total k-dominating set $\mathcal{D}$ and a maximal k-independent set S in a connected graph with at least $n \geq 2k + 1$ vertices. These sets $\mathcal{D}$ and S are shown in [27] to satisfy the inequality $|S| + k|\mathcal{D}| \leq n$. Using this result, we may establish the following result from [27].

Proposition 12.51 *If G is a connected graph on $n \geq 2k + 1$ vertices, then $i_k(G) + k\gamma_k^t(G) \leq n$.*

Note that Proposition 12.51 is not an immediate consequence of Proposition 12.50. For suppose T is a spanning tree of a connected graph G. Then any total k-dominating set of T is also a total k-dominating set of G, so $\gamma_k^t(G) \leq \gamma_k^t(T)$. However, a k-independent set of T is not necessarily a k-independent set of G (see Proposition 12.31).

That the bound in Proposition 12.51 is best possible may be seen by considering the graph G of order n obtained from a star $K(1, r)$, $r \geq 1$, by subdividing each edge $2k$ times. Then $n = (2k + 1)r + 1$, $\gamma_k^t(G) = 2r$ and $i_k(G) = r + 1$, so that $i_k(G) + k\gamma_k^t(G) = n$.

12.6.3 Bounds relating γ_k and γ_k^t

Since $\gamma_k(G) \leq i_k(G)$ for all graphs G, and since $\gamma_k(G) \leq \gamma_k(T)$ and $\gamma_k^t(G) \leq \gamma_k^t(T)$ for any spanning tree of a connected graph G, an immediate corollary of Proposition 12.50 now follows.

Corollary 12.52 *If G is a connected graph on $n \geq 2k+1$ vertices, then $\gamma_k(G) + k\gamma_k^t(G) \leq n$, and this bound is best possible.*

Corollary 12.52 also follows immediately from Proposition 12.51. Tuza [50] provides an elegant proof of Corollary 12.52. That the bound given in Corollary 12.52 is best possible, may be seen by considering the graph G obtained from a connected graph H by attaching a path of length k to each vertex of H. Then $\gamma_k(G) = \gamma_k^t(G) = |V(H)|$ and $\gamma_k(G) + k\gamma_k^t(G) = (k+1)|V(H)| = |V(G)|$. However, as pointed out by Tuza [50], the bound is best possible in a much stronger sense as well; namely, its left-hand side cannot be replaced by $(1 - \epsilon)\gamma_k(G) + (k + \epsilon)\gamma_k^t(G)$, for any $\epsilon > 0$. To see this, take $r - 1$ $(r \geq 1)$ vertex-disjoint paths $T_1, \ldots, T_{r-1}$ of length k and one path of length $2k - 1$. Joining a new vertex v with one endvertex of each T_i, we obtain a tree T of order $n = (r + 1)(k + 1) - 1$, with $\gamma_k(G) = r$ and $\gamma_k^t(G) = r + 1$, hence $\gamma_k(T) + k\gamma_k^t(T) = n$ while $(1 - \epsilon)\gamma_k(T) + (k + \epsilon)\gamma_k^t(T) = n + \epsilon$.

As an immediate consequence of Corollary 12.52, we have the following result which was first established in [37].

Corollary 12.53 *If G is a connected graph on $n \geq 2k+1$ vertices, then $\gamma_k(G) + \gamma_k^t(G) \leq (2n)/(k+1)$, and this bound is best possible.*

12.6.4 Bounds relating ir_k and γ_k

In this subsection we investigate various relations involving ir_k and γ_k. By Proposition 12.1, every minimal k-dominating set is an k-irredundant set. Hence the parameters ir_k and γ_k are related as follows.

Proposition 12.54 *For every graph G, $ir_k(G) \leq \gamma_k(G)$.*

It is quite possible that $ir_k(G)$ is the best overall lower bound for $\gamma_k(G)$. The graph G_k described in Section 12.5 is a graph for which $ir_k(G) = 4$ and $\gamma_k(G) = 5$. Hence strict inequality can occur in Proposition 12.54. The next result from [31] is a generalization of a result due to Cockayne and Hedetniemi [17].

Proposition 12.55 *If $G = (V, E)$ is a graph, then $D \subseteq V$ is minimal k-dominating if and only if D is k-irredundant and k-dominating.*

The following result appears in [32].

Proposition 12.56 *Let X be a maximal k-irredundant set of vertices in a graph G. If u is a vertex of G not k-dominated by X, then for some $x \in X$, $PN_k(x) \subseteq N_k(u)$.*

In [31] various properties of graphs that have a set of vertices U not k-dominated by a maximal k-irredundant set X are exhibited, and relationships between $ir_k(G)$ and $\gamma_k(G)$ are deduced. In particular, sufficient conditions for $ir_k(G) = \gamma_k(G)$ for a graph G are presented.

For a graph G, let $\rho_k(G) = ir_k(G)/\gamma_k(G)$. The following result in [32] extends a result due to Allan and Laskar (see [2]) and Bollobás and Cockayne (see [8]) for the case $k = 1$.

Proposition 12.57 *For $k \geq 1$ an integer, $\inf\{ \rho_k(G) \,|\, G \text{ is a graph}\} = \frac{1}{2}$ with no graph attaining the infimum.*

If we restrict our attention to trees, then we have the following lower bound on all quotients $\rho_k(T)$ in which T is a tree.

Proposition 12.58 *If T is a tree, then $\rho_k(T) \geq \frac{2}{3}$, and this bound is sharp.*

Proposition 12.58 extends a result due to Damaschke [20], namely that if T is a tree, then $\rho(T) > \frac{2}{3}$. Note that the result of Proposition 12.58 differs from the case $k = 1$, where the infimum, namely $\frac{2}{3}$, is not attained. That the bound in Proposition 12.58 is sharp, may be seen as follows. For $k \geq 2$, let T be the tree obtained from a path $v_0, v_1, \ldots, v_{2k+4}$ by attaching a path $v_{k+2}, u_1, u_2, \ldots, u_{k-1}$ of length $k - 1$ to the vertex v_{k+2}. Since $\{v_{k+1}, v_{k+3}\}$ is an k-irredundant set of T, while $\{v_k, v_{k+4}, u_{k-1}\}$ is an k-dominating set of G, it is easy to verify that $ir_k(T) = 2$, while $\gamma_k(G) = 3$. It follows that $\rho_k(T) = \frac{2}{3}$.

12.6.5   Bounds relating ir_k and γ_k^t

In [31], the following relation between ir_k and γ_k^t is established. This generalizes a result due to Allan, Laskar, and Hedetniemi [3].

Proposition 12.59 *For $k \geq 1$ an integer, and any connected graph G, $\gamma_k^t(G) \leq 2ir_k(G)$, and this bound is sharp.*

That the bound in Proposition 12.59 is sharp, may be seen considering a path $v_1, v_2, \ldots, v_{4k+2}$ on $4k + 2$ vertices. Note that $\{v_{2k+1}, v_{2k+2}\}$ is an ir_k-set and $\{v_{k+1}, v_{2k+1}, v_{2k+2}, v_{3k+2}\}$ is a γ_k^t-set of this path.

Since any k-independent set is also k-irredundant, it is evident that $ir_k(G) \leq i_k(G)$ for all graphs G. Hence as a consequence of Proposition 12.51 we have the following result.

Proposition 12.60 *If G is a connected graph on $n \geq 2k + 1$ vertices, then $ir_k(G) + k\,\gamma_k^t(G) \leq n$.*

That the bound in Proposition 12.60 is sharp, may be seen considering the graph G obtained from a tree T by attaching a path of length $2k$ to every vertex of T. Then $|V(G)| = n = (2k+1)|V(T)|$, $\gamma_k^t(G) = 2|V(T)|$, and $ir_k(G) = |V(T)|$ (since $V(T)$ is an ir_k-set of G), so that $ir_k(G) + k\gamma_k^t(G) = n$ in this case.

Bibliography

[1] R. B. Allan and R. Laskar, On domination and independent domination numbers of a graph. *Discrete Math.* 23 (1978) 73–76.

[2] R. B. Allan and R. Laskar, On domination and some related concepts in graph theory. *Utilitas Math.* 21 (1978) 43–56.

[3] R. B. Allan, R. Laskar and S. T. Hedetniemi, A note on total domination. *Discrete Math.* 49 (1984) 7–13.

[4] G. Bascó and Z. Tuza, Dominating cliques in P_5-free graphs. *Period. Math. Hungar.* 21 (1990) 303–308.

[5] G. Bascó and Z. Tuza, A characterization of graphs without long induced paths. *J. Graph Theory* 14 (1990) 455–464.

[6] T. Bean, M. A. Henning, and H. C. Swart, On the integrity of distance domination in graphs. *Australas. J. Combin.* 10 (1994) 29–43.

[7] L. Beineke and M. A. Henning, Some extremal results on independent distance domination in graphs. *Ars Combin.* 37 (1994) 223–233.

[8] B. Bollobás and E. J. Cockayne, Graph-theoretic parameters concerning domination, independence, and irredundance. *J. Graph Theory* 3 (1979) 241–249.

[9] J. A. Bondy and Geng-hau Fan, A sufficient condition for dominating cycles. *Discrete Math.* 76 (1987) 205–208.

[10] K. S. Booth and J. H. Johnson, Dominating sets in chordal graphs. *SIAM J. Comput.* 11 (1982) 191–199.

[11] R. C. Brigham, P. Z. Chinn, and R. D. Dutton, Vertex domination-critical graphs. *Networks* 18 (1988) 173–179.

[12] G. J. Chang, *k-domination and graph covering problems.* Ph.D. Dissertation, School of OR and IE, Cornell University, Ithaca, N.Y. (1982).

[13] G. J. Chang and G. L. Nemhauser, The k-domination and k-stability problems on sunfree chordal graphs. *SIAM J. Algebraic Discrete Methods* 5 (3) (1984) 332–345.

[14] G. J. Chang and G. L. Nemhauser, R-domination of block graphs. *Oper. Res. Lett.* 1 (6) (1982) 214–218.

[15] G. J. Chang and G. L. Nemhauser, *The k-domination and k-stability problem on graphs.* Technical Report 540, School of Operations Res. and Industrial Eng., Cornell Univ. (1982).

[16] E. J. Cockayne and S. T. Hedetniemi, Towards a theory of domination in graphs. *Networks* 7 (1977) 247–261.

[17] E. J. Cockayne and S. T. Hedetniemi, Independence graphs. In: *Proceedings of Fifth Southeastern Conference on Combinatorics, Graph Theory and Computing* (Utilitas Mathematica, Winnipeg, 1974) 471–491.

[18] E. J. Cockayne, S. T. Hedetniemi, and D. J. Miller, Properties of hereditary hypergraphs and middle graphs. *Canad. Math. Bull.*
21 (1978) 461–468.

[19] D. G. Corneil and Y. Perl, Clustering and domination in perfect graphs. *Discrete Appl. Math.* 9 (1984) 27–40.

[20] P. Damaschke, Irredundance number versus domination number. *Discrete Math.* 89 (1991) 101–104.

[21] G. S. Domke, S. T. Hedetniemi, and R. Laskar, Generalized packings and coverings of graphs. *Congr. Numer.* 62 (1988) 259–270.

[22] M. Farber, Applications of linear programming duality to problems involving independence and domination. Technical Report 81-13, Department of Computer Science, Simon Fraser University, Canada (1981).

[23] A. Finbow, B. Hartnell, and R. Nowakowski, Well-dominated graphs: a collection of well-covered ones. *Ars Combin.* 25 (1988) 5–10.

[24] P. Fraisse, A note on distance dominating cycles. *Discrete Math.* 71 (1988) 89–92.

[25] G. Fricke, S. T. Hedetniemi, and M. A. Henning, Distance independent domination in graphs. *Ars Combin.* 40 (1995) 1–12.

[26] G. Fricke, S. T. Hedetniemi, and M. A. Henning, Asymptotic results on distance independent domination in graphs. *J. Combin. Math. Combin. Comput.* 17 (1995) 160–174.

[27] G. Fricke, M. A. Henning, O. R. Oellermann, and H. C. Swart, An efficient algorithm to compute the sum of two distance domination parameters. *Discrete Appl. Math.* 68 (1996) 85–91.

[28] M. R. Garey and D. S. Johnson, *Computers and Intractability: A Guide to the Theory of NP-Completeness.* W. H. Freeman and Company, New York (1979).

[29] J. Gimbel and M. A. Henning, Bounds on an independent distance domination parameter. *J. Combin. Math. Combin. Comput.* 20 (1996) 193–205.

[30] J. H. Hattingh and M. A. Henning, A characterization of block graphs that are well-k-dominated. *J. Combin. Math. Combin. Comput.* 13 (1993) 33–38.

[31] J. H. Hattingh and M. A. Henning, Distance irredundance in graphs. *Graph Theory, Combinatorics, and Applications*, John Wiley & Sons, Inc. 1 (1995) 529–542.

[32] J. H. Hattingh and M. A. Henning, The ratio of the distance irredundance and domination numbers. *J. Graph Theory* 18 (1994) 1–9.

[33] T. Haynes, private communication.

[34] S. T. Hedetniemi and R. C. Laskar, Bibliography on domination in graphs and some basic definitions of domination parameters. *Discrete Math.* 86 (1990) 257–277.

[35] M. A. Henning, O. R. Oellermann, and H. C. Swart, Bounds on distance domination parameters. *J. Combin. Inform. System Sci.* 16 (1991) 11–18.

[36] M. A. Henning, O. R. Oellermann, and H. C. Swart, Distance domination critical graphs. To appear in *J. Combin. Inform. System Sci.*

[37] M. A. Henning, O. R. Oellermann, and H. C. Swart, Relationships between distance domination parameters. *Math. Pannon.* 5 (1) (1994) 69–79.

[38] M. A. Henning, O. R. Oellermann, and H. C. Swart, The diversity of domination. *Discrete Math.* 161 (1996) 161–173.

[39] M. A. Henning, O. R. Oellermann, and H. C. Swart, Relating pairs of distance domination parameters. *J. Combin. Math. Combin. Comput.* 18 (1995) 233–244.

[40] F. Jaeger and C. Payan, Relations du type Nordhaus-Gaddum pour le nombre d'absorption d'un graphe simple. *C.R. Acad. Sci. Ser. A* 274 (1972) 728–730.

[41] D. Lichtenstein, Planar satisfiability and its uses. *SIAM J. Comput.* 11 (2) (1982) 329–343.

[42] A. Meir and J. W. Moon, Relations between packing and covering numbers of a tree. *Pacific J. Math.* 61 (1975) 225–233.

[43] Z. Mo and K. Williams, (r, s)-domination in graphs and directed graphs. *Ars Combin.* 29 (1990) 129–141.

[44] E. A. Nordhaus and J. W. Gaddum, On complementary graphs. *Amer. Math. Monthly* 63 (1956) 175–177.

[45] O. Ore, *Theory of Graphs*. Amer. Math. Soc. Publ., Providence (1962).

[46] P. J. Slater, R-domination in graphs. *J. Assoc. Comput. Mach.* 23 (1976) 446–450.

[47] D. P. Sumner, Critical concepts in domination. *Discrete Math.* 86 (1990) 33–46.

[48] B. C. Tansel, R. L. Francis, and T. J. Lowe, Location on networks: a survey. I. The p-center and p-median problems. *Management Sci.* 29 (1983) 282–297.

[49] J. Topp and L. Volkmann, On packing and covering numbers of graphs. *Discrete Math.* 96 (1991) 229–238.

[50] Z. Tuza, Small n-dominating sets. *Math. Pannon.* 5 (2) (1994) 271–273.

[51] Xin He and Y. Yesha, Efficient parallel algorithms for r-dominating set and p-center problems on trees. *Algorithmica* 5 (1990) 129–145.

Chapter 13

Domatic Numbers of Graphs and Their Variants: A Survey

Bohdan Zelinka
Department of Discrete Mathematics and Statistics
Technical University
Liberec, Czech Republic

13.1 Introduction

This paper is a survey of the author's and others' results concerning the domatic number of a graph and its variants. These results were published in the listed references; most of the author's results are contained as well in his doctoral dissertation [13] in Czech and also in his survey lecture [35] at the Symposium on Graph Theory in Prachatice in 1990. Therefore many of the theorems are presented without proofs; only selected proofs are included.

When the reader meets the word "domatic", he probably expects something concerning crystallographics; in this science the word "domatic" is frequently used. But the word "domatic" used here means something quite different. In the same way as the word "smog" was created from "smoke" and "fog", the word "domatic" was created from the words "dominating" and "chromatic". The domatic number of a graph is defined using the concept of dominating set; on the other hand, this concept is somewhat analogous to the chromatic number of a graph.

The most usual numerical equivalent of a graph G based on the concept of dominating set is the dominating number $\gamma(G)$ of G. The domatic number $d(G)$ of G is related to $\gamma(G)$ in the same way as the chromatic number is to the independence number.

Let G be a graph. A partition $\mathcal{D}$ of its vertex set $V(G)$ is called a *domatic partition* of G if each class of $\mathcal{D}$ is a dominating set in G. The maximum number of classes of a domatic partition of G is called the *domatic number of G* and is denoted by $d(G)$. Note that $d(G)$ is always well-defined. In every graph G, there exists at least one domatic partition, namely the partition $\{V(G)\}$ consisting of one class. Therefore there exists a domatic partition with the maximum number of classes.

The domatic number was introduced in [3] by E. J. Cockayne and S. T. Hedetniemi. Later these authors together with R. M. Dawes introduced the total domatic number [2] based on the concept of a total dominating set. For a graph G without isolated vertices, a subset D of $V(G)$ is called total dominating, if for each vertex $x \in V(G)$ there exists a vertex $y \in D$ which is adjacent to x. A partition $\mathcal{D}$ of $V(G)$ is called a *total domatic partition of G* if each class of $\mathcal{D}$ is a total dominating set of G. The maximum number of classes of a total domatic partition of G is the *total domatic number of G* and is denoted by $d_t(G)$. This concept is well-defined only for graphs without isolated vertices. If a graph contains an isolated vertex, then it has no total dominating set and thus does not have a total domatic partition. Obviously for every graph G without isolated vertices each total dominating set is a dominating set. Therefore each total domatic partition is a domatic partition and hence $d_t(G) \leq d(G)$.

Sometimes we shall consider infinite graphs instead of finite ones. Then, instead of the maximum number of classes, we take the supremum of the numbers of classes of all domatic (or total domatic) partitions.

Many properties of $d(G)$ and $d_t(G)$ are analogous, so it is reasonable to treat these concepts together. This will be done in Section 13.2. Section 13.3 will concern some numerical invariants of graphs which may be considered as variants of the domatic number.

13.2   Domatic and Total Domatic Numbers

13.2.1   Connectivity

We compare the domatic number $d(G)$ with the numerical invariants concerning the connectivity.

Let G be a connected graph. A vertex cutset of G is a subset S of $V(G)$ such that the subgraph of G induced by the set $V(G) - S$ is disconnected. An edge cutset of G is a subset R of $E(G)$ such that the factor of G with the edge set $E(G) - R$ is disconnected. The minimum number of edges of an edge cutset of G is the edge connectivity number $\kappa_1(G)$ of G. If G is not a complete graph, then the minimum number of vertices of a vertex cutset of G is the vertex connectivity number $\kappa_0(G)$ of G. If G is the complete graph K_n with n vertices, then by definition $\kappa_0(G) = n - 1$. It is necessary to distinguish between these two cases, because a complete graph contains no vertex cutset; all of its induced subgraphs are connected.

The following theorems compare $d(G)$ with these numbers.

Theorem 13.1 [15] *Let p,q be positive integers, p < q. Then there exists a graph G such that $\kappa_0(G) = p$, $d(G) = q$.*

A graph G consisting of two cliques with q vertices each, while the intersection of these cliques has p vertices, satisfies the conditions of Theorem 13.1.

Theorem 13.2 [15] *Let p,q be positive integers, p < q. Then there exists a graph G such that $\kappa_1(G) = p$, $d(G) = q$.*

A graph consisting of two disjoint cliques with q vertices each and of p independent edges joining vertices of distinct cliques has the properties of Theorem 13.2.

Theorem 13.3 [15] *Let h be a positive integer. Then there exists a graph G such that $\kappa_0(G) - d(G) = \kappa_1(G) - d(G) = h$.*

Proof. Consider the cartesian product $K_{h+2} \times K_{h+2}$. We may consider its vertex set as the set of all ordered pairs (p, q), where p, q are integers, $1 \leq p \leq h + 2$, $1 \leq q \leq h + 2$. Two distinct vertices (p_1, q_1), (p_2, q_2) are adjacent if and only if either $p_1 = p_2$ or $q_1 = q_2$. It is easy to prove that $\kappa_0(K_{h+2} \times K_{h+2}) = \kappa_1(K_{h+2} \times K_{h+2}) = 2h+2$. Now let S be a subset of $V(K_{h+2} \times K_{h+2})$, $|S| < h+2$. Then there exist numbers p_0, q_0 from $1, ..., h + 2$ such that S contains no vertex (p_0, q) for $q \in \{1, ..., h + 2\}$ and no vertex (p, q_0) for $p \in \{1, ..., h + 2\}$. Then the vertex (p_0, q_0) is not in S and is adjacent to no vertex of S. Thus the set S is not a dominating set. We have proved that a dominating set in $K_{h+2} \times K_{h+2}$ must have at least $h + 2$ vertices. Since this graph has $(h + 2)^2$ vertices, we have $d(K_{h+2} \times K_{h+2}) \leq h+2$. If $D_i = \{(i, q) | q \in \{1, ..., h+2\}\}$ for $i = 1, ..., h+2$, then $\{D_1, ..., D_{h+2}\}$ is a domatic partition of $K_{h+2} \times K_{h+2}$ and $d(K_{h+2} \times K_{h+2}) = h+2$. □

13.2.2 Cliques and clique graphs

A clique in a graph G is a subgraph of G which is complete and is not a proper subgraph of another complete subgraph of G. The minimum number of vertices of a clique in G is denoted by $c(G)$. The clique graph $K(G)$ of G is a graph whose vertices are cliques of G and in which two vertices are adjacent if and only if they have (as cliques of G) a non-empty intersection.

Theorem 13.4 [14] *There exists a graph G whose clique graph $K(G)$ is a cycle of odd length and for which $c(G) > d(G)$ holds.*

Theorem 13.4 gives a solution to a problem of E. J. Cockayne and S. T. Hedetniemi. We shall describe the mentioned graph G. Its vertex set consists of vertices $v(i, j)$ for $i \in \{1,2,3,4,5\}$ and $j \in \{1,2,3\}$ and of vertices $u(i)$ for

$i \in \{1,2,3,4,5\}$. For each $i \in \{1,2,3,4,5\}$ the set $U_i = \{v(i,1), v(i,2), v(i,3), u(i+1), v(i+1,1), v(i+1,2), v(i+1,3)\}$, where the sum $i + 1$ is taken modulo 5, induces a clique and G has no edges except the edges of these cliques. We have $c(G) = 7$ and $K(G)$ is the cycle of length 5. The graph G has 20 vertices. Therefore each partition of $V(G)$ with at least 7 classes contains a class with at most 2 vertices. But clearly no subset of $V(G)$ with at most 2 vertices is dominating in G; hence no domatic partition with at least 7 classes exists and $d(G) < 7 = c(G)$.

We conclude this subsection with the following related result.

Theorem 13.5 [14] *Let G be a graph whose clique graph $K(G)$ is a cycle of odd length and let G contain a clique with the property that each of its vertices is contained also in another clique of G. Then $c(G) \leq d(G)$.*

13.2.3 Degrees of vertices

In this section we shall show that even graphs having every vertex of large degree need not have a large domatic number. Here we shall consider also infinite graphs.

Theorem 13.6 [17] *For each non-zero cardinal number a there exists a graph G in which each vertex has degree at least a and whose domatic number is 2. If a is finite, then there exists both a finite graph and an infinite graph with this property.*

We shall describe this graph G. Choose a cardinal number $b > 3a$. Let A be a set of cardinality b, let B be the set of all subsets of A which have cardinality a. The vertex set of G is $A \cup B$. A vertex $u \in A$ is adjacent to a vertex $v \in B$ in G if and only if v is a set which contains the element u. No two vertices of A and no two vertices of B are adjacent. The degree of each vertex of B is a and the degree of a vertex of A cannot be less than a. Suppose that there exist three disjoint dominating sets D_1, D_2, D_3. First suppose that the intersections $F_1 = A \cap D_1$, $F_2 = A \cap D_2$ and $F_3 = A \cap D_3$ are all non-empty. Since $b > 3a$, at least one of the (disjoint) sets F_1, F_2, F_3, say F_1, has more than a elements. Let M be a subset F_1 of cardinality a. It is a vertex of B and is adjacent to no vertex of D_2 and to no vertex of D_3. It may be contained in at most one of the sets D_2 and D_3. Thus one of these sets is not dominating, a contradiction. Therefore at least one of the sets F_1, F_2, F_3, say F_3, is empty and then $D_3 \subseteq B$. But, as B is independent, we have $D_3 = B$ and thus $D_1 = F_1 \subseteq A$, $D_2 = F_2 \subseteq A$. Since A is also independent, the sets D_1, D_2, being non-empty and disjoint, are not dominating, which is a contradiction. On the other hand, $\{A, B\}$ is a domatic partition of G and thus $d(G) = 2$.

It may be proved that in this graph G no total domatic partition with more than one class exists. Therefore we have analogous theorem for the total domatic number.

Theorem 13.7 [15] *For each non-zero cardinal number a there exists a graph G in which each vertex has degree at least a and whose total domatic number is 1. If a is finite, then there exists both a finite graph and an infinite graph with this property.*

We add a further result, concerning infinite cardinal numbers.

Theorem 13.8 [13] *If one of the numbers $d(G)$, $d_t(G)$ for a graph G is infinite, then $d(G) = d_t(G)$.*

Proof. Let $d(G) = a$, where a is an infinite cardinal number. Then there exists a domatic partition $\mathcal{D}$ having a classes. The family $\mathcal{D}$ can be partitioned into two subfamilies $\mathcal{D}_1$, $\mathcal{D}_2$ which both have the cardinality a. There exists a bijection f of $\mathcal{D}_1$ onto $\mathcal{D}_2$. Let $\mathcal{D}_0$ be the family of all sets $D \cup f(D)$ for all $D \in \mathcal{D}_1$. This is evidently a total domatic partition of G having a classes and thus $d_t(G) \geq a = d(G)$. Since $d_t(G) \leq d(G)$, we have $d_t(G) = a$. Now assume that $d_t(G)$ is infinite. Then so is $d(G)$ and also $d_t(G) = d(G)$. □

As it was proved by E.J. Cockayne and S.T. Hedetniemi [3], always $d(G) \leq \delta(G) + 1$ and $d_t(G) \leq \delta(G)$. We conclude this subsection by listing some other results of this kind.

Theorem 13.9 [17] *Let G be a finite undirected graph with n vertices. Then*

$$d(G) \geq \lfloor n/(n - \delta(G)) \rfloor.$$

Theorem 13.10 [17] *Let G be an undirected graph whose vertex set has an infinite cardinality a and let $\overline{G}$ be its complement. If the supremum of the degrees of vertices of G is less than a, then $d(\overline{G}) = a$.*

Theorem 13.11 [18] *Let G be a finite undirected graph with n vertices. Then*

$$d_t(G) \geq \lfloor n/(n - \delta(G) + 1) \rfloor.$$

Theorem 13.12 [18] *Let G be an undirected graph whose vertex set has the infinite cardinality a and let $\overline{G}$ be its complement. If the supremum of degrees of vertices of G is less than a, then $d_t(\overline{G}) = a$.*

13.2.4 Total domatic number

Here we shall indicate some basic results concerning the total domatic number $d_t(G)$ of a graph G. All of them are from the paper [16].

Theorem 13.13 *Let G be an undirected graph without isolated vertices. Then*

$$d_t(G) \leq \lfloor n/2 \rfloor.$$

The minimum number of edges of the graph G for which the equality may occur is $\frac{1}{4}n^2$ for n even and $\frac{1}{4}(n^2 - 1)$ for n odd.

Proof. A total dominating set has at least two elements. Therefore a total domatic partition has at most $\lfloor n/2 \rfloor$ classes and $d_t(G) \leq \lfloor n/2 \rfloor$. Now suppose that $d_t(G) = \lfloor n/2 \rfloor$. If n is even, there exists a total domatic partition consisting of $n/2$ classes, each having two elements. There are $n/2$ edges joining the vertices of the same class. Further the vertices of any one class are incident to at least $n - 2$ edges joining vertices of distinct classes. Thus there are $\frac{1}{4}n(n - 2)$ such edges and the number of all edges is $n/2 + \frac{1}{4}n(n - 2) = n^2/4$. This may be realized by the cartesian product $K_{n/2} \times K_{n/2}$. If n is odd, then one of the classes has three elements and all the others have two. Thus there are $\frac{1}{4}(n^2 - 1)$ edges and this is realized by the graph obtained from $K_{(n-1)/2} \times K_{(n-1)/2}$ by adding one vertex and joining it with all the $(n - 1)/2$ vertices of one clique. $\square$

Theorem 13.14 *For any graph G without isolated vertices,*

$$\lfloor d(G)/2 \rfloor \leq d_t(G) \leq d(G).$$

The proof is based on the facts that every total dominating set is a dominating set and that the union of at least two disjoint dominating sets is a total dominating set. An example of a graph for which $d_t(G) = \lfloor d(G)/2 \rfloor$ is the complete graph K_n. An example of a graph with $d(G) = d_t(G)$ is the complete bipartite graph $K_{r,r}$.

For the next theorem, we shall again use the concept of cartesian product of graphs.

Theorem 13.15 *Let G be a graph without isolated vertices. Then*

$$d_t(G \times K_2) \leq d(G).$$

The following theorem concerns the special graph known as the wheel. The wheel W_n, where n is a positive integer, $n \geq 3$, is the graph obtained from a cycle C_n of length n by adjoining a new vertex and joining it by edges with all vertices of this cycle.

Theorem 13.16 *For every wheel W_n, $d_t(W_n) = 2$.*

13.2.5 Cacti

We present some theorems which concern the domatic number and the total domatic number of graphs of a special type, the so-called cacti. A cactus is a connected graph G with the property that each edge of G is contained in at most one cycle of G. If each edge of G is contained in exactly one cycle of G, then G is called a round cactus. Evidently each block of a cactus either is a cycle or consists of one edge. If a cactus is round, then each of its blocks is a cycle. Every tree is a cactus. If a cactus has at least two blocks, we call it non-trivial.

A terminal block of a graph G is a block which contains exactly one cutvertex of G.

In [20] a necessary and sufficient condition is given for a non-trivial cactus to have the domatic number equal to 3. We do not reproduce this theorem here, because it uses a special class of paths in a cactus whose description is complicated. Instead of it we state two theorems which are a little weaker and simpler.

Theorem 13.17 *Let G be a finite non-trivial cactus. Then either $d(G) = 2$ or $d(G) = 3$.*

Theorem 13.18 *Let G be a finite non-trivial cactus with $d(G) = 3$. Then there exits a domatic partition of G with three classes such that each vertex is adjacent to at most one vertex of the same class and each edge joining two vertices of the same class belongs to a cycle. Each terminal block of G is a cycle of length divisible by 3.*

A graph G is called *uniquely total domatic*, if there exists a unique total domatic partition of G with the maximum number of classes. The following theorems were proved in [21] and concern the total domatic numbers of cacti.

Theorem 13.19 *Let G be a non-trivial cactus in which at least one terminal block is a cycle of length congruent to 2 modulo 4. Then $d_t(G) = 1$.*

Theorem 13.20 *Let G be a round cactus with exactly one cutvertex. The total domatic number of G is 2 if and only if no block of G is a cycle of length congruent to 2 modulo 4 and either there exists at least one block of G which is a cycle of length divisible by 4, or there exists at least one block of G which is a cycle of length congruent to 1 modulo 4 and at least one block of G which is a cycle of length congruent to 3 modulo 4.*

Theorem 13.21 *Let G be a round cactus with exactly one cutvertex and let $d_t(G) = 2$. The graph G is uniquely total domatic if and only if it does not contain a cycle of length divisible by 4.*

Further in [21] the cacti with total domatic number 2 are characterized in the general case. This characterization is not reproduced here, because it requires large preparatory considerations.

13.2.6   Cubes

For every positive integer k the graph Q_k, the cube of dimension k, is defined as the graph whose vertex is the set of all k-dimensional Boolean vectors (i.e., vectors having k coordinates, each of which is 0 or 1) and in which two vertices are adjacent if and only if they differ in exactly one coordinate. We produce two theorems from [13] and [19].

Theorem 13.22 *Let k be a positive integer. Then the graph of the cube of dimension 2^k-1 and the graph of the cube of dimension 2^k both have the domatic number equal to 2^k.*

A graph is called *totally domatically full* if its total domatic number is equal to its minimum degree (it cannot be greater). Similarly a graph is called *domatically full* if its domatic number is equal to its minimum degree plus one.

Theorem 13.23 *Let k be a positive integer. Then the graph of the cube of dimension 2^k has total domatic number 2^k and is totally domatically full.*

13.2.7   Kneser graphs

The following results are from the paper [5] co-authored by J. Ivančo.

In addition to theorems concerning the domatic number and the total domatic number, we present also theorems concerning the domination number $\gamma(G)$ and the total domination number $\gamma_t(G)$, because they are closely related to the others.

Let k, n be two integers such that $2 \leq k \leq n$. Then the Kneser graph $K(n,k)$ is defined in the following way. Let M be a set, $|M| = n$. Let $V(K(n,k))$ be the set of all subsets of M which have the cardinality k. The vertex set of $K(n,k)$ is $V(K(n,k))$ and two vertices are adjacent in $K(n,k)$ if and only if they are disjoint (as sets). This concept was introduced by M. Kneser [6] and studied by L. Lovasz [8]. The particular case when $n = 2k+1$ was studied by H. M. Mulder [10] under the name "odd graph". Here we study only the case when $k = 2$.

We note that the Kneser graph $K(n,2)$ is the complement of the line graph of K_n.

Theorem 13.24 *The domination number of the Kneser graph $K(n,2)$ for every n is equal to 3. For any three two-element sets u_1, u_2, u_3, set $D = \{u_1, u_2, u_3\}$ is dominating in $K(n,2)$ for $n \geq 5$ if and only if either $u_1 \cap u_2 = u_1 \cap u_3 = u_2 \cap u_3 = \emptyset$, or $|u_1 \cup u_2 \cup u_3| = 3$.*

Proof. A dominating set in $K(n,2)$ is a set of pairs of elements of M such that each other pair is disjoint with at least one of them. For any two two-element sets there exists another two-element set which has a non-empty intersection with each of them. For three two-element sets u_1, u_2, u_3 another two-element set disjoint at least with one of them exists if and only if $u_1 \cap u_2 = u_1 \cap u_3 = u_2 \cap u_3 = \emptyset$ or $|u_1 \cup u_2 \cup u_3| = 3$.□

Theorem 13.25 *The domatic number of $K(5,2)$ is equal to 2.*

Theorem 13.26 *Let n be an integer, $n \geq 3, n \neq 5$. Then the domatic number satisfies*

$$d(K,n,2)) = \left\lfloor \frac{1}{6}n(n-1) \right\rfloor.$$

Note that the graph $K(n,2)$ has $n(n-1)/2$ vertices and its domination number is 3. Therefore, $d(K(n,2)) \le \lfloor \frac{1}{6}n(n-1) \rfloor$. The proof is based on the construction of the corresponding domatic partition.

Theorem 13.27 *For the total domination number of the Kneser graph $K(n,2)$ the following holds:*

$$\gamma_t(K(4,2)) = 6,$$
$$\gamma_t(K(5,2)) = 4,$$
$$\gamma_t(K(n,2)) = 3 \text{ for } n \ge 6.$$

The set $D = \{ u_1, u_2, u_3 \}$ is total dominating set in $K(n,2)$ if and only if $u_1 \cap u_2 = u_1 \cap u_3 = u_2 \cap u_3 = \emptyset$.

Remark. The graph $K(3,2)$ consists of three isolated vertices and therefore $\gamma_t(K(3,2))$ is not defined.

Theorem 13.28 *Let $n \ge 6$ be an integer. Then*

$$d_t(K(n,2)) = \left\lfloor \frac{1}{6}n(n-1) \right\rfloor.$$

Theorem 13.29 *The total domatic number of the graph $K(4,2)$ is equal to 1 and the total domatic number of $K(5,2)$ is equal to 2.*

13.2.8 The bichromaticity of a graph

The bichromaticity of a connected bipartite graph was introduced by F. Harary, D. Hsu and Z. Miller [4]. Here we shall compare it with the domatic number.

Let B be a connected bipartite graph on the bipartition classes C and D. A bicomplete homomorphism of B is a homomorphic mapping ϕ of B onto the complete bipartite graph $K_{r,s}$ (where r, s are positive integers) with the property that for any two vertices x, y of B the identity $\phi(x) = \phi(y)$ holds only if either $x \in C$ and $y \in C$, or $x \in D$ and $y \in D$. The maximum value $r + s$ for all graphs $K_{r,s}$ with the property that there exists a bicomplete homomorphism of B onto $K_{r,s}$ is called the *bichromaticity* of B. In this subsection only we use $\beta(B)$ to denote the bichromaticity of B.

If B is a finite bipartite graph with bipartition classes C and D then the *majority* of B is the number $\mu = max(|C|, |D|)$.

Theorem 13.30 [32] *For every connected finite bipartite graph B the inequality*

$$\beta(B) \ge \mu + \lfloor d(B)/2 \rfloor$$

holds. This inequality cannot be improved.

Corollary 13.31 *For every connected finite bipartite graph B with $d(B) \geq 3$ we have*

$$\beta(B) \geq \lfloor (3/2)d(B) \rfloor.$$

Note that if $d(B) = 2$, then the inequality need not hold. For instance, $\beta(K_{1,1}) = d(K_{1,1}) = 2$.

13.2.9 Domatically critical graphs

If we delete one edge from a graph G, then we obtain a graph G' for which the inequalities $d(G') \geq d(G) - 1$ and $d_t(G') \geq d_t(G) - 1$ hold. In other words, by deleting an edge from a graph its domatic number either is diminished by one, or remains unchanged. The same holds also for the total domatic number. If we start with a graph G and repeat the operation of deleting an edge whose deletion leaves the domatic number unchanged, then after a finite number of steps we obtain a graph in which no such edge exists; this graph is called *domatically critical*. Thus a graph G is *domatically critical* if by deleting an arbitrary edge from G a graph is obtained whose domatic number is less than that of G. The *totally domatically critical graph* is defined analogously.

We shall describe the structure of a domatically critical graph with a given domatic number.

Theorem 13.32 [22] *Let G be a domatically critical graph with the domatic number $d(G) = d$. Then the vertex set $V(G)$ of G is the union of d pairwise disjoint sets $V_1, \ldots, V_d$ with the property that for any two distinct numbers i, j from numbers $1, \ldots, d$ the subgraph $G_{i,j}$ of G induced by the set $V_i \cup V_j$ is a bipartite graph on the sets V_i, V_j, all of whose components are stars.*

Theorem 13.33 [22] *A regular domatically full graph G with n vertices and domatic number d exists if and only if d divides n; such a graph is also domatically critical. Its structure is the following: The vertex set $V(G) = \bigcup_{i=1}^{d} V_i$, $V_i \cap V_j = \emptyset$ for $i \neq j$, $|V_i| = n/d$ and the subgraph $G_{i,j}$ of G induced by the set $V_i \cup V_j$ is regular of degree 1 (for $i = 1, \ldots, d; j = 1, \ldots, d; i \neq j$).*

Analogous assertions hold for totally domatically critical graphs.

Theorem 13.34 [13] *Let G be a totally domatically full graph with n vertices and with the total domatic number $d_t(G) = d$. Then the vertex set $V(G)$ of G is the union of d pairwise disjoint sets $V_1, \ldots, V_d$ with the property that for any two distinct numbers i, j from numbers $1, \ldots, d$ the subgraph $G_{i,j}$ of G formed by all edges joining vertices from V_i with vertices from V_j (and by their endvertices) is a bipartite graph on the sets V_i, V_j, all of whose components are stars, and for each $i \in \{1, \ldots d\}$ the subgraph G_i of G induced by V_i is a forest, all of whose components are stars.*

Theorem 13.35 [13] *Let G be a totally domatically critical graph which is regular of degree d with the total domatic number d. Then its order n is an integral multiple of the number $2d$ and its vertex set $V(G)$ is the union of d pairwise disjoint sets $V_1, \ldots, V_d$ of equal cardinalities with the property that for each $i \in \{1, \ldots d\}$ the subgraph G_i of G induced by V_i is a regular graph of degree 1 and for any two distinct numbers i, j from the numbers $1, \ldots, d$ the subgraph $G_{i,j}$ of G formed by all edges joining vertices from V_i with vertices from V_j is also a regular graph of degree 1.*

Theorem 13.36 [13] *Let G be a regular totally domatically full graph. Then G has the structure described in 13.35.*

13.2.10 Hypergraphs

The concept of the domatic number may also be applied to hypergraphs. Let $H = (V, \mathcal{E})$ be a hypergraph. A dominating set in H is a subset D of $V(H)$ with the property that for each vertex $x \in V(H) - D$ there exists a vertex $y \in D$ such that there exists an edge of H which contains both x and y. Analogously as for a graph, we may define the domatic partition and the domatic number $d(H)$ for a hypergraph.

Here we shall consider only r-uniform hypergraphs (for $r \geq 2$). An *r-uniform hypergraph* is a hypergraph in which each edge contain exactly r vertices.

We shall use also the concept of a simple hyperforest. A hypergraph H is called a *simple hyperforest* if it has no isolated vertices and for any k edges $E_1, \ldots E_k$ of G, where $k \geq 3$ and $E_i \cap E_{i+1} \neq \emptyset$ for $i = 1, \ldots k - 1$, the intersection $E_1 \cap E_k = \emptyset$. A domatically full hypergraph is defined analogously to a domatically full graph and the following theorem is used.

Theorem 13.37 [23] *Let $H = (V, \mathcal{E})$ be an r-uniform hypergraph with domatic number $d(H)$ and minimum degree $\delta(H)$. Then*

$$d(H) \leq \delta(H)(r-1) + 1.$$

Proof. If a vertex x is incident with $\delta(H)$ edges, then it is adjacent to at most $\delta(H)(r-1)$ vertices. Any class of a domatic partition must contain either x, or a vertex adjacent to x. This implies the inequality. $\square$

Thus a *domatically full hypergraph* is a hypergraph H in which $d(H) = \delta(H)(r-1) + 1$.

Theorem 13.38 [23] *For each positive integer $r \geq 2$ there exists an r-uniform hypergraph without isolated vertices such that $d(H) = 2$.*

Theorem 13.39 [23] *Let H be a simple hyperforest and let H be r-uniform for $r \geq 2$. Then there exists a domatic partition $\{D_1, \ldots D_r\}$ of H such that $|D_i \cap E| = 1$ for each $i = 1, \ldots r$ and for each edge E of H.*

Theorem 13.40 [23] *Let H be a domatically full r-uniform hypergraph which is regular of degree $\delta(H)$. Then any two edges of H have at most one vertex in common.*

Proof. Let two edges E_1, E_2 have two common vertices x and y. Then the vertex x is adjacent to at most $\delta(H)(r-1) - 1$ vertices and thus $d(H) \leq \delta(H)(r-1)$, which is a contradiction to the assumption that H is domatically full. $\square$

Theorem 13.41 [23] *If there exists a domatically full r-uniform hypergraph with n vertices regular of degree $\delta(H)$, then there exists a bipartite graph B on the vertex sets U, V such that $|U| = n$, all vertices of U have degree δ, all vertices of V have degree r and B contains no cycle of length 4.*

Proof. The vertices of U are the vertices of H, the vertices of V are the edges of H and the adjacency in B is the incidence in H. By Theorem 13.40 two edges of H have at most one vertex in common and therefore no cycle of length 4 exists. $\square$

Theorem 13.42 [23] *Let H be a domatically full r-uniform hypergraph which is regular of degree $\delta(H)$ and let n be its number of vertices. Then n is divisible by $(r-1)\delta(H) + 1$ and by $r/(r, \delta(H))$, where $(r, \delta(H))$ is the greatest common divisor of the numbers r and $\delta(H)$.*

13.3   Variants of the Domatic Number

In this section we shall a number of variations of the domatic number of a graph. Some of them were introduced by the author of this chapter, some by other authors.

13.3.1   Adomatic number and total adomatic number

The adomatic number of a graph was introduced by E. J. Cockayne and S. T. Hedetniemi in [3], the total adomatic number by the present author in [13]. A (total) dominating set in a graph is called *indivisible* if it is not a union of two disjoint (total) dominating sets. The minimum number of classes of a partition of $V(G)$, all of whose classes are indivisible (total) dominating sets, is called the *(total) adomatic number* of G. The adomatic number of G is denoted by $ad(G)$ and the total adomatic number by $ad_t(G)$. Note that here we take the minimum number of classes, not the maximum. Note also that an indivisible total dominating set need not be an indivisible dominating set. The adomatic number depends on the connectivity and the metric properties of a graph.

Theorem 13.43 [24] *Let G be a disconnected graph without isolated vertices. Then $ad(G) = 2$.*

Proof. Let G be a disconnected graph with components $H_1, ..., H_k$. In each H_i for $i = 1, ..., k$ we choose a dominating set S_i with the minimum number of vertices. Let $T_i = V(H_i) - S_i$ for $i = 1, ..., k$. Suppose that T_i is not a dominating set of H_i. Then there exists a vertex $x \in V(H_i) - T_i = S_i$ which is adjacent to no vertex of T_i. But then $S_i - \{x\}$ is a dominating set in H_i, which is a contradiction with the minimality of S_i. Therefore T_i is a dominating set in H_i for $i = 1, ..., k$. Now we define the sets D_1, D_2. If $k = 2$, then $D_1 = S_1 \cup T_2$ and $D_2 = S_2 \cup T_1$. If $k \geq 3$, then $D_1 = S_1 \cup T_2 \bigcup_{i=3}^{k} S_i$, $D_2 = S_2 \cup T_1 \bigcup_{i=3}^{k} T_i$. The set D_1 contains a dominating set of each component of G, therefore it is dominating in G. Suppose that D_1 is the union of two disjoint sets R_1 and R_2, each of which is dominating in G. Let $Q_1 = V(H_1) \cap R_1$, $Q_2 = V(H_1) \cap R_2$. Then Q_1 and Q_2 are disjoint dominating sets in H_1, which contradicts the minimality of S_1. Therefore D_1 is an indivisible dominating set in G. Analogously we prove that D_2 is also and hence $\{D_1, D_2\}$ is the required partition. $\square$

Theorem 13.44 [24] *Let G be a connected graph whose diameter is at least 3. Then $ad(G) = 2$.*

Theorem 13.45 [24] *Let a, n be integers such that $2 \leq a \leq n - 2$ or $2 \leq a = n$. Then there exists a connected graph G with n vertices such that $ad(G) = a$.*

Theorem 13.46 [24] *If G is a connected graph with $n \geq 4$ vertices, then $ad(G) \neq n - 1$.*

Now we shall consider complete graphs. An indivisible dominating set in a complete graph obviously consists of one vertex and hence evidently $ad(K_n) = n$. We shall treat the total domatic number. A subset of $V(K_n)$ is total dominating if and only if it has at least two elements. This implies that indivisible total dominating sets in K_n are exactly those subsets of $V(K_n)$ which have two or three elements and hence the following theorem.

Theorem 13.47 [13] *The total adomatic number of the complete graph K_n with n vertices, $n \geq 2$, is equal to $\lceil n/3 \rceil$.*

We conclude this subsection with the adomatic and total adomatic numbers of complements of cycles.

Theorem 13.48 [13] *Let C_n be the cycle of length n and $\overline{C_n}$ be its complement. Then*

$$ad(\overline{C_n}) = \lceil n/3 \rceil,$$
$$ad_t(\overline{C_4}) = ad_t(\overline{C_5}) = 1,$$
$$ad_t(\overline{C_n}) = \lceil n/4 \rceil \text{ for } n \geq 6.$$

13.3.2   Idomatic number

A partition of the vertex set $V(G)$ of a graph, all of whose classes are simultaneously dominating and independent sets in G is called an *idomatic partition* of G. Note that an idomatic partition need not exist in a graph. For example in a cycle C_5 of length 5 the independent dominating sets are exactly all two-element sets consisting of non-adjacent vertices. Since C_5 has an odd number of vertices, its vertex set cannot be partitioned into such sets.

If there exists at least one idomatic partition of a graph G, then the idomatic number $id(G)$ of G is the maximum number of classes of an idomatic partition of G. If no idomatic partition of G exists, then by definition $id(G) = 0$. A graph for which $id(G) \neq 0$ is called *idomatic*. In [3] a problem was suggested to characterize idomatic graphs.

Theorem 13.49 [24] *Let c, d be integers such that $2 \leq c \leq d$. Then there exists a graph G such that $id(G) = c$ and $d(G) = d$.*

We describe the structure of such a graph. The vertex set of G is the union of three disjoint sets $X = \{x_1, \ldots x_{d-c+2}\}$, $Y = \{y_1, \ldots y_{d-c+2}\}$, $Z = \{z_1, \ldots z_{c-2}\}$. Two vertices of G are adjacent if and only if they belong neither both to X, nor both to Y. The idomatic partition of G with the maximum number of classes is $\{X, Y, \{z_1\}, \ldots \{z_{c-2}\}\}$, while the domatic partition with the maximum number of classes is $\{\{x_1, y_1\}, \ldots, \{x_{d-c+2}, y_{d-c+2}\}, \{z_1\}, \ldots \{z_{c-2}\}\}$.

We conclude this section with the statement of two theorems on idomatic graphs.

Theorem 13.50 [24] *Let G be an idomatic graph. Then*

$$ad(G) \leq id(G) \leq d(G).$$

Theorem 13.51 [24] *Let G be an idomatic graph. Then*

$$\chi(G) \leq id(G)$$

and consequently

$$\chi(G) \leq d(G)$$

where $\chi(G)$ is the chromatic number of G.

13.3.3 k-domatic number

The concept of k-domination was introduced by M. Borowiecki and M. Kuzak in [1]. Let G be a graph and k be a positive integer. A k-*dominating set* in a graph G is a subset D of $V(G)$ with the property that for each vertex $x \in V(G) - D$ there exists a vertex $y \in D$ such that $d(x, y) \leq k$. Here $d(x, y)$ denotes the distance between x and y; i.e., the minimum length of a path connecting x and y in G. (Note that a more general form of k-domination was introduced by Slater in 1976 [11]). This leads to a generalization of the concept of domatic number. A *k-domatic partition* of G is a partition of $V(G)$, all of whose classes are k-dominating sets in G. The maximum number of classes of a k-domatic partition of G is called the *k-domatic number* of G and denoted by $d_k(G)$. For $k = 1$ we have $d_k(G) = d(G)$.

First we shall compare k-domatic numbers for different values of k.

Theorem 13.52 [25] *Let k, l be positive integers, $k < l$. Let G be a graph. Then $d_k(G) \leq d_l(G)$.*

Theorem 13.53 [25] *Let G be a graph with n vertices and diameter $diam(G)$. Then $d_k(G) = n$ for each $k \geq diam(G)$.*

Theorem 13.54 [25] *Let G be a graph and G' be a spanning subgraph of G. Then $d_k(G) \geq d_k(G')$.*

Theorem 13.55 [25] *Let G be a connected graph with n vertices and let k be a positive integer. Then $d_k(G) \geq min(n, k+1)$.*

The next theorem shows that this bound is sharp.

Theorem 13.56 [25] *Let P_n be the path of order n and k be a positive integer. Then $d_k(P_n) = min(n, k+1)$.*

The next theorem states the existence of graphs with all intermediate values of $d_k(G)$.

Theorem 13.57 [25] *Let k, n be two positive integers and let $2 \leq k \leq n$. Then for each integer h such that $k+1 \leq h \leq n$ there exists a graph G_h with n vertices such that $d_k(G_h) = h$.*

Proof. Let $X = \{x_1, ..., x_k\}$, $Y = \{y_1, ..., y_{h-k}\}$ and $Z = \{z_1, ..., z_{n-h}\}$ be pairwise disjoint sets of vertices. If $h = n$, then $Z = \emptyset$. The vertex set of G_h is $X \cup Y \cup Z$. The subgraph of G_h induced by X is a path with the terminal vertices x_1, x_k. The subgraph induced by $Y \cup \{x_k\}$ is the star with the center x_k. The set Z induces a complete subgraph. Further each vertex of Y is adjacent to each vertex of Z. There are exactly h vertices whose distance from x_1 is less than or equal to k (including x_1 itself), namely all vertices of $X \cup Y$. Every k-dominating set must contain at least one of them and hence $d_k(G_h) \leq h$. Let $D_1 = \{x_1\} \cup Z$, $D_i = \{x_i\}$ for $i = 2, ..., k$, $D_i = \{y_j\}$ for $j = 1, ..., h - k$. Evidently $\{D_1, ..., D_h\}$ is a k-domatic partition of G_h and therefore $D_k(G_h) = h$. $\square$

The last theorem of this subsection concerns cycles.

Theorem 13.58 [25] *Let C_n be a cycle of length n and k be a positive integer. Then*

$$d_k(C_n) = \left\lfloor \frac{n}{\lceil n/(2k+1) \rceil} \right\rfloor$$

13.3.4 k-ply domatic number

Another way to generalize the domatic number is the following. Let G be a graph and k be a positive integer. A subset D of $V(G)$ is called k-ply *dominating*, if for each $x \in V(G) - D$ there exist k vertices $y_1, \ldots y_k$ from D which are adjacent to x. A k-ply *domatic partition* of G is a partition $V(G)$, all of whose classes are k-ply dominating sets in G. The maximum number of classes of a k-ply domatic partition of G is called the k-ply *domatic number* of G and denoted by $d^k(G)$. Here k is no exponent, only an upper index; this notation is used for distinguishing from the k-domatic number. For $k = 1$ we have $d^k(G) = d(G)$, the domatic number of G. We shall reproduce only two theorems from [26]; others may be found in that paper.

Theorem 13.59 *Let G be a graph and k be a positive integer. Then*

$$d^k(G) \leq \lfloor \delta(G)/k \rfloor + 1$$

The second theorem describes the k-ply domatically critical graphs (defined analogously to domatically critical graphs). A graph G is called k-ply *domatically critical* (for a given positive integer k), if $d^k(G') < d^k(G)$ for each proper factor (spanning subgraph) G' of G.

Theorem 13.60 *Let G be a k-ply domatically critical graph for a positive integer k, let $d^k(G) = d$. Then the vertex set $V(G)$ of G is the union of pairwise disjoint sets $V_1, \ldots V_d$ such that for any i, j from the numbers $\{1, \ldots d\}$ such that $i \neq j$ the subgraph G_{ij} of G induced by the set $V_i \cup V_j$ is a bipartite graph on the sets V_i, V_j with the property that each vertex of G_{ij} has degree at least k in it and each edge of G_{ij} is incident with at least one vertex of degree k in G_{ij}.*

13.3.5 Edge-domatic number and edge-adomatic number

In addition to dominating sets of vertices, dominating sets of edges were also studied, *eg.* in [9]. An *edge-dominating set* in a graph G is a subset D of the edge set $E(G)$ of G with the property that for each edge $e \in E(G) - D$ there exists at least one edge $f \in D$ having an endvertex in common with e. The minimum number of edges of an edge-dominating set in G is the edge-domination number $\gamma'(G)$ of G.

An *edge-domatic partition* of G is a partition of $E(G)$, all of whose classes are edge-dominating sets in G. The maximum number of classes of an edge-partition of G is called the edge-domatic number of G and denoted by $d'(G)$. A total edge-dominating set in a graph G is a subset D of the edge set $E(G)$ of G with the property that for each edge $e \in E(G)$ there exists at least one edge $f \in D$ having an endvertex in common with e. The *total-edge domatic partition* and the *total edge-domatic number* $d'_t(G)$ are defined quite analogously to the preceding definitions.

Theorem 13.61 [27] *Let K_n be the complete graph with $n \geq 2$ vertices. If n is even, then*

$$d'(K_n) = n - 1.$$

If n is odd, then

$$d'(K_n) = n.$$

Proof. Let D be an edge-dominating set of K_n. Then there exists at most one vertex of K_n which is incident to no edge of D. If there were two, then the edge joining them would have no endvertex in common with any edge of D, a contradiction. Therefore $|D| \geq \lceil (n-1)/2 \rceil$, i.e., $|D| \geq (n-1)/2$ for n odd and $|D| \geq n/2$ for n even. The graph K_n with n even can be decomposed into $n - 1$ pairwise edge-disjoint linear factors; their edge sets form an edge-domatic partition of K_n and thus $d'(K_n) = n - 1$. The graph K_n with n odd can be decomposed into n pairwise edge-disjoint factors, each of which has $(n-1)/2$ independent edges; again their edge sets form an edge-domatic partition of K_n and $d'(K_n) = n$. $\square$

Theorem 13.62 [27] *Let $K_{r,s}$ be a complete bipartite graph. Then*

$$d'(K_{r,s}) = max(r, s).$$

Theorem 13.63 [27] *Let C_n be a cycle of length n. If n is divisible by 3, then $d'(C_n) = 3$. Otherwise $d'(C_n) = 2$.*

The next theorem concerns the hypercube Q_k of dimension k. Such graphs were mentioned in Section 13.2.

Theorem 13.64 [36] *Let k be a positive integer divisible by 3. Then*

$$d'(Q_k) \geq 4k/3.$$

The degree of an edge is the number of edges which have an endvertex in common with it. The minimum edge degree of G is denoted $\delta_e(G)$. Next we present an inequality involving minimum vertex degree $\delta(G)$ and minimum edge degree $\delta_e(G)$.

Theorem 13.65 [27] *For each graph G we have $\delta(G) \leq d'(G) \leq \delta_e(G) + 1$. These bounds cannot be improved.*

Theorem 13.66 [27] *For any tree T, $d'(T) = \delta_e(T) + 1$.*

Corollary 13.67 *The edge-domatic number of a path is equal to 2 and edge-domatic number of a star is equal to the number of its edges.*

If $d'(G) = \delta_e(G) + 1$, the graph G is called *edge-domatically full*. Odd graphs are a particular case of Kneser graphs and were introduced by H. M. Mulder in [10]. Let k be an integer, $k \geq 2$. Let $\mathcal{M}_k$ be the family of all subsets of the number set $\{1, \ldots, 2k-1\}$ which have the cardinality $k-1$. The odd graph O_k is the graph whose vertex set is $\mathcal{M}_k$ and in which two vertices are adjacent if and only if they are disjoint (as sets). The graph O_k is regular of degree k. For example, the graph O_2 is the triangle and O_3 is the Petersen graph. We define a related graph. Let the graph B_k^j be the bipartite graph whose bipartition classes are the family $\mathcal{M}_k^j$ of all subsets of the number set $\{1, \ldots, j\}$ which have the cardinality k and the family $\mathcal{M}_{j-k-1}^j$ of such subsets which have the cardinality $j-k-1$ and in which $X \in \mathcal{M}_k^j$ is adjacent to $Y \in \mathcal{M}_{j-k-1}^j$ if and only if $X \cap Y = \emptyset$. For example, the graph B_0^j is the star with j edges.

We shall define yet two more concepts. The concept of multiple cover of a graph G is a generalization of double cover introduced by D. A. Waller in [12]. Let G be a graph and k be a positive integer. To each vertex $v \in V(G)$ we assign a set $S(v)$ such that $|S(v)| = k$ for each vertex $v \in V(G)$ and $S(v_1) \cap S(v_2) = \emptyset$ for $v_1 \neq v_2$. Let H be a graph with the vertex set $S = \bigcup_{v \in V(G)} S(v)$ and with the following properties:

1. Each set $S(v)$ is independent in H.

2. If v_1, v_2 are adjacent vertices of G, then $S(v_1) \cup S(v_2)$ induces a linear subgraph of H.

3. If v_1, v_2 are non-adjacent vertices in G, then $S(v_1) \cup S(v_2)$ is an independent set in H.

Then H is called a *k-cover* of G. For $k = 1$ the unique k-cover of G is G itself. The k-covers of G for all k are called *multiple covers* of G.

Theorem 13.68 [33] *Let G be a finite connected edge-regular graph of degree r. Then the following two assertions are equivalent:*

1. *G is edge-domatically full.*

2. *G is a multiple cover of the odd graph $O_{r/2+1}$ or of the graph B_{s-1}^{r+1} for some s, $1 \leq s \leq r/2 - 1$.*

Theorem 13.69 [33] *Let G be a regular finite connected graph, let H be the subdivision graph of G (i.e., the obtained from G by inserting a vertex onto each edge). The graph G is domatically full if and only if H is edge-domatically full.*

The following theorem concerns the usual domatic number, but it is related to the preceding theorems.

Theorem 13.70 [33] *Let G be a finite connected regular graph of degree r. Then the following two assertions are equivalent.*

1. *G is edge-domatically full.*

2. *G is a multiple cover of the complete graph K_r.*

The next theorem gives the edge-domatic number for an odd graph.

Theorem 13.71 [29] *The edge-domination number of the odd graph O_k is equal to $\binom{(1/2)k-1}{2k-2}$ and its edge-domatic number is equal to $2k-1$ for every positive integer k.*

We conclude this section with some theorems from [37] concerning the total edge-domatic number, d'_t.

Theorem 13.72 *Let C_n be the cycle of length n. If n is divisible by 4, then $d'_t(C_n) = 2$; otherwise $d'_t(C_n) = 1$.*

Theorem 13.73 *Let $K_{r,s}$ be the complete bipartite graph. Then*

$$d'_t(K_{r,s}) = \lfloor s/2 \rfloor \text{ for } r = 1, \ s \geq 2;$$

$$d'_t(K_{r,s}) = max(r,s) \text{ for } r \geq 2, \ s \geq 2.$$

Theorem 13.74 *Let G be a finite connected edge-regular graph of degree r. Then G is totally edge-domatically full, if and only if it is obtained from a regular graph G' of degree r having a decomposition into r linear factors by inserting one vertex onto each edge.*

Theorem 13.75 *The total edge-domatic number of a wheel is 3.*

Theorem 13.76 *Let K_n be a complete graph with n vertices. Then $d'_t(K_n) \leq 3n/4$.*

13.3.6   Complementarily domatic number

The following concept relates the graph and its complement. A subset D of its vertex set $V(G)$ of a graph G is called *complementarily dominating* if for each vertex $x \in V(G) - D$ there exists vertices y, z from D such that x is adjacent to y and non-adjacent to z. A *complementarily domatic partition* of G is a partition of G, all of whose classes are complementarily dominating sets in G. The maximum number of classes of a complementarily domatic partition of G is called the *complementarily domatic number* of G and denoted by $d_{cp}(G)$. The complementarily domatic number of a graph was introduced and studied in [28].

Theorem 13.77 *Let G be a disconnected graph. Then $d_{cp}(G) = d(G)$.*

Proof. Every complementarily dominating set is a dominating set and therefore $d_{cp}(G) \leq d(G)$ for every graph G. Let G be disconnected with components $H_1, ..., H_k$. Let D be a dominating set in G and let $D_i = V(H_i) \cap D$ for $i = 1, ..., h$. Then D_i is a dominating set in H_i for $i = 1, ..., h$. Now let $x \in V(G) - D$ and $x \in V(H_i)$. If $1 \leq j \leq k$, $i \neq j$, then in G no vertex of H_i is adjacent to a vertex of H_j and thus x is not adjacent to a vertex $y \in D_j \subseteq D$. We have proved that every dominating set D in G is a complementarily dominating set in G and therefore $d_{cp}(G) = d(G)$. □

A saturated vertex is vertex which is adjacent to all other vertices of the graph. The next two theorems follow from Theorem 13.77.

Theorem 13.78 *Let a graph G contain either an isolated vertex or a saturated vertex. Then $d_{cp}(G) = 1$.*

Theorem 13.79 *Let G be a disconnected graph without isolated vertices. Then $d_{cp}(G) \geq 2$.*

The next theorem concerns the diameter.

Theorem 13.80 *Let G be a connected graph with the diameter at least 4. Then $d_{cp}(G) \geq 2$.*

We conclude this section with existence theorems.

Theorem 13.81 *There exist connected graphs G_1, G_2 with diameter 2 such that their complements have also the diameter 2 and $d_{cp}(G_1) = 1$, $d_{cp}(G_2) \geq 2$.*

Theorem 13.82 *There exist connected graphs G_1, G_2 with the diameter 3 such that their complements have also the diameter 3 and $d_{cp}(G_1) = 1$, $d_{cp}(G_2) \geq 2$.*

Theorem 13.83 *For each integer $k \geq 5$ there exists a graph G with $4k$ vertices such that $d(G) = k+1$, $d_{cp}(G) = k$.*

13.3.7 Connected domatic number

The connected domatic number of a graph was introduced by R. Laskar and S. T. Hedetniemi in [7]. A dominating set in a graph G is called connected if the subgraph of G induced by this set is connected. A *connected domatic partition* of G is a partition of $V(G)$, all of whose classes are connected dominating sets in G. The maximum number of classes of a connected domatic partition of G is called the *connected domatic number* of G and denoted by $d_c(G)$.

First we compare $d_c(G)$ with the vertex connectivity number $\kappa_0(G)$.

Theorem 13.84 [31] *Let G be a connected graph which is not complete. Then $d_c(G) \le \kappa_0(G)$.*

Now some existence theorems follow.

Theorem 13.85 [31] *For an arbitrary positive integer q, there exists a graph G such that $d(G) = d_c(G) = q$.*

Theorem 13.86 [31] *Let $n \ge 3$ an integer. For each integer k such that $1 \le k \le n - 2$ or $k = n$ there exists a graph G with n vertices such that $d_c(G) = k$. For $k = n - 1$ such a graph does not exist.*

Proof. For $k = n$ the required graph is the complete graph K_n. For $k = 1$ it is a path with n vertices. For $2 \le k \le n - 2$ it is obtained from a path with $n - k + 1$ vertices and a complete graph with $k - 1$ vertices by joining all vertices of one graph with all vertices of the other. The connected domatic partition with k classes consists of the vertex set of the path and $k - 1$ one-element sets of the vertices of the complete graph. Now suppose that there exists a graph G with n vertices and with the connected domatic number $n - 1$. Let $\mathcal{D}$ be a connected domatic partition of G with $n - 1$ classes. Then exactly one class of $\mathcal{D}$ has the cardinality 2, all the others have the cardinality 1. Let $\{u, v\}$ be the class of $\mathcal{D}$ of cardinality 2. Let $x \in V(G) - \{u, v\}$. As $\{x\}$ is a dominating set of G, the vertex x is saturated and u and v are adjacent to x. Since x was chosen arbitrarily, these vertices are adjacent to all vertices of $V(G) - \{u, v\}$. As $\{u, v\}$ is a connected dominating set, they are also adjacent to each other. Thus graph G is complete and $d_c(G) = n$, which is a contradiction. $\square$

Now some inequalities follow.

Theorem 13.87 [31] *Let G be a connected graph with order n and let n_0 be the number of its saturated vertices. Then $d_c(G) \le (n + n_0)/2$.*

Theorem 13.88 [31] *Let G be a connected graph with at least three vertices. Let e be an edge of G which is not a bridge and let G' be the graph obtained from G by deleting e. If e joins two saturated vertices of G, then $d_c(G') \ge d_c(G) - 2$, otherwise $d_c(G') \ge d_c(G) - 1$.*

Therefore we may define a *connectively domatically critical graph* as a graph G such that $d_c(G') < d_c(G)$ for each proper factor (spanning subgraph) G' of G.

Theorem 13.89 [31] *Let G be a connectively domatically critical graph and let $d_c(G) = d$. Then the vertex set of G is the union of pairwise disjoint sets $D_1, \ldots, D_d$ such that*

1. *the subgraph G_i of G induced by D_i is a tree for each $i = 1, \ldots, d$.*

2. *the subgraph G_{ij} of G with the vertex set $D_i \cup D_j$ and the edge set consisting of all edges joining a vertex from D_i with a vertex from D_j is a forest, each of whose components is a star for any i, j from the set $\{1, \ldots, d\}$, $i \neq j$.*

Our next theorem uses the notation of Theorem 13.89.

Theorem 13.90 [31] *Let G be a connectively domatically critical graph for which $d_c(G) = d$. If G is regular of degree d, then $G_i \cong K_2$ for each $i \in \{1, \ldots, d\}$ and G_{ij} consists of two components isomorphic to K_2 for any i and j from the set $\{1, \ldots, d\}$, $i \neq j$. If G is regular of degree $d-1$, then it is the complete graph K_d.*

13.3.8   Semidomatic numbers of directed graphs

In this section we consider directed graphs and introduce an analogous concept to the domatic number for them.

A subset D of the vertex set $V(G)$ of a directed graph G is an *inside-semidominating (outside-semidominating)* set in G, if for each vertex $x \in V(G) - D$ there exists a vertex $y \in D$ such that the edge xy (yx, respectively) is an edge of G. A set which is simultaneously inside-semidominating and outside-semidominating is said to be dominating. An *inside-semidomatic (outside-semidomatic) partition* of G is a partition of $V(G)$, all of whose classes are inside-semidominating (or outside-semidominating) sets in G. The maximum number of classes of an inside-semidomatic (or outside-semidomatic) partition of G is called the *inside-semidomatic (or outside-semidomatic respectively) number* of G. The inside-domatic number of G is denoted by $d^-(G)$, the outside domatic number of G by $d^+(G)$. Now we present some theorems from [30].

Theorem 13.91 *Let d_1, d_2, n be positive integers such that $d_1 \leq n/2$, $d_2 \leq n/2$. Then there exists a tournament T with n vertices such that $d^-(T) = d_1$, $d^+(T) = d_2$.*

Theorem 13.92 *Let G be a directed graph. Then the following two assertions are equivalent:*

1. *G contains a factor G_0 which is bipartite and has no sink.*

2. $d^-(G) \geq 2$.

The following theorem is dual to this one.

Theorem 13.93 *Let G be a directed graph. Then the following two assertions are equivalent:*

1. *G contains a factor G_0 which is bipartite and has no source.*

2. $d^+(G) \geq 2$.

13.3.9 Antidomatic number

We return to undirected graphs and finish with a section in which "everything is conversely".

An *antidomatic partition* of a graph G is a partition of $V(G)$ such that none of its classes is a dominating set in G. The minimum number of classes of an antidomatic partition of G is called the *antidomatic number* of G and denoted by $\bar{d}(G)$. Note that in a graph with a saturated vertex, each subset of the vertex set which contains a saturated vertex is dominating, so no antidomatic partition exists. Hence in all the theorems graphs without saturated vertices are considered. Note also that $\bar{d}(G) \geq 2$ because the partition $\{V(G)\}$ of $V(G)$ is always domatic.

We shall present some theorems which are contained in [13] and will be published in [34].

Theorem 13.94 *Let G be a disconnected graph. Then $\bar{d}(G) = 2$.*

Proof. Let G be a disconnected graph and let its components be $H_1, ..., H_k$. Let $D_1 = V(H_1)$ and $D_2 = \bigcup_{i=2}^{k} V(H_i)$. None of these sets is domatic in G and therefore $\{D_1, D_2\}$ is an antidomatic partition of G and $\bar{d}(G) = 2$. □

Theorem 13.95 *Let G be a connected graph without saturated vertices with diameter at least 3. Then $\bar{d}(G) = 2$.*

Theorem 13.96 *Let G be a connected graph without saturated vertices and minimum degree $\delta(G)$. Then $\bar{d}(G) \leq \delta(G) + 2$.*

Let $G_1 \oplus G_2$ denotes the Zykov sum of graphs, i.e., the graph obtained from the disjoint union of graphs G_1, G_2 by adding all edges joining each and every vertex of G_1 with each and every vertex of G_2.

Theorem 13.97 *Let G_1, G_2 be two graphs without saturated vertices. Then $\bar{d}(G_1 \oplus G_2) = \bar{d}(G_1) + \bar{d}(G_2)$.*

Next we give some existence theorems.

Theorem 13.98 *Let G be a connected graph of order n without saturated vertices. The antidomatic number of G is equal to n if and only if n is even and G is obtained from the complete graph with n vertices by deleting edges of a linear factor (matching).*

Proof. Let G be a graph without saturated vertices. Suppose that there exists a vertex u in G whose degree is less than or equal to $n - 3$. Then there are at least two vertices to which u is not adjacent, say v and w. Then $\{v, w\}$ is not a dominating set in G, because no vertex of this set dominates u. Since G has no saturated vertex, no one-vertex subset of $V(G)$ is dominating in G. Consider a partition $\mathcal{D}$ of $V(G)$, one of whose classes is $\{v, w\}$ and all others are one-element sets. Then $\mathcal{D}$ is an antidomatic partition of G with $n - 1$ classes and $\bar{d}(G) \leq n - 1$. Therefore if $\bar{d}(G) = n$, then $\delta(G) = n - 2$. A vertex of degree $n - 1$ would be saturated; therefore G is regular of degree $n - 2$ and is obtained from K_n for n even by deleting edges of a linear factor. $\square$

Theorem 13.99 *Let n, k be integers, $n \geq 4$. Let $2 \leq k \leq n - 2$. Then there exists a graph G with n vertices such that $\bar{d}(G) = k$.*

We conclude by stating some theorems which concern special graphs.

Theorem 13.100 *Let C_n be a cycle of length $n \geq 4$. Then $\bar{d}(C_4) = 4$, $\bar{d}(C_5) = 3$, $\bar{d}(C_n) = 2$ for $n \geq 6$.*

Theorem 13.101 *Let C_n be a cycle of length n and $\overline{C_n}$ be its complement. Then $\bar{d}(\overline{C_n}) = n/2$ for $n \equiv 0 \,(mod\,4)$, $\bar{d}(\overline{C_n}) = (n + 1)/2$ for n odd, $\bar{d}(\overline{C_n}) = n/2 + 1$ for $n \equiv 2 \,(mod\,4)$.*

Theorem 13.102 *Let G be a complete k-partite graph without saturated vertices. Then $\bar{d}(G) = 2k$.*

Bibliography

[1] M. Borowiecki and M. Kuzak, On the k-stable and k-dominating sets of graphs. *Graphs, Hypergraphs and Block Systems, Proc. Symp. Zielona Góra 1976*, Eds. M. Borowiecki, Z. Skupień and L. Szamkolowicz, Univ. Zielona Góra (1976).

[2] E. J. Cockayne, R. M. Dawes, and S. T. Hedetniemi, Total domination in graphs. *Networks* 10 (1980) 211–219.

[3] E. J. Cockayne and S. T. Hedetniemi, Towards a theory of domination in graphs. *Networks* 7 (1977) 247–261.

[4] F. Harary, D. Hsu, and Z. Miller, The bichromaticity of a tree. *Theory and Applications of Graphs*, Springer Verlag, Berlin - Heidelberg - New York (1978).

[5] J. Ivančo and B. Zelinka, Domination in Kneser graphs. *Math. Bohem.* 118 (1993) 147–152.

[6] M. Kneser, Aufgabe 300. *Jahrber. Deutsch. Math. Verein* 58 (1978).

[7] R. Laskar and S. T. Hedetniemi, *Connected domination in graphs*. Technical Report No. 414, Clemson Univ. (1983).

[8] L. Lovasz, Kneser's conjecture, chromatic number and homotopy. *J. Combin. Theory A* 25 (1978) 319–324.

[9] S. Mitchell and S. T. Hedetniemi, Edge domination in trees. In: *Proceedings of Eighth Southeastern Conference on Combinatorics, Graph Theory, and Computing* (Utilitas Mathematica, Winnipeg, 1977) 489–511.

[10] H. M. Mulder, The interval function of a graph. *Math. Centrum Amsterdam* (1980).

[11] P. J. Slater, R-domination in graphs. *J. Assoc. Comput. Mach.* 23 (1976) 446–450.

[12] D. A. Waller, Double covers of graphs. *Bull. Austral. Math. Soc.* 14 (1976) 233–248.

[13] B. Zelinka, *Některé číselné invarianty grafu. (Some numerical invariants of graphs, Czech)*. Ph.D. Dissertation, Prague (1988).

[14] B. Zelinka, On domatic numbers of graphs. *Math. Slovaca* 31 (1981) 91–95.

[15] B. Zelinka, Some remarks on domatic numbers of graphs. *Časop. pěst. mat.* 106 (1981) 373–375.

[16] B. Zelinka, Total domatic number of a graph. *Proc. Math. Liberec* (1994).

[17] B. Zelinka, Domatic number and degrees of vertices of a graph. *Math. Slovaca* 33 (1983) 145–147.

[18] B. Zelinka, Total domatic number and degrees of vertices of a graph. *Math. Slovaca* 39 (1989) 7–11.

[19] B. Zelinka, Domatic numbers of cube graphs. *Math. Slovaca* 32 (1982) 117–119.

[20] B. Zelinka, Domatic number and linear arboricity of cacti. *Math. Slovaca* 36 (1986) 41–54.

[21] B. Zelinka, Total domatic number of cacti. *Math. Slovaca* 38 (1988) 207–214.

[22] B. Zelinka, Domatically critical graphs. *Czech. Math. J.* 30 (1980) 486–489.

[23] B. Zelinka, Domatic number of uniform hypergraphs. *Arch. Math. Brno.* 21 (1985) 129–134.

[24] B. Zelinka, Adomatic and idomatic numbers of graphs. *Math. Slovaca* 33 (1983) 99–103.

[25] B. Zelinka, On k-domatic numbers of graphs. *Czech. Math. J.* 33 (1983) 309–313.

[26] B. Zelinka, On k-ply domatic numbers of graphs. *Math. Slovaca* 34 (1984) 313–318.

[27] B. Zelinka, Edge-domatic number of a graph. *Czech. Math. J.* 33 (1983) 107–110.

[28] B. Zelinka, Complementarily domatic number of a graph. *Math. Slovaca* 38 (1988) 27–32.

[29] B. Zelinka, Odd graphs. *Arch. Math. Brno.* 21 (1985) 181–188.

[30] B. Zelinka, Semidomatic numbers of directed graphs. *Math. Slovaca* 34 (1984) 371–374.

[31] B. Zelinka, Connected domatic number of a graph. *Math. Slovaca* 36 (1986) 387–391.

[32] B. Zelinka, Bichromaticity and domatic number of a bipartite graph. *Časop. pěst. mat.* 110 (1985) 113–115.

[33] B. Zelinka, Edge-domatically full graphs. *Math. Slovaca* 40 (1990) 359–365.

[34] B. Zelinka, Antidomatic number of a graph. To appear in *Arch. Math. Brno.*

[35] B. Zelinka, Domatic number of a graph and its variants (extended abstract). *Fourth Czech. Symp. on Combinatorics, Graphs and Complexity*, Eds. M. Fiedler, J. Nešetřil, Elsevier Science Publishers (1992).

[36] B. Zelinka, Domination in cubes. *Math. Slovaca* 41 (1991) 17–19.

[37] B. Zelinka, Total edge-domatic number of a graph. *Math. Bohemica* 116 (1991) 96–100.

Chapter 14

Domination-Related Parameters

E.J. Cockayne
Department of Mathematics and Statistics
University of Victoria
P.O. Box 3045
Victoria, BC V8W 3P4
Canada

C.M. Mynhardt
Department of Mathematics
University of South Africa
P.O. Box 312
Pretoria 0003
South Africa

Abstract

We discuss domination related parameters such as generalized minimal domination and maximal independence numbers, including domination sequences, Nordhaus-Gaddum type results and a generalization of Gallai's Theorem for these parameters. We also survey results on lower Ramsey numbers and irredundant Ramsey numbers and conclude with some open problems and directions for further research.

14.1 Introduction

Many types of generalizations of dominating sets and the domination number of a graph are discussed in various contributions to this volume. In this chapter we discuss a generalization which is not so much a generalization of domination concepts as of the concepts of minimality and maximality; the related generalized domination, independence and irredundance parameters arise naturally from these concepts.

We begin with the six basic parameters, namely the upper and lower domination, independence and irredundance numbers as defined in the introduction to this volume and related by the inequality chain

$$ir \leq \gamma \leq i \leq \beta \leq \Gamma \leq IR \tag{I}$$

379

(where we use β for the vertex independence number β_0.) These parameters can be considered to be the "basic building blocks" of the theory of domination, independence and irredundance. The mathematical complexity of these concepts is amply illustrated by the computational complexity of the parameters (see "Fundamentals of Domination in Graphs" [32]) and by the richness of the mathematical theory involved. One aspect which has received considerable attention is the characterization of classes of graphs for which two or more of the above numbers are equal. For example, results concerning well-dominated graphs ($\gamma = \Gamma$), graphs for which $i = \gamma$ and graphs which have all three lower (or upper) parameters equal are surveyed in [32]. None of these classes of graphs have been characterized but various sufficient conditions for equality exist. In Section 14.2 we discuss a construction which illustrates that all inequalities among the six parameters which hold for all graphs and which do not involve other parameters are already known.

The generalization of minimality and maximality is given in Section 14.3. The related inequality chains are discussed and the values of these parameters are given for paths and cycles. We also explain how this generalization leads to a generalization of the theorem of Gallai which relates maximal independent sets and minimal vertex covers of a graph.

Section 14.4 is devoted to Nordhaus-Gaddum type results, that is, results concerning the sum or product of a given parameter for a graph and its complement. The corresponding results for the classical domination parameters are surveyed in [32]. Lower Ramsey numbers (which involve the independent domination number i as well as generalized maximal independent numbers) are discussed in Section 14.5. For the sake of completeness a brief survey of irredundant Ramsey numbers is given in Section 14.6; a comprehensive survey can be found in [37]. We conclude with some open problems and directions for further research in Section 14.7.

14.2   Dii-sequences

A sequence $m_1, m_2, \cdots, m_6$ of six positive integers is called a *dii-sequence* (dii for domination, independence, irredundance) if for some graph G,

$$ir(G) = m_1, \quad \gamma(G) = m_2, \quad i(G) = m_3, \quad \beta(G) = m_4,$$

$$\Gamma(G) = m_5 \text{ and } IR(G) = m_6.$$

The investigation of the existence of dii-sequences was begun by Cockayne, Favaron, Payan and Thomason [12] and these sequences were characterized in [22].

Theorem 14.1 [22] *A sequence* $m_1, m_2, \cdots, m_6$ *of positive integers is a dii-sequence if and only if:*

(i) $m_1 \leq m_2 \leq \cdots \leq m_6$;

(ii) $m_1 = 1$ *implies that* $m_3 = 1$;

(iii) $m_4 = 1$ *implies that* $m_6 = 1$;

(iv) $m_2 \leq 2m_1 - 1$.

The necessity of (i) follows directly from the inequality chain (I) while (ii) and (iii) are trivial. The inequality $\gamma(G) \leq 2ir(G) - 1$ for any graph G was established independently in [1] and [2] and thus (iv) is necessary. The sufficiency of these four conditions follows from a construction given in [22]. For a given sequence $m_1 \leq m_2 \leq \cdots \leq m_6$, a graph G having this sequence as its inequality chain was constructed by first constructing three graphs F, H and J, and then joining F and J if $m_1 = m_2$ and H and J if $m_1 < m_2$ in a variety of ways, depending on the difference between m_3 and m_4. We give a brief description of the construction.

Let the vertex set $U(F)$ of $F = F(a, b)$ consist of disjoint subsets A and $B_1, \cdots, B_a$, where $2 \leq a \leq b$ and

$$
\begin{aligned}
A &= \{u_1, \cdots, u_a\}, \\
B_1 &= \{u_{11}, \cdots, u_{1(b-a+2)}\}, \\
B_2 &= \{u_{21}, \cdots, u_{2(b-a+2)}\} \quad \text{and} \\
B_i &= \{u_{i1}\} \quad \text{for each} \quad i = 3, \cdots, a.
\end{aligned}
$$

Add edges such that the subgraph $\langle A \rangle$ of F induced by A is complete. Also join each u_i to every vertex in B_i, $i = 1, \cdots, a$; join u_1 to u_{21} and u_2 to u_{11}, and finally join u_{1j} to u_{2j} for every $j = 1, \cdots, b - a + 2$. (See Figure 14.1.) Then

$$2 \leq a = ir(F) = \gamma(F) \leq i(F) = \beta(F) = \Gamma(F) = IR(F) = b.$$

Let the vertex set $W(H)$ of $H = H(c, d)$ consist of disjoint subsets C, $W_1, \cdots, W_{c-1}$, $Y_1, \cdots, Y_c$ and $Z_1, \cdots, Z_c$, where

$$
\begin{aligned}
C &= \{w_1, \cdots, w_c\}, \\
W_j &= \{w_{jk} \mid k = 1, \cdots, c - j\}, \quad j = 1, \cdots, c - 1, \\
Y_j &= \{y_{j1}, \cdots, y_{j(d-2c+3)}\} \text{ for } j = 1, 2 \text{ and } Y_j = \{y_{j1}\}, \quad j = 3, \cdots, c, \\
Z_j &= \{z_{j1}, \cdots, z_{j(d-2c+3)}\} \text{ for } j = 1, 2 \text{ and } Z_j = \{z_{j1}\}, \quad j = 3, \cdots, c.
\end{aligned}
$$

Join the vertices of H in such a way that $\langle C \rangle \cong K_c$ and $\langle W_j \rangle \cong K_{c-j}$, $j = 1, \cdots, c - 1$. Also join every vertex in Y_j to every vertex in Z_j, $j = 1, \cdots, c$, every vertex in Y_1 (Z_1, respectively) to every vertex in Y_2 (Z_2, respectively), and w_j to every vertex in Y_j, $j = 1, \cdots, c$. Add all edges $y_{j\ell} z_{k\ell}$ for $j, k \in \{1, 2\}$, $j \neq k$, and $\ell = 1, \cdots, d - 2c + 3$. Finally, add the edges $w_j w_{jk}$ and $w_{j+k} w_{jk}$ for each $j = 1, \cdots, c - 1$ and each $k = 1, \cdots, c - j$ to complete the construction of $H = H(c, d)$. (See Figure 14.2.) Then

$$ir(H) = c, \quad \gamma(H) = 2c - 1 \quad \text{and} \quad i(H) = \beta(H) = \Gamma(H) = IR(H) = d.$$

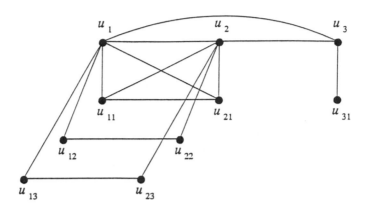

Figure 14.1: $F(a, b)$ with $a = 3$, $b = 4$.

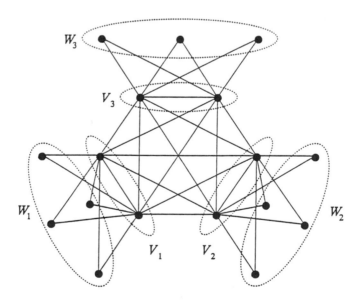

Figure 14.2: $H(c, d)$ with $c = 3$, $d = 5$.

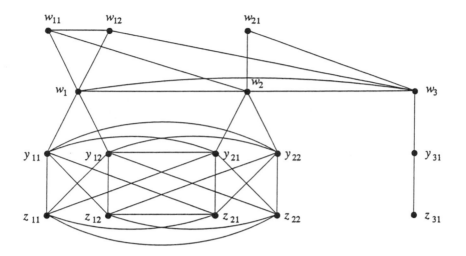

Figure 14.3: $J(r,s,t)$.

Let the vertex set of $J = J(r,s,t)$, where $r \geq 2$, consist of the disjoint subsets $V_1, \cdots, V_6$, where

$$V_i = \{v_{i1}, \cdots, v_{i(s-r+2)}\} \text{ for } i = 1, 2, 3;$$
$$V_i = \{v_{i1}, \cdots, v_{i(t-r+2)}\} \text{ for } i = 4, 5 \text{ and}$$
$$V_6 \text{ consists of } r - 2 \text{ independent vertices.}$$

Let $M = V_1 \cup V_2 \cup V_4$ and $N = V_3 \cup V_5$. Join the vertices of J in such a way that $\langle M \rangle$, $\langle V_1 \cup V_5 \rangle$ and $\langle N \rangle$ are complete. Add the edges $\{v_{1j}v_{3j} \mid j = 1, \cdots, s-r+2\}$ and $\{v_{4j}v_{5j} \mid j = 1, \cdots, t-r+2\}$ to complete the construction of $J = J(r,s,t)$. (See Figure 14.3.) Then

$$ir(J) = \gamma(J) = i(J) = \beta(J) = r, \quad \Gamma(J) = s \text{ and } IR(J) = t.$$

Consider any sequence $m_1, \cdots, m_6$ satisfying the conditions of Theorem 14.1. If $m_4 = 1$, then $G \cong K_1$ satisfies the requirements, thus we may assume that $m_4 \geq 2$. If $m_1 = 1$, then the graph obtained by joining a new vertex to every vertex of $J(m_4, m_5, m_6)$ satisfies the requirements and we may therefore also assume that $m_1 \geq 2$. Suppose firstly that $m_1 = m_2$. If $m_4 - m_3 \geq 2$, let $G = G(m_1, m_3, m_4 - m_3, m_5 - m_3, m_6 - m_3)$ be the graph obtained by joining every vertex of $J(m_4 - m_3, m_5 - m_3, m_6 - m_3)$ to every vertex in the subset A of the vertex set of $F(m_1, m_3)$. Then $m_1, \cdots, m_6$ is the dii-sequence of G. Let $G^* = G^*(m_1, m_3, 2, m_5 - m_3 + 1, m_6 - m_3 + 1)$ be the graph obtained from $G(m_1, m_3, 2, m_5 - m_3 + 1, m_6 - m_3 + 1)$ by joining every vertex of $J(2, m_5 - m_3 + 1, m_6 - m_3 + 1)$ to both u_{11} and u_{21}. Also, let $G^{**} = G^{**}(m_1, m_3, 2, m_5 - m_3 + $

$2, m_6 - m_3 + 2)$ be obtained from $G(m_1, m_3, 2, m_5 - m_3 + 2, m_6 - m_3 + 2)$ by joining every vertex in $M = V_1 \cup V_2 \cup V_4$ ($N = V_3 \cup V_5$, respectively) to every vertex in $\{u_{11}, u_{12}, u_{21}\}$ ($\{u_{11}, u_{21}, u_{22}\}$, respectively). If $m_4 - m_3 = 1$ ($m_4 - m_3 = 0$, respectively), then $m_1, \cdots, m_6$ is the dii-sequence of G^* (G^{**}, respectively).

Now suppose $m_1 < m_2$. In this case, $m_2 \leq 2m_1 - 1$. Let $p = 2m_1 - m_2 - 1$ and let $m'_j = m_j - p$, $j = 1, 2, 3$ and $m'_j = m_j - p - m'_3$ for $j = 4, 5, 6$; note that $m'_2 = 2m'_1 - 1$. If $m'_4 \geq 2$, let $G = G(m'_1, m'_3, m'_4, m'_5, m'_6)$ be the graph obtained by joining every vertex of $J(m'_4, m'_5, m'_6)$ to each vertex in the subset C of the vertex set of $H(m'_1, m'_3)$. Then $m_1, \cdots, m_6$ is the dii-sequence of $G \cup \bar{K}_p$. Let $G^* = G^*(m'_1, m'_3, 2, m'_5 + 1, m'_6 + 1)$ be obtained from $G(m'_1, m'_3, 2, m'_5 + 1, m'_6 + 1)$ by joining every vertex of $J(2, m'_5 + 1, m'_6 + 1)$ to every vertex in $Y_1 \cup Y_2 \cup \{z_{11}, z_{21}\}$, while $G^{**} = G^{**}(m'_1, m'_3, 2, m'_5 + 2, m'_6 + 2)$ is obtained from $G(m'_1, m'_3, 2, m'_5 + 2, m'_6 + 2)$ by joining every vertex in M (N, respectively) to every vertex in $Y_1 \cup Y_2 \cup \{z_{11}, z_{12}, z_{21}\}$ ($Y_1 \cup Y_2 \cup \{z_{11}, z_{21}, z_{22}\}$, respectively). If $m'_4 = 1$ ($m'_4 = 0$, respectively), then $m_1, \cdots, m_6$ is the dii-sequence of $G^* \cup \bar{K}_p$ ($G^{**} \cup \bar{K}_p$, respectively). This completes the construction of the graphs G, G^* and G^{**}. We note that this construction was amended in [40] to yield slightly smaller graphs in some cases.

Theorem 14.1 shows that the only inequalities among the six parameters which are true for all graphs and which do not involve other parameters (*e.g.* number of vertices, maximum/minimum degree) are those given in Theorem 14.1. It may be interesting to characterize the dii-sequences of particular classes of graphs, such as bipartite, chordal, planar, regular graphs, trees, etc.

14.3   Generalized Parameters

The concepts of minimality and maximality were generalized in [3] as follows. Let S be a set and P a property associated with some of the subsets of S. A subset of S with (without, respectively) property P is called a *P-set* (*$\bar{P}$-set*, respectively). A subset X of S is called a *k-minimal P-set* if X has the property P, but for all ℓ-subsets U of X, where $1 \leq \ell \leq k$, and all $(\ell - 1)$-subsets R of S, $(X - U) \cup R$ is a $\bar{P}$-set. Note that 1-minimality is the usual concept of minimality. In the case (such as in this paper) where S is the vertex set V of a graph and $X \subseteq V$ has property P if and only if X is dominating, k-minimal P-sets become *k-minimal dominating sets* of the graph. We use the acronym kMDS for a k-minimal dominating set.

Similarly, $X \subseteq S$ is a *k-maximal P-set* if X is a P-set, but for all ℓ-subsets W of X, where $0 \leq \ell \leq k - 1$, and all $(\ell + 1)$-subsets T of $S - X$, $(X - W) \cup T$ is a $\bar{P}$-set. Again, 1-maximality is the usual concept of maximality, and if S is the vertex set V of a graph, where $X \subseteq V$ is a P-set if and only if X is independent, the k-maximal P-sets are precisely the *k-maximal independent sets* of the graph. The acronym kMIS is used to denote a k-maximal independent set.

Let $\Gamma_k(G)$ ($\beta_k(G)$, respectively) denote the largest (smallest, respectively) cardinality of a kMDS (kMIS, respectively) of the graph G. Since a kMDS is also a $(k-1)$MDS, and a kMIS also a $(k-1)$MIS for $k \geq 2$, the following inequalities, first given in [3] and [34] respectively, are obvious for any graph G:

$$\Gamma(G) = \Gamma_1(G) \geq \Gamma_2(G) \geq \cdots \geq \Gamma_k(G) \geq \cdots \geq \gamma(G) \qquad \text{(II)}$$

and

$$i(G) = \beta_1(G) \leq \beta_2(G) \leq \cdots \leq \beta_k(G) \leq \cdots \leq \beta(G). \qquad \text{(III)}$$

Although $\Gamma_k(G)$ and $\beta_k(G)$ are defined for all positive integers k, it is clear that the sequences (II) and (III) contain only finitely many distinct terms. Indeed, the following results were obtained in [21] and [34], respectively.

Proposition 14.2 [21] *If $\gamma(G) = r$, then $\Gamma_k(G) = r$ for all $k \geq r+1$.*

Proposition 14.3 [34] *If $\beta(G) = t$, then for all $k = 1, 2, \cdots, t$, $\beta_k(G) \geq k$, and for all $k \geq t$, $\beta_k(G) = t$. Furthermore, if G is the complete t-partite graph $K_{1,2,\cdots,t}$, then $\beta_k(G) = k$ for all $k = 1, 2, \cdots, t$.*

The sequence

$$R : m_1 \geq m_2 \geq \cdots \geq m_k \geq \cdots \geq m_r \geq r$$

of positive integers is a *domination sequence* if there exists a graph G such that

$$\Gamma_k(G) = m_k \text{ for each } k = 1, 2, \cdots, r \text{ and } \gamma(G) = r.$$

Similarly, the sequence

$$T : p_1 \leq p_2 \leq \cdots \leq p_k \leq \cdots \leq p_t = t$$

is an *independence sequence* if there exists a graph G such that

$$\beta_k(G) = p_k \text{ for each } k = 1, 2, \cdots, t \text{ and } \beta(G) = t.$$

Domination and independence sequences were studied in [21] and [34] respectively. To characterize the former turned out to be a very tough problem and the best result obtained is the following sufficient condition.

Theorem 14.4 [21] *Consider the sequence $R : m_1 \geq m_2 \geq \cdots \geq m_k \geq \cdots \geq m_r \geq r$. Let $K = \{1, \cdots, r\}$ and $J' = \{j \in K - \{1\} : m_j > m_{j+1}\}$. If $J' = \emptyset$, define J as*

$$J = \begin{cases} \emptyset & \text{if } m_1 = m_2 \\ \{1\} & \text{otherwise.} \end{cases}$$

If $J' \neq \emptyset$, let $a_q = \max_{j \in J'}\{j\}$; let $m_1 \geq m_2 + a_q - 2$ and define J as

$$J = \begin{cases} J' & \text{if } m_1 = m_2 + a_q - 2 \\ J' \cup \{1\} & \text{otherwise}; \end{cases}$$

say $J = \{a_1, \cdots, a_q\}$ with $a_1 \leq \cdots \leq a_q$. If $m_{a_\ell} > m_{a_{\ell+1}} + a_{\ell-1}$ for all $2 \leq a_\ell \in J - \{a_q\}$, then R is a domination sequence.

Corollary 14.5 [21] *Any sequence R of the form $m_1 \geq m_2 \geq m_3 = \cdots = k$ is a domination sequence.*

The following characterization of independence sequences was obtained in [34].

Theorem 14.6 [34] *The sequence $T : p_1 \leq p_2 \leq \cdots \leq p_k \leq \cdots \leq p_t = t$ is an independence sequence if and only if $p_k \geq k$ for each $1 \leq k \leq t$.*

The necessity of this condition follows from Proposition 14.3 and the sufficiency was established by explicitly constructing a graph $G = G(T)$ for each such sequence T. Since the construction of $\bar{G}$ is easier to explain, we give the construction of $\bar{G}$ instead. Suppose the sequence $T : p_1 \leq p_2 \leq \cdots \leq p_k \leq \cdots \leq p_t = t$ with $p_k \geq k$ is given. Let $K = \{1, 2, \cdots, t\}$ and let $I = \{1\} \cup \{j \in K - \{1\} \mid p_j > p_{j-1}\}$; say $I = \{a_1, a_2, \cdots, a_q\}$ with $1 = a_1 < a_2 < \cdots < a_q$. Let $V_1, V_2, \cdots, V_{q-1}$ be disjoint sets of vertices such that $|V_j| = p_{a_j}$ for each $j \in J = \{1, 2, \cdots, q-1\}$. Note that

$$p_{a_j} \leq p_{a_{j+1}} - 1$$

by definition of $p_{a_{j+1}}$. Let r_j be the binomial coefficient given by

$$r_j = \binom{p_{a_j}}{a_{j+1} - 1}$$

and let

$$d_j = p_{a_j} - a_{j+1} + 1$$

for each $j \in J$. Construct the graph G_j as follows: Let $V(G_j) = V_j \cup W_{j1} \cup \cdots \cup W_{jr}$ (disjoint union), where $r = r_j$ and where $|W_{jk}| = t - d_j$ for each $k = 1, 2, \cdots, r$. Add edges such that the subgraphs induced by V_j and by each of the W_{jk} are complete. Note that V_j has r subsets of cardinality d_j; for each of these d_j-subsets $X_{j1}, X_{j2}, \cdots, X_{jr}$ of V_j, join every vertex of X_{jk} to every vertex of W_{jk}. The resulting graph is G_j, and $\bar{G}$ is the graph $G_1 \cup G_2 \cup \cdots \cup G_{q-1}$. The construction is illustrated in Figure 14.4 for the sequence $T : p_1 \leq p_2 \leq \cdots \leq p_5$ with $p_1 = 3$, $p_2 = p_3 = 4$ and $p_4 = p_5 = 5$.

The domination and independence sequences of paths and cycles were determined in [3, 17, 16].

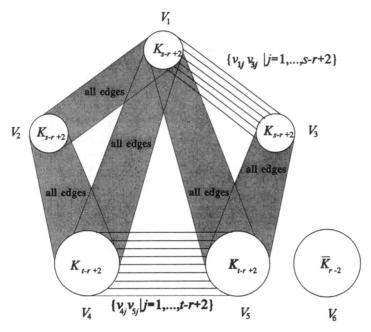

Figure 14.4: The graph $G(T)$ where $p_1 = 3$, $p_2 = p_3 = 4$ and $p_4 = p_5 = 5$.

Theorem 14.7 [3] *For all $k, n \geq 1$,*

$$\Gamma_k(P_n) = \begin{cases} \lceil (k+1)n/(3k+1) \rceil & \text{if } n \equiv 3\ell + 1 \text{ (mod } 3k+1) \\ \lfloor (k+2)(n+)/(3k+1) \rfloor & \text{otherwise} \end{cases}$$

Theorem 14.8 [17] *For all $k \geq 1$ and $n \geq 3$,*

$$\Gamma_k(C_n) = \begin{cases} \lfloor (k+1)n/(3k+1) \rfloor & \text{if } n \equiv 3\ell \text{ (mod } 3k+1) \\ \quad \text{for some } \ell \in \{0, \cdots, k\} \\ \quad \text{or } n \geq 3k+1 \text{ and } n \equiv 3\ell + 1 \text{ (mod } 3k+1) \\ \quad \text{for some } \ell \in \{0, \cdots, k-1\} \\ \lceil (k+1)n/(3k+1) \rceil & \text{if } n \equiv 3\ell + 1 \text{ (mod } 3k+1) \\ \quad \text{for some } \ell \in \{0, \cdots, k-1\} \\ \lfloor (k+1)(n+1)/(3k+1) \rfloor & \text{if } n \equiv 3\ell + 2 \text{ (mod } 3k+1) \\ \quad \text{for some } \ell \in \{0, \cdots, k-1\}. \end{cases}$$

Theorem 14.9 [16] *Let $n = q(2k+1) + r$, where $0 \leq r < 2k+1$. Then*

$$\beta_k(C_n) = \begin{cases} \lfloor n/2 \rfloor & \text{if } n < 4k+2 \\ kq + \lceil r/2 \rceil & \text{if } n \geq 4k+2 \end{cases}$$

Theorem 14.10 [16] *For any n and k, $\beta_k(P_n) = \beta_k(C_{n+2k+1}) - k$.*

The vertex covering number $\alpha_0(G)$ or simply $\alpha(G)$ of G (*i.e.*, the smallest number of vertices which cover all the edges) and the independence number $\beta(G)$ are related by the well-known result of Gallai [28] which states that $\alpha(G) + \beta(G) = n$ for any n-vertex graph G. We conclude this section by generalizing this theorem to obtain a corresponding result for k-minimal and k-maximal parameters. For this purpose we define a subset Y of the set S to be a Q-*set* if it intersects every $\bar{P}$-set X of S (*i.e.*, it is a transversal of the family of $\bar{P}$-sets of S). Property P is said to be *hereditary* if each subset of a P-set is also a P-set.

Theorem 14.11 [3] *Let P be a hereditary property. Then X is a k-maximal P-set if and only if $S - X$ is a k-minimal Q-set.*

If $\alpha_k(S, P)$ and $\beta_k(S, P)$ denote the smallest cardinality of a k-minimal Q-set and the largest cardinality of a k-maximal P-set respectively, we immediately obtain the following collorary.

Corollary 14.12 [3] *If P is a hereditary property on the subsets of S, then*

$$\alpha_k(S, P) + \beta_k(S, P) = |S|.$$

In particular, if $S = V(G)$, where G is an n-vertex graph, and P-sets are independent sets of vertices of G, then

$$\alpha_k(G) + \beta_k(G) = n.$$

We conclude this section by remarking that the k-maximal irredundance parameters IR_k and irredundance sequences are defined analogous to the numbers β_k and independence sequences. (Also see Section 14.7.)

14.4   Nordhaus-Gaddum Type Results

Let ξ be any graph-theoretical parameter. We sometimes denote $\xi(\bar{G})$ as $\bar{\xi}$. For an n-vertex graph G, results of the type

$$\xi(G)\xi(\bar{G}) \le (\ge) \ f(n) \quad \text{or} \quad \xi(G) + \xi(\bar{G}) \le (\ge) \ g(n)$$

are called Nordhaus-Gaddum type results in reference to the work of these authors on the sum and product of the chromatic numbers of G and $\bar{G}$ (see [38]). As far as domination related parameters are concerned, Jaeger and Payan [33] proved that $\gamma(G)\gamma(\bar{G}) \le n$ for any n-vertex graph G, while Chartrand and Schuster [7] proved that $\beta(G)\beta(\bar{G}) \le \lceil (n^2 + 2n)/4 \rceil$. Both these results have been improved – the former bound also holds if γ is replaced by Γ_2 ([3]) and the latter holds if β and $\bar{\beta}$ are replaced by IR and $\overline{IR}$ respectively ([18]).

Theorem 14.13 [3] *For any n-vertex graph G, $\Gamma_2(G)\gamma(\bar{G}) \leq n$.*

Theorem 14.14 [18]

(a) *For any n-vertex graph G,*

$$IR(G)IR(\bar{G}) \leq \lceil (n^2 + 2n)/4 \rceil.$$

(b) *G attains the bound in (a) if and only if G or its complement consists of*

(i) *a set X of $\lfloor (n+1)/2 \rfloor$ independent vertices,*

(ii) *a set Y of $\lceil (n+1)/2 \rceil$ vertices, where $\langle Y \rangle$ is complete and*

$X \cap Y = \{x\}$, *and*

(iii) *any set S of edges joining vertices of $X - \{x\}$ to vertices of $Y - \{x\}$.*

Corollary 14.15 [18] *For any n-vertex graph G, $IR(G) + IR(\bar{G}) \leq n$.*

Corollary 14.16 [18] *For any n-vertex graph G, the products $i\bar{i}$, $\gamma\bar{\beta}$, $\beta\bar{\beta}$, $\gamma\bar{\Gamma}$, $\Gamma\bar{\Gamma}$ (for example) are all bounded above by $\lceil (n^2 + 2n)/4 \rceil$. This bound is best possible in all the above cases except the case $i\bar{i}$.*

Let $f(n,k)$ $(h(n,k)$, respectively) denote the maximum value of $\beta_k(G)\beta_k(\bar{G})$ $(IR_k(G)IR_k(\bar{G})$, respectively) among all n-vertex graphs and note that $f(n) = f(n,1) = \max\left(i(G)i(\bar{G})\right)$ $(h(n) = h(n,1) = \max\left(ir(G)ir(\bar{G})\right)$, respectively), where the maximum is taken over all n-vertex graphs G. For $k > 1$ the extremal graphs in Theorem 14.14 show that $f(n,k) = h(n,k) = \lceil (n^2 + 2n)/4 \rceil$ (see [16]). The extremal graphs for the inequality $\gamma(G)\gamma(\bar{G}) \leq n$ were characterized by Payan and Xuong [39] and these graphs show that $h(n) = n$. However, the problem of calculating $f(n)$ turned out to be surprisingly difficult. The best upper bound so far is given in [11], namely

$$f(n) \leq \min\left\{(n+3)^2/8, \quad (n+8)^2/10.8\right\}.$$

If $f_r(n)$ is the restriction of $f(n)$ to regular graphs, then this bound can be improved, as shown by Haviland [31], to

$$f_r(n) \leq (n+14)^2/12.68.$$

The best lower bound (for $f(n)$, $n \geq 4$) is given in [18] (amended in [13]), namely

$$f(n) \geq \lfloor (n+4)^2/16 \rfloor \qquad (n \text{ even})$$

$$f(n) \geq \lfloor (n+3)^2/16 \rfloor \qquad (n \text{ odd}).$$

The following asymptotic bound can be found in [13]. Although this result implies that the above lower bound is good asymptotically, it is still an open problem to determine whether it is in fact the exact value of $f(n)$.

Theorem 14.17 [13] *Let $0 < k < 16$. There exists an integer n_0 such that for all $n \geq n_0$ and any graph G with n vertices, $i(G)i(\bar{G}) \leq n^2/k$.*

Corollary 14.18 [13] $\lim\limits_{n\to\infty} (f(n)/n^2) = 1/16$.

We now discuss Nordhaus-Gaddum type results for k-minimal domination numbers. Let $g(n, k)$ denote the maximum value of $\Gamma_k(G)\Gamma_k(\bar{G})$ among all n-vertex graphs G. Some of the extremal graphs of Payan and Xuong [39] are also extremal graphs for the inequality $\Gamma_2(G)\gamma(\bar{G}) \leq n$. This together with Corollary 14.16 and the sequence (II) immediately gives

$$n \leq g(n, k) \leq \lceil (n^2 + 2n)/4 \rceil.$$

The determination of $g(n, k)$ appears to be a formidable task and remains an open problem. It was shown by explicit construction in [19] that for any $k \geq 2$, $g(n, k) > n$. The process involved the construction of self-complementary graphs whose domination numbers exceed any given integer $k > 1$:

Let p be a prime such that $p \equiv 1 \pmod 4$ and let $V(G_{p,k}) = \{0, \cdots, p-1\} = Z_p$. Two vertices a, b are adjacent if and only if $a - b$ is not a quadratic residue of p. The graph $G_{p,k}$ is called a Paley graph (see *e.g.* [6, p. 345]). It is easily seen, using the transformation $x \to \lambda x$ where λ is any quadratic non-residue, that $G_{p,k}$ is self-complementary and it was shown in [19] that for $k \geq 2$ and $p > k^2 2^{2k-2}$, $\gamma(G_{p,k}) > k$.

Now let G and H be graphs of order m and n respectively, with $V(G) = \{v_1, \cdots, v_m\}$. The *composition* (*cf.* [29]) of G and H, denoted by $G[H]$, is the graph obtained by replacing each vertex v_i of G by a copy H_i of H and each edge v_iv_j of G by a copy of $K_{q,q}$, where the edges of $K_{q,q}$ join the vertices of H_i to the vertices of H_j. Let $V(H_i) = \{v_{i1}, \cdots, v_{iq}\}$, $i = 1, \cdots, m$ and let $G * H$ be the graph obtained from $G[H]$ in the following way: If $v_iv_j \in E(G)$ ($v_iv_j \notin E(G)$ respectively), remove (add, respectively) the set $E_{ij} = \{v_{i\ell}v_{j\ell} \mid \ell = 1, \cdots, q\}$ of edges from (to, respectively) $G[H]$, for $i, j = 1, \cdots, m$. Observe that if G and H are self-complementary graphs, then so is $G * H$.

Theorem 14.19 [19] *Let $k \geq 2$ and let p be any prime such that $p \equiv 1 (mod\ 4)$ and $p > (k2^{k-1})^2$. Then, for any self-complementary graph H of order $q > p$, the n-vertex graph $G = G_{p,k} * H$ satisfies $\Gamma_k(G)\Gamma_k(\bar{G}) > n$.*

It can be shown that $\gamma(G_{p,k} * H) > k$. However, if $\gamma(F) < k$ for $k \geq 2$ and some n-vertex graph F, then no dominating set of F with more than $\gamma(F)$ vertices can be a k-minimal dominating set and hence $\Gamma_k(G) = \gamma(G)$. By using the fact that $\gamma(G)\Gamma_2(\bar{G}) \leq n$ (see Theorem 14.13), it follows that in this case $\Gamma_k(G)\Gamma_k(\bar{G}) \leq n$.

14.5   Lower Ramsey Numbers

A *clique* of a graph $G = (V, E)$ is a subset Q of V such that the subgraph $\langle Q \rangle$ of G induced by Q is complete. The clique number $\omega(G)$ is the largest cardinality amongst the cliques of G, while $\mu(G)$ denotes the smallest cardinality amongst the maximal cliques of G. Clearly, Q is a clique of G if and only if Q is an independent set of $\bar{G}$ and hence $\mu(G) = i(\bar{G})$ and $\omega(G) = \beta(\bar{G})$. Further, Q is a *k-maximal clique* (kMC) of G if and only if Q is a kMIS of $\bar{G}$. Let $\omega_k(G)$ denote the smallest cardinality amongst the kMCs of G – obviously $\omega_k(G) = \beta_k(\bar{G})$ and thus we have the sequence

$$\mu(G) = \omega_1(G) \leq \omega_2(G) \leq \cdots \leq \omega_k(G) \leq \cdots \leq \omega(G). \qquad \text{(IV)}$$

Let $s_k(\ell, m)$ denote the largest integer such that every graph of order $s_k(\ell, m)$ has a kMIS of cardinality at most ℓ or a kMC of cardinality at most m. The numbers $s_k(\ell, m)$ are called *lower k-Ramsey numbers*, or, for $k = 1$, simply *lower Ramsey numbers*, denoted by $s(\ell, m)$, for graphs. In [34] these numbers were shown to exist for all k and $s_k(\ell, m)$ was determined for all $k \geq 2$ and bounded for $k = 1$.

Theorem 14.20 [34] *For any positive integers ℓ, m and $k \geq 2$, the number $s_k(\ell, m)$ exists and is equal to $\ell + m$.*

Theorem 14.21 [34] *For any positive integers ℓ and m, the number $s(\ell, m)$ exists, and*

$$\ell + m < s(\ell, m) < 2(\ell + m).$$

This upper bound was improved in [36]. For ℓ and m fixed, let

$$N_m = \left\{ (a, b) \in Z^+ \times Z^+ \mid a \leq b,\ ab \leq m \ \text{ and } \ a(b + 1) \geq m \right\}$$

(where Z^+ denotes the set of all positive integers) and define $q : N_m \to Z^+$ by

$$q(a, b) = (a + 1)(b + 1) + (m - ab) + \lceil (m - ab)/(a + 1) \rceil.$$

Construct the graph $G = G(\ell, m)$ as follows: Let $V(G) = V_1 \cup \cdots \cup V_{a+1} \cup W_1 \cup \cdots \cup W_{a+1}$ (disjoint union), where $|V_j| = \ell$ for $j = 1, \cdots, a + 1$ and

$$|W_j| = \begin{cases} b & \text{if } m = ab \text{ and } j \in \{1, \cdots, a + 1\}, \text{ or} \\ & \text{if } m > ab \text{ and } j \in \{m - ab + 2, \cdots, a + 1\}; \\ b + 1 & \text{if } m > ab \text{ and } j \in \{1, \cdots, m - ab + 1\}. \end{cases}$$

Add edges such that $\langle V_1 \cup \cdots \cup V_{a+1} \rangle$ is complete. Furthermore, for each $j \in \{1, \cdots, a+1\}$, join every vertex of W_j to every vertex of V_j. The resulting graph

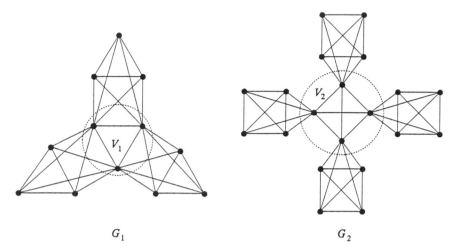

G_1 G_2

Figure 14.5: $G(2,7)$ where $a = 2$ and $b = 3$.

is $G(\ell, m)$. Note that since $a(b+1) \geq m$ it follows that $m - ab < a + 1$ and hence

$$\lceil (m - ab)/(a + 1) \rceil = \begin{cases} 0 & \text{if } m = ab, \\ 1 & \text{if } m > ab. \end{cases}$$

It thus follows that $|V(G(\ell, m))| = q(a, b)$. The construction is illustrated in Figure 14.5 for $\ell - 2$, $m = 7$, $a = 2$ and $b = 3$.

Theorem 14.22 [36] *For any positive integers ℓ and m and $(a, b) \in N_m$, the graph $G(\ell, m)$ satisfies*

$$\mu(G(\ell, m)) = \ell + 1 \quad and \quad i(G(\ell, m)) = m + 1.$$

Corollary 14.23 [36] *For any positive integers ℓ and m,*

$$s(\ell, m) \leq \min\{q(a, b)\} - 1.$$

Note that this bound is smaller than $2(\ell + m) - 1$ if $\ell < \lfloor m/2 \rfloor$. for ℓ close to $\lfloor m/2 \rfloor$ this upper bound is only a slight improvement over $2(\ell+m)-1$. However, if the ratio m/ℓ is large, then the ratio $2(\ell + m)/q(a, b)$ becomes close to two for those values of a and b for which $q(a, b)$ is minimum. (This can be shown to be the case when $|a\ell - b|$ is minimum – see [35,36].)

Using the existing bounds for $s(\ell, m)$, one obtains $s(1, 1) = 3$, $s(1, 2) = 5$, $s(2, 2) = 7$, $s(1, 3) = 7$, $s(1, 4) = 8$ and $s(2, 3) = 9$ (to mention but a few). These values are equal to the respective upper bounds and it is therefore reasonable to expect that the lower bound may not be particularly good. Indeed, better lower bounds were determined by Faudree, Gould, Jacobson and Lesniak [27] and these together with Corollary 14.23 determine $s(1, m)$ exactly. If $\ell = 1$, then Corollary 14.23 becomes the following.

Corollary 14.24 [36] *Let $m = n^2 + r$ for some $0 \le r \le 2n$. Then*

$$s(1, m) \le m + 2n + \lceil r/n \rceil.$$

Since it was proved in [27] that $s(1, m) \ge m + 2n + \lceil r/n \rceil$, it now follows immediately that

$$s(1, m) = m + 2n + \lceil r/n \rceil.$$

It can also be deduced from the upper bound given in Corollary 14.23 that, unless $\ell = 2$ and $m = 3$,

$$s(\ell, m) \le m + 2\sqrt{\ell m} + 2\ell,$$

while it was shown in [27] that

$$s(\ell, m) \ge m + 2\sqrt{\ell m} - 2\sqrt{\ell} - 1.$$

Hence

$$m + 2\sqrt{\ell m} - 2\sqrt{\ell} - 1 \le s(\ell, m) \le m + 2\sqrt{\ell m} + 2\ell$$

and we see that the range for $s(\ell, m)$ is quite small, particularly for small ℓ. Finally, if $\ell \ge m/2$ and particularly if $\ell = m$, the upper bound $s(\ell, m) \le 2(\ell + m) - 1$ may be close to the true value of $s(\ell, m)$, as indicated by the following result of [27].

Theorem 14.25 [27] *Let ℓ and m be positive integers with $\ell = cm$, where $0 \le c \le 1$. If*

$$\epsilon > 1 - \sqrt{4c/(c+1)^2}$$

and m is sufficiently large, then

$$s(\ell, m) \ge (2 - \epsilon)(\ell + m).$$

In particular, if $c = 1$, then for any $\epsilon > 0$ and m sufficiently large,

$$s(m, m) \ge (4 - \epsilon) m.$$

14.6   Irredundant Ramsey Numbers

Another variation of classical Ramsey numbers is obtained by noting that since cliques in a graph correspond to independent sets in its complement, Ramsey numbers can also be defined by using independent sets instead of cliques. Thus irredundant Ramsey numbers are defined by replacing independent sets in the definition of Ramsey numbers by irredundant sets. Let $G_1, G_2, \cdots, G_t$ be an arbitrary t-edge colouring of K_p, where for each $i \in \{1, 2, \cdots, t\}$, G_i is the spanning subgraph of K_p consisting of all edges coloured with colour i. The *irredundant Ramsey number* $S(q_1, q_2, \cdots, q_t)$ is defined as the smallest p such that for all t-edge colourings of K_p, there is an $i \in \{1, 2, \cdots, t\}$ for which $IR(\bar{G}_i) \geq q_i$. Accordingly, if $t = 2$, the irredundant Ramsey number $S(m, n)$ is the smallest integer p such that for every graph G of order p, $IR(G) \geq n$ or $IR(\bar{G}) \geq m$. Since any independent set is irredundant, the irredundant Ramsey numbers exist by Ramsey's theorem and satisfy $S(q_1, q_2, \cdots, q_t) \leq r(q_1, q_2, \cdots, q_t)$ for all $q_1, q_2, \cdots, q_t$. The *mixed Ramsey number* $t(m, n)$ is the smallest p such that for every graph G of order p, $IR(\bar{G}) \geq m$ or $\beta(G) \geq n$. Clearly, $S(m, n) \leq t(m, n) \leq r(m, n)$ for all m, n. Irredundant Ramsey numbers were introduced in [4] and mixed Ramsey numbers in [14]. A comprehensive survey of irredundant and mixed Ramsey numbers is given in [37] and we only briefly summarize the results here. Note that unlike the case for $S(m, n)$ and $r(m, n)$, $t(m, n) \neq t(n, m)$ in general.

m/n	3	4	5	6	7
3	6 [4]	8 [4]	12 [4]	15 [5]	18 [9,15]
4		13 [10]	$S(3,3,3) = 13$ [20,23]		

Table 14.1: Irredundant Ramsey numbers $S(m, n)$ and $S(3, 3, 3)$.

m/n	3	4	5	6
3	6	9	12	≤ 16
4	8			
5	13			

Table 14.2: Mixed Ramsey numbers $t(m, n)$ as given in [14].

As noted in [4], the same recurrence inequality which holds for $r(m, n)$ also holds for $S(m, n)$ and $t(m, n)$.

Proposition 14.26 [4] *For all integers* $m, n \geq 2$,

$$S(m, n) \leq S(m - 1, n) + S(m, n - 1)$$

and

$$t(m, n) \leq t(m - 1, n) + t(m, n - 1),$$

while strict inequality holds if $S(m - 1, n)$ *and* $S(m, n - 1)$ *(or* $t(m - 1, n)$ *and* $t(m, n - 1)$*) are both even.*

Since the difficulty of obtaining exact values for irredundant Ramsey numbers is comparable to that of the corresponding problem for classical Ramsey numbers, attempts were soon made to obtain asymptotic bounds for $S(m, n)$. Using the probabilistic method and an adaptation of Erdös's proof [24] for the corresponding result for $r(n, n)$, the following result was obtained.

Theorem 14.27 [8] *For all sufficiently large* n,

$$S(n, n) > 2^{n/2} \sqrt{n/3}.$$

Using a probabilistic inequality of Lovász [26] known as the "Lovász local lemma" and emulating the techniques of Spencer [41], one obtains a lower bound for $S(m, n)$.

Theorem 14.28 [8,25] *For each* $m \geq 3$ *there is a positive constant* C_m *such that*

$$S(m, n) > C_m (n/\log n)^{(m^2 - m - 1)/2(m-1)}.$$

Finally, using induction and a result of [30], an upper bound for $S(3, n)$ and $t(3, n)$ is obtained.

Theorem 14.29 [8] *For all* $n \geq 1$,

$$S(3, n) \leq t(3, n) \leq \tfrac{1}{2} n^{3/2} \sqrt{10}.$$

14.7　Open Problems and Directions for Further Research

We conclude by listing some open problems and discussing directions for further research.

1. Let $\mathcal{G}$ be a particular class of graphs, such as bipartite, chordal, interval, planar, regular, k-degenerate, perfect graphs, trees, etc. Characterize the dii-sequences of $\mathcal{G}$.

2. Consider the corresponding problem for the extended inequality chain

$$er \leq ir \leq \gamma \leq i \leq \beta \leq \Gamma \leq IR \leq ER,$$

(see [32]), including the case where $\mathcal{G}$ is the class of all graphs.

The generalizations of minimality and maximality given in Section 14.3 can also be applied to other domination related concepts, such as connected and total domination, irredundance, external redundance, etc., and to other graph-theoretical concepts such as covering and colouring. For example, it is easy to see that Proposition 14.3 remains true if $\beta(G)$ is replaced by $IR(G)$ and $\beta_k(G)$ by $IR_k(G)$. We suspect that the conditions which characterize independence sequences will also characterize irredundance sequences. However, there exist graphs G with $ir(G) < i(G)$ and $IR(G) > \beta(G)$ and the irredundance sequences of these graphs differ from their independence sequences.

3. Determine the irredundance sequences of special classes of graphs.

4. Consider a graph G with $ir(G) < i(G)$ and $IR(G) > \beta(G)$. How do the independence and irredundance sequences of G differ? Do certain sections (in the middle) coincide?

   Recall that $f(n) = f(n,1) = \max\left(i(G)i(\bar{G})\right)$ (see Section 14.4).

5. Determine the exact value of $f(n)$.

It is clear that the determination of the lower Ramsey numbers becomes tricky even for relatively small values of $\ell \geq 2$ and m and is a hard problem in general. Although there is some merit in calculating $s(\ell, m)$ for more small values of ℓ and m, it is more important to improve the known bounds and to find a formula for $s(\ell, m)$, perhaps for $\ell = 2$ if the general case proves to be too difficult. Also, it is easy to see that $s(\ell, m+1) \geq s(\ell, m) + 1$ (see [36]) and in view of the known upper bounds and the exact values of $s(\ell, m)$ for $\ell = 1$, or $\ell \geq 2$ and m small, it is reasonable to expect that $s(\ell, m+1) \leq s(\ell, m) + 2$. We therefore formulate the following problem.

6. Prove or disprove:
$$s(\ell, m + 1) \leq s(\ell, m) + 2.$$

The exact determination of $f(n)$ (Problem 5) is directly related to the problem of determining $s(\ell, m)$ as indicated in [34]: If $s(\ell, \ell) < 4\ell - 1$ for some integer ℓ and G is a graph of order $n \leq 4\ell - 1$ such that $i(G) \geq \ell + 1$ and $i(\bar{G}) \geq \ell + 1$, then $i(G)i(\bar{G}) > (n+4)^2/16$. Conversely, if G is an n-vertex graph such that $i(G)i(\bar{G}) > (n+4)^2/16$, where $i(G) = \ell + 1$ (say) and $i(\bar{G}) = m + 1$, then $n \leq 2(\ell + m) - 1$ and hence $s(\ell, m) \leq 2(\ell + m) - 2$. In view of Theorems 14.17 and 14.25 we therefore formulate the following conjecture.

Conjecture 14.30 *For positive integers ℓ and m with $\lfloor m/2 \rfloor \leq \ell \leq m$,*
$$s(\ell, m) = 2(\ell + m) - 1.$$

We also conjecture that the upper bound for $s(\ell, m)$ (Corollary 14.23) is best possible if $2 \leq \ell < \lfloor m/2 \rfloor$.

Conjecture 14.31 *Let ℓ and m satisfy $2 \leq \ell < \lfloor m/2 \rfloor$ and define*

$$N_m = \left\{ (a, b) \in Z^+ \times Z^+ \mid a \leq b, \ ab \leq m \ and \ a(b+1) \geq m \right\}.$$

Choose $(\alpha, \beta) \in N_m$ such that $|\alpha \ell - \beta| \leq |c \ell - d|$ for all $(c, d) \in N_m$. Then

$$s(\ell, m) = \ell + m + \beta + \alpha \ell + \begin{cases} 0 & \text{if } m = \alpha\beta, \\ 1 & \text{if } m > \alpha\beta. \end{cases}$$

The following problem has not been investigated at all.

7. Lower irredundant Ramsey numbers $q_k(\ell, m)$ may be defined similar to lower Ramsey numbers. Determine $q_k(\ell, m)$ for some values of k, ℓ and m and find upper and lower bounds and recurrence inequalities. Is $q_k(\ell, m) = s_k(\ell, m)$ for $k \geq 2$? (We suspect that this may well be true.)

Finally, we mention some open problems concerning irredundant and mixed Ramsey numbers.

8. Investigate irredundant-type Ramsey numbers for other irredundance parameters.

9. Determine further exact values of $S(m, n)$ and $t(m, n)$.

10. Determine constructive lower bounds for $S(n, n)$ or even $S(m, n)$.

11. [8] It can be shown that

$$\lim_{n \to \infty} \frac{t(3, n)}{r(3, n)} = 0.$$

This suggests the following problem: Prove or disprove that for each fixed $m \geq 3$,

$$\lim_{n \to \infty} \frac{t(m, n)}{r(m, n)} = 0.$$

Acknowledgement.
     This paper was written while E.J. Cockayne was visiting the Department of Mathematics of the University of South Africa in 1995. Financial support from the University of South Africa, the FRD (South Africa) and NSERC (Canada) is gratefully acknowledged.

Bibliography

[1] R. B. Allan and R. C. Laskar, On domination and some related concepts in graph theory. *Congr. Numer.* 21 (1978) 43-58.

[2] B. Bollobás and E. J. Cockayne, Graph-theoretic parameters concerning domination, independence and irredundance. *J. Graph Theory* 3 (1979) 241-249.

[3] B. Bollobás, E. J. Cockayne and C. M. Mynhardt, On generalised minimal domination parameters for paths. *Discrete Math.* 86 (1990) 89-97.

[4] R. C. Brewster, E. J. Cockayne and C. M. Mynhardt, Irredundant Ramsey numbers for graphs. *J. Graph Theory* 13 (1989) 283-290.

[5] R. C. Brewster, E. J. Cockayne and C. M. Mynhardt, The irredundant Ramsey number $s(3,6)$. *Quaestiones Math.* 13 (1990) 141-157.

[6] P. J. Cameron, Strongly regular graphs. *Selected Topics in Graph Theory*, Eds. L. Beineke and R. J. Wilson, Acad. Press (1978) 337-360.

[7] G. Chartrand and S. Schuster, On the independence numbers of complementary graphs. *Trans. New York Acad. Sci.* (2) 36 (1974) 247-251.

[8] G. Chen, J. H. Hattingh and C. C. Rousseau, Asymptotic bounds for irredundant and mixed Ramsey numbers. *J. Graph Theory* 17 (1993) 193-206.

[9] G. Chen and C. C. Rousseau, The irredundant Ramsey number $s(3,7)$. *J. Graph Theory* 19 (1995) 263-270.

[10] E. J. Cockayne, G. Exoo, J. H. Hattingh and C. M. Mynhardt, The irredundant Ramsey number $s(4,4)$. *Utilitas Math.* 41 (1992) 119-128.

[11] E. J. Cockayne, O. Favaron, H. Li and G. MacGillivray, On the product of the independent domination numbers of a graph and its complement. *Discrete Math.* 90 (1991) 313-317.

[12] E. J. Cockayne, O. Favaron, C. Payan and A. G. Thomason, Contributions to the theory of domination, independence and irredundance in graphs. *Discrete Math.* 33 (1981) 249-258.

[13] E. J. Cockayne, G. Fricke and C. M. Mynhardt, On a Nordhaus-Gaddum type problem for independent domination. *Discrete Math.* 138 (1995) 199–205.

[14] E. J. Cockayne, J. H. Hattingh, J. Kok and C. M. Mynhardt, Mixed Ramsey numbers and irredundant Turán numbers for graphs. *Ars Combin.* 29C (1990) 57–68.

[15] E. J. Cockayne, J. H. Hattingh and C. M. Mynhardt, The irredundant Ramsey number $s(3,7)$. *Utilitas Math.* 39 (1991) 145–160.

[16] E. J. Cockayne, G. MacGillivray and C. M. Mynhardt, Generalised maximal independence parameters for paths and cycles. *Quaestiones Math.* 13 (1990) 123–139.

[17] E. J. Cockayne and C. M. Mynhardt, k-Minimal domination numbers of cycles. *Ars Combin.* 23A (1987) 195–206.

[18] E. J. Cockayne and C. M. Mynhardt, On the product of upper irredundance numbers of a graph and its complement. *Discrete Math.* 76 (1989) 117–121.

[19] E. J. Cockayne and C. M. Mynhardt, On the product of k-minimal domination numbers of a graph and its complement. *J. Combin. Math. Combin. Comput.* 8 (1990) 118–122.

[20] E. J. Cockayne and C. M. Mynhardt, On the irredundant Ramsey number $s(3,3,3)$. *Ars Combin.* 29C (1990) 189–202.

[21] E. J. Cockayne and C. M. Mynhardt, Domination sequences of graphs. *Ars Combin.* 33 (1992) 257–275.

[22] E. J. Cockayne and C. M. Mynhardt, The sequence of upper and lower domination, independence and irredundance numbers of a graph. *Discrete Math.* 122 (1993) 89–102.

[23] E. J. Cockayne and C. M. Mynhardt, The irredundance number $s(3,3,3) = 13$. *J. Graph Theory* 18 (1994) 595–604.

[24] P. Erdös, Some remarks on the theory of graphs. *Bull. Amer. Math. Soc.* 53 (1947) 292–294.

[25] P. Erdös and J. H. Hattingh, Asymptotic bounds for irredundant Ramsey numbers. *Quaestiones Math.* 16 (1993) 319–331.

[26] P. Erdös and L. Lovász, Problems and results on 3-chromatic hypergraphs and some related questions. *Infinite and Finite Sets*, Eds. A. Hajnal, R. Rado and V. T. Sòs, North-Holland, Amsterdam (1975) 609–628.

[27] R. Faudree, R. J. Gould, M. S. Jacobson and L. Lesniak, Lower bounds for lower Ramsey numbers. *J. Graph Theory* 14 (1990) 723–730.

[28] T. Gallai, Über extreme Punkt-und-Kantenmengen. *Ann. Univ. Sci. Budapest, Eötvös Sect. Math.* 2 (1959) 133–138.

[29] F. Harary, *Graph Theory.* Addison Wesley, Reading, Mass. 1969.

[30] J. H. Hattingh, On irredundant Ramsey numbers for graphs. *J. Graph Theory* 14 (1990) 437–441.

[31] J. Haviland, Independent domination in regular graphs. To appear in *Discrete Math.*

[32] T. W. Haynes, S. T. Hedetniemi and P. J. Slater, *Fundamentals of Domination in Graphs*, Marcel Dekker, Inc., 1997.

[33] F. Jaeger and C. Payan, Relations du type Nordhaus-Gaddum pour le nombre d'absorption d'un graphe simple. *C.R. Acad. Sci. Paris, Series A* 274 (1972).

[34] C. M. Mynhardt, Generalised maximal independence and clique numbers of graphs. *Quaestiones Math.* 11 (1988) 383–390.

[35] C. M. Mynhardt, On a Ramsey type problem for independent domination parameters of graphs. Technical Report 57/88 (2), University of South Africa (1988).

[36] C. M. Mynhardt, Lower Ramsey numbers for graphs. *Discrete Math.* 91 (1991) 69–75.

[37] C. M. Mynhardt, Irredundant Ramsey numbers for graphs: a survey. *Congr. Numer.* 86 (1992) 65–79.

[38] E. A. Nordhaus and J. W. Gaddum, On complementary graphs. *Amer. Math. Monthly* 63 (1956) 175–177.

[39] C. Payan and N. H. Xuong, Domination balanced graphs. *J. Graph Theory* 6 (1982) 23–32.

[40] J. C. Schoeman, *'n Ry boonste en onderste dominasie -, onafhanklikheids- en onoorbodigheidsgetalle van 'n grafiek.* Master's Thesis, Rand Afrikaans University (1992).

[41] J. Spencer, Asymptotic lower bounds for Ramsey functions. *Discrete Math.* 20 (1977) 69–76.

Chapter 15

Topics on Domination in Directed Graphs

J. Ghoshal, R. Laskar, D. Pillone

Department of Mathematical Sciences
Clemson University
Clemson, SC 29634   USA

15.1   Introduction

Domination and other related concepts in undirected graphs are well studied. Recently, two special volumes appeared containing research papers dealing only with these topics [49], [58]. The need for a concise survey of the many types of domination models is long overdue. In this book such topics are dealt with primarily for undirected graphs. Although domination and related topics have been extensively studied, the respective analogs on digraphs have not received much attention. Such studies in the directed case have concentrated mainly on kernels, because of its applications to game theory and other areas, as discussed in Sections 15.3 and 15.8. In this chapter we present results concerning domination, kernels, and solutions in digraphs along with applications in game theory.

The pioneering work in digraphs in this area can be ascribed to Berge, Harary, König, Grundy and Richardson, amongst others. This chapter detailing the results on digraphs has been naturally influenced by the book *Graphs and Hypergraphs* by Berge and the survey of recent results on kernels by Berge and Duchet [8]. The concepts of kernels and solutions in digraphs have their origin in the concepts of imputations and domination developed by Von Neumann and Morgenstern [107]. We approach this topic by dividing it into four major areas; an exposition of results on kernels; some results on domination in digraphs; the concept and results concerning solutions in a digraph and the application of some of these ideas to game theory. The significant works in these areas by Blidia, Duchet, Galeana-Sanchez, Kwasnik, Meyneil, Neumann-Lara, Roth, Smith, Topp and others are recorded in this endeavor. We begin this journey with definitions of the major concepts.

15.2   Definitions

Perhaps no other area of domination has as great a need to standardize definitions and notation as that of directed domination. Different terms are chosen for the same concept and the same term is occassionally chosen for different concepts. We have tried to clarify the situation by giving common alternate terms and pointing out differences in definitions. For this chapter, unless otherwise mentioned, a digraph $D = (V, A)$ consists of a finite vertex set V and an arc set $A \subseteq P$, where P is the set of all ordered pairs of distinct vertices of V. That is, D has no multiple loops and no multiple arcs (but pairs of opposite arcs are allowed). For this chapter we assume that the underlying graph of the digraph D is *connected*. In the terminology of Berge we are considering connected 1-graphs without loops. Let $D = (V, A)$ be such a digraph. If $A = P$ then the digraph is *complete*. If $(x, y) \in A$ then the arc is directed from x to y and is denoted by $x \rightarrow y$. The vertex x is called *a predecessor of y* and y is called *a successor of x*. If the reversal (y, x) of an arc (x, y) of D is also present in D we say that (x, y) is a *reversible (symmetrical)* arc. If $(x, y) \in A$ but $(y, x) \notin A$ then (x, y) is an *asymmetric arc*.

A *symmetric digraph* is a digraph for which $(x, y) \in A$ implies $(y, x) \in A$. A *tournament* describes a digraph such that for all $x, y \in V$ where $x \neq y$ either $(x, y) \in A$ or $(y, x) \in A$ but not both. The *complement* of a digraph $D = (V, A)$ is the digraph $\overline{D} = (V, P - A)$.

The sets $O(u) = \{v : (u, v) \in A\}$ and $I(u) = \{v : (v, u) \in A\}$ are called the *outset* and *inset* of the vertex u. Likewise, $O[u] = O(u) \cup \{u\}$ and $I[u] = I(u) \cup \{u\}$. If $S \subseteq V$ then $O(S) = \cup_{s \in S} O(s)$ and $I(S) = \cup_{s \in S} I(s)$. Similarly $O[S] = \cup_{s \in S} O[s]$ and $I[S] = \cup_{s \in S} I[s]$. The *indegree* of a vertex u is given by $id(u) = |I(u)|$ and the *outdegree* of a vertex is $od(u) = |O(u)|$. If for some n we have a sequence $\sigma = x_1, x_2, \ldots x_n$ of vertices such that every x_{i+1} is a successor of x_i then σ is a *directed walk* from x_1 to x_n of length $n - 1$. If all the x_i's are different then σ is a *directed path*. If $x_1 = x_n$ then σ is a *circuit (directed cycle)*. The notation $d(u, v)$ is the length of a shortest directed path from u to v.

A set $S \subseteq V$ is *independent* (or *stable* or *internally stable*) if for all $x, y \in S$, $(x, y) \notin A$. The *coefficient of stability* of the digraph D is the maximum cardinality of an independent set. Since this is equivalent to the maximum cardinality of an independent set in the underlying graph we will adopt the notation $\beta(D)$ for the coefficient of stability. Following Berge, a subset $S \subseteq V$ is *absorbant* if for every vertex $x \notin S$ there is a vertex $y \in S$ such that y is a successor of x. We define a set $S \subseteq V$ of a digraph D to be a *dominating set* of D if for all $v \notin S$, v is a successor of some vertex $s \in S$. With this terminology we now define a *kernel* of D to be a subset S which is independent and absorbant. Note that the digraph of Figure 15.1 has no kernel. Moreover, every undirected graph G has a kernel. If, however, a digraph D contains a kernel K, then it follows from the definitions that $|K| \leq \beta(D)$. The concept of a kernel is important to the theory of digraphs because it arises naturally in applications such as cooperative n-person games, Nim-type games, and logic, to name a few. These aspects are taken up in Sections 15.3 and 15.8. A set $S \subseteq V$ is a *solution* of a digraph if S is both independent and dominating. Given a digraph $D = (V, A)$, the digraph D^{-1} is called the *reversal of D*, where $V(D) = V(D^{-1})$ and $(x, y) \in A(D^{-1})$ if and only if $(y, x) \in A(D)$. Observe that a solution of a digraph D is a kernel of D^{-1}.

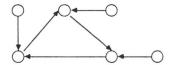

Figure 15.1: A digraph with a solution but no kernel.

A digraph is *strongly connected* if for each pair of vertices x and y, there exists a directed path from x to y and also a directed path from y to x. A digraph D is *transitive* if for any triple $x, y, z \in V$, $(x, y) \in A$, $(y, z) \in A$ implies $(x, z) \in A$ (cf. Figure 15.2). It is important to note here that an undirected graph is called a *comparability* graph if there exists a *transitive* orientation of its edges. An *orientation* of an undirected graph G is an assignment of directions to its edges. In particular an orientation may contain reversible arcs.

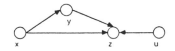

Figure 15.2: A transitive digraph.

15.3 Motivation

To motivate the study of kernels and solutions in digraphs we now present some examples where these concepts are used.

Example 1. *Game Theory* (Von Neumann, Morgenstern [107])

Suppose that n players, denoted by $(1), (2), \ldots, (n)$ can discuss together to select a point x from a set X (the "situations"). If player (i) prefers situation a to situation b, we shall write $a \geq^i b$. The individual preferences might not be compatible, and consequently it is necessary to introduce the concept of *effective preference*. The situation a is said to be *effectively preferred to* b, or $a > b$, if there is a set of players who prefer a to b and who are all together capable of enforcing their preference for a. However, effective preference is not transitive; i.e., $a > b$ and $b > c$ does not necessarily imply that $a > c$.

Consider the digraph $D = (V, A)$ where $O(x)$ denotes the set of situations effectively preferred to x. Let S be a kernel of D. Von Neumann and Morgenstern suggested that the selection be confined to the elements of S. Since S is independent, no situation in S is effectively preferred to any other situation in S. Since S is absorbant, for every situation $x \notin S$, there is a situation in S that is effectively preferred to x, so that x can be immediately discarded.

Example 2. *Problem in Logic* (Berge, Rao [10])

Let us consider a set of properties $P = \{p_1, p_2, \ldots\}$ and a set of theorems of the type: "property p_i implies property p_j". These theorems can be represented by a directed graph $D = (V, A)$ with vertex set P, where (p_i, p_j) is an arc if and only if it follows from one or more of the existing theorems that p_i implies p_j. Suppose we want to show that no arc of the complementary graph $\overline{D}$ is good to represent a true implication of that kind: more precisely, with each arc (p, q) with $p \neq q$ and $(p, q) \notin A$, we assign a student who has to find an example where p is fulfilled but not q (i.e., a counter-example to the statement that p implies q).

In [10] they determined the minimum number of students needed to show that all the possible (pairwise) implications are already represented in the digraph D. It was found that this number corresponded to the cardinality of the unique kernel of the transitive digraph under study.

Example 3. *Facility Location*

Let $D = (V, A)$ be a digraph where the vertices represent "locations" and there is an arc from location u to location v if location v can be "reached" from location u. Assume that each "location" has a weight associated with it which represents some parameter pertinent to the study.

Choose a subset of "locations" such that those outside the set have an arc incident from a member of the set, which means that all the "locations" can be "serviced" by the members of the set S. Let $w(S)$ denote the sum of the weights of the members of S. The problem of finding such a set S such that $w(S)$ is minimized. The relevant graph theoretic concept is that of directed domination.

15.4   Kernels in Digraphs

For this study of kernels it is appropriate to divide the results into four major areas. In the first section we look at the existence of kernels in various families of digraphs. Many papers focus on the connections between perfect graphs and kernels; this is discussed in the second and third section. The fourth section deals with the many generalizations of the concept of kernels in digraphs.

The next lemma follows directly from the definition of a kernel.

Lemma 15.1 (Berge [5]) *If S is a kernel then S is a maximal independent set and a minimal absorbant set.*

Algorithms to construct all kernels of a digraph D have been presented by Roy in 1970 [94] and Rudeanu in 1966 [95]. Other than these, very few algorithmic results exist concerning kernels; most papers deal with the existence of a kernel. Loukakis in [72] has given algorithms for finding independent sets in digraphs using tree search approach and dynamic programming. Fraenkel [35] has determined that deciding whether a finite digraph D has a kernel K or a Grundy function g (see Section 15.5) is NP-complete even if D is a cyclic planar digraph with degree constraints $od(x) \leq 2$, $id(x) \leq 2$ and $od(x) + id(x) \leq 3$. These results are best possible (if $P \neq NP$) in the sense that if any of the constraints are tightened, there are polynomial algorithms which either compute K and g or show that they do not exist. The proof uses a simple reduction from planar 3-satisfiability for both problems. See [22], [36], [71], [19], [37], [81] for other algorithmic results.

15.4.1   Results on the existence of kernels

The concept of kernels has many applications some of which we will discuss in Section 15.8. First we notice that some digraphs have no kernels, some have several kernels, and some have a unique kernel as illustrated, respectively, by the examples in Figure 15.3.

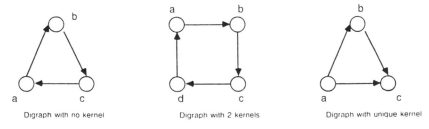

Figure 15.3: Examples of digraphs.

Thus a natural question arises: *Which structural properties of a digraph imply the existence of a kernel?*

The following theorem due to Berge gives a necessary and sufficient condition for a set $S \subseteq V$ to be a kernel. Recall that the characteristic function $\phi_s(x)$ of S is defined by $\phi_s(x) = 1$ if $x \in S$ and $\phi_s(x) = 0$ if $x \notin S$. If $O(x) = \emptyset$ we define $max_{y \in O(x)} \{\phi_s(y)\} = 0$.

Theorem 15.2 (Berge [5]) *A necessary and sufficient condition for a set $S \subseteq V(D)$ to be a kernel of a digraph $D = (V, A)$ is that its characteristic function $\phi_s(x)$ satisfies $\phi_s(x) = 1 - max\{\phi_s(y) : y \in O(x)\}$.*

Proof. Let S be a kernel in D and suppose that ϕ_s is the characteristic function defined on it. If $x \in S$ then $\phi_s(x) = 1$, and for all $y \in O(x), \phi_s(y) = 0$. This implies $max\{\phi_s(y) : y \in O(x)\} = 0$ and hence, $\phi_s(x) = 1 - max\{\phi_s(y) : y \in O(x)\}$. If $x \notin S$ then $\phi_s = 0$, and since S is absorbant, there exists a $y \in \phi_s(x)$ such that $\phi_s(y) = 1$. Hence, $max\{\phi_s(y) : y \in O(x)\} = 1$. So $\phi_s(x) = 1 - max\{\phi_s(y) : y \in O(x)\}$ for any $x \in V - S$.

To prove the converse, consider a set $S \subseteq V(D)$ for which there exists a characteristic function ϕ_s on S satisfying the given condition. Then $x \in S \implies \phi_s(x) = 1 \implies max\{\phi_s(y) : y \in O(x)\} = 0 \implies S \cap O(x) = \emptyset$. So S is independent. Now $x \in V - S \implies \phi_s(x) = 0 \implies max\{\phi_s(y) : y \in O(x)\} = 1 \implies S \cap O(x) \neq \emptyset$. Hence S is absorbant. Combining we see that S is a kernel. $\square$

The existence of kernels in transitive digraphs is demonstrated by König in the following theorem and its corollary.

Theorem 15.3 (König [62]) *If $D = (V, A)$ is a transitive digraph then each minimal absorbant set of D has the same cardinality. Furthermore, a set $S \subseteq V$ is a kernel if and only if S is a minimal absorbant set.*

Corollary 15.4 *A transitive digraph has a kernel and all of its kernels have the same cardinality.*

When a digraph is symmetric (equivalent to an undirected graph) this result is well known.

Lemma 15.5 (Berge [5]) *If D is a symmetric digraph then D has a kernel. Furthermore, a set $S \subseteq V(D)$ is a kernel if and only if S is a maximal independent set.*

See [5], [17], [25], and [34] for results and problems on kernels. Although not all digraphs have kernels, it has been shown that almost all digraphs have one. Let $D(n, p)$ denote a random digraph on n vertices, where each directed edge (x, y) is present with probability p. De la Vega has established the following result.

Theorem 15.6 (De la Vega [30]) *Let p be fixed $0 \leq p \leq 1$. The probability that the random digraph $D(n,p)$ possesses a kernel tends to 1 as $n \to \infty$.*

The same result is noted by Tomescu in [102] wherein he shows that almost all digraphs D of order n contain only kernels such that $log(n) - log(log(n)) - 1.43 \leq |K| \leq log(n) - log(log(n)) + 2.11$ and the number $K(D)$ of kernels of D satisfies $n^{0.913+O(1)} < K(D) < n^{1+O(1)}$ as $n \to \infty$.

The majority of existence results on kernels are proved under the stronger condition that the digraph D is kernel perfect. This important area of research is discussed next.

15.4.2 Kernel perfect graphs and related concepts

A digraph D having the property that all induced subgraphs have kernels is called *kernel perfect*; otherwise, D is *kernel imperfect*. This concept was motivated by the famous perfect graph conjecture (due to Berge)[1] and the search for a relationship between kernel perfect and perfect graphs. Note that if D is kernel perfect then D has a kernel. Thus, instead of exhibiting the existence of kernels, graph theorists attempted to define conditions which imply the stronger result of a digraph being kernel perfect.

A graph property, P, is *hereditary* if for any digraph D having property P, every induced subdigraph has property P. Properties like transitivity and being acyclic are hereditary. So in order to show that a digraph D with a hereditary property P is kernel perfect it suffices to show that D has a kernel. Using this idea, the following classical results can be proved.

Theorem 15.7 (Von Neumann [107]) *A digraph without circuits is kernel-perfect and has a unique kernel.*

Richardson [89] improved Von Neumann's result in Theorem 15.7. We give a proof (by V. Neumann) of this theorem as mentioned in Berge. This proof requires the concept of a semikernel and a lemma, which we give below. If $S \subseteq V$ is a kernel of the subdigraph of D induced by S and all its neighbors, then S is a *semikernel (local kernel)* of D. Equivalently, S is a semikernel of D if S is an independent set of vertices such that for all $z \in V - S$ for which there exists an Sz-arc, there is also a zS-arc. A kernel is a semikernel but not every semikernel is a kernel as Figure 15.4 illustrates. Observe that in Figure 15.4, the set $S = \{c\}$ is a semikernel but not a kernel.

Lemma 15.8 (Berge [5]) *If for each nonempty subset $V_o \subseteq V(D)$, the subdigraph induced by V_0 has a semikernel, then D has a kernel.*

---

[1] A graph G is perfect if and only if G contains neither an induced odd cycle nor its complement of order 5 or greater.

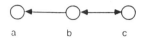

Figure 15.4: A digraph with a semikernel which is not a kernel.

Proof. [As in [5]] Let S be a maximal semikernel of D and let $V_o = V - S - O(S) - I(S)$. If $V_o = \emptyset$ then S is a kernel. If $V_o \neq \emptyset$ then $< V_o >$ has a semikernel T. There are no arcs between S and T hence $S \cup T$ is independent. Each $x \notin S \cup T$ that is adjacent from $S \cup T$ has a successor in $S \cup T$. Thus $S \cup T$ is a semikernel of D, which contradicts the maximality of S. $\square$

A nice result that uses this lemma to characterize kernel perfectness in terms of semikernels is due to Neumann-Lara.

Theorem 15.9 (Neumann-Lara [84]) *A digraph D is kernel perfect if and only if every induced subdigraph of D has a semikernel.*

We now state and prove Richardson's Theorem using Lemma 15.8.

Theorem 15.10 (Richardson) *A digraph without odd circuits is kernel perfect.*

Proof. [As in [5]] Since every induced subdigraph of D has no odd circuits it suffices to show that $D = (V, A)$ has a semikernel. Since D is strongly connected there exists at least one strongly connected component X_1 such that $O(X_1) \subseteq X_1$ (cf. [5]). If $|X_1| = 1$ then X_1 is a semikernel. If $|X_1| > 1$ and if $x_0 \in X_1$ then let x be a vertex in X_1 distinct from x_0. Then all paths from x_0 to x remain "inside" X_1 and have the same parity since otherwise an odd circuit could be formed with a path from x to x_0. Let S denote the set of all $x \in X_1$ such that all x_0 to x paths are even. Set S is independent. If $z \in V$ is adjacent from S then each successor of z belongs to S. Thus, S is a semikernel. $\square$

Many extensions of Richardson's Theorem have been found over the years. An important one which has since been studied and generalized is due to Duchet. Others results are given below.

Theorem 15.11 (Duchet [24]) *If every odd circuit possesses at least two reversible arcs then the digraph is kernel perfect.*

Theorem 15.12 *A digraph D is kernel perfect if*

1. (Duchet [24]) *Every circuit of D has at least one symmetric arc.*

2. (Duchet, Meyniel [26]) *Every odd circuit $\{v_1, v_2, \ldots, v_{2k+1}, v_1\}$ has two chords of the type (v_i, v_{i+2}) and (v_{i+1}, v_{i+3}).*

3. (Neumann-Lara, Galeana-Sanchez [42]) *If every odd circuit has 2 chords whose heads are consecutive vertices of the circuit.*

However, it is false that a digraph D all of whose odd circuits have two chords is kernel-perfect [42].

Chvatal and Berge conjectured that if D is kernel perfect then D^{-1} is kernel perfect. A counterexample was found by Duchet and Meyniel [27]. Likewise Kwasnik [66],[67] and Marcu [79], [80] obtained results concerning strongly connected digraphs which are generalizations of Richardson's Theorem. This is given in Section 15.4.4. See [97] for a counterexample to a generalization to Richardson's theorem.

Since not every digraph has a kernel it is frequently of interest to examine digraphs without kernels. A digraph D is *kernel perfect critical* (or *KP-critical* or *critical kernel imperfect*) if D does not have a kernel but every proper induced subdigraph of D has a kernel. Note that a cycle on three vertices is kernel perfect critical but not kernel perfect. The following is a very interesting result concerning kernel perfect critical graphs and strongly connected digraphs.

Theorem 15.13 (Berge, Duchet [8]) *A kernel perfect critical graph is strongly connected.*

Proof. [As given by Berge and Duchet] Suppose D is a kernel perfect critical graph which is not strongly connected. There exists a strong component C_1 of D such that $O(C_1) \subseteq C_1$. Let S_1 be a kernel of $< C_1 >$. Consider the subdigraph induced by $C_2 = V - S_1 - \{x : x \in V, x \text{ has a successor in } S_1\}$. Clearly, C_2 is a proper subset of V; consequently, $< C_2 >$ has a kernel S_2. The set $S_1 \cup S_2$ is independent, because no arc goes from S_1 to S_2 (because S_2 does not meet the terminal component C_1), and no arc goes from S_2 to S_1 (by the definition of C_2). Therefore, the set $S_1 \cup S_2$, which is also absorbant for D, is a kernel of D which contradicts the assumption that D is kernel perfect critical. $\square$

It is important to note that *any kernel perfect digraph is an induced subdigraph of some kernel perfect critical digraph.* While a classification of kernel perfect critical digraphs appears hopeless, the problem seems easier for the "minimal kernel imperfect" digraphs and hence the following definition. If D is a digraph with no kernel but for all $v \in V$, $< V - \{v\} >$ has a kernel then D is *kernel critical* (or *K-critical*).

Duchet conjectured that a kernel critical digraph is strongly connected. This was later shown to be false by Duchet and Meyniel [27]; their counterexample appears in Figure 15.5.

Note that kernel perfect critical implies kernel critical but the converse is not true, as illustrated by Figure 15.5. This digraph is not kernel perfect critical because it contains a directed 3-cycle as an induced subdigraph.

Constructions of kernel perfect critical digraphs can be found in [27], [41], [43]. After the initial surge of research on kernel perfectness, other researchers

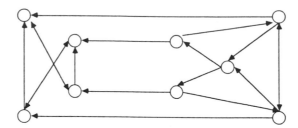

Figure 15.5: A digraph which is kernel critical but not strongly connected.

have made significant contributions. Notable among them are Galeana-Sanchez and Neumann-Lara. They define a digraph to be an *R-digraph* if and only if every non-empty induced subdigraph of D has a nonempty semikernel. It is then proven that an R-digraph is equivalent to a kernel perfect digraph.

In [43] these two authors investigate other sufficient conditions for a digraph to be kernel perfect. They have shown that $Asym(D)$, the subdigraph spanned by the asymmetric arcs in D, of any kernel perfect critical digraph D is strongly connected and a new method of constructing kernel perfect critical graphs is presented. We present some of their results, along with a result of Jacob [61], replacing the term R-digraph with the equivalent term kernel perfect. These and other results provide evidence that by using ordinary graph compositions the family of kernel perfect digraphs can be extended. Note that the other graph compositions can be tested for kernel perfectness.

Theorem 15.14 (Galeana-Sanchez, Neumann-Lara [43]) *If $V(D)$ has a partition $\{V_1, V_2\}$ such that every arc from V_1 to V_2 in D is symmetric and $< V_1 >$ and $< V_2 >$ are kernel perfect, then D is kernel perfect.*

Theorem 15.15 (Galeana-Sanchez, Neumann-Lara [43]) *A digraph D is kernel perfect if and only if for every strong component α of $Asym(D)$, $< V(\alpha) >$ is kernel perfect.*

Corollary 15.16 *If for every strong component α of $Asym(D)$, $< V(\alpha) >$ is bipartite then D is kernel perfect.*

Theorem 15.17 (Jacob [61]) *Let D_1, D_2 and D be digraphs such that $V(D_1) \cap V(D_2) = \{v\}$ and $D = D_1 \cup D_2$. Then D is an kernel perfect if and only if D_1 and D_2 are kernel perfect.*

In [20] K. Chilakamarri and P. Hamburger present a construction of a class of digraphs in which each of the digraphs is either kernel perfect or kernel perfect critical. Within this class they establish criteria to distinguish kernel perfect digraphs from kernel perfect critical digraphs. They rely on the following result proven by Galeana-Sanchez and Neumann-Lara [43], Duchet and Myniel [27], and by Duchet [24].

Theorem 15.18 (Neumann-Lara, Galeana-Sanchez [43]) *If Asym(D) is acyclic, then D is a kernel perfect graph.*

Construction: Let n, r, and s be integers with $r \geq 1, s \geq 0$ and $2r + 2s + 1 \leq n$. Let $\vec{C}_n$ be a circuit of length n on vertices $\{x_0, x_1, \ldots, x_{n-1}\}$, the arcs being $(x_0, x_1), (x_1, x_2), \ldots, (x_{n-2}, x_{n-1}), (x_{n-1}, x_0)$. Let the digraph $D(n, r, s)$ be constructed from $\vec{C}_n$ by adding the arcs (x_i, x_j) and (x_j, x_i) whenever the indices i and j satisfy the following condition:

$$\pm (i - j) \equiv r + 1, r + 2, \ldots, r + s(\text{mod } n). \qquad (15.1)$$

We give two examples that appear in [20]. The first digraph is $D(7, 2, 1)$ and the second is $D(7, 1, 1)$. It is also noted that if $D(n, r, s)$ has a kernel, then it is kernel perfect else it is kernel perfect critical.

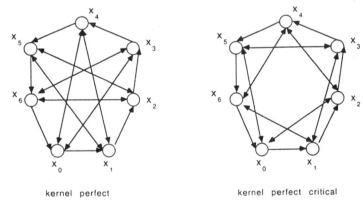

kernel perfect   kernel perfect critical

Figure 15.6: Kernel perfect and kernel perfect critical digraphs.

Theorem 15.19 (Chilakamarri, Hamburger [20]) *If r is an odd integer and $s \geq r$, then $D(n, r, s)$ is a kernel-perfect digraph if and only if $n = t(1+r+s)+2R$ for some positive integer t and for some nonnegative integer R with $2R \leq t(r-1)$. If r is even, then the digraph $D(n, r, s)$ is a kernel-perfect graph.*

Corollary 15.20 (Galeana-Sanchez, Neumann-Lara [43]) *If $r = 1, s \geq 1$, and $(s + 2)$ does not divide n, then $D(n, 1, s)$ is a kernel perfect critical graph.*

Corollary 15.21 *If r is odd, $s \geq r$ and $n = 1 + 2r + 2s$, then $D(n, r, s)$ is a kernel perfect critical graph.*

Corollary 15.22 *If n is odd, r is odd, $s \geq r$, and s is even, then $D(n, r, s)$ is a kernel-perfect-critical graph.*

15.4.3   Normal orientations and quasi perfect graphs

The attempt to link kernels with perfect graphs led to the definition of orientations of undirected graphs mainly for the class of perfect graphs. A graph G is *perfect* if for every induced subgraph H of G, $\chi(H) = \omega(H)$. The most interesting conjecture which led to extensive research in this area was one by Berge and Duchet [8].

Conjecture 15.23 (Berge, Duchet [8]) *A graph G is perfect if and only if any normal orientation of G is kernel perfect.*

Before we study these orientations we survey some of their properties beginning with some important definitions. A *clique* is a digraph whose underlying graph is complete. A digraph D is *normal* (*admissible*) if every clique in D has a kernel. The underlying graph G is then said to have a normal orientation. Any digraph can then be viewed as an orientation of its underlying graph. In particular an orientation may contain reversible arcs. In a normal digraph, D, any clique C has a *receiver*, i.e., a vertex which is a successor of all other vertices of C. Thus the elements of C can be ordered in a sequence $v_1, v_2, \ldots, v_p$ such that each v_i is a receiver of the clique $\{v_i, v_{i+1}, \ldots, v_p\}$. Any sequence of this form is called a *normal ordering* of C in D. The last vertex v_p of any normal ordering is a predecessor of all other elements of clique C (we say v_p is an *emitter* of C). To any family $(\leq_C)$ of linear orders on maximal cliques of an undirected graph G we associate a digraph D, where (x, y) is present whenever $x \leq_C y$. Digraphs which are obtained in such a way are called *principal directions* of the graph G.

Lemma 15.24 (Berge, Duchet [8]) *For any digraph D, the following conditions are equivalent:*

1. *D is normal.*

2. *D^{-1} is normal.*

3. *D contains a principal direction of the underlying digraph.*

4. *Every clique of D contains a receiver.*

5. *Every clique of D contains an emitter.*

6. *Every circuit of a clique of D possesses at least one reversible arc.*

An undirected graph G is said to be *quasi-perfect* if *every normal orientation* of G is kernel perfect. The complete undirected graph K_n is a quasi-perfect graph. The three succeeding results show a strong link between perfect graphs and quasi-perfect graphs [9].

Theorem 15.25 (Berge, Duchet [9]) *The graph C_{2k+1}, with $k \geq 2$, is not quasi-perfect.*

Theorem 15.26 (Berge, Duchet [9]) *The graph $\overline{C}_{2k+1}$, with $k \geq 2$, is not quasi-perfect.*

Theorem 15.27 (Berge, Duchet [9]) *Every induced subgraph of a quasi-perfect graph is quasi-perfect.*

It follows from the above that a quasi perfect graph cannot have an induced C_{2k+1} nor an induced $\overline{C}_{2k+1}$ as subgraphs, and hence the name for these graphs. Many classes of perfect graphs have been studied for quasi-perfectness and some of the important classes are given below. Refer to [9] for definitions of these classes.

1. (Maffray [75]) If G is chordal then G is quasi-perfect.

2. (Maffray [76]) If G is i-triangulated then G is quasi-perfect.

3. (Blidia, Duchet, Maffray [13]) Complements of strongly perfect graphs are quasi-perfect. This implies that complements of comparability and complements of Meyneil graphs are also quasi-perfect.

4. (Maffray [75]) Perfect graphs with no induced $K_{1,3}$ nor K_4 - an edge e are quasi-perfect; thus perfect line graphs are are also quasi-perfect.

Quasi-perfectness has yet to be determined for many subfamilies of perfect graphs for example, the class of weakly chordal graphs, the Meyneil graphs and parity graphs to name a few. Some work has been done on these families. Blidia in [12] and [11] has shown that a parity digraph has a kernel. Maffray has shown that i-triangulated digraphs have a kernel, and using some results from Jacob [61] it can be inferred that these are quasi-perfect. Another important area of research which is related to this study is the concept of M-solvability.

A digraph D has an *M-orientation* if and only if every 3-circuit has at least two symmetric arcs. A graph G is said to be *M-solvable* if every M-orientation of G is kernel-perfect. The concept of M-solvability received a lot of attention because of the following conjecture.

Conjecture 15.28 (Meyneil [16]) *Every perfect graph is M-solvable.*

Christopher Champetier [16] provides a proof of this conjecture of Meyniel in the case of comparability graphs. Any orientation of a comparability graph such that all 3-directed cycles have at least 2 reversible arcs is kernel-perfect. Parity graphs have been shown to be M-solvable by Blidia [23]. This result is generalized to Meyneil graphs by Blidia, Duchet and Maffray [13]. Here too as in the case of quasi-perfectness there are many open questions.

15.4.4   Generalizations of kernels

The attempt to generalize the concept of a kernel stems from the fact that not every digraph has a kernel. One of the first endeavors in this direction was by Chvatal and Lovasz and motivated by a famous result concerning tournaments. It is known that not every tournament contains a *transmitter* but every tournament contains a *king*. A *transmitter* is a vertex u in a tournament T of order n, such that $|O(u)| = n-1$. A *king* is a vertex u in a tournament T such that $\forall\, v \in V(T)$, $d(u,v) \leq 2$. The result specified above led them to define a *"semikernel"* in a digraph as follows:

1. If $u, v \in S$, where $u \neq v$, then $d(u,v) \geq 2$.

2. For any $v \notin S$ there exists a vertex $u \in S$ such that $d(u,v) \leq 2$.

We should warn the reader that this definition of semikernel is different from the concept introduced earlier and shall be denoted by *"semikernel"*.

Theorem 15.29 (Chvatal, Lovasz [21]) *Every digraph has a "semikernel".*

This idea of a "semikernel" was further generalized to the concept of a (k, l) kernel by Kwasnik [64], [65], [66], [68]. In this connection see Kwasnik et al [69] and Galeana-Sanchez [44]. She defined a (k, l) kernel in a digraph as a subset $J \subseteq V(D)$, $k \geq 2$ and $l \geq 1$, such that

1. If $u, v \in J$, where $u \neq v$, then $d(u,v) \geq k$,

2. For any $v \notin J$ there exists a vertex $u \in J$ such that $d(u,v) \leq l$.

For $k = 2$ and $l = 1$ we get a kernel in the sense of Berge and $k = 2$ and $l = 2$ then we get a "semikernel" in the sense of Chvatal and Lovasz. The existence of (k, k) was pursued in [68] where the following result was shown.

Theorem 15.30 (Kwasnik [68]) *Every digraph has a (k, k) kernel, for all $k \geq 2$.*

Also, existence results for (k, l) kernels are shown for strongly connected digraphs and some of their graph products [65],[66]. Furthermore, the concept of a k-kernel was defined as a $(k, k-1)$ kernel. For other results on strongly connected digraphs see Marcu [77], [78].

Theorem 15.31 (Kwasnik [68]) *Let D be a strongly connected digraph such that every directed cycle of D has length $\equiv 0 \pmod k$, $k \geq 2$. Then D has a k-kernel.*

Galeana-Sanchez and Neumann-Lara [45] generalized this as follows.

Theorem 15.32 (Galeana-Sanchez, Neumann-Lara [45]) *Let D be a digraph such that Asym(D) is strongly connected. Furthermore suppose that for every directed cycle C such that l(C) ≢ 0(mod k), where l denotes length of the cycle C, either of the following are satisfied:*

(a) Every arc of C is a symmetrical arc of D.

(b) C has at least k symmetrical arcs.

Then D has a k-kernel.

Note that when $l(C) \leq k - 1$ then C must satisfy (a). This result was again generalized for (k, l) kernels.

In another attempt at generalization, H. Galeana-Sanchez, L. Pastrana Ramirez and H. A. Rincon-Mejia discussed the term *quasikernel* [47] (see [86] for other results). A *quasikernel* Q of D is an independent set of vertices such that $V = Q \cup I(Q) \cup I(I(Q))$. Since a "semikernel" is a quasi kernel, every digraph will have a quasikernel. The *line digraph* of $D = (V, A)$ is the digraph $L(D) = (A, W)$ with the set of vertices the set of arcs of D and for any $h, k \in A$ there exists a $(h, k) \in W$ if and only if corresponding arcs h, k induce a directed path in D; i.e., the terminal endpoint of h is the initial endpoint of k. It is then proven [47] that:

Theorem 15.33 *If D is a digraph such that every vertex has indegree at least one then the number of semikernels of D is less than or equal to the number of semikernels of L(D).*

Theorem 15.34 (Galeana-Sanchez, Ramirez, Rincon-Mejia [47]) *If D is a digraph such that every vertex has indegree at least one then the number of quasikernels of D is less than or equal to the number of quasikernels of L(D).*

For other results on generalizations refer to [83], [105], [45] and [46].

15.5 Kernels and Grundy Functions

In 1939, P. M. Grundy [51] defined a nonnegative integer function on V satisfying certain conditions, later called a Grundy function. A nonnegative integer function $g : V \rightarrow Z^{+} \cup \{0\}$ is called a *Grundy function* on D if for every vertex x, $g(x)$ is the smallest nonnegative integer which does not belong to the set $O(x)$. This concept was introduced by Grundy for digraphs without circuits. Berge and Schützenberger [7] extended this concept to arbitrary digraphs.

A Grundy function can also be defined as a function $g(x)$ such that,

1. $g(x) = k > 0$ implies that for each $j < k$ there is a $y \in O(x)$ with $g(y) = j$

2. $g(x) = k$ implies that each $y \in O(x)$ satisfies $g(y) \neq k$

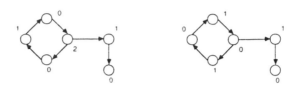

Figure 15.7: A digraph with more than one Grundy function.

It should be pointed out that not all digraphs have Grundy functions; for example a directed 5-cycle. Figure 15.7 illustrates that some digraphs have more than one Grundy function.

There is a close relationship between a Grundy function and a kernel. As a matter of fact, if D has a Grundy function g, then D has a kernel, since the set $S = \{x : x \in V, g(x) = 0\}$ satisfies simultaneously (as given in Berge):

1. $x \in S \Longrightarrow g(x) = 0 \Longrightarrow$ for $y \in O(x)$, that min $g(y) > 0 \Longrightarrow O(x) \cap S = \emptyset$

2. $x \notin S \Longrightarrow g(x) > 0 \Longrightarrow$ for $y \in O(x)$, that min $g(y) = 0 \Longrightarrow O(x) \cap S \neq \emptyset$.

and hence S is a kernel.

The converse is not true. It can be verified that the graph in the Figure 15.8 has a kernel $\{d\}$ but does not have a Grundy function.

Figure 15.8: A digraph with a kernel but no Grundy function.

The next result shows that if a digraph is kernel perfect then it has a Grundy function.

Theorem 15.35 (Berge, Schützenberger [7]) *C. Berge and M. S. Schützenberger. If D is a digraph such that each induced subgraph has a kernel then D has a Grundy function.*

Corollary 15.36 (Berge, Schützenberger [7]) *A transitive digraph has a Grundy function.*

Corollary 15.37 (Berge, Schützenberger [7]) *A digraph without odd circuits has a Grundy function.*

Theorem 15.38 (Grundy [51]) *A digraph without circuits possesses a unique Grundy function $g(x)$. Moreover, for each x, $g(x)$ does not exceed the length of the longest path from x.*

15.6 Solutions in Digraphs

The concept of domination of imputations in n-person games was developed in the now classical book by Von Neumann and Morgenstern [107]. In that formulation, each imputation was represented by an element a,b,c ... of some universal set P. If a dominates b then $a > b$. The *relation* $>$ defined on the elements of P was assumed to be *irreflexive* , i.e., $a \not> a$ for any a in P.

The subset S of P is a *solution* of the relation $>$ if the following holds:
(i) For any two elements a and b in S, $a \not> b$.
(ii) For any $a \in P - S$, $\exists\, b \in S$ such that $b > a$.
An irreflexive relation is *solvable* if it has at least one solution.

This formulation can be visualized in terms of a digraph D whose vertex set $V(D)$ represents the set P and there exists an arc from vertex a to vertex b if and only if $a > b$. Note that D has no loops since the relation $>$ is irreflexive. The problem of finding a *solution* of a relation is equivalent to finding an *independent* and *dominating* set in the digraph D, which we henceforth call a *solution* in the digraph. Note that the concepts of *solutions* and *kernels* are *directional duals* [56] and these sets coincide for ordinary graphs.

In the work by Von Neumann and Morgenstern, it is shown in the language of relations that an *acyclic* digraph has a solution (Theorem 15.9). This result also shows that the solution is unique. Richardson, in a series of papers ([87],[88],[89],[90]), generalizes the concept to arbitrary digraphs, finite or infinite. He shows in [89] that every digraph with no odd circuits has at least one solution. While this establishes the existence of solutions in some cases which are not acyclic, it is of little interest from the point of view of the theory of games, because the assumption that there is no odd cycle implies that the relation $>$ is transitive. For the theory of games, neither transitivity nor intransitivity can be assumed. The purpose of [88] is to establish sufficient conditions for the existence of solutions in certain cases where intransitivity is not required and which is not necessarily acyclic.

An extension of the concept of a solution is proposed in [90]. Richardson defines the relevant ideas as follows:

Let D be any digraph (system) and D_o be a subgraph (subsystem) and suppose S is a solution of D and S_o is a solution of D_o. Solution S is called an *extension* of S_o if $S \cap V(D_o) = S_o$, in which case S_o is called the relativization of S. In [90], he finds some sufficient conditions for the existence of extensions and relativizations of solutions.

The major questions in this field of study are proposed by Richardson and Harary in [54], where the following appears:

I. Characterize solvable and uniquely solvable irreflexive relations or equivalently, characterize those loopless digraphs which have a solution or those which have a unique solution.

II. Find all solutions of an irreflexive relation (loopless digraph).

Their paper supplies an algorithm for the second question in the case of finite irreflexive relations and develops some properties of a generalization of the concept of solution. In addition, some sufficient conditions for the existence of solutions in specific digraphs are mentioned. Define a *transmitter* to be a vertex with indegree 0.

Lemma 15.39 (Harary, Richardson [54]) *Every transmitter in a digraph lies in every solution.*

Lemma 15.40 (Harary, Richardson [54]) *If v is a transmitter of D then D is solvable if and only if D-v-$O(v)$ has a solution.*

Lemma 15.41 (Harary, Richardson [54]) *If a vertex v in a digraph has the property that it lies in every solution, then D is solvable if and only if D-v-$O(v)$ is solvable.*

The procedure for finding all solutions of an irreflexive relation is based on an observation concerning cliques [55]. Note that the complement of an independent set of vertices is a clique. So a matrix algorithm is developed to test all the cliques of the complementary relation.

At about the same time, the same subject matter is mentioned in the book *Theorie der endlichen und unendlichen Graphen*, by Konig [62]. He uses the term *a point basis of the second kind* to define a solution in a digraph. From that definition Harary and Richardson generalized the concept to a r-basis. A subset S of the set $V(D)$ of a digraph D is an *r-basis* if the following holds:

(i) For any two distinct vertices u,v in S, the distance from each other is greater than r.

(ii) For any vertex v in $V(D) - S$ there exists a vertex u in S such that the directed distance from u to v is no greater than r.

In these terms, a solution in a digraph D is a *1-basis*. Also in the case of tournaments there exists a famous theorem which can be worded in terms of an r-basis.

Theorem 15.42 (Harary, Richardson [54]) *Every tournament has a singleton 2-basis.*

Another result concerning tournaments which completely characterizes solvability among tournaments is also given in [54].

Theorem 15.43 (Harary, Richardson [54]) *A tournament T is solvable if and only if it has a transmitter.*

Proof. If T has a transmitter then it is solvable. Now suppose that T has a solution $S \subseteq V(T)$. Since a tournament has an arc between every pair of vertices, it is evident that $|S| \not> 1$. This says that $|S| = 1$. If this singleton has to be dominating then it must be a transmitter. □

The same concept received an independent and interesting treatment by Berge in his book [5]. He devotes a chapter to the study of kernels and their relation with Grundy functions. Solutions in digraphs are not dealt with in detail but mentioned as being the concept of a kernel in the reversal digraph.

Question **I**, as proposed by Harary and Richardson [54] is answered to some extent by Harary and Behzad.

Theorem 15.44 (Harary, Behzad [53]) *A digraph D has a solution if and only if D does not have an induced subdigraph E with no solution such that the indegree of each vertex in E is the same in E as in D.*

Theorem 15.45 (Harary, Behzad [53]) *Let D be a digraph with $n \geq 3$ vertices and at least $n^2 - 2n + 1$ arcs. Then D has a solution and a kernel. Moreover there exists a digraph with $n^2 - 2n$ arcs which has neither a solution nor a kernel, so this bound is sharp.*

The construction of this extremal digraph is as follows:
Consider the complete digraph with n vertices $v_1, v_2, \ldots, v_n$ where $n \geq 3$. Remove the arcs $(v_1, v_2), (v_2, v_3), \ldots, (v_{n-1}, v_n)$ and (v_n, v_1) which form a hamiltonian circuit. The resulting digraph has neither a solution nor a kernel.

Since acyclic digraphs have a solution, Harary and Behzad concentrated on unicyclic digraphs and obtained some interesting results. They stated that the questions **I** and **II** given above could be answered for special families of digraphs such as Eulerian, Unipathic, Strong, Flexible, Strictly Unilateral, Strictly Weak, etc. Other than tournaments, very few classes of digraphs have been studied for solvability. In addition to finding solutions, they state a result about the number of solutions in a digraph D. Let $S(D)$ denote the number of solutions in a digraph D.

Theorem 15.46 (Harary, Behzad [53]) *For each arbitrary positive integer m, there exists a connected digraph D of order n such that $S(D) > mn$.*

Harminic also studied the number of solutions in digraphs.

Theorem 15.47 (Harminic [57]) *The number of solutions (kernels) of a digraph is equal to the number of solutions (kernels) of its line graph.*

He also demonstrates how to construct solutions of the line graph from the solutions of the original digraph. A nice result concerning the characterization of digraphs having the same set as a solution and kernel is in [53].

Theorem 15.48 (Harary, Behzad [53]) *Let L be an independent set of vertices in D. Then L is both a solution and a kernel for D if and only if D contains a set K of directed paths or cycles containing all vertices not in L, which start and end with a vertex of L and alternately contain vertices of L.*

Lee, in his doctoral thesis [70], works on the solvability of digraphs for some special families of acyclic digraphs. He obtains bounds on the cardinality of solution for oriented trees, binary trees and the expected cardinality of the solution for a random binary tree.

Although the concept of domination, and therefore of independent domination, has been extensively studied in graphs, very little work has been done on solvability of digraphs. A host of open problems remain in the area, some of which are mentioned below.

15.7   Domination in Digraphs

Although the concept of domination in graphs has received extensive attention as evidenced by this volume, the same concept has been somewhat sparsely studied for digraphs. Even bounds for undirected graphs have not been considered and compared with their counterparts for digraphs. In terms of applications, the questions of *Facility Location, Assignment Problems* etc. are very much related to the idea of domination or independent domination on digraphs. There have been over the years a few papers on the domination number of digraphs. These and other related concepts are presented below. We use the notation $\gamma(D)$ to represent the domination number of a digraph, i.e., the *minimum* cardinality of a set $S \subseteq V(D)$ which is *dominating*.

15.7.1   Existing work

The first Ph.D. dissertation dedicated to the study of the domination numbers in digraphs was as recent as 1994. C. Lee [70] has surveyed some of the bounds on families of undirected graphs and proposed corresponding ones for digraphs. N. Alon and J. Spencer [1] have shown that for any graph G with minimum degree δ , $\gamma(G) \leq \frac{1+ln(\delta+1)}{\delta+1}|V(G)|$.

Theorem 15.49 (Lee [70]) *Let D be a digraph of order n and minimum indegree $\delta^- \geq 1$. Then, we have $1 \leq \gamma(D) \leq \frac{\delta^-+1}{2\delta^-+1} n$.*

In [70], bounds for various families of digraphs such as Unilateral and Contrafunctional are given. The domination number of a random digraph is as follows.

Theorem 15.50 (Lee [70]) *For p fixed, $0 < p < 1$, a random digraph D_n has domination number either $\lfloor k + 1 \rfloor$ or $\lfloor k + 2 \rfloor$ almost surely, where $k = \log n - 2\log\log n + \log\log e$, where log denotes the logarithm with base $\frac{1}{1-p}$.*

The previous result is analogous to the one proved by Weber [109] for undirected graphs.

The algorithmic results on the domination number of digraphs should parallel those for undirected graphs. The question of finding an efficient dominating set in a digraph was shown to be NP-complete by Barkauskas and Host [3] and a linear algorithm was given for finding the minimum number of vertices in an efficient dominating set for oriented trees. The existence of a dominating set which does not contain a path of length k in a digraph was shown to be NP-complete by Bar-Yehuda and Vishkin in [4].

In the study of domination on undirected graphs, the concepts of *strong* and *weak* domination were introduced by Sampathkumar and Latha [96]. It is shown in [48] that the study of these parameters on undirected graphs is equivalent to the study of the domination number of certain types of digraphs created from the graphs given.

15.7.2 Domination concepts in tournaments

Tournaments are the most studied class of digraphs because of the close linkage to various applications. In this case too there have been many domination related studies in tournaments. In his thesis on domination of digraphs Lee [70] exhibited bounds for various classes notably tournaments.

Theorem 15.51 (Lee [70]) *For any tournament T of order n. Then $1 \leq \gamma(T) \leq \lfloor \log_2(n + 1) \rfloor$.*

It is not at all obvious whether for arbitrary positive integers k, there exists a tournament T with $\gamma(T) = k$. Erdös [29] provides a partial answer using probabilistic methods. The construction of such tournaments is detailed in a paper by Graham and Spencer.

Theorem 15.52 (Graham, Spencer [50]) *For every $\epsilon > 0$, there is an integer K such that for every $k \geq K$ there exists a tournament T_k with no more than $k^2 2^k (\log 2 + \epsilon)$ vertices such that $\gamma(T) > k$.*

The problem of a dominating set for a tournament is also considered by Meggido and Vishkin [82], who show the problem of finding a minimum dominating set in a tournament can be solved in $n^{O(\log n)}$ time. It is shown that if this problem has a polynomial time algorithm then for every constant C, there is also a polynomial time algorithm for the satisfiability problem of boolean formulas in conjunctive normal form with m clauses and $C \log^2 m$ variables. Let us consider the probability space T_n consisting of random tournaments on the vertex set

$V = \{v_1, v_2, \ldots, v_n\}$. By a *random tournament* we mean here a tournament on vertex set V obtained by choosing for each $1 \leq i, j \leq n$, independently either the arc (v_i, v_j) or the arc (v_j, v_i) where each of the choices is equally likely.

Theorem 15.53 (Lee [70]) *A random tournament $T \in T_n$ has domination number either $\lfloor k_* \rfloor + 1$ or $\lfloor k_* \rfloor + 2$, where $k_* = log_2(n) - 2log_2(log_2(n)) + log_2(log_2(e))$.*

Domination graphs of tournaments were introduced in connection with competition graphs of tournaments. See Lundgren [74] for details on competition graphs. The *domination graph dom(D)* of a digraph D has the same vertex set as the digraph with an edge between two vertices if those two vertices dominate every other vertex in the digraph. Fisher, Lundgren, Merz, and Reid [31] showed that the domination graph of a tournament is either an odd cycle with or without isolated and/or pendant vertices or a forest of caterpillars.

Theorem 15.54 (Fisher, Lundgren, Merz, Reid [31]) *For a tournament T, dom(T) is either an odd cycle, with or without isolated and/or pendant vertices, or a forest of caterpillars. Further, any graph G consisting of an odd cycle with or without isolated and/or pendant vertices is the domination graph of some tournament.*

Later in [32], Fisher, Lundgren, Merz and Reid extended the concept to oriented graphs. They show that any caterpillar is the domination graph of some digraph, but a path on four or more vertices is not the domination graph of any tournament. Another paper which deals with connected domination graphs is [33].

15.7.3   New results

In this section we explore some domination related results on digraphs analogous to those of undirected graphs. First we look at some common bounds for $\gamma(D)$. One of the earliest bounds for the domination number for any undirected graph was proposed by Ore.

Theorem 15.55 (Ore [85]) *For any graph G without isolates, $\gamma(G) \leq \frac{n}{2}$, where n is the number of vertices.*

This result does not hold for directed graphs; a counterexample is the digraph $K_{1,n}$, $n \geq 2$, with its arcs directed from the endvertices towards the central vertex. The general bound which holds for digraphs is not very good for a majority of digraphs. We assume our digraphs to be those whose underlying graphs are connected.

Observation 15.56 *For any digraph D with n vertices, $\gamma(D) \leq n - 1$.*

This bound is sharp because the domination number of the digraph $K_{1,n}$, for $n \geq 2$ with its arcs directed from the endvertices towards the central vertex is n. Since very few graphs agree with this bound we find other bounds which are tighter for a significant number of digraphs.

Theorem 15.57 *For any digraph D on n vertices, $\frac{n}{1+\Delta(D)} \leq \gamma(D) \leq n - \Delta(D)$, where $\Delta(D)$ denotes the maximum outdegree.*

Proof. For the upper bound we form a dominating set of D by including the vertex v of maximum outdegree and all the other vertices in the digraph which are not dominated by v. This set is clearly a dominating set and has cardinality $n - \Delta(D)$.

Note that any vertex in D can dominate at most $1 + \Delta(D)$ vertices. In a minimum dominating set S of D there are $\gamma(D)$ vertices, so they can dominate at most $\gamma(D)(1 + \Delta(D))$ vertices. Since S is dominating this number has to be at least n. Thus we get the lower bound. $\square$

To get another bound we look for certain characteristics in a digraph.

Observation 15.58 *For any digraph D on n vertices, which has a hamiltonian circuit, $\gamma(D) \leq \lceil \frac{n}{2} \rceil$.*

Proof. Let D contain a hamiltonian circuit C. To dominate the vertices of D it suffices to dominate the cycle C. We know that the domination number of a circuit is bounded above by $\lceil \frac{n}{2} \rceil$ and so the same holds for the digraph D. $\square$

Theorem 15.59 [70] *For a strongly connected digraph D on n vertices, $\gamma(D) \leq \lceil \frac{n}{2} \rceil$.*

In addition to $\gamma(D)$ we introduce some domination related parameters in digraphs, in particular, the *irredundance number*, the *upper irredundance number* and the *upper domination number*, analogous to those for undirected graphs. Recall that a set $S \subseteq V(D)$ of a digraph D is a *dominating set* if for all $v \notin S$, v is a successor of some vertex in S. A dominating set S is a *minimal dominating set* if for every $v \in S$, $O[v] - O[S - v] \neq \emptyset$. If $u \in O[v] - O[S - v]$, then u will be called a *private outneighbor* (*pon*) of v with respect to S. See [38] for another characterization of minimal dominating sets in digraphs.

Let $\Gamma(D)$, the upper domination number, denote the maximum cardinality of a minimal dominating set. As in the undirected case, we define an *irredundant set* $S \subseteq V(D)$ to be a set such that every $v \in S$ has a private outneighbor. The *irredundance number $ir(D)$* and the *upper irredundance number $IR(D)$* are, respectively, the minimum and maximum cardinalities of a maximal irredundant set.

The notion of a solution also yields parameters which are new to the field of digraphs. Let $i(D)$ and $\beta(D)$ denote respectively the minimum and maximum cardinalities of an independent dominating set. It must be pointed out that not all digraphs have independent dominating sets. As these are special cases of solutions, these exist in digraphs which admit at least one solution. It must be mentioned here that due to the concepts defined above the following string of inequalities hold for any digraph D with a solution,

$$ir(D) \leq \gamma(D) \leq i(D) \leq \beta(D) \leq \Gamma(D) \leq IR(D). \qquad (15.2)$$

Researchers interested in domination theory for undirected graphs are quite familiar with the corresponding inequality chain. This chain raises some interesting questions about the structural properties of digraphs D (having a solution), for which

1. $\gamma(D) = i(D)$,

2. $ir(D) = \gamma(D)$,

3. $\beta(D) = \Gamma(D) = IR(D)$, or

4. $i(D) \neq \beta(D)$.

The following theorem is an interesting result for transitive digraphs.

Theorem 15.60 *For a transitive digraph D, we have $\gamma(D) = i(D) = \beta(D) = \Gamma(D) = IR(D)$.*

Proof. Note that if D is a transitive digraph so is its reversal D^{-1}. It is then known that a solution exists in D. Moreover, from Berge's theorem, we see that in D, every minimal absorbant set is independent and the kernel is unique. This implies that $\gamma(D) = i(D) = \beta(D) = \Gamma(D)$.

To show $\beta(D) = IR(D)$, suppose that S is an irredundant set with $|S| = IR(D)$. We will call such a set an IR-set. Amongst all IR-sets let S contain the minimum number of arcs in it. If S has no arcs, then certainly S is independent and $\beta(D) \geq IR(D)$ implying $\beta(D) = IR(D)$. So suppose that $< S >$ contains an arc (x, y). Since S is irredundant y must have a private outneighbor $y_1 \notin S$. But D is a transitive digraph, so (x, y_1) must be an arc, contradicting that y_1 is a private neighbor of x. Hence S is independent and the result follows. $\square$

In the spirit of the discussion above, a natural area of research would be to look at the smallest cardinality of a kernel, $k(D)$, and the maximum cardinality of a kernel, $K(D)$. Observe that by König's result (Corollary 15.4), for any transitive digraph D, $k(D) = K(D)$.

Recent work on domination in digraphs has been produced by researchers at Western Michigan University. In [18] the authors define the *lower orientable domination number, dom(G)*, as the minimum domination number among all orientations of G, while the *upper orientable domination number, DOM(G)*, is the maximum domination number among all orientations of G. Some results that the authors have presented are, for any graph, G, $dom(G) = \gamma(G)$ and for every integer c with $dom(G) \leq c \leq DOM(G)$, there exists an orientation D of G such that $\gamma(D) = c$. Other work in [60] extends this concept by defining, for a positive integer k, a *k-step dominating set* for a digraph D to be a set S of vertices of D such that for every v of D there is a vertex $u \in S$ such that the directed distance from u to v is k.

15.8   Applications in Game Theory

The theory of games and strategies involved for one, two, three and n-player games is studied in detail in the book *Theory of Games and Economic Behavior* by Von Neumann and Morgenstern. This pioneering work establishes the situations and variabilities of games and economic situations in terms of mathematical ideas and concepts. The analysis is done by representing the various situations in a game by elements in a set and defining a relation amongst them. The set of *imputations*, as they call it, led to the definition of a *solution* in terms of the relation defined on that set. After the connection between relations and digraphs is established by Richardson, much interest is generated in studying the *winning* strategies of a game in terms of the underlying digraph. These studies expressed the connection between the *winning* moves in a game in terms of a *kernel*. The concept of a game in a digraph has been extensively studied. For a list of references see [63] and [5], [6], [99], [100]. We now survey the development of this area of research starting with a famous work by C.A.B. Smith [99] which details the creation of a digraph from a given game.

We concentrate our attention on two-player games without chance like Chess, Nim or Tic-Tac-Toe. Let the vertices of a digraph D represent all the *situations* in the game being played. Vertices u and v are connected by an arc if there is a legal *move* for a player which changes situation u to v. Let the two players involved be denoted by A and B. In this digraph, the vertices with outdegree 0 are called *terminal*. The *class* of all terminal vertices is divided by the rules of the game into three classes: T_A, meaning win for A; T_B, meaning win for B and T_O, which represents a draw. Playing the game would mean choosing an initial vertex v_o (by player A) which represents a starting situation in the game. Player B would then have to choose a vertex v_1 such that $v_1 \in O(v_o)$ in the digraph. Player A then has to choose a vertex v_2 where $v_2 \in O(v_1)$. In this manner, the players alternately generate a *directed walk* in the digraph and try to reach one of the terminal classes. The different walks through the digraph represent different *strategies* for the two players. It is possible for a walk to be arbitrarily long, e.g., by going around in a cycle. Since most of the games generate acyclic digraphs, this problem does not arise often.

In some games all moves from a situation u are legal for both players, but in some the players have different moves from the same vertex in the digraph. The first case defines an *impartial* or *symmetric* game while the latter case is called partial or asymmetric. There exist *compound* games where the players A and B simultaneously play a number of impartial games. In a *conjunctive* compound game a player makes a move in every component game and in a *disjunctive* compound, a player selects makes a move in only one of the components.

Let us now look at some popular games and study the connection between graph theoretic concepts and winning strategies. We start with some examples of *Nim game* as given in [5] and [7].

Example 1: There are p piles of matches. Two players alternately selects a pile and removes at least one match from it. The player that removes the last match wins.

Example 2: There are p piles each with a different number of matches. In this variation of the game above, there are two moves available to a player. A player either selects a pile and removes at least one match from it or removes an equal number of matches from all piles. The player that removes the last match wins.

Example 3: Each of two players alternately place a tile that covers three squares in a straight line on an $n \times n$ chessboard. No square can be covered twice. The last player to place a tile wins.

Example 4: (Poison game) Players A and B play alternately on the vertices of a digraph D. Player A starts by selecting a vertex of his choice and then each player selects in turn a vertex which is a successor of the vertex previously chosen by his opponent. Player B "poisons" the vertices on which he plays. Player B wins if A cannot select at some stage of the game, a healthy vertex. To the contrary, Player A wins if he is not poisoned or if B is unable to move.

Let us consider the first game in terms of its underlying digraph. Suppose we have two piles of matches with 5 matches each. The vertices in the digraph represent the points (i, j), with i and j ranging from 0 to 5. The starting vertex is the one which represents the point $(5, 5)$. The first player to reach the terminal point $(0, 0)$ wins. The rules of the game dictate that from the vertex (i, j) one can go to the vertices $(i - 1, j)$, $(i - 2, j)$, $(i - 3, j)$, ..., $(0, j)$ or to the vertices $(i, j - 1)$, $(i, j - 2)$, ..., $(i, 0)$. This digraph contains no directed cycles and hence by Theorem 15.9, has a unique kernel. In fact it was shown for general Nim games and their underlying digraphs that:

Theorem 15.61 (Berge, Schützenberger [7]) *If the digraph D possesses a kernel S and if a player chooses a vertex in S this choice assures him of a win or a draw.*

Proof. If player A chooses a vertex x_1 in S, then either $O(x_1)$ is empty (win for A) or player B has to choose a vertex x_2 in $V(D) - S$. Player A can then choose a vertex x_3 in S, since x_2 was in $V(D) - S$ and S is a kernel (hence an absorbant set). The terminal points of the game must be contained in the kernel S since they have outdegree 0. Because S is also independent, it is impossible for B to move to a vertex in S and hence player B must lose. A draw will result if there is a circuit in the digraph in which case the game can continue indefinitely. □

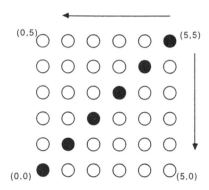

Figure 15.9: A digraph for example 1.

It is easy to check that the vertices $(0,0)$,$(1,1)$, $(2,2)$,$(3,3)$,$(4,4)$,$(5,5)$ form a kernel in the digraph above. The validity of the theorem can be checked by playing the game on the digraph given above.

Let us now play the second game with two piles of matches with 5 and 6 matches, respectively. The digraph representing this game is given below. Note that because of the additional rule this digraph contains the arcs along the diagonals but remains acyclic. The unique kernel in this digraph is given by $(0,0)$, $(1,2)$,$(2,1)$, $(3,5)$ and $(5,3)$. In this version of the game which was first shown by Withoff [110], it is possible for the first player to win by moving from the starting position of $(5,6)$ to an element of the kernel, either $(1,2)$ or $(5,3)$.

Regarding the Poison game in Example 4, Duchet and Meyniel looked at digraphs with a possibly infinite vertex and arc set. They insist that the digraphs are *outwardly finite*; i.e., $|O(x)|$ is finite for every vertex of the digraph or *progressively finite* (no vertex is the origin of an infinite path). The ability to survive the game is not dependent on the entire digraph having a kernel. What is important is the local structure of the digraph; where the players are making their moves. The condition of being outwardly finite allows player A to survive by occupying the semikernel. The main results follow.

Theorem 15.62 (Duchet, Meyniel [28]) *Let D be a progressively and outwardly finite digraph. Player A can survive the poison game on D if and only if D has a semikernel.*

Theorem 15.63 (Duchet, Meyniel [28]) *An outwardly finite digraph is kernel perfect if and only if every finite induced subdigraph has a kernel.*

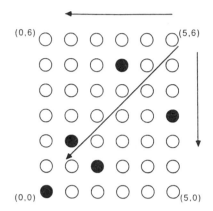

Figure 15.10: A digraph for example 2.

The games described above are all impartial or symmetric games, i.e., the moves for the two players are the same. In many works on this subject, partial or asymmetric games have been studied. Some of them are discussed in [93], [99]. J. Topp in [104] characterizes the winning moves for asymmetric games in terms of kernels and bikernels. Note that in the case of an asymmetric game there are two sets of arcs; one constituting a legal set of moves for player A and one for player B. Hence the digraph is denoted by (D, A_1, A_2).

An ordered pair (S_1, S_2) of subsets of vertices of the vertex set V is defined to be a *bikernel* of a digraph (D, A_1, A_2) if each S_i is a kernel of the digraph (D, A_i). Let $BK(D, A_1, A_2)$ denote the family of all bikernels of the digraph (D, A_1, A_2). Also let $BK_1(D, A_1, A_2)$ be the family of all subsets of V such that for every $S \in BK_1(D, A_1, A_2)$ there exists a set S' in V and $(S, S') \in BK_1(D, A_1, A_2)$. In a similar manner $BK_2(D, A_1, A_2)$ is defined.

A winning vertex for player A is one for which A is guaranteed a win irrespective of the strategies of B; losing and drawing vertices for A are defined analogously. Let W_1, L_1, D_1 (W_2, L_2, D_2) be the set of winning, losing and drawing vertices for player A (player B). Then it is shown that:

Theorem 15.64 (Topp [104]) *For every digraph* (D, A_1, A_2) *of some game,*

$W_i = V - \{S : S \in BK_i(D, A_1, A_2)\};$

$L_i = \cap \{S : S \in BK_i(D, A_1, A_2)\};$

$D_i = \cup \{S : S \in BK_i(D, A_1, A_2)\} - \cap \{S : S \in BK_i(D, A_1, A_2)\}.$

Many other concepts of games and their relations with graph theoretic properties have also been studied in [92], [103], [106], [108], [110].

15.9   Conclusions and Open Problems

Domination and other related topics in undirected graphs are extensively studied, both theoretically and algorithmically. However, the corresponding topics on digraphs have not received much attention, even though digraphs come up more naturally in modelling real world problems. With this view in mind, we have made an attempt to survey some of the existing results on domination related concepts on digraphs. We have also introduced some parameters on digraphs analogous to domination parameters on undirected graphs. All along this brief survey, we have tried to mention some open problems in the area. We have just scratched the surface however, and the interested reader can see that many areas remain open. As a matter of fact, it seems that almost all domination related problems on undirected graphs, if they make sense in digraphs, may be investigated. Algorithmic aspects of these problems on digraphs will be another good area of research.

Acknowledgement : The authors would like to thank the anonymous referee for the comments regarding additions to the survey and for pointing out the new developments in this field.

Bibliography

[1] N. Alon and J. H. Spencer, *The Probabilistic Method.* Wiley, New York, 1992.

[2] E. W. Adams and D. C. Benson, Nim type games. Technical Report No. 13, Carnegie Institute of Technology (1953).

[3] A. Barkauskas and L. Hunt, Finding efficient dominating sets in oriented graphs. *Congr. Numer.* 98 (1993) 27-32.

[4] R. Bar-Yehuda and U. Vishkin, Complexity of finding k-path-free dominating sets in graphs. *Inform. Process. Lett.* 14(5) (1982) 228-232.

[5] C. Berge, *Graphs and Hypergraphs.* North Holland Publishing Co. North Holland, New York, 1973. Chapter 14.

[6] C. Berge, Topological games with perfect information. *Contributions to the Theory of Games (3). Ann. of Math. St.* 39 (1957) 189-199.

[7] C. Berge and M. S. Schützenberger, Jeux de Nim et solutions. *C. R. Acad. Sci. Paris* 242 (1956) 16-72.

[8] C. Berge and P. Duchet, Perfect graphs and kernels. *Bull. Inst. Math. Acad. Sinica* 16 (1988) 263–274.

[9] C. Berge and P. Duchet, Recent problems and results about kernels in directed graphs. In *Applications of Discrete Mathematics*, SIAM, Philadelphia, PA, 1988. 200-204. also in: *Discrete Math.* 86 (1990) 27-31.

[10] C. Berge and A. Ramachandra Rao, A combinatorial problem in logic. *Discrete Math.* 17 (1977) 23-26.

[11] M. Blidia, Parity graphs with an orientation condition have a kernel. Submitted for publication.

[12] M. Blidia. A parity digraph has a kernel. *Combinatorica* 6 (1986) 23–27.

[13] M. Blidia, P. Duchet and F. Maffray, Meyniel graphs are kernel-M-solvable. *Res. RepÙniv. Paris* 6 (1988).

[14] G. Bucan and L. Varvak, On games on a graph *Algebra and Math. Logic.* Kiev (1966) 122-138.

[15] V. E. Cazanescu and S. Rudeanu, Independent sets and kernels in graphs and hypergraphs. *Ann. Fac. Sci. Univ. Nat. Zaire (Kinshasa) Sect. Math.-Phys.* 4 (1978) 37–66.

[16] C. Champetier, Kernels in some orientations of comparability graphs. *J. of Combin. Theory B* 47 (1989) 111-113.

[17] C. Chao, On the kernels of graphs. Technical Report RC-685, IBM (1962).

[18] G. Chartrand, D. W. VaderJagt and B. Quan Yue, Orientable domination in graphs. Presented at the *27th S.E. Conference on Combinatorics, Graph Theory, and Computing*.

[19] G. Chaty and Jayme L. Szwarcfiter, Enumerating the kernels of a directed graph with no odd circuits. *Inform. Process. Lett.* 51 (1994) 149-153.

[20] K. B. Chilakamarri and P. Hamburger, On a class of kernel-perfect and kernel-perfect- critical graphs. *Discrete Math.* 118 (1993) 253–257.

[21] V. Chvatal and L. Lovasz, Every directed graph has a semi-kernel. Springer-Verlag, *Lecture Notes in Mathematics* 411 (1974).

[22] V. Chvatal, On the computational complexity of finding a kernel. Technical Report No. CRM-300, Centre de Rechereches Mathematiques, University de Montreal (1973).

[23] P. Duchet, Parity graphs are kernel-M-solvable. *J. Combin. Theory B* 43 (1987) 121–126.

[24] P. Duchet, Graphes noyau-parfaits. *Ann. Discrete Math.* 9 (1980) 93-101.

[25] P. Duchet, Two problems in kernel theory. *Ann. Discrete Math.* 9 (1980) 302.

[26] P. Duchet, A sufficient condition for a graph to be kernel perfect. *J. Graph Theory* 11 (1987) 81–86.

[27] P. Duchet and H. Meyniel, A note on kernel-critical graphs. *Discrete Math.* 33 (1980) 93–101.

[28] P. Duchet and H. Meyniel, Kernels in directed graphs: a poison game. *Discrete Math.* 115 (1993) 273–276.

[29] P. Erdös, On a problem in graph theory. *Math. Gaz.* 47 (1963) 220–223.

[30] W. Fernandez De la Vega, Kernels in random graphs. *Discrete Math.* 82 (1990) 213-217.

[31] David Fisher, J. Richard Lundgren, S. Merz and K. B. Reid, The domination and competition graphs of a tournament. To appear in *J. Graph Theory*.

[32] David Fisher, J. Richard Lundgren, S. Merz and K. B. Reid, Domination graphs of tournaments and digraphs. *Congr. Numer.* 108 (1995) 97-107.

[33] David Fisher, J. Richard Lundgren, S. Merz and K. B. Reid, Connected domination graphs of tournaments. Submitted for publication.

[34] A. Frank, Kernel systems of directed graphs. *Acta Sci. Math. (Szeged)* 4 (1979) 63–76.

[35] A. Fraenkel, Planar kernel and Grundy with $d \leq 3$, $d_{out} \leq 2$, $d_{in} \leq 2$ are NP-complete. *Discrete Appl. Math.* 3 (1981) 257-262.

[36] A. S. Fraenkel and Y. Yesha, Complexity of problems in games, graphs, and algebraic equations. *Discrete Appl. Math.* 1 (1979) 15-30.

[37] C. N. Frangakis, A backtracking algorithm to generate all kernels of a directed graph. *Internat. J. Comput. Math.* 10(1) (1981/82) 35-41.

[38] Yumin Fu, Dominating set and converse dominating set of a directed graph. *Amer. Math. Monthly* (1968) 861-863.

[39] H. Galeana-Sánchez, A counterexample to a conjecture of Meyniel on kernel-perfect graphs. *Discrete Math.* 41 (1982) 105–107.

[40] H. Galeana-Sánchez, A theorem about a conjecture of H. Meyniel on kernel-perfect graphs. *Discrete Math.* 59 (1986) 35–41.

[41] H. Galeana-Sánchez and L. V. Neumann-Lara, Extending kernel perfect digraphs to kernel-perfect critical digraphs. To appear in *Discrete Math.*

[42] H. Galeana-Sánchez and L. V. Neumann-Lara, On kernels and semikernels of digraphs. *Discrete Math.* 48 (1984) 67–76.

[43] H. Galeana-Sánchez and L. V. Neumann-Lara, On kernel perfect critical digraphs. *Discrete Math.* 59 (1986) 257–265.

[44] H. Galeana-Sánchez, On the existence of (k,l)-kernels in digraphs. *Discrete Math.* 85(1) (1990) 99-102.

[45] H. Galeana-Sánchez and V. Neumann-Lara, Orientations of graphs in kernel theory. *Discrete Math.* 87 (1991) 271-280.

[46] H. Galeana-Sánchez and V. Neumann-Lara, New Extensions of kernel perfect digraphs to kernel imperfect critical digraphs. *Graphs Combin.* 10(4) (1994) 329-336.

[47] H. Galeana-Sánchez, L. Pastrana-Ramirez and H. A. Rincon-Mejia, Semikernels, quasikernels, and Grundy functions in line digraphs. *SIAM J. Discrete Math.* 4(1) (1991) 80-83.

[48] J. Ghoshal, R. Laskar and D. Pillone, Strong bondage and strong reinforcement numbers of graphs. To appear in *Congr. Numer.*

[49] Combinatorial Algorithms, Optimization, and Computers. Eds. M. C. Golumbic and R. C. Laskar, *Discrete Appl. Math.* 44(1-3) July 19, 1993.

[50] R. L. Graham and J. H. Spencer, A constructive solution to a tournament problem. *Canad. Math. Bull.* 14(1) (1971) 45-48.

[51] P. M. Grundy, Mathematics and games. *Eureka* 2 (1939) 6-8.

[52] P. M. Grundy and C. A. B. Smith, Disjunctive games with the last player losing. *Proc. Cambridge Philos. Soc.* 52 (1936) 527-533.

[53] F. Harary and M. Bezhad, On the problem of characterizing digraphs with solutions and kernels. *Sociometry* 20 (1957) 205-215.

[54] F. Harary and M. Richardson, A matrix algorithm for solutions and r-bases of a finite irreflexive relation. *Nav. Res. Log. Quart.* 6 (1959) 307-314.

[55] F. Harary and I. C. Ross, A procedure for clique detection using the group matrix. *Bull. Iranian Math. Soc.* 3 (1975).

[56] F. Harary, R. Z. Norman and D. Cartwright, *Structural Models*, Wiley, New York, 1965.

[57] M. Harminic, Solutions and kernels of a directed graph. *Math. Slovaca* 3, 32 (1982) 263-267.

[58] Topics on Domination. Eds. S. T. Hedetniemi and R. C. Laskar. *Discrete Math.* 86 (1990) (1-3).

[59] J. C. Holladay, Cartesian product of termination games. *Contributions to the Theory of Games 3, Ann. of Math. Stud.* 39 (1957) 189-199.

[60] L. Holley, Yung-Ling Lai and B. Quan Yue, Orientable step domination numbers of graphs. Presented at *the 27th S.E. Conference on Combinatorics, Graph Theory, and Computing* (1996).

[61] H. Jacob, Etude theorique du Noyau d' un graphe, These Universite Pierre et Marie Curie, Paris, VI, 1979.

[62] D. König, *Theorie der endlichen undendlichen Graphen.* Reprinted from Chelsea Publishing Company (1950).

[63] B. Kummer, *Spiele auf Graphen.* Deutscher Verlag der Wissenschaften, Berlin, 1979.

[64] M. Kwasnik, Charakteritiche Funktion, k-Grundy Funktion, Ordinalfunktion und k-Kern. *Zeszyty Naukowe WSInz.*, Nr. 55, Matematyka-Fizyka, Zielona Gora, 1980.

[65] M. Kwasnik, Die Kerne in der Summe und Komposition der Graphen. Manuscript.

[66] M. Kwasnik, On (k; l)-kernels of exclusive disjunction, Cartesian sum and normal point product of two directed graphs. Manuscript.

[67] M. Kwasnik, The generalization of Richardson's theorem. Manuscript.

[68] M. Kwasnik, On the (k,l)-kernels. *Lecture notes in Mathematics. Graph Theory*, Lagow 1981. Springer-Verlag (1983) 114-121.

[69] M. Kwasnik, A. Wloch and I. Wloch, Some remarks about (k,1)-kernels in directed and undirected graphs. *Discuss. Math.* 122 (1993) 29-37.

[70] Changwoo Lee, *On the Domination Number of a Digraph.* Ph.D. Dissertation, Michigan State University (1994).

[71] J. van Leeuwen, Having a Grundy-numbering is NP-complete. Technical Report No. 207, Computer Science Dept., Pennsylvania State University, University Park, P. A. (1976).

[72] E. Loukakis, Two algorithms for generating and ranking the family of solutions of a finite irreflexive relation. *Math. Soc. Sci.* 6(1) (1983) 75-86.

[73] R. D. Luce and H. Raiffa, *Games and Decisions.* Wiley, N. Y., 1957.

[74] J. R. Lundgren, Food webs, competition graphs, competition-common enemy graphs, and niche graphs. *Applications of Combinatorics and Graph Theory to the Biological and Social Sciences.* Springer-Verlag, 17 (1989).

[75] F. Maffray, *Sur l' existence des noyaus dans les graphes parfaits.* Thesis, University of Paris, 6 (1979).

[76] F. Maffray, On kernels in *i*-triangulated graphs. *Discrete Math.* 61 (1986) 247-251.

[77] D. Marcu, Kernels of strongly connected digraphs. *Iasi Sect. I a Mat.* 26(2) (1980) 417-418.

[78] D. Marcu, On the existence of kernels in a strong connected digraph. *Bul. Inst. Politehn. Iasi Sect. I* 25(1-2) (1979) 35-37.

[79] D. Marcu, Some remarks concerning the kernels of a strong connected digraph. *An. Stiint. Univ. "Al. I. Cuza" Iasi Sect. I a Mat. (N.S.)* 26(2) (1980) 417-418.

[80] D. Marcu, On the existence of a kernel in a strong connected digraph. *Bul. Inst. Politehn. Iasi Sect. I* 25(2) (1979) 35-37.

[81] D. Marcu, On finding a kernel in a symmetrical digraph without loops. *Polytech. Inst. Bucharest Sci. Bull. Mech. Engrg.* 54(3-4) (1992) 55-57.

[82] N. Megiddo and U. Vishkin, On finding a minimum dominating set in a tournament. *Theoret. Comput. Sci.* 61(2-3) (1988) 307-316.

[83] Z. Mo and K. Williams, (r, s)-domination in graphs and directed graphs. *Ars Combin.* 29 (1990) 129–141.

[84] V. Neumann-Lara, Seminucleos de una digrafica. *Anales del Instituto de Matematicas 2,* Universidad Nacional Autonoma de Mexico (1971).

[85] O. Ore, *Theory of Graphs.* Amer. Math. Soc. Collq. Publ. 38, Amer. Math. Soc., Providence 1962.

[86] V. Phillippe, Quasi-kernels of minimum weakness in a graph. *Discrete Math.* 20 (1977) 187–192.

[87] M. Richardson, Solutions of irreflexive relations. *Ann. Math.* 58 (1953) 573-580.

[88] M. Richardson, Extensions theorems for solutions of irreflexive relations. *Proc. Nat. Acad. Sci. U.S.A.* 39 (1953) 649-651.

[89] M. Richardson, On weakly ordered systems. *Bull. Amer. Math. Soc.* 52 (1946) 113-116.

[90] M. Richardson, Relativization and extension of solutions of irreflexive systems. *Pacific J. Math.* 50 (1955) 551-584.

[91] Z. Romanowicz and K. Wozniak, Games on a graph. *Graphs, Hypergraphs, and Block Systems.* Eds. M. Borowiecki, Z. Skupien and L. Szamkolowicz, Zielona Gora (1976) 231-237.

[92] A. Roth, Two person games on graphs. *J. Combin. Theory* 24 (1978) 238-241.

[93] A. Roth, A note concerning asymmetric games on graphs. *Nav. Res. Log. Quart.* 25 (1978) 365-367.

[94] B. Roy, *Algebre moderne et theorie des Graphes.* Volume 2, Chapter VI, Dunod, Paris, 1970.

[95] S. Rudeanu, Notes sur l' existence et l' unicite du noyau d' un graphe. *Revue Francaise Rech. Operat.* 33 (1964) 20-26.

[96] E. Sampathkumar and P. Latha, Strong, weak domination and domination balance in a graph. Manuscript.

[97] A. Sanchez-Flores, A counterexample to a generalization of Richardson's theorem. *Discrete Math.* 65 (1987) 319-320.

[98] G. Schmidt and T. Strohlein, On kernels of graphs and solutions of games: a synopsis based on relations and fixedpoints. *SIAM J. Algebraic Discrete Meth.* 6 (1985) 54–65.

[99] C. A. B. Smith, Graphs and composite games. *A Seminar on Graph Theory* (F. Harary, L. Beineke), Holt-Rinehart. Winston, N. Y. (1967) 86-111.

[100] H. Steinhaus, Definitions for a theory of games and pursuit. *Mysl Akad. Lvov* 1(1) (1925) 13-14; reprinted in *Nav. Res. Log. Quart.* 7 (1960) 105-108.

[101] I. Tomescu, Almost all digraphs have a kernel. *Discrete Math.* 2, 84 (1990) 181-192.

[102] I. Tomescu, Almost all digraphs have a kernel. *Random Graphs '87* Eds. M. Karonski, J. Jaworski and A. Rucinski. John Wiley and Sons Ltd. (1990) 325-340.

[103] J. Topp, Kernels of digraphs formed by some unary operations from other graphs. *Rostock. Math. Kolloq.* 21 (1982) 73-81.

[104] J. Topp, Asymmetric games on graphs. *Lecture notes in Mathematics.* Graph Theory, Lagow 1981. Springer-Verlag (1983) 260-265.

[105] L. P. Varvak, Generalization of a kernel of a graph. *Ukrainian Math. J.* 25 (1973) 78–81.

[106] L. P. Varvak, Games on a sum of graphs (Russian). *Kiber.* 4(1) (1968) 63-66.

[107] J. Von Neumann and O. Morgenstern, *Theory of Games and Economic Behaviour.* Princeton University Press, Princeton, 1944.

[108] C. P. Welter, The theory of a class of games on a sequence of squares, in terms of the advancing operation in a special group. *Proc. Amer. Math. Soc.* 13 (1962) 47-52.

[109] K. Weber, Domination number for almost every graph. *Rostock. Math. Kolloq.* 16 (1981) 31-43.

[110] N. Y. Withoff, A modification of the game of Nim. *Nieuw Arch. voor Wiskunde* 7 (1907) 199-203.

Chapter 16

Graphs Critical with Respect to the Domination Number

David P. Sumner
Department of Mathematics
University of South Carolina
Columbia, SC   USA 29208

Ewa Wojcicka
Department of Mathematics
University of Charleston
Charleston, SC 29424   USA

Ewa Wojcicka, the driving force behind this chapter, died January 14, 1996. She had suffered multiple severe injuries from an automobile accident near Savannah, Georgia. She was recovering remarkably well when yet another accident occurred in the rehabilitation clinic, and she died of the consequences. Ewa was responsible for much of the progress in domination critical graphs. A memorial page for her can be viewed on the World Wide Web at *http://www.math.sc.edu/ sumner/ewa.html.*

16.1   Introduction and Preliminaries

If a graph-theoretic property is worth studying at all, it is worthwhile to investigate those graphs that are extremal with respect to that property. This is particularly true in discrete mathematics where arguments by induction abound. There is a multitude of ways in which a graph can be extremal with respect to some property P. Typically, a graph is said to be *critical* with respect to P if G has P, but for every vertex v, $G - v$ fails to have P. Also, G is often said to be *minimal* with respect to a property P if G has P, but $G - e$ fails to have P for every edge of G. However, these are not always the most productive ways of indicating that G is extremal with respect to P. In the case of the domination number, there is a multitude of possible definitions for extremal concepts.

For example, consider edge deletion. Graphs that are minimal in this sense (i.e., for every edge e of G, $\gamma(G-e) > \gamma(G)$) have been characterized by Bauer, Harary, Nieminen and Suffel [5] where they show that such graphs are disjoint unions of stars.

In this chapter we concentrate our attention on two classes of graphs that are critical with respect to the domination number: graphs whose domination number drops whenever an edge is added - the *edge-critical* graphs, and graphs whose domination number drops whenever a vertex is deleted - the *vertex-critical* graphs.

These two concepts are quite distinct, but have more in common than first meets the eye. Although we devote the great majority of this chapter to these two properties, we also touch on some of the work that has been done on other ways in which graphs may be extremal with respect to the domination number. In particular we discuss the local-edge-critical graphs recently defined by Henning, Oellermann and Swart [28], and the edge-*i*-critical and vertex-*i*-critical graphs studied by Ao [3] and Ao and MacGillivray [4]. A graph G in which no two distinct vertices have the same closed neighborhood is said to be *point distinguishing* (see [40, 41]).

When no confusion is possible, we will not distinguish between a set A of vertices of a graph G, and the subgraph induced by A. In particular, if A is a subset of $V(G)$, then we write just $\gamma(A)$ for the domination number of the subgraph induced by A. The *nucleus* of a graph G is the set $G^* = \{v : \gamma(G-v) = k\}$. For examples, see Figures 16.1, 16.2, and 16.3.

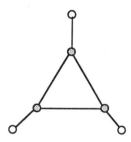

Figure 16.1: The smallest, connected, 3-edge-critical graph. The shaded vertices form the nucleus.

16.1.1   Edge-critical graphs

A graph G is *k-edge-domination-critical* (or just *k-edge-critical*) if $\gamma(G) = k$, and for every nonadjacent pair of vertices v and u, $\gamma(G + vu) = k - 1$; thus an edge-critical graph is one for which the domination number drops whenever a missing edge is added.

It is easy to characterize the *k*-edge-critical graphs for $k \leq 2$. A graph is 1-edge-critical if and only it is complete. The next theorem characterizes 2-edge-critical graphs.

Theorem 16.1 [38] *A graph G is 2-edge-critical if and only if it is the comple-ment of a union of stars; i.e., $\overline{G} = \bigcup_{i=1}^{t} K_{1,p_i}$ for some $t \geq 1$.*

Note that a disconnected graph is 2-edge-critical if and only if it is the disjoint union of K_n and an isolated vertex. Theorem 16.1 shows that the 2-edge-critical graphs are complements of the domination critical graphs as defined by Bauer, Harary, Nieminen and Suffel [5]. Walikar and Acharya [46] also studied this class of graphs.

Consequently, we focus our attention on k-edge-critical graphs having $k \geq 3$.

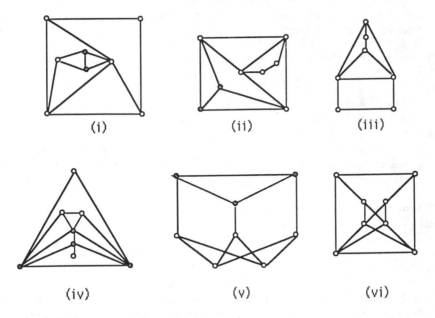

Figure 16.2: Small examples of 3-edge-critical graphs. The shaded vertices comprise the nucleus. (Graphs (iii) and (vi) have an empty nucleus.)

16.1.2   The arrow notation

If v and u are nonadjacent vertices in a k-edge-critical graph G, then there must exist a set S of cardinality $k - 2$ such that either $S \cup \{v\}$ dominates all of $G - u$, or $S \cup \{u\}$ dominates all of $G - v$. In the first case we write $[v, S] \to u$, and in the second we write $[u, S] \to v$. In particular, when we write $[v, S] \to u$, it is understood that u is *not* dominated by $S \cup \{v\}$. We write $v \to u$ to indicate that such a set S exists with $[v, S] \to u$.

Note:

(i) $v \rightarrow u$ means that there exists a $(k-1)$-element set $S \cup \{v\}$ that contains v and dominates all of $G - u$.

(ii) Clearly, if G is edge-critical, then for any two nonadjacent vertices v and u, either $v \rightarrow u$ or $u \rightarrow v$.

For graphs in general, removing a vertex may produce a dramatic increase in the domination number. For example, the graph $K_{1,n-1}$ has domination number 1, but removing the central vertex yields a graph with domination number $n - 1$. However, the next result shows that edge-critical graphs are much better behaved in this regard. Removing a vertex from a domination-critical graph cannot increase the domination number.

Theorem 16.2 [38] *If v is any vertex of an k-edge-critical graph G, then $\gamma(G - v) \leq k$.*

Thus, if v is any vertex of an edge-critical graph G that does not belong to the nucleus G^*, then $\gamma(G - v) = k - 1$. We define a vertex v to be *critical* if $\gamma(G - v) < \gamma(G)$. If G is an edge-critical graph, then for every pair of nonadjacent vertices x and y, either $x \rightarrow y$ or $y \rightarrow x$ and hence at least one of x and y is critical. Moreover, this means that at most one of x and y can belong to the nucleus. Hence, the following result.

Theorem 16.3 [18] *For any connected, k-edge-critical graph G, if G^* is nonempty, then G^* is a complete graph.*

Generally speaking, most of the difficulties involved with investigating the k-edge-critical graphs appear in the case $k = 3$ (but with lesser complexity than in the case for $k \geq 4$). For that reason, more attention has been given to the case $k = 3$ than to the general case. However, as we will see later, there are a few properties of 3-edge-critical graphs that do not seem to generalize to the case $k \geq 4$.

16.1.3   Vertex-critical graphs

A graph is *k-domination-vertex-critical* (or just *k-vertex-critical*) if $\gamma(G) = k$ and for every vertex v of G, $\gamma(G - v) = k - 1$; i.e., every vertex of G is critical. Vertex-critical graphs are distinct from edge-critical graphs. The cycle C_7 is vertex-critical, but not edge-critical. The graph in Figure 16.1 is edge-critical, but not vertex-critical. The graph in Figure 16.4 is both vertex- and edge-critical, but most graphs are neither. However, there are several relationships between these two classes of graphs that will become apparent from the results in this survey.

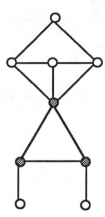

Figure 16.3: A 4-edge-critical graph. The darkened vertices form the nucleus.

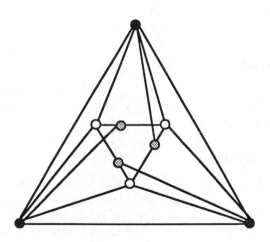

Figure 16.4: The graph Q_3. The solid vertices are from U, the shaded vertices from V and the clear vertices are from W.

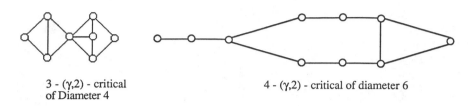

3 - (γ,2) - critical   4 - (γ,2) - critical of diameter 6
of Diameter 4

Figure 16.5: A 3-$(\gamma, 2)$-critical graph of diameter 4 and a 4-$(\gamma, 2)$-critical graph of diameter 6.

Theorem 16.4 [9] *The only 1-vertex-critical graph is the graph on a single vertex. The 2-vertex-critical graphs are the graphs $K_{2t} - F$ where F is a perfect matching.*

Note that every 2-vertex-critical graph is also a 2-edge-critical graph, but not conversely.

We use the term *domination-critical* to refer to graphs that are either vertex-critical or edge-critical. The following simple observation is frequently useful.

Observation. If G is any domination-critical graph, then every vertex of G belongs to some minimum dominating set.

16.1.4   Closely related concepts

While the vertex-critical and edge-critical graphs are the main focus of this survey, there are two recently studied concepts that are closely related and deserve recognition here, too.

Locally-edge-critical graphs

It is reasonable to consider those graphs whose domination number drops whenever one of a set of special edges is added. Henning, Oellermann and Swart [28] defined such a variant on the edge-critical theme. They call a graph k-(γ, r)-*critical* if $\gamma(G) = k$, and $\gamma(G + vu) < k$ for all pairs of nonadjacent vertices v and u such that $d(v, u) \leq r$. Such graphs are referred to as *local-edge-domination-critical*. It is important to point out that while every k-edge-critical graph is also k-(γ, r)-critical, the converse need not be true. The authors give examples of 3-$(\gamma, 2)$ graphs that are not 3-edge-critical and also examples of 4-$(\gamma, 2)$ graphs that are not 4-edge-critical (see Figure 16.5). They demonstrate bounds on the diameter of 3-$(\gamma, 2)$ and 4-$(\gamma, 2)$-critical graphs that are similar to those for edge-critical graphs.

Henning et al. also characterized the 2-$(\gamma, 2)$-critical graphs as follows. A *double star* $S(r, s)$ $(r, s \geq 1)$ is obtained from the stars $K_{1,r}$ and $K_{1,s}$ by joining their central vertices by an edge.

Theorem 16.5 [28] *A graph G is 2-$(\gamma, 2)$-critical if and only if it is the complement of a disjoint union of stars or the complement of a double star.*

They give examples of 3-$(\gamma, 2)$-critical graphs that are not 3-edge-critical, however there is no known characterization of such graphs.

It would be valuable to exhibit general classes of k-(γ, r) graphs that are not k-edge-critical. Also, Henning et al. concentrated primarily on the diameter of locally-edge-critical graphs. It is natural to wonder to what extent do the other known results for k-edge-critical graphs extend to locally-edge-critical graphs.

The independent domination number

Ao [2] and Ao and MacGillivray [4] defined a graph to be k-*vertex-i-critical* if $i(G) = k$ and $i(G - v) < k$ for every vertex v. They also defined a graph G to be k-*edge-i-critical* if $i(G) = k$ and $i(G + e) = k - 1$ for every edge e missing from G.

Ao [2] studied edge-i-critical graphs and related classes of graphs in some detail. In particular she showed that most of the results in [38] for 3-edge-critical graphs hold also for the class of 3-edge-i-critical graphs. Moreover, the arguments are exactly similar in most instances. It should be noted that the edge-i-critical graphs are distinct from the edge-critical graphs. For example, the graph in Figure 16.4 is 3-edge-critical, but not 3-edge-i-critical (joining a vertex of degree 5 to some vertex of degree 3 does not cause $i(G)$ to decrease), and the complement of the Cartesian product $K_t \times K_t$ $(t \geq 4)$ is t-edge-i-critical, but not edge-critical. However, Ao shows that a graph is 2-edge-i-critical if and only if it is 2-edge-critical. Ao also showed that many of the results previously known to hold for vertex-critical graphs hold as well for the class of vertex-i-critical graphs. The classes of vertex-i-critical and edge-i-critical graphs appear to be quite interesting and deserve to be further investigated.

16.2 Examples

It is always helpful to have as many examples handy as possible when investigating a particular class of graphs. In this section we discuss several additional classes of edge-critical and vertex-critical graphs in addition to those that arise naturally in later sections. We also provide a few constructive techniques for generating domination-critical graphs.

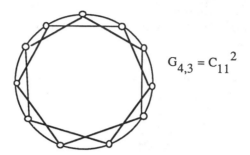

$$G_{4,3} = C_{11}{}^2$$

Figure 16.6: Graph $G_{4,3} = C_{11}{}^2$.

16.2.1   Vertex-critical examples

There are many examples of k-vertex-critical graphs for $k \geq 3$, and several ways to generate them.

Let A and B be two graphs with $a \in A$ and $b \in B$. The *coalescence* $A._{ab}B$ (or just $A.B$ if a and b are understood) of A and B via a and b, is the graph obtained from the disjoint union of A and B by identifying the vertex a and b. (The graph in Figure 16.17 is formed by a coalescence of three 4-cycles.) Brigham, Chinn and Dutton [10] discuss this operation and prove the following key result.

Theorem 16.6 [10] *A coalescence of two graphs A and B is vertex-critical if and only if both of A and B are vertex-critical.*

Theorem 16.7 [10] *If H is a coalescence of A and B, then $\gamma(A) + \gamma(B) - 1 \leq \gamma(H) \leq \gamma(A) + \gamma(B)$.*

In [9] the authors provide a class of k-vertex-critical graphs $G_{\Delta,k}$ (where $\Delta \geq 2$ is even, and $k \geq 2$), which is simply the $\frac{\Delta}{2}$ power of the cycle on $(k-1)(\Delta+1) + 1$ vertices. For example, the graph in Figure 16.6 is $G_{4,3}$, the square of a cycle on 11 vertices.

Also constructed in [9] and [10], are the k-vertex-critical graphs Q_k for $k \geq 3$. The vertex set of Q_k consists of the disjoint union of $U \cup V \cup W$ where $U = \{u_0, u_1, \cdots, u_{k-1}\}$, $V = \{v_0, v_1, \cdots, v_{k-1}\}$, and $W = \{w_0, w_1, \cdots, w_{k-1}\}$. The neighborhoods of the vertices are given by: $N(u_i) = \{u_{i-1}, u_{i+1}, v_{i-1}, v_i, w_i\}$, $N(v_i) = \{u_i, u_{i+1}, w_{i-1}, w_i\}$, and $N(w_i) = \{u_i, v_i, v_{i+1}\}$ where all indices are modulo k. The graph Q_3 shown in Figure 16.4 is also edge-critical.

Ao [2] defined the class of t-vertex-critical graphs O_t for each $t \geq 1$, to be the graph on $3t$ vertices formed by taking two cycles C_{2t} and identifying those vertices having odd subscripts. The graph in Figure 16.7 is O_3.

Figure 16.7: Graph O_3.

16.2.2   Edge-critical examples

Figure 16.2 shows some small examples of 3-edge-critical graphs (on $n \leq 8$ vertices).

We next give a simple construction from Favaron, Sumner and Wojcicka [18] that frequently makes it possible to extend a k-critical graph to a larger one. Let G be a graph, $v \in V(G), v' \notin V(G)$. Then $< G, v, v' >$ will denote the graph with $V(< G, v, v' >) = V(G) \cup \{v'\}$, and $E(< G, v, v' >) = E(G) \cup \{v'u : u \in N[v]\}$. Thus $< G, v, v' >$ is obtained from G by adding a new vertex v' that has the same closed neighborhood as v. We will call $< G, v, v' >$ the *expansion of G via v (by v')*. If the expansion of the k-critical graph G via v is also k-critical then v will be called *expandable*. In Figure 16.2 the graphs (i) and (ii) are both expansions of (iii). Also all the vertices in (iii) of Figure 16.2 are expandable except for the two vertices of degree 4.

Theorem 16.8 [42] *If v is a vertex of G, then v is expandable if and only if $v \to u$ for every vertex $u \notin N[v]$.*

So, in particular, every vertex in G^* is expandable. Also, as a direct consequence of this result, if v is expandable and not an element of G^* then v must be adjacent to every vertex of G^*. Also note that if v is an expandable vertex, then both v and v' belong to the nucleus of the expanded graph $< G, v, v' >$.

Conversely, it is straightforward to check that if v and u are vertices of a k-edge-critical graph and $N[v] = N[u]$, then $G - v$ is k-edge-critical, and G is the expansion of $G - v$ by u. So, it follows that if $N[v] = N[u]$ in a k-edge-critical graph, then v and u must both belong to the nucleus of G. Thus if $|G^*| \leq 1$ for a k-edge-critical graph, then G is point distinguishing.

16.2.3   Edge-extensions

Any graph G with domination number k, can be extended to a k-edge-critical graph by adding edges. The following algorithm from [39] does the job.

Let G be a graph with $\gamma(G) = k$, and let
$e_1, e_2, \cdots, e_t$ be the edges missing from G.
For $i = 1$ to t
       if $\gamma(G + e_i) = \gamma(G)$ then add the edge e_i to G

After one pass through this loop the resulting graph is k-edge-critical. We refer to any such graph as an *edge extension* of G. If G is vertex-critical, and e is any edge missing from G such that $\gamma(G + e) = \gamma(G)$, then $G + e$ is also vertex-critical.

Theorem 16.9 [9] *Any edge extension of a vertex-critical graph is both vertex and edge-critical.*

Along similar lines, Brigham, Chinn and Dutton [9] showed the following.

Theorem 16.10 [9] *For every graph G there is a vertex-critical graph H such that G is an induced subgraph of H.*

The k-edge-critical graphs in Figure 16.8 consist of a partition $A_1, A_2, \cdots, A_k$ of a complete graph, together with an independent set of vertices $\{x_1, x_2, \cdots, x_k\}$ with $N(x_k) = A_k$. We refer to these graphs as full graphs. Note that every full graph is the expansion of a graph in the same class with each A_i a singleton.

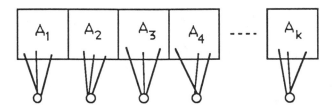

Figure 16.8: The full graphs.

16.2.4   3-edge-critical examples

The graph (v) in Figure 16.2 is a special case of a general class of 3-edge-critical graphs. Here we take any 3-edge-critical graph G with a nonempty nucleus and which has the property that no vertex in $A = G - G^*$ is adjacent to all of G^*. Let H be any 2-vertex-critical graph. Join each vertex in H to each vertex in A. The resulting graph is also 3-edge-critical. (See Figure 16.10 where G is the graph in Figure 16.1.)

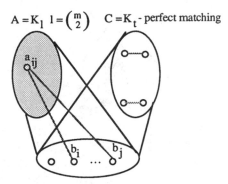

$A = K_1 \; 1 = \binom{m}{2}$ $C = K_t$ - perfect matching

a_{ij}

b_i b_j

B = Independent set on m vertices

Figure 16.9: $N(b_{i,j}) = B - \{b_i, b_j\}$.

A similar class of 3-edge-critical graphs is obtained by letting $V(G)$ be $A \cup B \cup C$, where $A = \{a_{i,j} : 1 \le i < j \le p\}$ is a complete graph on $\binom{p}{2}$ vertices, $B = \{b_1, b_2, \cdots, b_p\}$ is an independent set on p vertices, and C is a 2-vertex-critical graph. Each vertex of C is adjacent to each vertex of B and each vertex $a_{i,j}$ of A is adjacent to all of $B - \{b_i, b_j\}$. (See Figure 16.9)

Another class of 3-edge-critical graphs consists of $A \cup B \cup \{v\}$ where A and B are 2-vertex-critical graphs, with $B = \{x_1, x_2, \cdots, x_p, y_1, y_2, \cdots, y_p\}$. For each $i = 1, 2, \cdots, p$, x_i and y_i are adjacent to all of B except for one another, and the vertex v is adjacent to every vertex in A. Also, every vertex of A is adjacent to every x_i in B.

Ao [2] notes that for $p \ge 1$, the Cartesian product $K_t \times K_t$ is both t-vertex-critical and t-edge-critical. Also, for $t \ge 4$, the subdivision graph of K_t is t-edge-critical but not t-vertex-critical.

16.2.5 i-critical examples

Many of the known classes of vertex-critical and edge-critical graphs are also vertex-i-critical or edge-i-critical. Ao [2] provides numerous additional examples of both edge-i-critical and vertex-i-critical graphs. In particular, she defines for each $r \ge s \ge 3$, a graph $P_{r,s}$ which is s-edge-i-critical and has maximum independence number r. The definition of $P_{r,s}$ follows. Let $A = \{x_1, x_2, \cdots, x_r\}$ be a set of r independent vertices. Let $p = \binom{r}{r-s+1}$ and let $A_1, A_2, \cdots, A_p$ be an arrangement of the $r - s + 1$ element subsets of A. Let $B = \{y_1, y_2, \cdots, y_p\}$ be a complete graph on p vertices. Then form $P_{r,s}$ from $A \cup B$ by adding the edges from x_i to all the vertices in A_i, for each $i = 1, 2, \cdots, p$.

Also Ao [2] observes that every vertex-i-critical graph can be extended to a

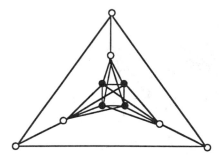

Figure 16.10: The graph induced by the black vertices is 2-vertex-critical. The graph induced by the white vertices is 3-edge-critical, with G^* forming the outer triangle.

graph that is both edge-i-critical and vertex-i-critical by an algorithm exactly similar to that in Section 16.2.3.

Ao shows, too, that if $G_1, G_2, \cdots, G_p$ are k-vertex-i-critical (resp. k-edge-i-critical) graphs, then their join is also a k-vertex-i-critical (resp. k-edge-i-critical) graph. Also, each full graph (see Section 16.2.3) is also an edge-i-critical graph. The Cartesian product $K_t \times K_t$ is both t-vertex-i-critical and t-edge-i-critical.

16.3   Perfect Matchings and Hamiltonicity

A *perfect matching M* in a graph is set of independent edges such that every vertex of G is incident with one edge of M. Tutte [44] characterized graphs with perfect matchings in a now classic theorem. For a set S of vertices of G, denote by $O(S)$ the number of components of $G - S$ that have odd cardinality.

Theorem 16.11 [44] *A graph G of even order has a perfect matching if and only if for every nonempty set S of vertices of G, $O(G - S) \leq |S|$.*

It is useful to notice that if there is a set S such that $O(G - S) > |S|$, then a simple parity argument shows that, in fact, $O(G - S) \geq |S| + 2$.

Chvátal [12] investigated the possibility of determining an analogous condition for Hamiltonicity. He defined a graph to be t-tough $(t > 0)$ if and only if for every set S of vertices of G, $C(G - S) \leq \frac{|S|}{t}$ where $C(G - S)$ denotes the number of components of $G - S$.

Chvátal conjectured that every 2-tough graph was Hamiltonian. A large value of toughness does tend to suggest the possibility that a graph is Hamiltonian. Note that by Tutte's Theorem, connected, even-order graphs that are 1-tough always have a perfect matching.

Sumner and Blitch [38] showed that 3-edge-critical graphs satisfy a toughness-like condition that is stronger than that of Tutte's Theorem.

Theorem 16.12 [38] *If G is a connected, 3-edge-critical graph and S is any set of vertices in G, then the number of components of G − S is at most |S| + 1.*

As a consequence of this result and Tutte's Theorem, connected, even-order, 3-edge-critical graphs have a perfect matching. Moreover, Theorem 16.12 shows that every connected, 3-edge-critical graph is $\frac{1}{2}$-tough, and every 2-connected, 3-edge-critical graph is $\frac{2}{3}$-tough. The proof of Theorem 16.12 is an induction argument that relies heavily on the following lemma.

Lemma 16.13 [38] *Let S be an independent set of t ≥ 4 vertices in the connected, 3-edge-critical graph G, then it is possible to order the vertices of S as $S = \{v_1, v_2, \cdots, v_t\}$ in such a way that there exists a path $X = x_1, x_2, \cdots x_{t-1}$ such that*

(i) $[v_i, \{x_i\}] \to v_{i+1}$ for all $i = 1, 2, \cdots, t-1$, and

(ii) $X \subseteq V(G) - S$.

It is natural to hope to use a similar inductive argument to prove a generalization of Theorem 16.12. Unfortunately, it is not known to what extent Lemma 16.13 generalizes to the case $k > 3$.

Question. Is it true that every connected edge-critical graph has the property that $C(G - S) \leq |S| + 1$?

As we see in the next section, this is true if $|S| = 1$, but little is known about the case $|S| > 1$ for $k > 3$. Sumner and Wojcicka [42] show that it is true for $k = 4$ and $|S| \leq 2$.

Moreover, it is not known if connected even-order 4-edge-critical graphs always have perfect matchings. There is some evidence that this may be true.

Conjecture 16.14 [42] *Every even-order, connected, k-critical graph for $k \geq 3$ has a perfect matching.*

Theorem 16.12 does not hold for vertex-critical graphs. For example, the graph in Figure 16.11 is 4-vertex-critical, and yet $G - v$ has three components. However, Ao [2] shows that the conditions of Theorem 16.12 hold for edge-*i*-critical graphs.

Considering that connected 3-edge-critical graphs tend to have few vertices of small degree (a concept made more precise in Section 16.7), and that such graphs of even order always have perfect matchings, it is not too unreasonable to suspect that such graphs are always at least traceable. Supported by several heuristics and computer-generated data, Sumner conjectured that this was so. Wojcicka [47] settled this conjecture.

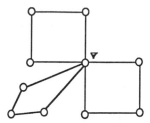

Figure 16.11: A vertex-critical graph having a cutvertex v with $G - v$ having three components.

Theorem 16.15 [47] *If G is a connected, 3-edge-critical graph on $n > 6$ vertices, then G is traceable.*

Wojcicka conjectured that more is true.

Conjecture 16.16 [47] *Every connected, 3-edge-critical graph without cutvertices has a Hamiltonian cycle.*

Edge-critical graphs need not be traceable for $k > 3$. In fact, it is easy to construct non-traceable, k-edge-critical graphs for all $k \geq 4$. For example, any full graph (see Figure 16.8) with three of the A_i's consisting of single vertices. Also, any traceable, edge-extension of the graph in Figure 16.12 is k-edge-critical, but not traceable. (Since any traceable edge-extension must be spanned by a path on $3(k-1)$ vertices, and hence has domination number at most $(k-1)$.)

However, Wojcicka conjectures that 3-connected, 4-critical graphs are Hamiltonian and that perhaps, in general, $(k-1)$-connected, k-edge-critical graphs are Hamiltonian. One problem in extending the approach in [47] to k-edge-critical graphs with $k \geq 4$ is that the following key result from [47] does not generalize to values of k larger than 3.

Theorem 16.17 [47] *If G is a connected, 3-edge-critical graph, then*

(i) G has a cycle whose vertices form a dominating set, and

(ii) if A is the set of endvertices of G, then $G - A$ is 2-connected.

Sumner and Wojcicka [42] show that if G is 3-edge-critical with a nonempty set A of endvertices, then $G - A$ is Hamiltonian.

Note that Henning, Oellermann and Swart [28] showed that part (i) of Theorem 16.17 does not hold for their local-3-edge-critical graphs. They ask, however, if local-edge-critical graphs must contain a dominating path.

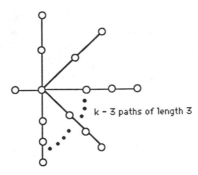

k - 3 paths of length 3

Figure 16.12: No edge-extension of this graph can be traceable.

Hanson [24] approached the Hamiltonian problem for 3-edge-critical graphs by defining a closure operation. Such an approach has been successful in other circumstances. For example, the closure operator of Bondy and Chvátal [8] can be used to deduce many of the classical conditions for a graph to be Hamiltonian. They defined the closure of a graph G to be the graph, $cl(G)$, obtained from G by recursively adding edges between pairs of nonadjacent vertices whose degrees sum to at least n. They showed that G is Hamiltonian if and only if $cl(G)$ is Hamiltonian. So, in particular if $cl(G)$ is complete, then G must be Hamiltonian.

Theorem 16.18 [24] *Let v be a vertex with $\delta(v) \geq 3$ in a 2-connected, 3-edge-critical graph G, and let a and b be nonadjacent vertices of G such that $[a, \{b\}] \to v$. Then $G + ab$ is Hamiltonian if and only if G is Hamiltonian.*

Now, given a 2-connected, 3-critical graph G, repeat the following recursive operation as long as possible:

For each pair of nonadjacent vertices a and b, if there exists a vertex v with $\delta(v) \geq 3$ such that $[a, \{b\}] \to v$, then add the edge ab to G. Denote the resulting graph by $D^*(G)$.

Finally, if G is any graph that has a spanning 2-connected, 3-critical subgraph H, then define $D^*(G)$ to be the graph with vertex set $V(G)$ and $E(D^*(G)) = E(G) \cup E(D^*(H))$. The key property of $D^*(G)$ is reflected by the next result.

Theorem 16.19 [24] *If G has a spanning subgraph that is 2-connected and 3-critical, then $D^*(G)$ is Hamiltonian if and only if G is Hamiltonian.*

Hanson notes that for the graph (v) in Figure 16.2, $D^*(G)$ is not complete, but the degrees of the vertices of $D^*(G)$ are sufficiently large that $cl(D^*(G))$ *is* complete. Hence, $D^*(G)$ is Hamiltonian as a consequence of the Bondy and Chvátal Theorem and so G is Hamiltonian as a result of Theorem 16.19.

While Hanson's result is not strong enough to show that 2-connected 3-edge-critical graphs are always Hamiltonian, it still suggests an attractive line of attack. Hanson notes that the condition $\delta(v) \geq 3$ may not be necessary and that its use in the proof of Theorem 16.18 can be avoided in most cases.

Recently, Ao and MacGillivray [4] defined a similar closure operation to that of Hanson's for the class of i-edge-critical graphs, and used it to establish some stronger results for this class of graphs in the case $i(G) = 3$. They also show that their results do not extend for $i(G) > 3$.

Theorem 16.20 [4]

(i) If G is a 2-connected, 3-edge-i-critical graph, then G is Hamiltonian.

(ii) If G is a connected, 3-edge-i-critical graph on more than 6 vertices, then G is traceable.

Another approach to the general Hamiltonian problem that has a great deal of success in recent years in that of neighborhood unions. For example, Faudree, Gould, Jacobson, and Schelp [16] give a sufficient condition for Hamiltonicity involving the size of neighborhood unions. Perhaps this approach could be useful in this context as well.

16.4   Components and Block Structure

Suppose that G is a disconnected graph having components $A_1, A_2, \cdots, A_t$. Then it is easy to see that G is k-vertex-critical if and only if each A_i is vertex-critical, and G is k-edge-critical if and only if all the A_i's are edge-critical and all but at most one are vertex-critical. Consequently, we generally confine our attention to connected domination-critical graphs.

Vertex-critical graphs are quite well-behaved when it comes to cutvertices and blocks. The next result is a direct consequence of Theorem 16.6.

Theorem 16.21 [10] *A graph G is vertex-critical if and only if each of its blocks is vertex-critical.*

Theorem 16.22 [10] *If G is a connected vertex-critical graph having blocks $B_1, B_2, \cdots, B_k$, then $\gamma(G) = 1 + \sum_{i=1}^{k} (\gamma(B_i) - 1).$*

Things are not quite so nice with respect to the edge-critical graphs, but still quite a bit can be said about the structure of edge-critical graphs with cutvertices. Note that for $k \leq 2$, no k-edge-critical graph can contain a cutvertex. In [38] it was observed that every cutvertex of a 3-edge-critical graph is adjacent to an endvertex.

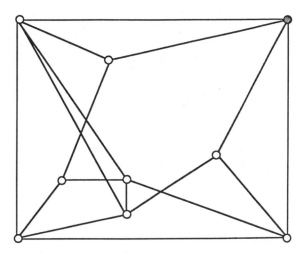

Figure 16.13: A 3-edge-critical graph with $|G^*| = 1$ (G^* consists of the single gray vertex).

Paris, Sumner and Wojcicka [33] studied the structure of connected k-edge-critical graphs containing cutvertices.

Recall that the nucleus of an edge-critical graph is the set G^* of vertices whose removal leaves the domination number unchanged. The nucleus of G plays a vital role in the structure of edge-critical graphs with cutvertices. The next theorem gives some insight into why this is so.

Theorem 16.23 [33] *If v is a cutvertex of the connected, k-critical graph G, then $v \in G^*$.*

It is an immediate consequence of the previous theorem that if $x \rightarrow y$ for some vertices x and y of a connected, k-critical graph G, then y cannot be a cutvertex of G.

Theorem 16.24 [33] *If G is a connected, k-edge-critical graph, and v is a cutvertex of G, then v belongs to exactly two blocks A and B. One of these, say A, is such that $G^* \cap (A - \{v\}) = \emptyset$ and hence $G^* \subseteq B$.*

The next result gives a partial extension to Theorem 16.12 for $k \geq 4$ and suggests that more may be true.

Corollary 16.25 [33] *If G is a connected k-edge-critical graph with $k \geq 3$ and v is a cutvertex for G, then $G - v$ consists of exactly two components.*

Corollary 16.26 [33] *If G is a connected edge-critical graph with at least one cutvertex then $|G^*| \geq 2$.*

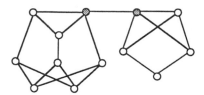

Figure 16.14: Edge-critical blocks.

Note that there are graphs with $|G^*| = 0$ (e.g., Figure 16.4) and $|G^*| = 1$ (Figure 16.13).

Since an edge-critical graph is vertex-critical if and only if it has an empty nucleus, graphs which are both vertex-critical and edge-critical cannot contain cutvertices.

Corollary 16.27 [33] *If G is both edge-critical and vertex-critical, then G is 2-connected.*

Let G be a connected, edge-critical graph with nonempty nucleus. Then from Theorem 16.24, the nucleus of G is contained in exactly one block of G. This block is called the *nuclear block* of G. The remaining blocks of G are called, naturally enough, the *non-nuclear blocks* of G. The structure of edge-critical graphs having cutvertices can be summarized by the following result.

Theorem 16.28 [33] *Let G be a connected, edge-critical graph with at least one cutvertex. Let C be the set of cutvertices of G, and let N be the nuclear block of G. Then $C \subseteq G^*$ and $G^* = N[c] \cap N$ for each $c \in C$. Moreover, exactly one non-nuclear block is attached to each cutvertex. (See Figure 16.15).*

Theorem 16.29 [33] *If G is a connected, edge-critical graph, and $A_1, A_2, \cdots, A_r$ are the non-nuclear blocks of G, then $\gamma(G) \geq \sum_{i=1}^{r} \gamma(A_i)$.*

Equality holds in Theorem 16.29 when G^* consists precisely of the set of cutvertices of G. Looking at the graphs in Figures 16.3 and 16.14, you might notice that the non-nuclear blocks in each case are themselves edge-critical graphs. In [33] it was conjectured that this is always the case.

Conjecture 16.30 [33] *If G is an edge-critical graph that contains cutvertices, then the non-nuclear blocks of G are edge-critical.*

Note that the nuclear block of the graph in Figure 16.3 is complete and hence trivially edge-critical. However, Sumner and Wojcicka [42] note that it is frequently the case that the nuclear block is *not* edge-critical. They conjecture that the nuclear block is edge-critical only if it is complete.

Theorem 16.31 [42] *If G is a connected, k-edge-critical graph having exactly one cutvertex, then the nuclear block of G is not edge-critical.*

For any k, and $1 \leq t \leq k$, there is a k-edge-critical graph with exactly t cutvertices and whose nuclear block is as large as desired. For example, simply choose a full graph (see Figure 16.8) with t of the A_i's being singletons and the rest as large as desired.

It should be noted that if A is a non-nuclear block corresponding to the cutvertex v, then there is a minimum dominating set for A that contains v. On the other hand, no minimum dominating set for the nuclear block can contain v. This property plays a significant role in many of the arguments dealing with the non-nuclear blocks.

In [42] 3-edge-critical graphs with cutvertices are completely characterized and Ao [2] does the same for 3-edge-i-critical graphs with cutvertices.

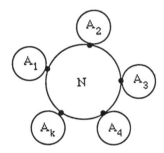

Figure 16.15: The block structure of a k-edge-critical graph.

16.5 Diameter

In general the diameter of a connected graph with domination number k can be as large as $3k - 1$ (just consider a path on $3k$ vertices). However, edge-critical graphs are much more restrictive. It was shown in [38] that the diameter of a 3-edge-critical graph is at most 3, and Blitch [6] showed that the diameter of a k-edge-critical graph was at most $3k - 5$. Paris [32] showed that this could be reduced to $3k - 6$, and Favaron, Sumner, and Wojcicka [18] showed that, in fact, the diameter of a k-edge-critical graph is at most $2k - 2$.

Theorem 16.32 [18] *If G is a k-edge-critical graph, then the diameter of G is at most $2k - 2$.*

This is still not best possible. In particular, the same paper shows that every 4-edge-critical graph has diameter at most 5. It would be interesting and potentially useful to characterize the 4-edge-critical graphs of diameter 5.

Also Favaron et al. constructed a class of examples that showed that for each $k \geq 3$, there exists a k-edge-critical graph that has diameter $\left[\frac{3}{2}k - 1\right]$ (where $[x]$ indicates the greatest integer in x). Moreover, they conjectured that this latter bound actually gives the best possible result.

Theorem 16.33 [18] *For every $k \geq 2$, there is a k-edge-critical graph G_k having diameter $\left[\frac{3}{2}k - 1\right]$.*

Conjecture 16.34 [18] *The diameter of a connected k-edge-critical graph is at most $\left[\frac{3}{2}k - 1\right]$.*

Figure 16.16 depicts a 4-edge-critical graph with diameter 5.

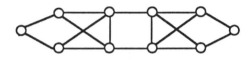

Figure 16.16: A 4-edge-critical graph with diameter 5.

The graphs G_k constructed in [18] have the property that $\gamma(G) = k$, and for every two nonadjacent vertices x, y of G, both $x \to y$ and $y \to x$. Such a graph G is said to be *k domination symmetric*. Note that as a consequence of the comment following Theorem 16.23, domination-symmetric graphs must be 2-connected. So far there has been no detailed study of this class of graphs.

Remark. It is clear that every domination-symmetric graph is both vertex and edge-critical. However, the converse does not hold, as the graph (vi) in Figure 16.2 shows (the interior vertices of degree 3 are not expandable, and hence the graph is not domination-symmetric).

It is curious that a similar bound on the diameter holds for vertex-critical graphs as Fulman, Hanson and MacGillivray [21] showed.

Theorem 16.35 [21] *If G is a k-vertex-critical graph, then the diameter of G is at most $2k - 2$.*

However, while Theorem 16.32 is not best possible for edge-critical graphs, by contrast it is shown in [21] that the bound in Theorem 16.32 is best possible for vertex-critical graphs. The graph H_k obtained by "stringing together" a series of $(k - 1)$ 4-cycles (see Figure 16.17 for the case $k = 4$) is k-vertex-critical with diameter $2k - 2$. In Fulman et al. [21], the authors characterize the 3,4-vertex-critical graphs that achieve the extremal diameter, and Ao [2] characterized, in general, the vertex-critical graphs with maximum diameter. A *linear coalescence* G_1 & G_2 & $\cdots$ & G_k of the graphs $G_1, G_2, \cdots, G_k$ is defined inductively as follows: (i). G_1 & G_2 is any coalescence of G_1 and G_2 (ii). For $k \geq 2$, G_1 & G_2 & $\cdots$ & $G_{k+1} = (G_1$ & G_2 & $\cdots$ & $G_k)_{uv}G_{k+1}$ where u is the unique vertex of G_k not adjacent to the vertex of attachment of G_1 & G_2 & $\cdots$ & G_{k-1} and G_k, and v is any vertex of G_{k+1}.

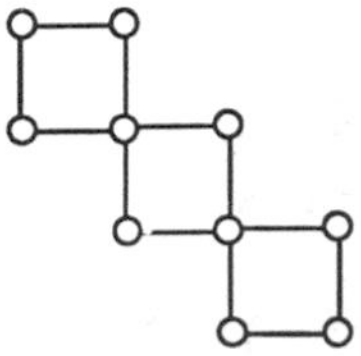

Figure 16.17: A 4-vertex-critical graph of diameter 6.

Theorem 16.36 [2] *If G is a connected, k-vertex-critical graph having diameter $2k - 2$, then G is a linear coalescence of some $k - 1$, 2-vertex-critical graphs.*

Ao [2] establishes essentially the same bounds on the diameter of vertex-i-critical graphs.

While 3-edge-critical graphs have diameter at most 3 and 4-edge-critical graphs have diameter at most 5, the corresponding bounds on local-edge-domination number are greater by 1 in each case.

Theorem 16.37 [28] *The diameter of a 3-$(\gamma, 2)$-critical graph is at most 4.*

Theorem 16.38 [28] *The diameter of a 4-$(\gamma, 2)$-critical graph is at most 6.*

Both these bounds are best possible. For example, the graph in Figure 16.5 (ii) is 4-$(\gamma, 2)$-critical with diameter 6.

Henning, Oellermann and Swart [28] ask for a characterization of the 3-$(\gamma, 2)$ graphs that have diameter 3. Such a characterization, together with their results, would give a complete characterization of 3-$(\gamma, 2)$ graphs. They also ask if it is true that the independence number and domination number are always the same for local-edge-critical graphs.

16.6   The Independent Domination Number

Many people have studied relationships between the domination number and various measures of independence. In particular, the independent dominating number, $i(G)$, has been studied by many people (See for example [1, 7, 20, 23,

37]). Evidently, $\gamma(G) \leq i(G)$ for any graph G. Allan and Laskar [1] proved that these two parameters are equal in the class of claw-free graphs. A graph is *claw-free* if it does not contain an induced subgraph isomorphic to $K_{1,3}$.

Theorem 16.39 [1] *If G is claw-free, then $\gamma(G) = i(G)$.*

A great deal of heuristic and computer-generated data suggests that the following conjecture from [38] is true:

Conjecture 16.40 [38] *If G is a 3-edge-critical graph, then $\gamma(G) = i(G)$.*

In [38], the authors actually made an even stronger (albeit false) conjecture, namely that if G is a k-edge-critical graph, with $k \geq 3$, then $\gamma(G) = i(G)$. However, Ao [2] gives a counterexample to this conjecture for $k = 4$, and in Ao, Cockayne, MacGillivray and Mynhardt [3], the authors provide an elegant construction that gives for each $k > 3$ a k-edge-critical graph that has $\gamma(G) < i(G)$.

However, the conjecture for $k = 3$ is still unsettled, and we still believe it to be true. The main support for it comes from a prolonged computer search for a counterexample that tended to show that maybe the conjecture is even too weak. In fact, it appeared at one point that it might be true that in a 3-edge-critical graph every vertex belonged to an independent dominating set of size 3. We now know this is not the case, but counterexamples seem to be rare.

There are several special cases in which the conjecture for $k = 3$ is easily seen to be true. For instance, if G is 3-edge-critical and has diameter 3, or $\delta(G) \leq 2$, or G has a cutvertex then $\gamma(G) = i(G)$.

Brigham, Chinn and Dutton [10] asked if $\gamma(G) = i(G)$ for vertex-critical graphs. This was shown not to be true in general in Fulman, Hanson and MacGillivray [21].

It is tempting to think that, for each k, there may be an absolute bound on the size of a maximum independent set in k-edge-critical graphs. This is not the case, and in fact for 3-edge-critical graphs Trotter [43] provided a class of 3-edge-critical graphs in which the independence number can be as large as desired. Blitch [6] improved this construction a bit.

Theorem 16.41 [6] *For every $t \geq 3$, there exists a 3-edge-critical graph with $n = 3t$ vertices and containing an independent set of size t.*

Theorem 16.42 [6] *If G is a 3-edge-critical graph with $\beta_0(G) = t$ and having as few vertices as possible, then $2t \leq |V(G)| \leq 3t$.*

Ao, Cockayne, MacGillivray and Mynhardt [3] construct an example of k-edge-critical graphs whose independence number exceeds k. Also Ao [2] constructs a class, $T_{k,t}$ of k-edge-i-critical graphs that have independence number t.

16.7   Degrees and Edges

It is reasonable to expect that a 3-edge-critical graph on n vertices, where n is large, cannot have many vertices of small degree. This is so, and in fact letting d_r denote the number of vertices in G having degree at most r, it was shown in Sumner and Blitch [38] that $d_r \leq 3r$.

Moreover, when n is sufficiently larger than r, d_r must be bounded by a simple linear function of r.

Theorem 16.43 [38] *If G is 3-edge-critical on $n >> r$ vertices then $d_r \leq r+1$.*

So in particular, a 3-edge-critical graph can have at most three endvertices and for $n \geq 7$, no 3-edge-critical graph can have more than two endvertices. Moreover, if G has two endvertices, then G is a full graph. In any k-edge-critical graph, no vertex can be adjacent to two endvertices. Theorem 16.43 is best possible for r odd. However, in Sumner [39] it was shown that for $n >> 2$, $d_2 \leq 2$. Perhaps it is the case that $d_r \leq r$ for r even.

Note that a domination-critical graph can have at most $(k-1)(\Delta+1)+1$ vertices, and this is best possible for vertex-critical graphs. It was conjectured in Brigham, Chinn and Dutton [10] that every vertex-critical graph on this many vertices had to be regular. Fulman, Hanson and MacGillivray [21] proved this conjecture.

Theorem 16.44 *Let G be a connected, vertex-critical graph with order n. Then*

(i) [10] $n \leq (k-1)(\Delta+1)+1$, and

(ii) [21] if G has exactly $(k-1)(\Delta+1)+1$ vertices, then G is regular.

In fact, [10] shows that the bound in part (i) of the previous theorem holds for any graph that contains at least one critical vertex. Stronger bounds than these seem to hold for edge-critical graphs. Ao [2] shows an exactly similar result for vertex-i-critical graphs.

The earliest bound on the number of edges in a graph with domination number k is due to Vizing [45].

Theorem 16.45 [45] *The maximum number of edges in a graph having n vertices and domination number k is*

$$\left\lfloor \frac{(n-k)(n-k+2)}{2} \right\rfloor.$$

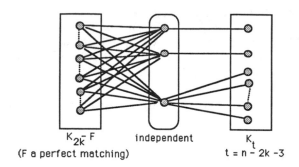

$$K_{2k}^{-} \; F \qquad \text{independent} \qquad K_t$$
$$\text{(F a perfect matching)} \qquad\qquad t = n - 2k - 3$$

Figure 16.18: Edge-critical graphs.

If G has as many edges as possible for a connected graph on n vertices with domination number k, then G is a k-edge-critical. Blitch [6] showed that this was $\binom{n-2}{2}$ for $k = 3$ and that the full graphs of Figure 16.8 achieve this value. Sanchis [36] generalized this result and showed that for $k \geq 3$, the maximum number of edges is given by $\binom{n-k+1}{2}$, and the value is also achieved for the graphs in Figure 16.8. Fulman [22], improved Sanchis' bound in terms of the maximum of degree of G.

It is still an open question as to what is the minimum number of edges in a k-edge-critical graph on n vertices. This value has only been determined for a few small values of 3-edge-critical graphs. For the 3-edge-critical graphs we conjecture that the minimum number of edges is achieved by the graphs in Figure 16.18. Thus, the conjectured minimum number of edges in a k-edge-critical graph on n vertices is: $\left\lceil \frac{(n-3)(n+2)}{4} \right\rceil$.

It is curious that each graph in the class of graphs that achieves the maximum number of edges for a k-edge-critical graph, and also each graph in the class of graphs that achieves the conjectured minimum number of edges, contains vertices with the same closed neighborhoods. Such graphs are obtainable from smaller edge-critical graphs by expanding vertices as explained in the preliminary section. So graphs that are point distinguishing are most primitive in this sense. Thus it would be worthwhile to determine the maximum and minimum number of edges in a point distinguishing k-edge-critical graph on n vertices.

Vertex-critical graphs must always be point distinguishing as observed by Brigham, Chinn, and Dutton [10] (although they did not use that terminology) – in fact, they note that vertex-critical graphs do not contain a pair of vertices v and u with $N[v] \subseteq N[u]$.

Brigham, Chinn and Dutton [10] show that vertex-critical graphs with order n and size m satisfy the following relation between m, γ, Δ, and n.

Theorem 16.46 [10] *For any vertex-critical graph G,*

$$n \leq \frac{2m + 3\gamma - \Delta}{3}.$$

16.8 Future Progress/Other Directions

What remains to be done? Obviously, there is quite a lot still unknown in regards to all of vertex-critical, edge-critical, locally-edge-critical, edge-i-critical, and vertex-i-critical. We recap here a few of the most outstanding problems in these areas.

1. For $k \geq 4$, do connected k-edge-critical graphs of even order have a perfect matching?

2. Is a non-nuclear block of a edge-critical graph always itself edge-critical?

3. To what extent do the known properties of edge-critical graphs carry over to locally-edge-critical graphs?

4. What is the best bound on the diameter of a k-edge-critical graph? What about the diameter of edge-i-critical graphs?

5. What can be said about the class of graphs that are both vertex-critical and edge-critical?

6. What about the class of domination-symmetric graphs? They have not been studied at all, yet appear to be quite interesting.

7. Is $\gamma(G) = i(G)$ for 3-edge-critical graphs? What about locally-edge-critical graphs? What about the same question for edge-i-critical graphs?

8. What is the minimum number of edges in a k-edge-critical graph on n vertices?

9. Is it true that every connected, 3-edge-critical graph without cutvertices has a Hamiltonian cycle?

10. Is every 3-connected (2-connected?), 4-edge-critical graph Hamiltonian?

11. Is it possible to characterize the 3-edge-critical graphs with diameter three?

12. Is there a good characterization for the 3-$(\gamma, 2)$-critical graphs with diameter three?

13. Which graphs are both edge-critical and edge-i-critical?

In this chapter we have only touched on a few of the most-studied ways in which a graph can be critical with respect to the domination number. Ao [2] does a good job of comparing several of the other possible ways that a critical graph might be defined.

We mention just two last directions here. First, the so-called edge *domination insensitive graphs* – defined by Dutton and Brigham [15] – those for which the domination number does not change whenever an edge is deleted; i.e., $\gamma(G-e) = \gamma(G)$ for every edge e of G. Note that these graphs are the flip side of the edge-minimal graphs studied by Bauer, Harary, Nieminen and Suffel [5] and Walikar and Acharya [46].

More generally, define a graph to be *k-domination-insensitive* (or just (γ, k)-*insensitive*) if the domination number does not change whenever any set of k edges is deleted from G. Let $E_k(n, \gamma)$ denote the smallest number of edges required for any (γ, k)-insensitive graph on n vertices. Haynes, Brigham and Dutton [26] studied this parameter. Their principal result follows.

Theorem 16.47 [26] *When $k + 1 \leq \gamma \leq 2k$, $E_k(n, \gamma)$ is equal to $(k + 3)n/2$ as n approaches infinity.*

It might be worthwhile to study the other graph theoretic properties of k-domination-insensitive graphs. Closely related to these ideas is that of the bondage number of a graph. The *bondage number* $b(G)$ is defined to be the minimum number of edges whose removal from the graph increases the domination number. The bondage number has been studied by J. E. Fink, M. S. Jacobson, L. F. Kinch, and J. Roberts [19] as well as B. L. Hartnell and D. F. Rall [25]

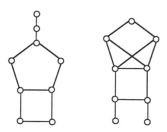

Figure 16.19: A 4-conservative graph that is not 4-critical.

Finally, a graph is *k-conservative* if it satisfies the conditions to be k-edge-critical, but has domination number $k - 1$. More accurately, say that A *precisely dominates* B if A dominates B but does not dominate any vertex in $G-B$. Then G is k-conservative if $\gamma(G) = k - 1$, and for every nonadjacent pair of vertices x and y there is a $k - 2$ element set S such that either $S \cup \{x\}$ precisely dominates $G - y$ or $S \cup \{y\}$ precisely dominates $G - x$. Figures 16.19 and 16.20 show examples of 4-conservative graphs.

Note that a 4-conservative graph can have a diameter larger than that possible for a 4-edge-critical graph. So we ask, how much of what is known about k-edge-critical graphs is also true for k-conservative graphs; i.e., how much of what is true about the k-edge-critical graphs really depends on the value of the domination number? In particular, what is the best bound on the diameter of k-conservative graphs?

Figure 16.20: A 4-conservative graph of diameter 6.

For additional references to work on critical concepts in domination not surveyed here, see Ao [2], Haynes, Brigham and Dutton [26], Carrington, Harary, and Haynes [11], Rice [35], Walikar and Acharya [46], and Gunther, Hartnell and Rall [23]. See also Chapter 17 of this book and Chapter 5 of [27].

Bibliography

[1] R. B. Allan and R. Laskar, On domination and independent domination numbers of a graph. *Discrete Math.* 23 (1978) 73-76.

[2] S. Ao, *Independent domination critical graphs.* Masters Thesis, University of Victoria, Victoria, BC, Canada (1994).

[3] S. Ao, E. J. Cockayne, G. MacGillivray and C. M. Mynhardt, Domination critical graphs with higher independent domination numbers. *J. Graph Theory* 22 (1) (1996) 9-14.

[4] S. Ao and G. MacGillivray, Hamiltonian properties of independent domination critical graphs. Preprint (1996).

[5] D. Bauer, F. Harary, J. Nieminen and C. Suffel, Domination alteration sets in graphs. *Discrete Math.* 47 (1983) 153-161.

[6] P. Blitch, *Domination in Graphs.* Ph.D. Dissertation, University of South Carolina (1983).

[7] B. Bollobás and E. J. Cockayne, Graph theoretic parameters concerning domination, independence, and irredundance. *J. Graph Theory* 3 (1979) 241-249.

[8] J. A. Bondy and V. Chvátal, A method in graph theory. *Discrete Math.* 15 (1976) 111-136.

[9] R. C. Brigham, P. Z. Chinn and R. D. Dutton, A study of vertex domination critical graphs. Technical Report, University of Central Florida (1984).

[10] R. C. Brigham, P. Z. Chinn and R. D. Dutton, Vertex domination critical graphs. *Networks* 18 (1988) 173-179.

[11] J. R. Carrington, F. Harary and T. W. Haynes, Changing and unchanging the domination number of a graph. *J. Combin. Math. Combin. Comput.* 9 (1991) 57-63.

[12] V. Chvátal, Tough graphs and Hamiltonian circuits. *Discrete Math.* 2 (1973) 215-228.

[13] E. J. Cockayne, O. Favaron, C. Payan and A. G. Thomason, Contributions to the theory of domination, independence and irredundance in graphs. *Discrete Math.* 33 (1981) 249-258.

[14] E. J. Cockayne and S. T. Hedetniemi, Disjoint independent dominating sets in graphs. *Discrete Math.* 15 (1976) 213-222.

[15] R. D. Dutton and R. C. Brigham, An extremal problem for edge domination insensitive graphs. *Discrete Appl. Math.* 20 (1988) 113-125.

[16] R. J. Faudree, R. J. Gould, M. S. Jacobson and R. H. Schelp, Neighborhood conditions and Hamiltonian properties in graphs. *J. Combin. Theory Ser. B* 46 (1989) 1-20.

[17] O. Favaron, Stability, domination and irredundance in a graph. *J. Graph Theory* 10 (1986) 429-438.

[18] O. Favaron, D. Sumner and E. Wojcicka, The diameter of domination-critical graphs. *J. Graph Theory* 18 (1994) 723-734.

[19] J. E. Fink, M. S. Jacobson, L. F. Kinch and J. Roberts, The bondage number of a graph. *Discrete Math.* 86 (1990) 47-57.

[20] J. Fulman, A note on the characterization of domination-perfect graphs. *J. Graph Theory* 7 (1993) 47-51.

[21] J. Fulman, D. Hanson and G. MacGillivray, Vertex domination-critical graphs. *Networks* 25 (1995) 41-43.

[22] J. Fulman, Generalization of Vizing's theorem on domination. *Discrete Math.* 126 (1994) 403-406.

[23] G. Gunther, B. Hartnell and D. F. Rall, Graphs whose vertex independence number is unaffected by single edge addition or deletion. *Discrete Appl. Math.* 46 (1993) 167-172.

[24] D. Hanson, Hamilton closures in domination critical graphs. *J. Combin. Math. Combin. Comput.* 13 (1993) 121-128.

[25] B. L. Hartnell and D. F. Rall, Bounds on the bondage number of a graph. *Discrete Math.* 128 (1994) 173-177.

[26] T. W. Haynes, R. C. Brigham and R. D. Dutton, Extremal graphs domination insensitive to the removal of k edges. *Discrete Appl. Math.* 44 (1993) 295-304.

[27] T. W. Haynes, S. T. Hedetniemi and P. J. Slater, *Fundamentals of Domination in Graphs.* Marcel Dekker, Inc., 1997.

[28] M. A. Henning, O. R. Oellermann and H. C. Swart, Local edge domination critical graphs. Preprint (1996).

[29] A. Meir and J. W. Moon, Relations between packing and covering numbers of a tree. *Pacific J. Math.* 61 (1975) 225-233.

[30] L. Nebesky, On the existence of 1-factors in partial squares of graphs. *Czechoslovak Math. J.* 29 (1979) 349-352.

[31] J. Nieminen, Two bounds for the domination number of a graph. *J. Inst. Math. Appl.* 14 (1974) 183-187.

[32] M. Paris, The diameter of edge domination critical graphs. *Networks* 24 (1994) 261-262.

[33] M. Paris, D. Sumner and E. Wojcicka, Edge-domination critical graphs with cut-vertices. Submitted for publication.

[34] D. Rall, Domatically critical and domatically full graphs. *Discrete Math.* 86 (1990) 81–87.

[35] T. W. (Haynes) Rice, *On k-γ-insensitive domination.* Ph.D. Dissertation, University of Central Florida (1988).

[36] L. A. Sanchis, Maximum number of edges in connected graphs with a given domination number. *Discrete Math.* 87 (1991) 65-72.

[37] N. Seifter, Domination and independent domination numbers of graphs. *Ars Combin.* 38 (1994) 119-128.

[38] D. Sumner and P. Blitch, Domination critical graphs. *J. Combin. Theory Ser. B* 34 (1983) 65-76.

[39] D. Sumner, Critical concepts in domination. *Discrete Math.* 86 (1990) 33-46.

[40] D. Sumner, Point determination in graphs. *Discrete Math.* 5 (1973) 179-187.

[41] D. Sumner, The Nucleus of a point determining graph. *Discrete Math.* 14 (1976) 91-97.

[42] D. Sumner and E. Wojcicka, Cut-vertices in k-edge-domination-critical graphs. Preprint (1996).

[43] W. T. Trotter, personal communication.

[44] W. T. Tutte, The factorization of linear graphs. *J. London Math. Soc.* (2) 13 (1976) 351-359.

[45] V. G. Vizing, A bound on the external stability number of a graph. *Dokl. Akad. Nauk.* 164 (1965) 729-731.

[46] H. B. Walikar and B. D. Acharya, Domination critical graphs. *Nat. Acad. Sci. Lett.* 2 (1979) 70-72.

[47] E. Wojcicka, Hamiltonian properties of domination-critical graphs. *J. Graph Theory* 14 (1990) 205-215.

Chapter 17

Bondage, Insensitivity, and Reinforcement

Jean E. Dunbar
Department of Mathematics
Converse College
Spartanburg, SC 29302 USA

Teresa W. Haynes
Department of Mathematics
East Tennessee University
Johnson City, TN 37614 USA

Ulrich Teschner
Lehrstuhl II fur Mathematik
RWTH Aachen
52056 Aachen, Germany

Lutz Volkmann
Lehrstuhl II fur Mathematik
RWTH Aachen
52056 Aachen, Germany

17.1 Introduction

Given an arbitrary graph, a new graph can be obtained by deleting a vertex, or adding or deleting an edge. Much work has been done concerning these graph alterations and their effects on the domination number of a graph. A survey of the literature in this area is found in [13], Chapter 5. Here we are concerned with generalizations of the alterations involving edges.

Graphs for which the domination number changes upon the removal of an arbitrary edge were first investigated by Walikar and Acharya in [30]. These graphs satisfy $\gamma(G - e) = \gamma(G) + 1$ for every edge $e \in E(G)$. Bauer, Harary, Nieminen, and Suffel [1] characterized such a graph as a union of stars. On the other hand, graphs for which the domination number does not change, i.e., $\gamma(G - e) = \gamma(G)$ for every edge $e \in E(G)$, are called γ-insensitive and were first studied by Dutton and Brigham [7].

Similarly, a graph can be altered by adding an edge. If $\gamma(G + e) = \gamma(G) - 1$ for every edge $e \in E(\overline{G})$, then G is called an *edge domination critical* graph. These graphs are discussed by Sumner and Wojcicka in the previous chapter of this book. A counterpart concerns graphs for which the domination number does not change upon the addition of an arbitrary edge. These graphs were

471

characterized by Carrington, Harary, and Haynes [4].

We consider generalizations of three of the above problems. In particular, we modify a graph by adding or deleting a number of edges and investigate the effect on the domination number of the resulting graph.

In Section 17.2 we consider the smallest number of edges whose removal from a graph G causes the modified graph to have a larger domination number than G. This number, defined by Fink, Jacobson, Kinch, and Roberts [8] is the *bondage number* of a graph. In Section 17.3 we summarize work done by Haynes, Brigham, and Dutton [14] on graphs whose domination number remains unchanged when any k edges are removed from the graph. These graphs are said to be (γ, k)-*insensitive*. Finally, in Section 17.4 we examine the minimum number of edges that must be added to a graph G in order to cause the domination number of the resulting graph to be smaller than that of G. This number, defined by Kok and Mynhardt [16], is the *reinforcement number* of a graph.

17.2   Bondage

Fink et al. [8] introduced the bondage number of a graph in 1990 with the following application in mind. "Among the various applications of the theory of domination that have been considered, the one that is perhaps most often discussed concerns a communications network. This network consists of existing communication links between a fixed set of sites. The problem at hand is to select a smallest set of sites at which to place transmitters so that every site in the network that does not have a transmitter is joined by a direct communication link to one that does have a transmitter. This problem reduces to finding a minimum dominating set in the graph, corresponding to this network, that has a vertex corresponding to each site, and an edge between two vertices if and only if the corresponding sites have a direct communications link joining them."

"We now carry the foregoing example further and examine a question concerning the vulnerability of the communications network under link failure. In particular, suppose that someone (a saboteur) does not know which sites in the network act as transmitters, but does know that the set of such sites corresponds to a minimum dominating set in the related graph. What is the fewest number of communication links that he must sever so that at least one additional transmitter would be required in order that communication with all sites be possible?"

The *bondage number* $b(G)$ of a nonempty graph G is the cardinality of a smallest set of edges whose removal from G results in a graph with domination number greater than $\gamma(G)$. Since the domination number of every spanning subgraph of a nonempty graph G is at least as great as $\gamma(G)$, the bondage number of a nonempty graph is well defined.

First results on the bondage number are found in an article of Bauer, Harary,

Nieminen, and Suffel [1] from 1983 (the term "domination line-stability" was used instead of bondage number). In [1] and [8] it was proved that any tree has bondage number 1 or 2, and the authors of [8] posed as an open problem classifying trees of bondage number 1. Two years later, Hartnell and Rall [11] presented a constructive characterization of all trees with bondage number 2.

In [8], the authors conjectured that $b(G) \leq \Delta(G) + 1$ for any nonempty graph G. Three years later, Teschner [22] disproved this conjecture and gave a necessary condition for all counterexamples. Moreover, Hartnell and Rall [12] and Teschner [23] independently proved $b(G_n) = \frac{3}{2}\Delta(G_n)$ for the cartesian product $G_n = K_n \times K_n$. This disproves the existence of an upper bound of the form $b(G) \leq \Delta(G) + c$ for any $c \in \mathbf{N}$. Teschner [24] showed that $b(G) \leq \frac{3}{2}\Delta(G)$ for any graph G satisfying $\gamma(G) \leq 3$.

Before presenting the details of these results, we digress briefly to include a related concept, *fractional bondage*. Chvátal and Cook [5] showed how to formulate computing $b(G)$ as an integer program. They defined a *whip* in a graph G as any spanning subgraph F of G such that each component of F is a star and F has precisely $\gamma(G)$ components; let $W(G)$ denote the set of all whips in G. Then $b(G)$ is expressable as the optimal value of the problem:

$$\text{minimize } \Sigma\{x_e \mid e \in E(G)\}$$
$$\text{subject to } \Sigma\{x_e \mid e \in E(F)\} \geq 1 \text{ for all } F \in W(G),$$
$$x_e \geq 0 \text{ for all } e \in E(G)$$
$$x_e \text{ an integer} \qquad \text{for all } e \in E(G).$$

Chvátal and Cook [5] defined the *fractional bondage number* $b^*(G)$ as the optimal value of the linear programming relaxation of the above problem.

$$\text{minimize } \Sigma\{x_e \mid e \in E(G)\}$$
$$\text{subject to } \Sigma\{x_e \mid e \in E(F)\} \geq 1 \text{ for all } F \in W(G),$$
$$x_e \geq 0 \qquad \text{for all } e \in E(G).$$

In light of the disproved conjecture relating bondage number and maximum degree, it is noteworthy that after relaxing the integer restriction Chvátal and Cook showed that $b^*(G) \leq \Delta(G)$.

17.2.1 Upper bounds

We begin with a straightforward observation.

Lemma 17.1 *If k edges can be removed from a graph G to yield a subgraph H with $b(H) = 1$, then $b(G) \leq k + 1$.*

Now let u and v be distinct vertices in a graph G. The *distance from u to v*, denoted $d(u, v)$, is the length of a shortest path from u to v. If no such path exists, we say $d(u, v) = \infty$. Let E_v be the set of edges incident with vertex v.

Lemma 17.2 [1, 8, 26] *If G is a graph, then $b(G) \leq deg(u) + deg(v) - 1$ for every pair u and v of vertices with $d(u,v) \leq 2$.*

Proof. Let G be a graph with vertices u, v such that $d(u,v) \leq 2$. If u and v are adjacent, let $H = G - (E_v \cup E_u - \{uv\})$. If $d(u,v) = 2$, then for some $w \in N(u) \cap N(v)$, let $H = G - (E_v \cup E_u - \{uw, vw\})$. In both cases $b(H) = 1$ and Lemma 17.1 yields the desired bound. □

Corollary 17.3 [1, 8] *If G is a graph with no isolated vertices, then $b(G) \leq \Delta(G) + \delta(G) - 1$.*

A generalization of Corollary 17.3 was found by Hartnell and Rall [12] (see also Teschner [26]).

Theorem 17.4 [12, 26] *If G has edge connectivity $\lambda(G) \geq 1$, then $b(G) \leq \Delta(G) + \lambda(G) - 1$.*

In our first conjecture, we present a possible generalization of Theorem 17.4.

Conjecture 17.5 *If G has vertex connectivity $\kappa(G) \geq 1$, then $b(G) \leq \Delta(G) + \kappa(G) - 1$.*

Hartnell and Rall [12] improved the bound of Lemma 17.2 for adjacent vertices.

Theorem 17.6 [12] *If G is a graph, then for every pair u and v of adjacent vertices $b(G) \leq deg(u) + deg(v) - 1 - |N(u) \cap N(v)|$.*

Bauer, Harary, Nieminen, and Suffel [1] made the following observation.

Lemma 17.7 [1] *If there exists a vertex $v \in V(G)$ for which $\gamma(G - v) \geq \gamma(G)$, then $b(G) \leq deg(v) \leq \Delta(G)$.*

Proof. Let G be a graph with vertex v such that $\gamma(G - v) \geq \gamma(G)$. Then it follows by the hypothesis that $\gamma(G - E_v) = \gamma(G - v) + 1 \geq \gamma(G) + 1 > \gamma(G)$. □

Corollary 17.8 *If $b(G) > \Delta(G)$, then $\gamma(G - v) < \gamma(G)$ for all vertices $v \in V(G)$.*

Brigham, Chinn, and Dutton [2, 3] defined vertex domination–critical graphs exactly in this way. A vertex v is *critical* if $\gamma(G - v) < \gamma(G)$, and G is vertex domination–critical if each vertex is critical. We refer to graphs with this property as *vc-graphs*. The concept of vertex domination–critical graphs plays an important role in the study of the bondage number. For instance, it immediately follows from Corollary 17.8 that each counterexample to the conjecture $b(G) \leq \Delta(G) + 1$ is necessarily a vc–graph. The bondage number of a vc–graph is examined in [21].

As already mentioned, a class of graphs disproving this conjecture was found independently by Hartnell and Rall [12] and Teschner [23] as follows.

Theorem 17.9 [12, 23] *Let G_t be the cartesian product $K_t \times K_t$ for a positive integer $t \geq 2$. Then $b(G_t) = 3(t-1) = \frac{3}{2}\Delta(G_t)$.*

Note that even the graphs without large cliques do not satisfy the conjecture $b(G) \leq \Delta(G) + 1$. It was shown in [12] that the bipartite graph G resulting from removing three vertex–disjoint 4-cycles from $K_{6,6}$ is 4-regular and $b(G) = 6$. Theorem 17.9 disproves the existence of an upper bound of the form $b(G) \leq \Delta(G) + c$ for any $c \in \mathbf{N}$. On the other hand, since every planar graph G has $\delta(G) \leq 5$, if G is a planar graph, then Corollary 17.3 yields $b(G) \leq \Delta(G) + 4$. Furthermore, for a planar graph G it is not very difficult to show that $b(G) \leq \Delta(G) + 3$ if $\Delta(G) \geq 8$, and $b(G) \leq \Delta(G) + 2$ if $\Delta(G) \geq 13$ or if the girth $g(G) \geq 4$, and $b(G) \leq \Delta(G) + 1$ if the girth $g(G) \geq 6$.

Conjecture 17.10 *If G is a planar graph, then $b(G) \leq \Delta(G) + 1$.*

Conjecture 17.11 *For any graph G, $b(G) \leq \frac{3}{2}\Delta(G)$.*

Teschner [24] has shown that Conjecture 17.11 is true for vc–graphs G with $\gamma(G) = 3$.

17.2.2 Lower bounds

Since the bondage number is defined as a minimum, each constructive method resulting in a bondage-edge-set leads to an upper bound of the bondage number. Although lower bounds are not as readily found, we mention a few here. For any graph G with no isolated vertices, any vertex cover is a dominating set, and hence $\gamma(G) \leq \alpha_0(G)$. We use this fact in the following proof.

Theorem 17.12 [26] *If $\alpha_0(G) = \gamma(G)$, then*
i) $b(G) \geq \delta(G)$
ii) $b(G) \geq \delta(G) + 1$, if G is vertex-critical.

Proof. Let G be a graph with $\alpha_0(G) = \gamma(G)$.
i) Without loss of generality, let $\delta(G) \geq 2$. If $X \subseteq E$ with $|X| \leq \delta(G) - 1$, then $\delta(G - X) \geq 1$ and therefore $\gamma(G - X) \leq \alpha_0(G - X) \leq \alpha_0(G) = \gamma(G)$. Hence $b(G) \geq \delta(G)$.
ii) If G is a vc-graph, then $\gamma(G - v) < \gamma(G)$ for all vertices $v \in V(G)$. Hence $b(G) \geq \delta(G) + 1$. □

For example, we construct a vc-graph G with $\alpha_0(G) = \gamma(G)$. Begin with $k \geq 2$ cycles C_4 and identify a vertex v of one of the cycles with a vertex of each remaining C_4, such that v is a cutvertex of the resulting vc-graph G. Then $\alpha_0(G) = \gamma(G) = k + 1$ and Theorem 17.12 yields $b(G) \geq 3$. Observe that the upper bound from Lemma 17.2 is also achieved. More about graphs with $\alpha_0(G) = \gamma(G)$ can be found in [29].

A result due to Sanchis [20] gives an upper bound on the size m of a graph G with a given domination number. We will use this result to develop the next lower bound on $b(G)$.

Theorem 17.13 [20] *For a graph G with $3 \leq \gamma(G) \leq n/2$ and no isolated vertices, $m \leq \binom{n-\gamma(G)+1}{2}$.*

Theorem 17.14 [26] *Let G be a graph with $2 \leq \gamma(G) \leq n/2 - 1$. Then*
i) $b(G) \geq min\{\delta(G), m - \binom{n-\gamma(G)}{2}\}$
ii) $b(G) \geq min\{\delta(G) + 1, m - \binom{n-\gamma(G)}{2}\}$, if G is a vc–graph.

Proof. Let G be a graph with $2 \leq \gamma(G) \leq n/2 - 1$. Remove a bondage-edge-set from G to create G'. Then $\gamma(G') = \gamma(G) + 1$ and $3 \leq \gamma(G') \leq n/2$.
Case 1 : G' has an isolated vertex. Then we have $b(G) \geq \delta(G)$ or $b(G) \geq \delta(G)+1$ if G is a vc–graph.
Case 2 : G' has no isolated vertices. According to Theorem 17.13

$$m(G') \leq \binom{n(G') - \gamma(G') + 1}{2} = \binom{n(G) - \gamma(G)}{2},$$

and therefore the theorem is proved. $\square$

The lower bound of Theorem 17.14 is sharp for the class of vc–graphs with domination number 2. Brigham, Chinn, and Dutton [3] showed that G belongs to this class if and only if G is the complete graph K_{2k}, $k \geq 2$, with a 1-factor removed. For $G = K_{2k}$, it follows from Theorem 17.14 that $b(G) \geq \delta(G) + 1 = \Delta(G) + 1$ while $b(G) \leq \Delta(G) + 1$ is straightforward.

Conjecture 17.15 *For a vc–graph G, $b(G) \geq \delta(G) + 1$.*

17.2.3   Selected families

We begin this section by giving exact values for the bondage number of selected graphs.

Proposition 17.16 [8] *For the complete graph of order $n \geq 2$, $b(K_n) = \lceil \frac{n}{2} \rceil$.*

Proposition 17.17 [8] *For the cycle of order n,*

$$b(C_n) = \begin{cases} 3 & if \ n \equiv 1 \ (mod \ 3), \\ 2 & otherwise. \end{cases}$$

Corollary 17.18 [8] *For the path of order $n \geq 2$,*

$$b(P_n) = \begin{cases} 2 & if \ n \equiv 1 \ (mod \ 3), \\ 1 & otherwise. \end{cases}$$

Proposition 17.19 [8] *Let $G = K_{n_1,n_2,...,n_t}$ be the complete t-partite graph such that $n_1 \leq n_2 \leq ... \leq n_t$ and $n_t > 1$. Then*

$$b(G) = \begin{cases} \lceil \frac{j}{2} \rceil & if \ n_j = 1 \ and \ n_{j+1} \geq 2, \ for \ some \ j, \ 1 \leq j < t, \\ 2t - 1 & if \ n_1 = n_2 = ... = n_t = 2, \\ \sum_{i=1}^{t-1} n_i & otherwise. \end{cases}$$

Proposition 17.20 *Let $G = C_t \times K_2$ with $t \geq 3$. Then*

$$b(G) = \begin{cases} 2 & if \ t \equiv 0, 1 \ (mod \ 4), \\ 3 & if \ t \equiv 3 \ (mod \ 4), \\ 4 & if \ t \equiv 2 \ (mod \ 4). \end{cases}$$

Next we consider bounds on the bondage number for trees.

Theorem 17.21 [1, 8] *If T is a nontrivial tree, then $b(T) \leq 2$.*

Proof. If T is a tree of order 2, then $b(T) = 1$. If T has at least 3 vertices, let $v_0, v_1, ..., v_k$ be a longest path in T. Observe that v_0 and v_k must necessarily be endvertices in T. Now either $deg(v_1) = 2$ or v_1 is adjacent to another endvertex $w \neq v_0$. In both cases, Lemma 17.2 yields the desired result. □

In [8] it was proven that a forbidden subgraph characterization to classify trees with bondage number 1 or 2, respectively, is not possible. However, characterizations have been determined [11, 26, 28]. Here we briefly describe the method, due to Hartnell and Rall [11], that inductively constructs all trees with bondage number 2. An important tree (F_t) in their construction is shown in Figure 17.1.

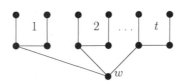

Figure 17.1: Tree F_t.

Terminology:
1. *Attach* a P_2 (P_3) to a vertex v of a tree means that the vertex v and one endvertex of the P_2 (P_3) are joined by an edge.
2. *Attach* F_t to a vertex v of a tree means that the vertex v and the vertex w of F_t are joined by an edge.

The four operations on a tree T:
Type 1: Attach a P_2 to $v \in V(T)$, where $\gamma(T - v) = \gamma(T)$ and v belongs to at least one γ-set of T.
Type 2: Attach a P_3 to $v \in V(T)$, where $\gamma(T - v) < \gamma(T)$, i.e., v is a critical

vertex.

Type 3: Attach F_1 to $v \in V(T)$, where v belongs to at least one γ-set of T.

Type 4: Attach F_t, $t \geq 2$, to $v \in V(T)$, where v can be any vertex of T.

Let $C = \{ T \mid T$ is a tree and $T = K_1$, $T = P_4$, $T = F_t$ for some $t \geq 2$, or T can be obtained from P_4 or F_t ($t \geq 2$) by a finite sequence of operations of Type 1, 2, 3 or 4 $\}$.

Theorem 17.22 [11] *A tree T has $b(T) = 2$ if and only if T belongs to the set C defined above.*

In [6], Cockayne, Goodman, and Hedetniemi presented a linear time algorithm for the domination number of a tree. Therefore, we can decide in $O(n^2)$ time by removing each edge, whether the bondage number of a given tree of order n is 1 or 2. In fact, Hartnell, Jorgensen, Vestergaard, and Whitehead [10] have produced a linear time algorithm to find the bondage number of a tree.

A *block graph* is a graph whose blocks are complete graphs. Each block in a *cactus graph* is either a cycle or a K_2. If each block of a graph G is either a complete graph or a cycle, then we call the graph a *block-cactus graph*. The *corona $G \circ H$* is the graph formed from a copy of G and $|V(G)|$ copies of H by joining the *ith* vertex of G to every vertex in the *ith* copy of H. Teschner and Volkmann [25, 27] found that vc–graphs were instrumental in determining bounds on the bondage number of cactus and block graphs. They showed that $b(G) \leq 3$ for non-trivial cactus graphs. For an example of the cactus graphs G which achieve this upper bound, let H be a nontrivial cactus graph without endvertices and G be the corona $H \circ K_1$. We pose the following open problem.

Problem: Characterize all cactus graphs with bondage number 3.

We conclude this section with selected results concerning the bondage number of block-cactus graphs.

Theorem 17.23 ([25, 27] *Let G be a connected block–cactus graph with at least two blocks. Then $b(G) \leq \Delta(G)$.*

Proof. Let B be an end block of G with the cut vertex x and let $v \in V(B) - \{x\}$. If B is complete, then $N[v] \subseteq N[x]$ and hence x is not critical. Thus Lemma 17.7 yields $b(G) \leq \Delta(G)$. If B is a cycle, then there exist two adjacent vertices of degree 2, and Lemma 17.2 yields $b(G) \leq 3 \leq \Delta(G)$. $\square$

We can apply the same proof techniques as in [25] and [27] to show the following result.

Theorem 17.24 *Let G be a connected block–cactus graph which is neither a cactus graph nor a block graph. Then $b(G) \leq \Delta(G) - 1$.*

17.3   Insensitivity

Pursuing further the effect of edge removal on the domination number, we consider graphs for which the domination number remains the same after the removal of an arbitrary set of edges. Dutton and Brigham [7] coined the term *edge domination insensitive* or just *γ-insensitive* to identify graphs for which the domination number is unchanged when any single edge is removed. These graphs were characterized by Walikar and Acharya [30].

Proposition 17.25 [30] *A graph G has $\gamma(G) = \gamma(G - e)$ for any edge $e \in E$ if and only if for each $e = uv \in E$, there exists a γ-set S such that one of the following conditions is satisfied:*

(i) $u, v \in S$

(ii) $u, v \in V - S$

(iii) $u \in S$ and $v \in V - S$ implies $|N(v) \cap S| \geq 2$.

Haynes, Brigham, and Dutton [14, 17] extended the notion of γ-insensitive graphs to (γ, k)-insensitive graphs by considering the removal of $k \geq 1$ arbitrary edges. Observe that if G is a (γ, k)-insensitive graph, the bondage number $b(G) \geq k + 1$ and equality holds when k is maximized for G. Since the bondage number of any tree T is at most 2, for a (γ, k)-insensitive tree T, k must be 1. Hence (γ, k)-insensitive trees are precisely the trees with bondage number 2 that were characterized in the previous section. (For an example of a $(\gamma, 1)$-insensitive tree, consider the corona of a path $P_t \circ K_1$, $t \geq 2$.) Note that in general for (γ, k)-insensitive graphs, k is arbitrary, but fixed and not necessarily maximum.

Research of (γ, k)-insensitivity has mostly been concerned with extremal graphs having the property. In this context, a connected graph of order n is extremal if it is (γ, k)-insensitive and has the minimum number of edges among all such graphs of order n.

Graphs with this property are useful in network design. For example, if graph G represents a communication network, a γ-set of G is a minimized core group that could function in a variety of ways, i.e., as "masters", fileservers, or repositories for a global database essential to the other computers in the network. It might be desirable that the number of processors in the core group stay the same even after k links (edges) fail. Then a network corresponding to an extremal (γ, k)-insensitive graph has minimum link cost (minimum number of edges) and the desired fault tolerant property. A special $(\gamma, 2)$-insensitive graph, called the "G-network" was introduced in [15] as a suitable design for communications and interconnection networks. Other applications of (γ, k)-insensitive graphs were investigated in [9, 15, 17, 19].

Dutton and Brigham [7] studied connected extremal $(\gamma, 1)$-insensitive graphs. The minimum number of edges in any (γ, k)-insensitive graph of order n is denoted $E_k(n, \gamma)$. Extremal graphs were found in [7, 14, 17] for the case when $\gamma(G) = 1$.

Theorem 17.26 [14] *Let G be a graph with $\gamma(G) = 1$ and $n > 2k$ such that $\gamma(G) = \gamma(G - F)$ for any subset $F \subseteq E$ such that $|F| = k \geq 1$. Then*

$$E_k(n, 1) = (2k + 1)(n - k - 1).$$

Proof. Let G be a $(1, k)$-insensitive graph. Suppose G has at most $2k$ vertices of degree $n - 1$. Then k edges can be selected such that each vertex of degree $n - 1$ is incident to at least one of them. No single vertex dominates the graph after these k edges are removed, contradicting that G is $(1, k)$-insensitive. Thus G has at least $2k + 1$ vertices with degree $n - 1$, so $E_k(n, 1) \geq (2k + 1)(n - k - 1)$. On the other hand, a connected graph with exactly $2k + 1$ vertices of degree $n - 1$ and no other edges is $(1, k)$-insensitive, so equality holds. $\square$

Dutton and Brigham [7] calculated $E_k(n, \gamma)$ for the case when $k = 1$. Note that Theorem 17.26 takes care of the case for which $\gamma(G) = 1$, so we summarize their results for $\gamma(G) \geq 2$.

Theorem 17.27 [7] *For a $(\gamma, 1)$-insensitive graph G with $\gamma(G) \geq 2$,*

$$E_1(n, \gamma) = \begin{cases} n - 1 & \text{if } n \leq 3\gamma(G) - 2, \\ n & \text{if } n = 3\gamma(G) - 1, \\ 2n - 3\gamma(G) & \text{if } n \geq 3\gamma(G). \end{cases}$$

The value of $E_k(n, \gamma)$ and extremal graphs were determined in [18] for the case when $\gamma(G) = k = 2$.

Theorem 17.28 [18] *For a $(2, 2)$-insensitive graph G with $n \geq 11$,*

$$E_2(n, 2) = \lfloor (5n - 10)/2 \rfloor.$$

The proof is long and tedious with many cases and is omitted here. However, we give an example of a family of extremal $(2, 2)$-insensitive graphs in Figure 17.2.

Although extremal (γ, k)-insensitive graphs and the exact value of $E_k(n, \gamma)$ have not been found for the cases $\gamma(G) \geq 3, k = 2$ and $\gamma(G) \geq 2, k \geq 3$, upper and lower bounds on $E_k(n, \gamma)$ were determined [17] and an asymptotic bound on $E_k(n, \gamma)$ for $k \geq 2$ was derived in [14]. Unless otherwise stated the remaining results in this section are from [14] and [17].

Theorem 17.29 *For $k + 1 \leq \gamma(G) \leq 2k$, $E_k(n, \gamma)$ is asymptotically equal to $(k + 3)n/2$ as n approaches infinity.*

Figure 17.2: A family of $(2,2)$-insensitive graphs.

We outline the proof of Theorem 17.29. First an upper bound was established by constructing a family of (γ, k)-insensitive graphs. Let $k + 1 \leq \gamma(G) \leq 2k$, $n \geq \gamma(G)(k + 1)$, $t = \lfloor (n - \gamma(G))/k \rfloor$, and $r = (n - \gamma(G)) \bmod k$. Note that $t \geq \gamma(G)$. Construct graph $G = (V, E)$ as follows:

(1) $V = A \cup B_1 \cup B_2 \cup ... \cup B_t$ where $A = \{a_1, a_2, ..., a_\gamma\}$, $B_i = \{b_{i1}, b_{i2}, ..., b_{ik}\}$ for $1 \leq i \leq t - 1$ and $B_t = \{b_{t1}, b_{t2}, ..., b_{t,k+r}\}$.

(2) Each B_i, $1 \leq i \leq t$, induces a complete subgraph.

(3) Each vertex b_{ij} is adjacent to exactly two vertices of A, one of which is a_1. The other is a_s, for $s \geq 2$, subject to the restriction that at least k distinct vertices a_s, for $s \geq 2$, are adjacent to each B_i.

(4) Every vertex a_s, $s \geq 2$, is adjacent to a vertex b_{ij} for at least $\gamma(G) - k$ distinct values of i.

Figure 17.3 shows a graph G having $n = 17$, $\gamma(G) = 4$, and $k = 3$ which has been constructed according to the above specifications.

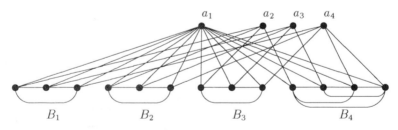

Figure 17.3: A $(4,3)$-insensitive graph.

Haynes, Brigham, and Dutton [14] proved that the graphs obtained from the construction are connected, (γ, k)-insensitive, and have

$$(k + 3)n/2 - [(k + 3)\gamma(G) - 2kr - r^2 + r]/2$$

edges. They also noted that the edge count is maximized when r has its largest value of $k - 1$, and hence the following bound is established.

Upper Bound. For a (γ, k)-insensitive graph G with $2 \leq k + 1 \leq \gamma(G) \leq 2k$ and $n \geq \gamma(G)(k + 1)$,

$$E_k(n, \gamma) \leq \frac{(k + 3)n - ((k + 3)\gamma(G) - 3k^2 + 5k - 2)}{2}.$$

Therefore, for fixed k and $\gamma(G)$, the upper bound is asymptotically equal to $(k + 3)n/2$.

To continue the proof of Theorem 17.29, we consider the lower bound. Recall that Dutton and Brigham [7] determined $E_1(n, \gamma) = 2n - 3\gamma(G)$ for $n \geq 3\gamma(G) \geq 6$. Since any (γ, j)-insensitive graph is also (γ, i)-insensitive for $i \leq j$, we have $E_i(n, \gamma) \leq E_j(n, \gamma)$ for $i \leq j$. Hence $E_k(n, \gamma) \geq 2n - 3\gamma(G)$ for $k \geq 1$. Considering the method used in [7] to obtain this edge count, Haynes, Brigham, and Dutton [14] used $2n - 3\gamma(G)$ as a basis and were able to increase this edge count for $k \geq 2$. Let S be a γ-set of a (γ, k)-insensitive graph G and $f(k)$ be the number of vertices in $V - S$ having degree at most k. The lower bound established in [14] follows.

Lower Bound. Let G be a (γ, k)-insensitive graph with $k \geq 2$, $\gamma(G) \geq 3$, and $n \geq \gamma^2 + 2\gamma + f(k)$. Then $E_k(n, \gamma) \geq (k+3)n/2 - [2(k+2)\gamma + (k-1)(\gamma^2 + f(k))]/2$.

Finally, they showed that $f(k)$ is bounded by an expression that is independent of n, and hence the lower bound is asymptotically equal to $(k + 3)n/2$ (the same asymptotic value as the upper bound), completing the proof of Theorem 17.29.

Open Problems

1. Determine $E_k(n, \gamma)$ for $k = 2$ and $\gamma(G) \geq 3$.

2. Determine $E_k(n, \gamma)$ and for $k \geq 3$ and $\gamma(G) \geq 2$.

3. Determine bounds on the diameter of (γ, k)-insensitive graphs.

4. Study properties of (γ, k)-insensitive graphs including maximum and minimum degree, clique number, and independent domination number.

5. Study (γ, k)-insensitive graphs which have the added property that G also remains connected when k arbitrary edges are removed.

17.4   Reinforcement

A concept which is, in a sense, dual to deleting edges from a graph G is that of adding edges to G. Clearly, when one or more edges are added to a graph, the domination number of the graph does not increase, but may decrease. The *reinforcement number* $r(G)$ is the smallest number of edges which must be added to G in order for the resulting graph G' to satisfy $\gamma(G') < \gamma(G)$. If $\gamma(G) = 1$, then the reinforcement number is defined to be $r(G) = 0$. Furthermore, if $\gamma(G + e) < \gamma(G)$, then $\gamma(G + e) = \gamma(G) - 1$.

The reinforcement number was introduced by Kok and Mynhardt in [16]. They noted that "The significance of this concept becomes apparent in the applications of domination theory. For example, in a situation where a γ-set is to represent costly facilities in a network N, it may be preferable to establish additional links between points of N rather than to construct facilities at all points of a γ-set of N."

All the results in this section are from [16]. Kok and Mynhardt showed that $r(G)$ may be used to provide a better upper bound for $\gamma(G)$ than the bound of $n - \Delta(G)$.

Proposition 17.30 [16] *For any graph G, $\gamma(G) \leq n - \Delta(G) - r(G) + 1$.*

Proof. If $\gamma(G) = 1$, the result is trivial. Thus assume $\gamma(G) \geq 2$ (and so $\Delta(G) < n - 1$). Let v be a vertex of G of maximum degree. Note that adding $n - (\Delta(G) + 1)$ edges to v would cause v to have degree $n - 1$. Thus $r(G) \leq n - \Delta(G) - 1$. Now let G' be a graph obtained from G by adding $r(G) - 1$ edges incident with v. (This is possible since $r(G) \leq n - \Delta(G) - 1$.) By definition of $r(G)$, $\gamma(G') = \gamma(G)$. Thus $\gamma(G) = \gamma(G') \leq n - \Delta(G') = n - (\Delta(G) + r(G) - 1)$ and the result follows. □

This bound is sharp, as shown by the corona $K_t \circ K_1$.

The inequality given above also yields an upper bound for the reinforcement number of a graph in terms of the domination number of the graph. In order to calculate the reinforcement number of a graph once the domination number is known, another (equivalent) parameter which is vertex-based is introduced. Let G be a graph with $\gamma(G) \geq 2$ and define

$$\eta(G) = min\{|V(G) - N[X]| : X \subseteq V(G), |X| = \gamma(G) - 1\}.$$

Intuitively, $\eta(G)$ gives the smallest deficiency of a $\gamma - 1$-set. Kok and Mynhardt [16] showed that $\eta(G) = r(G)$ if G has no vertex of degree $n - 1$.

Proposition 17.31 *If G is a graph with $\gamma(G) \geq 2$, then $r(G) = \eta(G)$.*

Proof. Let X be an η-set. That is, let X be a set of cardinality $\gamma(G) - 1$ such that $\eta(G) = |V(G) - N[X]|$. Then $\eta(G)$ vertices are not dominated by X. By joining each of these vertices to a vertex in X, we obtain a graph G' which is dominated by X and hence $\gamma(G') \leq |X| < \gamma(G)$, which implies that $r(G) \leq \eta(G)$.

Next let F be a reinforcement set of edges. That is, $|F| = r(G)$ and $\gamma(G + F) = \gamma(G) - 1$. Let D be a γ-set of $G + F$. For every $uv \in F$, we may assume that $u \in D$ and $D \cap N_G[v] = \emptyset$, for otherwise D dominates $G + F - \{uv\}$, contradicting the choice of F. Hence $r(G)$ vertices of G are not dominated in G by D. Thus $\eta(G) \leq r(G)$. $\square$

Using Proposition 17.31, the calculation of $r(G)$ is immediate if $\gamma(G) = 2$.

Corollary 17.32 *If G is a graph with $\gamma(G) = 2$, then $r(G) = n - \Delta(G) - 1$.*

Proposition 17.31 can also be used to calculate the reinforcement number for various classes of graphs.

Proposition 17.33 *Let $n \geq 4$ be an integer $n = 3k + i$, where $i \in \{1, 2, 3\}$. Then $r(P_n) = r(C_n) = i$.*

Proof. Let $(0, 1, ..., n - 1)$ be the vertex sequence in P_n. It is easy to see that $\{1, 4, ..., 3k - 2\}$ is a set of $k = \lceil n/3 \rceil - 1$ vertices of P_n which dominates all but i vertices, and that no subset of $V(P_n)$ of cardinality $\gamma(P_n) - 1$ leaves fewer than i vertices undominated. Hence $\eta(P_n) = i$ and the result follows from Proposition 17.31. The proof for cycles is identical and is omitted here. $\square$

Kok and Mynhardt [16] determined the reinforcement number of the cartesian product $K_s \times K_t$, and demonstrated that when $s = t$ the cartesian product is edge domination critical.

Proposition 17.34 *For $2 \leq s \leq t$, $r(K_s \times K_t) = t - s + 1$. Moreover,*

$$\gamma(K_s \times K_s + e) = \gamma(K_s \times K_s) - 1$$

for every edge $e \in E(\overline{K_s \times K_s})$.

Their next result gives reinforcement numbers for the union $G \cup H$, the join $G + H$, and the corona $G \circ H$.

Proposition 17.35 *Let G and H be graphs of order n and n', respectively. Then*

$$(i) \; r(G \cup H) = \begin{cases} \min\{n, n'\} & \text{if } \gamma(G) = \gamma(H) = 1, \\ \min\{n, r(H)\} & \text{if } \gamma(G) = 1 \text{ and } \gamma(H) \geq 2, \\ \min\{r(G), r(H)\} & \text{if } \gamma(G), \gamma(H) \geq 2. \end{cases}$$

$$(ii) \; r(G + H) = \begin{cases} 0 \text{ if } \min\{\gamma(G), \gamma(H)\} = 1, \\ \min\{n - \Delta(G) - 1, n' - \Delta(H) - 1\} & \text{otherwise.} \end{cases}$$

$$(iii) \; r(G \circ H) = \begin{cases} 0 & \text{if } n = 1, \\ n' + 1 & \text{if } G = \overline{K_n}, n \geq 2, \\ n' & \text{otherwise.} \end{cases}$$

Corollary 17.36 *If $t \geq 3$, then $K_t \circ K_1$ is edge domination critical.*

To determine an upper bound for the reinforcement number, another parameter is required. For a graph G, let $x \in X \subseteq V$. The *private neighborhood of* x *relative to* X consists of those vertices which are in the closed neighborhood of x, but not in the closed neighborhood of $X - \{x\}$. Thus, the private neighborhood of x relative to X, denoted $PN(x, X) = N[x] - N[X - \{x\}]$. Next let $\varepsilon(X) = min\{|PN(x, X)| : x \in X\}$, and finally we define the *private neighborhood number* of G, denoted $\varepsilon(G)$, to be the $min\{\varepsilon(D) : D \text{ is a } \gamma\text{-set of } G\}$. Kok and Mynhardt [16] proved that the reinforcement number $r(G)$ is bounded above by $\varepsilon(G)$.

Proposition 17.37 *For any graph G with $\gamma(G) \geq 2$, $r(G) \leq \varepsilon(G)$. Moreover, if $r(G) = 1$, then equality holds.*

Proof. Let D be a γ-set of G such that $\varepsilon(D) = \varepsilon(G)$ and let $v \in D$ satisfy $|PN(v, D)| = \varepsilon(D)$. Let $u \in D - \{v\}$ and let H be the graph obtained from G by joining u to every vertex in $PN(v, D)$. Clearly $D - \{v\}$ dominates H and hence $r(G) \leq \varepsilon(G)$.

Let $r(G) = 1$ and $\gamma(G + uv) = \gamma(G) - 1$. Consider any γ-set D of $G + uv$. Without loss of generality, we may assume that $u \in D$ and $N_G[v] \cap D = \emptyset$, for otherwise D dominates G which is impossible. Clearly, every vertex of G apart from v is dominated by D and consequently, $D' = D \cup \{v\}$ is a γ-set of G with the property $|PN(v, D')| = 1$. Hence $\varepsilon(G) = 1$. $\square$

Since any γ-set D of a graph G contains a vertex v such that $|PN(v, D)| \leq n/\gamma(G)$, the following bound is obtained for $r(G)$ and $\varepsilon(G)$.

Corollary 17.38 *If G is a graph with n vertices, then $r(G) \leq \varepsilon(G) \leq n/\gamma(G)$.*

Examples of graphs G with $\varepsilon(G) = r(G)$ include cycles C_{3k} and paths P_{3k}. On the other hand, Kok and Mynhardt [16] demonstrated that for any value of $\gamma(G) \geq 2$, the difference between $\varepsilon(G)$ and $r(G)$ can be made arbitrarily large for connected graphs.

Theorem 17.39 *For any integers r, s, t with $2 \leq r \leq s$ and $2 \leq t$, there exists a connected graph G with $r(G) = r$, $\varepsilon(G) = s$, and $\gamma(G) = t$.*

We omit the proof to the theorem, but give a construction for these graphs.

Construction
First observe that if $s = 2$, then $r = 2$ and the cycle C_{3t-1} has the desired property. Next let $s \geq 3$. Let F (H, respectively) be the graph obtained by joining every vertex of K_2 (K_{r-1}, respectively) to every vertex of $\overline{K}_{s-1}$. If $r = s$ ($r < s$, respectively), let Q be the graph obtained by taking $t - 2$ disjoint copies of H and one copy F ($K_{1,s-1}$, respectively), adding a new vertex v and joining

v to every vertex of $\overline{K}_{s-1}$ in each copy of H as well as to every vertex of $\overline{K}_{s-1}$ in F ($K_{1,s-1}$, respectively). The desired graph G is constructed by adding one copy of $K_{1,s-2}$ to Q, joining the vertex u of $K_{1,s-2}$ with degree $s-2$ as well as (if $r < s$) $s - r - 1$ other vertices of $K_{1,s-2}$, to v. A graph G formed by the construction with $r = 3$, $s = 4$, and $t = 2$ is shown in Figure 17.4.

We conclude this section and hence this chapter with open problems regarding the reinforcement number.

Open Problems

1. Characterize the graphs G for which $r(G) = \varepsilon(G)$.

2. Characterize the graphs G for which $r(G) = \varepsilon(G) = n/\gamma(G)$.

3. Study characteristics (for example, the diameter) of graphs having a given reinforcement number.

4. Determine additional upper and lower bounds for $r(G)$.

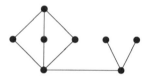

Figure 17.4: A graph G having $r(G) = 3$, $\varepsilon(G) = 4$, and $\gamma(G) = 2$.

Bibliography

[1] D. Bauer, F. Harary, J. Nieminen, and C. L. Suffel, Domination alteration sets in graphs. *Discrete Math.* 47 (1983) 153–161.

[2] R. C. Brigham and R. D. Dutton, A study of vertex domination critical graphs. Technical Report M-2, Department of Mathematics, University of Central Florida, (1984).

[3] R. C. Brigham, P. Z. Chinn, and R. D. Dutton, Vertex domination-critical graphs. *Networks* 18 (1988) 173–179.

[4] J. Carrington, F. Harary, and T. W. Haynes, Changing and unchanging the domination number of a graphs. *J. Combin. Math. Combin. Comput.* 9 (1991) 57–63.

[5] V. Chvàtal and W. Cook, The discipline number of a graph. *Discrete Math.* 86 (1990) 191–198.

[6] E. Cockayne, S. Goodman, and S. Hedetniemi, A linear algorithm for the domination number of a tree. *Inform. Proc. Lett.* 4 (1975) 41–44.

[7] R. D. Dutton and R. C. Brigham, An extremal problem for edge domination in insensitive graphs. *Discrete Appl. Math.* 20 (1988) 113–125.

[8] J. F. Fink, M. S. Jacobson, L. F. Kinch, and J. Roberts, The bondage number of a graph. *Discrete Math.* 86 (1990) 47–57.

[9] R. K. Guha and T. W. Haynes, Some remarks on k-insensitive graphs in network system design. *Sankhyā* 54 (1992) 177–187.

[10] B. L. Hartnell, L. K. Jorgensen, P. D. Vestergaard, and C. Whitehead, Edge stability of the k-domination number of trees. Submitted for publication.

[11] B. L. Hartnell and D. F. Rall, A characterization of trees in which no edge is essential to the domination number. *Ars Combin.* 33 (1992) 65–76.

[12] B. L. Hartnell and D. F. Rall, Bounds on the bondage number of a graph. *Discrete Math.* 128 (1994) 173–177.

[13] T. W. Haynes, S. T. Hedetniemi, and P. J. Slater, *Fundamentals of Domination in Graphs*, Marcel Dekker, Inc., 1997.

[14] T. W. Haynes, R. C. Brigham, and R. D. Dutton, Extremal graphs domination insensitive to the removal of k edges. *Discrete Appl. Math.* 44 (1993) 295–304.

[15] T. W. Haynes, R. K. Guha, R. C. Brigham, and R. D. Dutton, The G-network and its inherent fault tolerant properties. *Internat. J. Comput. Math.* 31 (1990) 167–175.

[16] J. Kok and C. M. Mynhardt, Reinforcement in graphs. *Congr. Numer.* 79 (1990) 225–231.

[17] T. W. Rice (Haynes), On (γ, k)-insensitive domination. Ph.D. Dissertation, University of Central Florida (1988).

[18] T. W. Rice (Haynes), R. C. Brigham, and R. D. Dutton, Extremal $(2, 2)$-insensitive graphs. *Congr. Numer.* 67 (1988) 158–166.

[19] T. W. Rice (Haynes) and R. K. Guha, A multi-layered G-network for massively paralled computation. *Proc. Second Symposium on the Frontiers of Massively Parallel Computation* (1989) 519–520.

[20] L. A. Sanchis, Maximum number of edges in connected graphs with a given domination number. *Discrete Math.* 87 (1991) 65–72.

[21] U. Teschner, Die Bondagezahl eines Graphen, Ph.D. Dissertation, RWTH Aachen (1995).

[22] U. Teschner, A counterexample to a conjecture on the bondage number of a graph. *Discrete Math.* 122 (1993) 393–395.

[23] U. Teschner, The bondage number of a graph G can be much greater than $\Delta(G)$. To appear in *Ars Combin.*

[24] U. Teschner, A new upper bound for the bondage number of graphs with small domination number. *Australas. J. Combin.* 12 (1995) 27–35.

[25] U. Teschner, On the bondage number of block graphs. To appear in *Ars Combin.*

[26] U. Teschner, New results about the bondage number of a graph. To appear in *Discrete Math.*

[27] U. Teschner and L. Volkmann, On the bondage number of cactus graphs. Submitted for publication.

[28] J. Topp and P. D. Vestergaard, α_k- and γ_k-stable graphs. Submitted for publication.

[29] L. Volkmann, On graphs with equal domination and covering numbers. *Discrete Appl. Math.* 51 (1994) 211–217.

[30] H. B. Walikar and B. D. Acharya, Domination critical graphs. *Nat. Acad. Sci. Lett.* 2 (1979) 70–72.

Index

Absorbant, 402
A-class graph, 169
Acyclic digraph, 417
Adomatic number, 362-363
Algorithms, 191-231
 cocomparability graphs, 199-203
 and complexity results, 36-37, 40-41, 325-326,332-333
  summary, 218-219
 dually chordal graphs, 204-206
 interval graphs, 209-216
 involving weighted rooks, 154-155
 parallel, 218
 permutation graphs, 194-199
 strongly chordal graphs, 206-209
Antidomatic number, 373-374
Arrow notation, 441-442
Asymmetric arc, 402
Attachable set, 174
Attachable vertex, 174
Attacher, 50
Automorphism class, 7-10

Bichromaticity of a graph, 359-360
Block-cactus graph, 478
Block graph, 45, 65, 479
Block structure, 450-453
Bondage
 fractional, 473
 insensitivity, and reinforcement, 471-489
 number, 77, 472
 and reinforcement, 77-79
Bottom Left Crossing List, 197
Bottom Right Crossing List, 198
Bounds
 on γ_s, 39-40
 relating i_k and γ_k, 340-342
 relating i_k and $\gamma_k{}^l$, 342
BREADTH-FIRST-SEARCH, 202

Cacti, 356-357, 478
Chordal, 80
 bipartite, 80

[Chordal]
 strongly, 80
Circle graphs, 199
Cliques, 353, 412
 and clique graphs, 353-354
Closed neighborhood, 8, 110, 262, 321
Closed-open irredundance trees, 254
Cocomparability graphs, 199-203
Coefficient of stability, 402
Combinatorial problems on chessboards, 133-162
 algorithms involving weighted rooks, 154
 domination on variant chessboards, 152-153
 first 10 values, 139-142
 N-queens problem, 142-146
 non-attacking queens, 147-152
 queens on a column and on a diagonal, 146-147
 total domination problems, 154
 updates to survey I, 135-138
Comparability graph, 200, 403
Complementable, 56
Complements, 12
 of a digraph, 402
Complement related families,
Complementarily domatic number, 370
Complementarily domatic partition, 370
Complementarity
 and duality, 10-14
Complementary slackness, 208
Complexity
 results, 233-269
  applying the proof, 238-243
  applying the variation, 246-258
  extending the result, 258-263
  generalized proof, 236-238
  perfect domination 235-236
  variation of the proof, 244-246
  summary, 218-219
Components and block structure, 450-453
Computational aspects, 84
Conjunctive compound, 425
Connected bipartite graph, 55
Connected domatic number, 371-372